PLANT MOLECULAR BIOLOGY

LABFAX

The LABFAX series

Series Editors:

B.D. HAMES Department of Biochemistry and Molecular Biology, University of Leeds, Leeds LS2 9JT, UK

D. RICKWOOD Department of Biology, University of Essex, Wivenhoe Park, Colchester CO4 3SQ, UK

MOLECULAR BIOLOGY LABFAX

CELL BIOLOGY LABFAX

CELL CULTURE LABFAX

BIOCHEMISTRY LABFAX

VIROLOGY LABFAX

PLANT MOLECULAR BIOLOGY LABFAX

Forthcoming titles

IMMUNOCHEMISTRY LABFAX

BACTERIOLOGY LABFAX

ENZYMOLOGY LABFAX

CELLULAR IMMUNOLOGY LABFAX

PLANT MOLECULAR BIOLOGY

LABFAX

EDITED BY

R.R.D. CROY

Department of Biological Sciences,
University of Durham, Science Laboratories,
South Road, Durham, DH1 3LE, UK

BIOS SCIENTIFIC PUBLISHERS

ACADEMIC PRESS

BIOS Scientific Publishers Limited,
St Thomas House, Becket Street, Oxford OX1 1SJ

ISBN 0 12 198370 6

A CIP catalogue entry for this book is available from the British Library.

This Edition published jointly in the United States of America by Academic Press, Inc. and BIOS Scientific Publishers Limited.

Distributed in the United States, its territories and dependencies, and Canada exclusively by Academic Press, Inc., 1250 Sixth Avenue, San Diego, California 92101 pursuant to agreement with BIOS Scientific Publishers Limited, St Thomas House, Becket Street, Oxford OX1 1SJ, UK.

The cover illustration is based on an original design by Marianne K. Croy

Typeset by Unicus Graphics Ltd, Horsham, UK.
Printed by Information Press Ltd, Oxford, UK.

PREFACE

Plant Molecular Biology Labfax is a companion text to its successful forerunner *Molecular Biology Labfax*, edited by Terry Brown from UMIST, and published 2 years ago. Plant molecular biology has undergone a spectacular expansion over the last 5 years both as an applied science and as an intellectually testing discipline in its own right. The recent appearance of several new journals specializing in the molecular aspects of plant biology and the widespread acceptance of plant molecular biology papers in the established plant journals, are testimony to this development. It is therefore timely for such a laboratory databook as *Plant Molecular Biology Labfax* to be produced in support of this technology. Plant molecular biology has evolved into a multi-disciplinary science utilizing a great many technologies and I hope that this is reflected in the specialized subject areas covered by the different chapters in *Plant Molecular Biology Labfax*. These range from simple, though essential, aspects of plant anatomy and classification, to practical details of plant transformation and data on plant nucleic acids and genes, through to highly specialized topics of study such as chloroplast molecular biology, plant gene introns and transposable elements.

Although plants have a number of unique biological features amongst living organisms, the manipulation of plant nucleic acids is not a special technology — DNA has more or less the same chemistry irrespective of its source; it can be manipulated using the same enzymes and cloned using the same vectors. Several sections in *Plant Molecular Biology Labfax* contain information and data of general utility to molecular biologists, for example, *Chapter 2* contains details of genomic DNA and RNA isolation techniques, and *Chapter 10* deals with PCR technology and gene expression. In this respect I hope that the scientific community will regard *Plant Molecular Biology Labfax* as a companion text which complements *Molecular Biology Labfax* and reserve a place for it on the laboratory bench.

Wherever practicable the accuracy of the data presented in this book has been checked and most of the information was up-to-date at the time of publication. Inevitably in such a large collection of data and information, mistakes can occur and data rapidly becomes out of date and so I would be grateful to readers for any corrections, additions and constructive comments which might be incorporated into future editions.

In editing and writing *Plant Molecular Biology Labfax*, I concur with Terry Brown on his point about needing to be slightly unbalanced to compile such a databook — I too found that the compilation and presentation of data was almost addictive — many of the one page entries and graphs took several days to arrange and present in a form which was deemed easier to use or clearer for the user to understand. All scientists should experience the challenge of collecting and compiling large amounts of data, if only to appreciate the substantial effort and stamina that is required for the production of such texts, which around the lab are very much taken for granted.

The production of this book would not have been possible without the suggestions and contributions from my fellow molecular biologists and also the stimulation and interest in plant molecular sciences imparted by my past supervisors, for which I am deeply indebted. My particular thanks go to my expert co-contributors listed in the following pages. On a sad note, the final stages of production of this book were marred by the untimely death of one of the contributors, Ray Ellis, who fell foul of the changeable weather conditions in the Cairngorm mountains — an area which he loved and visited often. Ray will be sadly missed by his many friends and colleagues and plant science will be all the poorer for this loss.

Finally, special thanks are due to my wife, Liz, for her constant encouragement and support during the lengthy preparation of this book, and who even resorted to writing several of the specialized contributions in order to speed up the submission of the final manuscript.

R.R.D. Croy

CONTENTS

Contributors xiii
Abbreviations xv

1. PLANT ANATOMY, HISTOLOGY AND TAXONOMY FOR MOLECULAR BIOLOGISTS 1

Histology: plant cell types 1
- Parenchyma 1
  - Parenchyma cells (Figure 1) 1
- Epidermis 1
  - Epidermis (Figure 2) 2
- Mesophyll 1
- Endodermis 2
- Pericycle 2
- Collenchyma 2
  - Collenchyma (Figure 3) 2
- Sclerenchyma 2
  - Sclerenchyma (Figure 4) 3
- Vascular bundles 2
  - Vascular bundle (Figure 5) 3
- Xylem 2
- Phloem 3
  - Phloem (Figure 6) 4
- Cambium 3

Anatomy 4
- Stem 4
  - Transverse section of a monocot stem (Figure 7) 4
  - Transverse section of a dicot stem (Figure 8) 5
- Root 4
  - Transverse section of a dicot root (Figure 9) 5
- Leaf 4
  - Transverse section of a dicot leaf (Figure 10) 6
- Flower 5
  - Diagram of a dicot flower (Figure 11) 6
  - Monocot flower: grass spikelet (Figure 12) 7
- Fruits and seeds 5
  - Monocot seed (caryopsis) (Figure 13) 7
  - Dicot seed in longitudinal section (Figure 14) 8

Methods for examination of plant tissues 8
- Embedding and sectioning 8
- Histochemical staining 8

Fluorescence microscopy 8
- General information 8
- Procedure 9
- Commonly used fluorochromes 9

Taxonomy 10
- Rules for the citation of plant names in scientific publications 10
- Nomenclature of major crop and experimental plants 11
  - Nomenclature: common name to Latin name and family (Table 1) 11
  - Nomenclature: Latin name to common name and family (Table 2) 15

References 20
Further reading 20

2. PLANT NUCLEIC ACIDS **21**

DNA isolation from plants 21
- Compilation of plant DNA isolation methods (Table 1) 22
- Key to species listed in *Table 1* (Table 2) 27
- Tissue types cited in *Table 1* (Table 3) 28

Isolation of *Agrobacterium* Ti plasmid DNA 21
- Methods for isolation of plasmid DNA from *Agrobacterium* (Table 4) 29

Calculation of gene copy number equivalents 30
- Plant genome sizes (Table 5) 31
- Formula 30
- Nomogram 30
  - Nomogram for estimation of gene copy equivalents (Figure 1) 32
- Computer program 31

RNA isolation from plants 32
- RNase inhibitors (Table 6) 33
- A compilation of plant RNA isolation methods (Table 7) 36

Codon usage 37
- Codon usage in plant nuclear genes (Table 8) 38
- Plant chloroplast codon usage and mitochondrial genes (Table 9) 43
- Key to organisms cited in *Tables 8* and *9* (Table 10) 46

References 47

3. PLANT GENE INDEX **49**

Introduction 49
- Sequence analysis program suppliers (Table 1) 51
- Example plant gene database (Table 2) 52
- ATP synthesis and electron transport (Table 3) 53
- Antigens (Table 4) 54
- Development- and tissue-specific expression (Table 5) 55
- DNA binding and associated proteins (Table 6) 57
- Structural DNA (Table 7) 60
- Enzymes (Table 8) 63
- Lectins (Table 9) 78
- Membrane-associated proteins (Table 10) 80
- Metal-associated proteins (Table 11) 81
- Miscellaneous (Table 12) 82
- Nodulins (Table 13) 83
- Photosynthesis (Table 14) 85
- Protein modification (Table 15) 95
- Structural RNA (Table 16) 98
- Specific ligand-binding proteins (Table 17) 104
- Storage proteins (Table 18) 105
- Stress and induction (Table 19) 112
- Structural proteins (Table 20) 116
- Transposons (Table 21) 118

References 120

4. PLANT TRANSFORMATION AND EXPRESSION VECTORS **121**

Introduction 121
- Abbreviated terms used in this chapter (Table 1) 122

Expression cassettes 121
- Nucleotide sequences at cloning junctions and 5′ ends (Table 2) 123
- Codes and accession numbers (Table 3) 125
- pRT103 121
  - Map for plasmid pRT103 (Figure 1) 125
- pJIT74 121
  - Map for plasmid pJIT74 (Figure 2) 126

pDH51 125
Map for plasmid pDH51 (Figure 3) 126
pJIT60 125
Map for plasmid pJIT60 (Figure 4) 127
pMA406 126
Map for plasmid pMA406 (Figure 5) 127
pKYLX3 127
Map for plasmid pKYLX3 (Figure 6) 128
pJIT117 128
Map for plasmid pJIT117 (Figure 7) 128
Reporter genes 129
gus gene 129
Map for plasmid pBI101.1, .2, .3 (Figure 8) 129
Map for plasmid pJIT75 (Figure 9) 129
Enzymatic β-glucuronidase assay (Figure 10) 130
Histochemical β-glucuronidase assay (Figure 11) 131
cat and *luc* genes 132
Map for plasmid pJIT26 (Figure 12) 132
Map for plasmid pJIT27 (Figure 13) 132
Map for plasmid pJIT52 (Figure 14) 133
Map for plasmid pJIT53 (Figure 15) 133
Chloramphenicol acetyltransferase assay (Figure 16) 134
Luciferase assay (Figure 17) 135
Other reporter genes 136
Selectable markers 136
Kanamycin resistance 136
Map for plasmid pJIT134 (Figure 18) 136
Map for plasmid pJIT161 (Figure 19) 137
Hygromycin resistance 137
Map for plasmid pJIT6 (Figure 20) 137
Map for plasmid pJIT72 (Figure 21) 138
Bialaphos resistance 138
Map for plasmid pIJ4104 (Figure 22) 138
Map for plasmid pJIT82 (Figure 23) 139
Sulfonamide resistance 139
Map for plasmid pJIT92 (Figure 24) 139
Map for plasmid pJIT118 (Figure 25) 140
Streptomycin/spectinomycin resistance 140
Map for plasmid pPM19 (Figure 26) 140
Map for plasmid pPM21 (Figure 27) 141
Chlorsulfuron resistance 141
Methotrexate resistance 141
Phenotypic markers 141
Anthocyanin regulatory genes 141
Map for plasmid pPHI443 (Figure 28) 142
Binary vectors 142
pBIN19 142
Map for plasmid pBIN19 (Figure 29) 143
pROK1 142
Map for plasmid pROK1 (Figure 30) 143
pGA482 143
Map for plasmid pGA482 (Figure 31) 144
pCGN1547, pCGN1548 144
Map for plasmid pCGN1547 (Figure 32) 144
Map for plasmid pCGN1548 (Figure 33) 145
pUCD2340 145
Map for plasmid pUCD2340 (Figure 34) 145
pC22 146
Map for plasmid pC22 (Figure 35) 146
References 146

5. PLANT SELECTABLE GENES, REPORTER GENES AND PROMOTERS 149

Introduction 149
Gene abbreviations and phenotypes (Table 1) 150
Aminoglycoside adenyltransferase gene [**S**] 149
Restriction map of the *aad* gene (Figure 1) 151
Bleomycin-resistance gene [**S**] 152
Restriction map of the *ble* gene (Figure 2) 152
Cauliflower mosaic virus 35S promoter [**P**] 152
Restriction map of the CaMV 35S promoter (Figure 3) 153
Cauliflower mosaic virus 35S promoter and terminator [**P,T**] 153
Restriction map of the CaMV 35S promoter and terminator (Figure 4) 154
Chloramphenicol acetyltransferase gene [**R**] 155
Restriction map of the *cat* gene (Figure 5) 156
Dihydrofolate reductase [**S**] 156
Restriction map of the bacterial *dhfr* gene (Figure 6) 157
Dihydropteroate synthase gene [**S**] 157
Restriction map of the *sul* gene (Figure 7) 158
Enolpyruvylshikimate phosphate synthase gene [**S**] 158
Restriction map of the *epsps* gene (Figure 8) 160
β-galactosidase gene [$\mathbf{R^H}$] 161
Restriction map of the *gal* gene (Figure 9) 164
β-glucuronidase gene [$\mathbf{R^H}$] 164
Restriction map of the *gus* gene (Figure 10) 166
Hygromycin phosphotransferase gene [**S**] 166
Restriction map of the *hyg* gene (Figure 11) 168
Luciferase gene [**R**] 168
Restriction map of the *luc* gene (Figure 12) 170
Luciferase A and B genes [**R**] 171
Restriction maps of the *luxA* and *luxB* genes (Figure 13) 173
Neomycin phosphotransferase gene [**S, R**] 174
Restriction map of the *npt* gene (Figure 14) 175
Nopaline synthase gene [**R, P, T**] 175
Restriction map of the *nos* gene (Figure 15) 178
Octopine synthase gene [**R, P, T**] 178
Restriction map of the *ocs* gene (Figure 16) 180
Phosphinothricin acetyltransferase gene [**S**] 180
Restriction map of the *pat* gene (Figure 17) 181
References 181

6. CHARACTERISTICS OF PLANT PRE-mRNA INTRONS AND TRANSPOSABLE ELEMENTS 183

Introduction 183
Compilation of monocot intron data (Table 1) 184
Compilation of dicot intron data (Table 2) 201
Plant splice site consensus (Table 3) 236
Occurrence of nucleotides at monocot splice sites (Table 4) 237
Occurrence of nucleotides at dicot splice sites (Table 5) 238
Size distribution of plant introns (Figure 1) 239
Mobile genetic elements in higher plants 239
Homology within terminal inverted repeats (Table 6) 240
Homology within terminal repeats (Table 7) 240
Transposable elements with retrotransposon-like structure (Table 8) 241
Identified transposable elements not in *Tables 6–8* (Table 9) 242
Transposable elements not yet studied (Table 10) 242
References 242

7. PLANT TISSUE CULTURE AND GENETIC TRANSFORMATION 253

Tissue culture media 253
Common basal media 253
Common basal media — inorganic salts (Table 1) 253
Common basal media — vitamins and organics (Table 2) 255
Media stock concentrations (Table 3) 255
Media supplements 253
Growth regulators: chemical names and molecular weights (Table 4) 256
Structures of growth regulators (Figure 1) 258
Other media additives (Table 5) 256
Gelling agents and concentrations (Table 6) 257
Media and methods for the culture of tobacco 257
Cultivars of tobacco commonly used for tissue culture (Table 7) 259
Sterilant concentrations for explants (Table 8) 259
Media for manipulation of tobacco in culture (Table 9) 260
Media for tobacco protoplast isolation, transformation and culture (Table 10) 261
Salts for tobacco protoplast culture media (Table 11) 261
Genetic transformation 262
Definitions of transformation (Table 12) 262
Gene transfer methods and their evaluation 262
Status of alternative gene delivery approaches (Table 13) 263
Criteria for assessing transformation (Table 14) 264
Segregation ratios for transformed phenotype (Table 15) 264
Stably transformed plant species 265
List of stably transformed plant species (Table 16) 265
Agrobacterium strains, media and methods 268
Wild-type (oncogenic) strains of *Agrobacterium* (Table 17) 269
Transconjugant C58 *Agrobacterium* strains (Table 18) 269
Derivation of disarmed Ti and Ri plasmids (Table 19) 270
Introducing vector plasmids into *Agrobacterium* (Table 20) 270
Culture media for *Agrobacterium* (Table 21) 271
Antibiotics for plant and bacterial culture work (Table 22) 272
Antibiotics to control *Agrobacterium* (Table 23) 276
Conditions for *Agrobacterium* virulence induction (Table 24) 276
Structure of acetosyringone (Figure 2) 276
Agroinfection of Gramineae with gemini viruses (Table 25) 277
Protoplast transformation 277
Enzymes for protoplast production (Table 26) 278
Chemicals facilitating DNA uptake by protoplasts (Table 27) 278
Particle bombardment of tissues 277
Propulsion methods of particle guns (Table 28) 278
DNA coating of microprojectiles (Table 29) 279
Examples of tissue-specific transient expression 279
Tissue-specific transient expression (Table 30) 279
References 279

8. PLANT ORGANELLAR TARGETING SEQUENCES 287

Nucleus 287
Peroxisomes/glyoxysomes (microbodies) 287
Secretion pathway 288
Mitochondria 288
Chloroplasts 289
Other plastids 291
References 291

9. CHLOROPLAST MOLECULAR BIOLOGY 293

Introduction 293
What cpDNA encodes 293
Nomenclature of genes and their products 293

Genes for the polypeptides of multi-component thylakoid complexes involved in electron transport 294
Sequential flow of electrons through thylakoid membrane (Figure 1) 294
Photosystem I (PSI) plastocyanin:ferredoxin oxidoreductase 294
Genes for PSI polypeptide components (Table 1) 295
Photosystem II (PSII) water:plastoquinone oxidoreductase 294
Genes for PSII polypeptide components (Table 2) 296
Cytochrome *b/f* complex (plastoquinol:plastocyanin oxidoreductase) 297
Genes for chloroplast electron transport components (Table 3) 298
ATP synthase (F_0F_1-ATPase, EC 3.6.1.34) 297
Genes for chloroplast ATP synthase polypeptide components (Table 4) 299
NADPH: plastoquinone oxidoreductaste (NADPH dehydrogenase) 297
Chloroplast genes for NAD(P):quinone oxidoreductase components (Table 5) 300
The chloroplast genetic system 297
Ribosomal RNA 297
Chloroplast ribosomal proteins 301
tRNA genes 301
Chloroplast genetic code — codons with matching tRNAs (Table 6) 301
RNA polymerase 302
Chloroplast gene expression 302
Transcription 302
Introns, splicing and processing 302
Chloroplast genome structure 302
Representative map of chloroplast genome (Figure 2) 303
Gene arrangements 302
Chloroplast transformation 304
Transformation of plastids of higher plants 304
Transformation of higher-plant chloroplasts (Table 7) 305
Transformation of the chloroplast of *Chlamydomonas reinhardtii* 304
Transformation of *Chlamydomonas* chloroplasts (Table 8) 306
Markers for selection of transformants in *C. reinhardtii* 304
Apparatus for particle bombardment 308
References 308

10. PLANT GENE EXPRESSION AND PCR TECHNIQUES 313

PCR strategies 313
General PCR (Table 1) 314
Genome analysis (Table 2) 325
PCR using cDNA after reverse transcription (Table 3) 328
PCR using degenerate oligonucleotides (Table 4) 332
Use of PCR to modify DNA sequences (Table 5) 336
Randomly amplified polymorphic DNA analysis (Table 6) 337
Detection and analysis of recombinant clones and transgenic plants (Table 7) 338
Evolutionary and archaeological PCR (Table 8) 340
Miscellaneous (Table 9) 342
PCR product cloning vectors (Figure 1) 343
PCR acronyms (Table 10) 345
Assay of plant gene expression 345
Introduction 345
Expression assays — RNA analyses (Table 11) 346
Expression assays — protein analyses (Table 12) 350
Assays for regulation of gene expression (Table 13) 357
References 365

INDEX 373

CONTRIBUTORS

J. BROWN
Scottish Crop Research Institute, Invergowrie, Dundee DD2 5DA, UK

E.J. CROY
Department of Biological Sciences, University of Durham, South Road, Durham DH1 3LE, UK

R.R.D. CROY
Department of Biological Sciences, University of Durham, South Road, Durham DH1 3LE, UK

T. DYER
Cambridge Laboratory, AFRC Institute of Plant Sciences Research, John Innes Centre, Colney, Norwich NR4 7UJ, UK

†J.R. ELLIS
ICI Seeds, Jealotts Hill Research Station, Bracknell RG12 6EY, UK

I.M. EVANS
Department of Biological Sciences, University of Durham, South Road, Durham DH1 3LE, UK

A. FORDHAM-SKELTON
Department of Biology, Colorado State University, Fort Collins, CO 80523, USA

T. FRANKLIN
Hybaid, 111–113 Waldegrave Road, Teddington TW11 8LL, UK

P. GATES
Department of Biological Sciences, University of Durham, South Road, Durham DH1 3LE, UK

†Deceased, 1993.

F. GUERINEAU

Department of Botany, University of Leicester, University Road, Leicester LE1 7RH, UK

D.L. HERRIN

Department of Botany, University of Texas at Austin, Austin, TX 8713, USA

C.J. HOWE

Department of Biochemistry, University of Cambridge, Cambridge CB2 1QW, UK

T. IKEMURA

National Institute of Genetics, Mishima, Shizuoka-ken 411, Japan

M. LATTERICH

Department of Molecular and Cell Biology, University of California at Berkeley, Berkeley, CA 94720, USA

D.J. LEADER

Scottish Crop Research Institute, Invergowrie, Dundee DD2 5DA, UK

P. MULLINEAUX

Department of Applied Genetics, John Innes Institute, Colney Lane, Norwich NR4 7UJ, UK

A. SHIRSAT

School of Biological Sciences, University of Wales at Bangor, Bangor, Gwynned LL57 2UW, UK

C.G. SIMPSON

Scottish Crop Research Institute, Invergowrie, Dundee DD2 5DA, UK

A. THOMPSON

Department of Molecular Biology, Horticultural Research International, Littlehampton, Sussex, UK

T.P. WALLACE

Scotgen Ltd, Aberdeen, UK

ABBREVIATIONS

aa	amino acid
AFLPs	amplification-fragment-length-polymorphisms
ALA	δ-aminolevulinic acid
ALS	acetolactate synthase
ANS	8-anilino-1-naphthalene sulfonic acid
APCR	anchored PCR
ATA	aurintricarboxylic acid
ATPase	adenosine triphosphatase
CaMV	cauliflower mosaic virus
CAT	chloramphenicol acetyltransferase
cDNA	complementary DNA
cpDNA	chloroplast DNA
DAPI	diamidino phenolindole
dc^7GTP	7-deaza-2′deoxyguanosine
DCMU	3-(3,4-dichlorophenyl)-1,1-dimethylurea
DEPC	diethylpyrocarbonate
DHFR	dihydrofolate reductase
dicot	dicotyledonous (plant)
DMSO	dimethylsulfoxide
DNA	deoxyribonucleic acid
DTPA	diethylenetriaminepentaacetic acid
EDTA	ethylenediaminetetraacetic acid
ELISA	enzyme-linked immunosorbent assay
EPSPS	5-enolpyruvylshikimate-3-phosphate synthase
ER	endoplasmic reticulum
EtBr	ethidium bromide
GUS	β-glucuronidase
IPCR	inverse PCR
LHC	light-harvesting complex
LM-PCR	ligation-mediated PCR
Luc	luciferase (firefly gene)
LUX	luciferase (bacterial gene)
MES	2-morpholinoethanesulfonic acid
MOPAC	mixed oligonucleotide primer amplification
mRNA	messenger RNA
monocot	monocotyledonous (plant)
MUG	4-methylumbelliferyl glucuronide
NPT	neomycin phosphotransferase
ORF	open reading frame
PAGE	polyacrylamide gel electrophoresis
PCR	polymerase chain reaction

PEG	polyethylene glycol
pre-mRNA	precursor mRNA
PSI	photosystem I
PSII	photosystem II
RACE	rapid amplification of cDNA ends
RAPD	randomly amplified polymorphic DNA
rDNA	ribosomal RNA genes
RFLP	restriction fragment length polymorphism
RHB	right-hand border
RT-PCR	reverse transcription-PCR
Rubisco	ribulose bisphosphate carboxylase/oxygenase
SDDTC	disodium diethyldithiocarbamate
SDS	sodium dodecyl sulfate
snRNAs	small nuclear RNAs
snRNPs	small nuclear ribonucleoprotein particles
T-DNA	transfer DNA
TINS	tri-isopropylnaphthalene 1,5 disulfonate
TLC	thin-layer chromatography
TMAC	tetramethylammonium chloride
UsnRNAs	uracil-rich small nuclear RNAs
UV	ultraviolet
VRCs	vanadyl ribonucleoside complex
w/v	weight/volume
X-Gal	5-bromo-4-chloro-3-indolyl-β-D-galactoside
X-Gluc	5-bromo-4-chloro-3-indolyl β-D-glucuronide
YAC	yeast artificial chromosome

CHAPTER 1
PLANT ANATOMY, HISTOLOGY AND TAXONOMY FOR MOLECULAR BIOLOGISTS

P. Gates

This brief guide provides an introduction to some basic elements of plant anatomy and histology, and information on simple methods for the examination of plant tissues. Users should be aware that the anatomical information is of a very generalized nature and that many variations are likely to be encountered on the basic ground plans of tissue anatomy described here. Detailed references to the tissue structure of various economic plants can be found in the references (1, 2).

1. HISTOLOGY: PLANT CELL TYPES

1.1. Parenchyma
Living cells (*Figure 1*). Usually isodiametric (*c.* 25–50 μm). Form the bulk of plant packing tissue. Metabolically active with large intercellular spaces and thin walls. Often modified for storage of lipids, proteins and starch.

1.2. Epidermis
Living, flattened and elongated cells (*Figure 2*). Cover the surface of the plant. Protective (often covered with cutin) and sometimes secretory. Epidermal cells are often modified into protective or secretory hairs (trichomes).

1.3. Mesophyll
Living cells, primarily active in leaf photosynthesis. Divided into a pallisade layer of columnar cells under the leaf adaxial surface (facing the stem) and the more irregular, roughly isodiametric spongy mesophyll cells above the leaf abaxial surface. The spongy mesophyll is loosely packed, with large intercellular air spaces (see *Figure 10*).

Figure 1. Parenchyma cells.

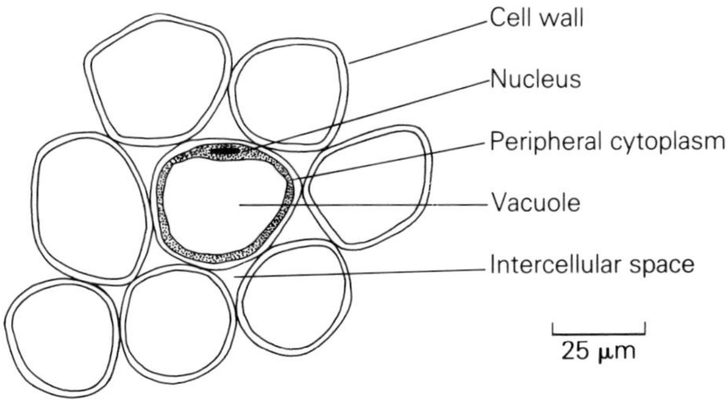

1.4. Endodermis

A ring of living cells with suberized walls (casparian strip) surrounding the stele of the root. Forms a selective barrier to the movement of water and mineral salts (see *Figure 9*).

1.5. Pericycle

A layer of meristematic cells between root central vascular tissue and the endodermis, where lateral roots and secondary thickening are initiated (see *Figure 9*).

1.6. Collenchyma

Long, tapering, living cells which provide support and mechanical strength (*Figure 3*). No intercellular spaces. Commonly found in the angles of stems and leaf midribs. Cells have pronounced cellulose thickening in the corners.

1.7. Sclerenchyma

Fibers are elongated dead cells, often in bundles, often with a small-diameter lumen, always highly lignified (*Figure 4*). They provide mechanical strength. Sclereids are short, often irregularly shaped sclerenchyma cells, functionally similar to fibers; often forming by lignification of parenchyma. They are found in the testas of seeds, e.g. *Pisum*.

1.8. Vascular tissues

These are organized into discrete vascular bundles (*Figure 5*).

1.9. Xylem

Dead, lignified water-conducting tissue. Vessels have open ends and various forms of thickening (spiral, annular, reticulate); tracheids are spindle-shaped and are linked by pits in the wall. Both cell types are often mixed with sclerenchyma fibers.

Figure 2. Epidermis.

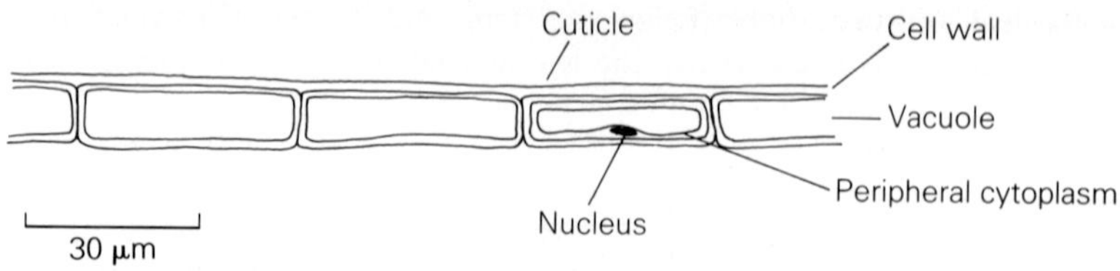

Figure 3. Collenchyma.

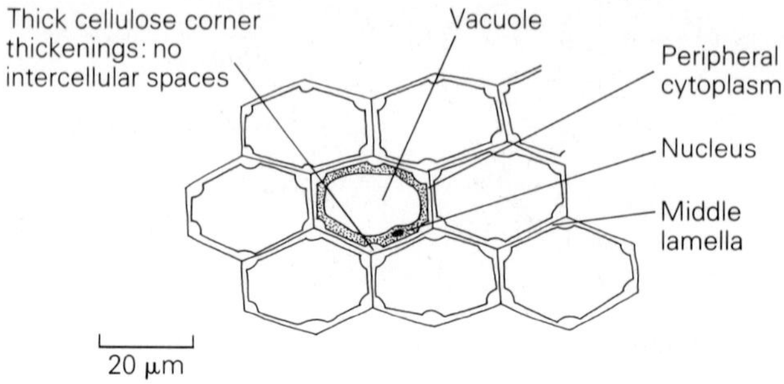

1.10. Phloem

Living cells conducting organic solutes (*Figure 6*). Sieve tubes are elongated and tubular, with cross walls that are perforated with sieve pores. Companion cells are elongated, narrow, densely cytoplasmic and are associated with sieve tubes. Both cell types may be mixed with sclereids and sclerenchyma fibers.

1.11. Cambium

Densely cytoplasmic cells which divide. Intrafascicular cambium lies within vascular bundles and differentiates into metaxylem, phloem, fibers, etc. Interfascicular cambium links vascular bundles, producing secondary thickening in dicotyledonous stems (see *Figure 8*).

Figure 4. Sclerenchyma.

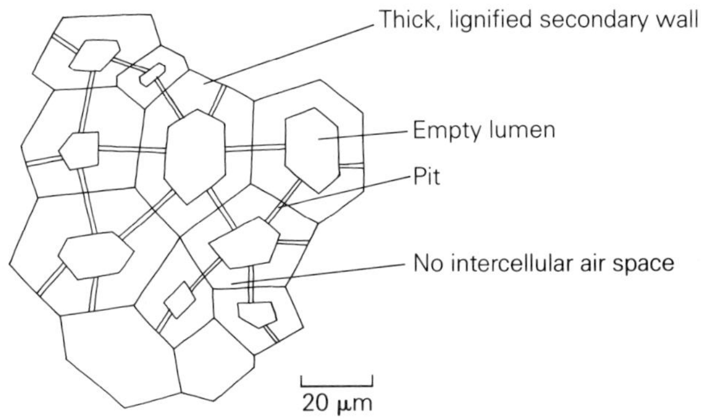

Figure 5. Vascular bundle (*Zea mays*).

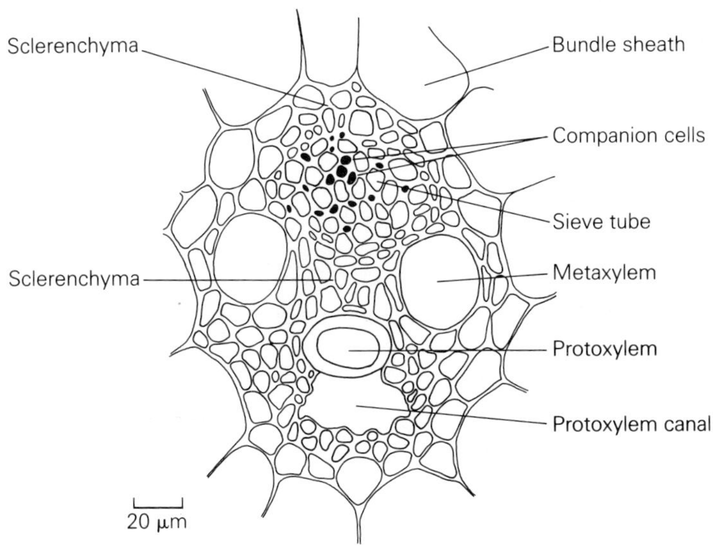

2. ANATOMY

Most molecular biological studies are conducted on short-lived annuals which do not show any pronounced secondary development of stem and root tissues. The following diagrams all refer to the primary structure of these organs. Details of secondary development can be found in refs 2 and 3.

2.1. Stem

Figures 7 and *8* show sections through a monocot and a dicot stem, respectively.

2.2. Root

Figure 9 shows a section through a dicot root.

2.3. Leaf

Figure 10 shows a section through a dicot leaf.

Figure 6. Phloem.

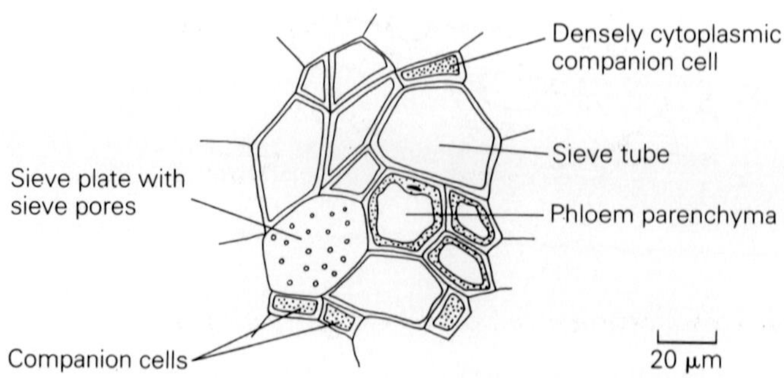

Figure 7. Transverse section of a monocot stem (*Zea mays*).

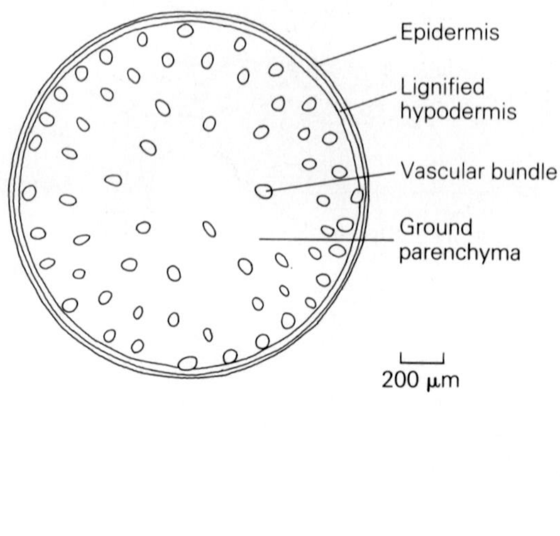

2.4. Flower

Figures 11 and *12* show details of a dicot and a monocot flower, respectively.

2.5. Fruits and seeds

Figures 13 and *14* show a monocot and a dicot seed, respectively.

Figure 8. Transverse section of a dicot stem (*Helianthus annuus*).

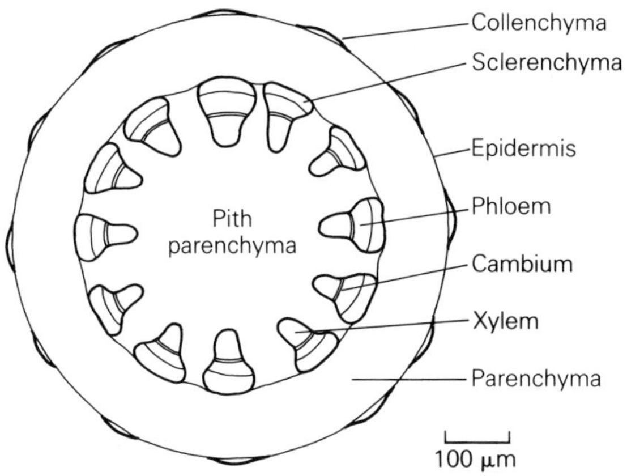

Figure 9. Transverse section of a dicot root, e.g. *Vicia faba*.

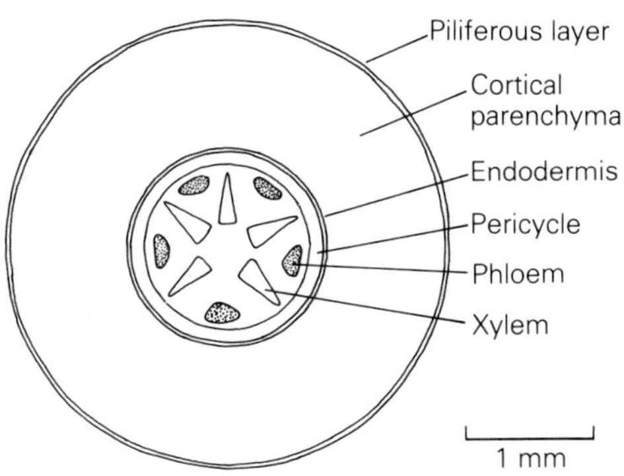

Figure 10. Transverse section of the central region of a dicot leaf.

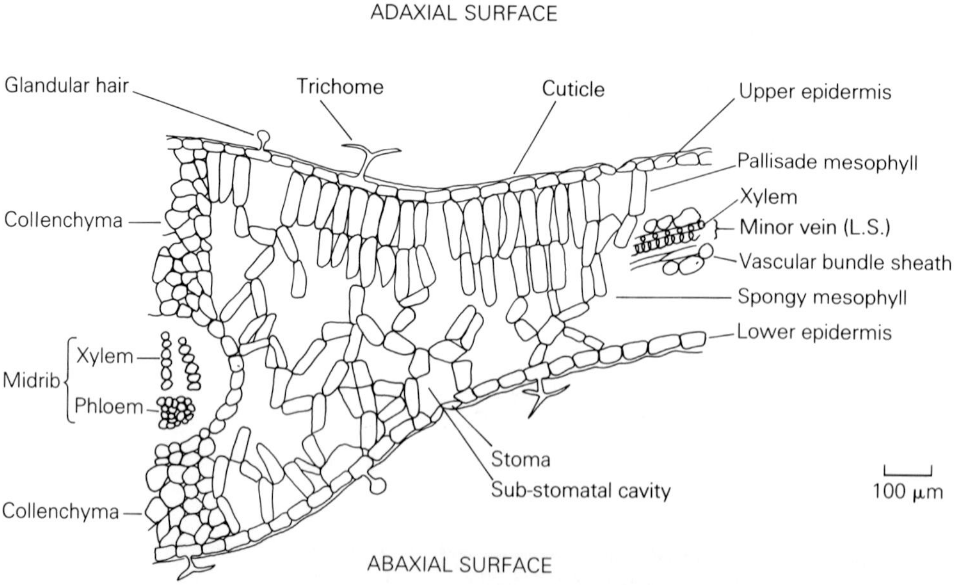

Figure 11. Diagram of a dicot flower (*Arabidopsis thaliana*).

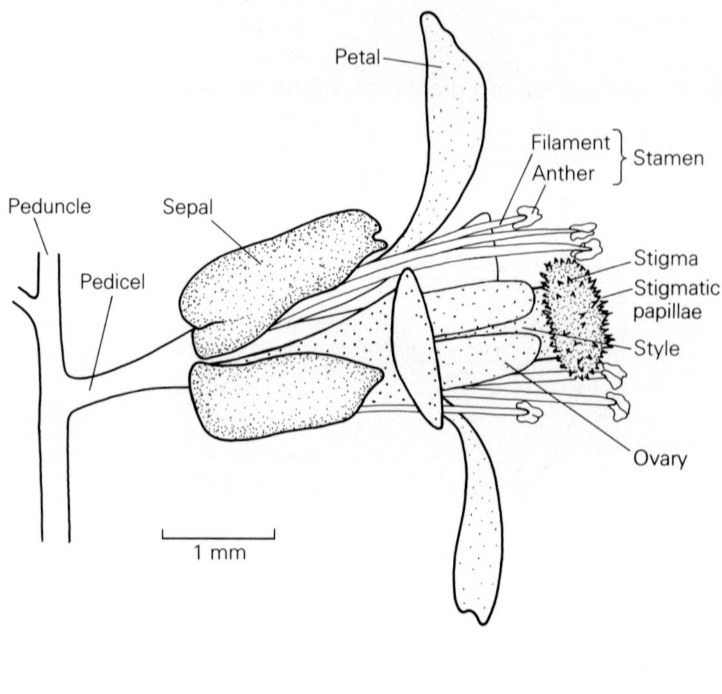

Figure 12. Monocot flower: exploded diagram of a grass spikelet.

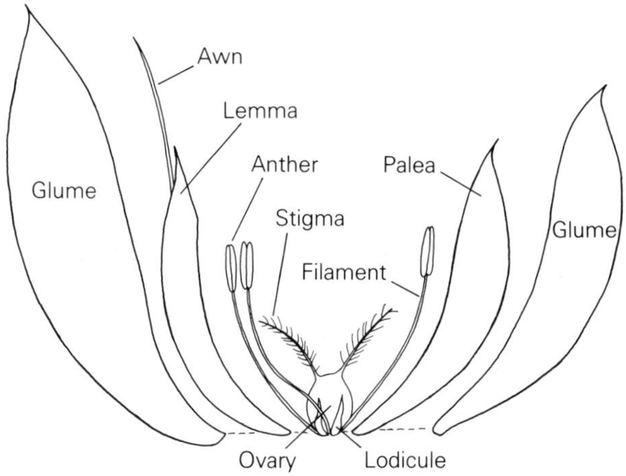

Figure 13. Monocot seed (caryopsis) (*Oryza sativa*).

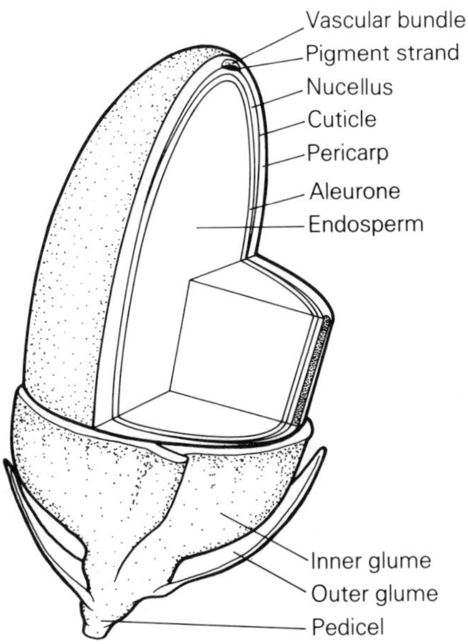

Figure 14. Dicot (*Phaseolus*) seed in longitudinal section.

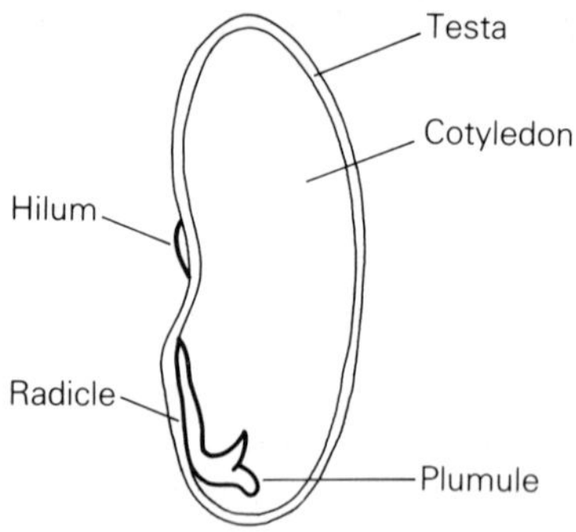

3. METHODS FOR EXAMINATION OF PLANT TISSUES

3.1. Embedding and sectioning

The conventional method for examining plant anatomical characteristics is to fix, dehydrate, embed, thin section, rehydrate, stain and mount sections. There are many variations on this methodology; sources of detailed information can be found in refs 4–8.

3.2. Histochemical staining

The most useful general histochemical stain is toluidine blue, used as a 1% (w/v) solution in 0.05% citrate–phosphate buffer, pH 4.0. This is a metachromatic stain. Lignified tissues are stained blue-green; chromatin, blue; nucleoli, purple; and unlignified cell walls, red. The stain will work adequately on fresh, hand-cut sections.

4. FLUORESCENCE MICROSCOPY

4.1. General information

Advantages

Conventional techniques of specimen preparation are slow and labor-intensive. For many applications, molecular biologists will need rapid methods which allow the examination of tissues with the minimum of preparation. Fluorescence microscopy is ideal for this purpose.

Equipment

Conventional microscopes equipped with fluorescence facilities are satisfactory but inverted microscopes are more convenient, since the cut surface of the specimen in contact with the microscope slide is always at right angles to the optical axis of the lens when the slide is viewed from beneath; this ensures that the specimen is uniformly in focus across the field of view, even at high magnifications. Ultraviolet, violet, blue-violet and blue excitation filters cover the most frequently used wavelengths.

Fluorochromes

These are molecules which do not fluoresce in aqueous solution, but which fluoresce when they are bound to biological molecules in tissue sections. Many are effective at very low concentrations and may not be phytotoxic.

Safety precautions
The high-pressure mercury vapor lamps used in fluorescence microscopes generate ozone, so these should only be used in well-ventilated surroundings. Unshielded mercury vapor lamps emit short-wave ultraviolet radiation which is injurious to eyes.

4.2. Procedure

(i) Cut through the tissue in the required plane of section. The thickness of the specimen is unimportant, since only the surface layer of cells will fluoresce; the tissue will appear to be one cell thick when viewed under the fluorescence microscope.
(ii) Briefly stain the tissue in an aqueous solution of the fluorochrome. In some (e.g. Calcofluor) the reaction is virtually instantaneous, in others longer staining times can be determined by experiment. For fluorochromes which stain cytoplasmic organelles (e.g. diamidino phenolindole (DAPI)) a mild fixation with buffered glutaraldehyde fixative may be helpful in allowing the stain to penetrate, but fixation is generally unnecessary for fluorochromes which stain cell walls.
(iii) Mount the specimen in a drop of distilled water and illuminate with the specific excitation wavelength.
(iv) For semi-permanent mounts, use a non-fluorescent mountant, such as Citifluor. Fluorescence may fade rapidly when high-intensity light sources and high-magnification objectives are used, but fading may be retarded with suitable treatments (3).

4.3. Commonly used fluorochromes

Calcofluor M2R
Use as 0.01% (w/v) aqueous solution (9). Cellulose cell walls fluoresce blue in ultraviolet light. Cell walls without secondary deposits, e.g. phloem and cambial cells, fluoresce brightest.

Decolorized aniline blue
Use as 0.01% (w/v) solution in 0.1 M tripotassium phosphate. Phloem sieve plates, pollen tubes and callose plugs fluoresce blue-green in ultraviolet light.

Auramine O
Use as a 0.01% (w/v) aqueous solution. Lignified cells and cuticle fluoresce yellow in blue light. Can also be used to stain living endomembrane systems (10).

8-anilino-1-naphthalene sulfonic acid (ANS)
Use as 0.1% (w/v) in pH 4.0 phosphate buffer. A non-specific fluorochrome which binds to hydrophobic molecules in tissue sections, which fluoresce yellow in blue light. Useful for localizing protein and lipid deposits in seeds and for detecting cuticles.

Acridine orange
Use as 0.01% (w/v) aqueous solution. Nuclei (double-stranded nucleic acids) fluoresce yellow-green and nucleoli (single-stranded nucleic acids) fluoresce red in blue light. Lignified cell walls also fluoresce yellow-orange.

Diamidino phenolindole (DAPI)
Use as 0.001% (w/v) aqueous solution. DNA in nuclei and organelles fluoresces bright blue in ultraviolet light. Mild fixation may aid penetration of the stain. Some phenolic compounds may also fluoresce.

5. TAXONOMY

5.1. Rules for the citation of plant names in scientific publications

(i) Plant names should be cited with generic names beginning with an upper-case letter and specific names with a lower-case letter.

(ii) When included in text and published tables, Latin names should be presented in italic typescript and indicated in manuscripts as underlined text or, if available, italic typescript.

Example: *Vicia faba*

(iii) The initial citation of the name should be followed by the abbreviation of the authority who coined the name.

Example: *Vicia faba* L.

In this case, "L." denotes that this name was coined by Linnaeus. Names cited in this way can be found in floras and reference texts on many economically important plants.

(iv) After the first citation in the text, the generic name may subsequently be abbreviated to its initial letter, unless this causes confusion with another genus that has a similar specific name.

Example: *V. faba*

(v) It is helpful if the family to which plants belong is also cited in paper titles, after initial citations, in abstracts and in key words. This additional information is often useful to researchers who are working on plant families, rather than on single species.

Example: *Vicia faba* L. (Leguminosae)

(vi) The term "subspecies" (abbreviation ssp.) is generally used to denote morphologically or genetically distinct populations which occur within the ecological or geographical range of a species.

Example: *Vicia faba* ssp. *eu-faba*

where, in this case, the subspecific epithet "*eu-faba*" refers to a subdivision of the species with particular leaflet morphology.

(vii) The term variety (abbreviation var.) is in common usage for ornamental plants and is more correctly applied to distinct subdivisions of species with particular morphological characteristics, without reference to geographical distribution.

Example: *Vicia faba* ssp. *eu-faba* var. *minor*

When domesticated plants are used in research projects, their varietal or cultivar names should be given. The term cultivar (derived from *culti*vated *var*iety) refers to crop or horticultural plants which do not occur in natural populations, and is abbreviated to cv.

Example: *Vicia faba* ssp. *eu-faba* var. *minor* cv. Herz Freya

where, in this case, the varietal epithet "*minor*" refers to small-seeded varieties which have been agriculturally selected for particular agronomic purposes, and "Herz Freya" is the commercial cultivar name.

(viii) The lowest rank commonly used in taxonomy is the forma specialis (abbreviation f.sp.), which refers to minor genetic variants, often based on characteristics such as leaf shape or flower color. This rank is most commonly encountered in the horticultural literature, where numerous forms of cultivated species are recognized.

(ix) The relevance of molecular biological research may often depend on the genetic constitution of the plant material used. For this reason it is essential that the methods sections of research papers should always list the original source of supply or natural location of the experimental material, so that other workers have access to the same genotypes.

(x) For full details of the rules of biological nomenclature, see ref. 11.

5.2. Nomenclature of major crop and experimental plants

Tables 1 and *2* provide useful lists of the species and common names of important crop and experimental plants. *Table 2* allows translation from a known Latin name to the common name and identification of the plant family. *Table 1* allows translation from the common name to the Latin name and identification of the plant family.

Table 1. Nomenclature of major crop and experimental plants: common name to Latin name and family

Common name	Latin name	Family
Adlay	*Coix lachryma-jobi*	Gramineae
Alfalfa	*Medicago sativa*	Leguminosae
Almond	*Prunus amygdalus*	Rosaceae
Apple	*Malus pumila*	Rosaceae
Apricot	*Prunus armeniaca*	Rosaceae
Arabica coffee	*Coffea arabica*	Rubiaceae
Asparagus	*Asparagus officinalis*	Liliaceae
Aspen	*Populus* spp.	Salicaceae
Aubergine	*Solanum melongena*	Solanaceae
Avocado	*Persea americana*	Lauraceae
Bamboo	*Bambusa* sp.	Gramineae
Banana	*Musa* spp.	Musaceae
Barley	*Hordeum vulgare*	Gramineae
Beetroot	*Beta vulgaris*	Chenopodiaceae
Bird's foot trefoil	*Lotus corniculatus*	Leguminosae
Blackcurrent	*Ribes nigrum*	Grossulariaceae
Blackberry	*Rubus* spp.	Rosaceae
Black mustard	*Brassica nigra*	Cruciferae
Black pepper	*Piper nigrum*	Piperaceae
Blueberry	*Vaccinium corymbosum*	Ericaceae
Breadfruit	*Artocarpus altilis*	Moraceae
Broad bean	*Vicia faba*	Leguminosae
Brown mustard	*Brassica juncea*	Cruciferae
Brussels sprout	*Brassica oleracea* var. *gemmifera*	Cruciferae
Buckwheat	*Fagopyrum* spp.	Polygonaceae
Bulrush millet	*Pennisetum typhoides*	Gramineae
Cabbage	*Brassica oleracea* var. *capitata*	Cruciferae
Calabrese	*Brassica oleracea* var. *italica*	Cruciferae
Cantaloup	*Cucumis melo*	Cucurbitaceae
Carnation	*Dianthus caryophyllus*	Caryophyllaceae
Carrot	*Daucus carota*	Umbelliferae
Cassava	*Manihot esculenta*	Euphorbiaceae
Castor bean	*Ricinus communis*	Euphorbiaceae

Table 1. Continued

Common name	Latin name	Family
Cauliflower	*Brassica oleracea* var. *botrytis*	Cruciferae
Celery	*Apium graveolens*	Umbelliferae
Chickpea	*Cicer arietinum*	Leguminosae
Chicory	*Chichorium intybus*	Compositae
Chilli pepper	*Capsicum annuum*	Solanaceae
Chinese cabbage	*Brassica campestris* ssp. *pekinensis*	Cruciferae
Chinese mustard	*Brassica campestris* ssp. *chinensis*	Cruciferae
Chrysanthemum	*Dendranthema indicum*	Compositae
Clove	*Eugenia caryophyllus*	Myrtaceae
Cocoa	*Theobroma cacao*	Sterculiaceae
Cock's foot (UK)	*Dactylis glomerata*	Gramineae
Coconut	*Cocos nucifera*	Palmae
Cocoyam	*Xanthosoma* spp.	Araceae
Common bean	*Phaseolus vulgaris*	Leguminosae
Corn	*Zea mays*	Gramineae
Cotton	*Gossypium hirsutum*	Malvaceae
Courgette	*Cucurbita pepo*	Cucurbitaceae
Cowpea	*Vigna unguiculata*	Leguminosae
Crab's eyes	*Abrus precatorius*	Leguminosae
Cranberry	*Vaccinium macrocarpon*	Ericaceae
Cucumber	*Cucumis sativus*	Cucurbitaceae
Date palm	*Phoenix dactylifera*	Palmae
Durum wheat	*Triticum durum*	Gramineae
Eddo	*Colocasia esculenta*	Araceae
Eggplant	*Solanum melongena*	Solanaceae
Ethiopian mustard	*Brassica carinata*	Cruciferae
Faba bean	*Vicia faba*	Leguminosae
Fennel	*Foeniculum vulgare*	Umbelliferae
Field bean (USA)	*Phaseolus vulgaris*	Leguminosae
Field bean (Eur)	*Vicia faba*	Leguminosae
Fig	*Ficus carica*	Moraceae
Finger millet	*Eleusine coracana*	Gramineae
Flax	*Linum usitatissimum*	Linaceae
Forage rape	*Brassica napus* (biennial types)	Cruciferae
Foxglove	*Digitalis* sp.	Scrophulariaceae
French bean	*Phaseolus vulgaris*	Leguminosae
Garlic	*Allium sativum*	Liliaceae
Giant Taro	*Alocasia macrorrhiza*	Araceae
Grain amaranth	*Amaranthus* spp.	Amaranthaceae
Grape	*Vitis vinifera*	Vitaceae
Grapefruit	*Citrus × paradisi*	Rutaceae
Groundnut	*Arachis hypogaea*	Leguminosae
Guayule	*Parthenium argentatum*	Compositae
Hairy vetch	*Vicia villosa*	Leguminosae
Haricot	*Phaseolus vulgaris*	Leguminosae
Hemp	*Cannabis sativa*	Moraceae
Hop	*Humulus lupulus*	Moraceae

Table 1. Continued

Common name	Latin name	Family
Horseradish	*Armoracia rusticana*	Cruciferae
Indian mustard	*Brassica juncea*	Cruciferae
Indian rape	*Brassica campestris* ssp. *dichotoma*	Cruciferae
Italian ryegrass	*Lolium multiflorum*	Gramineae
Jackbean	*Canavalia ensiformis*	Leguminosae
Jerusalem artichoke	*Helianthus tuberosus*	Compositae
Job's tears	*Coix lachryma-jobi*	Gramineae
Kalanchoe	*Kalanchoe blossfeldiana*	Crassulaceae
Kale	*Brassica oleracea* var. *acephala*	Cruciferae
Kidney bean	*Phaseolus vulgaris*	Leguminosae
Leek	*Allium ampeloprasum*	Liliaceae
Lemon	*Citrus limon*	Rutaceae
Lentil	*Lens culinaris*	Leguminosae
Lettuce	*Lactuca sativa*	Compositae
Lily	*Lilium auratum*	Liliaceae
Lima bean	*Phaseolus lunatus*	Leguminosae
Linseed	*Linum usitatissimum*	Linaceae
Lisianthus	*Eustoma grandiflorum*	Gentianaceae
Lucerne	*Medicago sativa*	Leguminosae
Lupin	*Lupinus albus*	Leguminosae
Madagascan periwinkle	*Catharanthus roseus*	Apocyanaceae
Maize	*Zea mays*	Gramineae
Mango	*Mangifera indica*	Anacardiaceae
Marrow	*Cucurbita pepo*	Cucurbitaceae
Mint	*Mentha × spicata*	Labiatae
Mistletoe	*Viscum album*	Viscaceae
Monkey flower	*Mimulus guttatus*	Scrophulariaceae
Morning glory	*Ipomoea purpurea*	Convolvulaceae
Morning glory	*Pharbitis nil*	Convolvulaceae
Moth bean	*Vigna aconitifolia*	Leguminosae
Mung bean	*Vigna radiata*	Leguminosae
Muskmelon	*Cucumis melo*	Cucurbitaceae
Narbonne vetch	*Vicia narbonensis*	Leguminosae
Navy bean	*Phaseolus vulgaris*	Leguminosae
Oats	*Avena sativa*	Gramineae
Oil palm	*Elaeis guinensis*	Palmae
Oilseed rape	*Brassica napus* (annual types)	Cruciferae
Okra	*Abelmoschus esculentus*	Malvaceae
Olive	*Olea europaea*	Oleaceae
Onion	*Allium cepa*	Liliaceae
Orchard grass (US)	*Dactylis glomerata*	Gramineae
Pak-choi	*Brassica campestris* ssp. *chinensis*	Cruciferae
Papaya	*Carica papaya*	Caricaceae
Paprika	*Capsicum annuum*	Solanaceae
Pasta wheat	*Triticum durum*	Gramineae
Pea	*Pisum sativum*	Leguminosae
Peach	*Prunus persica*	Rosaceae

Table 1. Continued

Common name	Latin name	Family
Peanut	*Arachis hypogaea*	Leguminosae
Pear	*Pyrus communis*	Rosaceae
Pearl millet	*Pennisetum typhoides*	Gramineae
Perennial ryegrass	*Lolium perenne*	Gramineae
Pe-tsai	*Brassica campestris* ssp. *pekinensis*	Cruciferae
Petunia	*Petunia hybrida*	Solanaceae
Pigeon pea	*Cajanus cajan*	Leguminosae
Pineapple	*Ananas comosus*	Bromeliaceae
Plum	*Prunus domestica*	Rosaceae
Poplar	*Populus* spp.	Salicaceae
Potato	*Solanum tuberosum*	Solanaceae
Pumpkin	*Cucurbita pepo*	Cucurbitaceae
Pyrethrum	*Chrysanthemum cinerariifolium*	Compositae
Quinine	*Chichona* spp.	Rubiaceae
Quinoa	*Chenopodium* spp.	Chenopodiaceae
Radish	*Raphanus sativus*	Cruciferae
Ragi	*Eleusine coracana*	Gramineae
Rape	*Brassica napus*	Cruciferae
Raspberry	*Rubus idaeus*	Rosaceae
Red clover	*Trifolium pratense*	Leguminosae
Rice (African)	*Oryza glaberrima*	Gramineae
Rice (Asian)	*Oryza sativa*	Gramineae
Robusta coffee	*Coffea canephora*	Rubiaceae
Rosary pea	*Arbus precatorius*	Leguminosae
Rose periwinkle	*Catharanthus roseus*	Apocyanaceae
Rubber	*Hevea brasiliensis*	Euphorbiaceae
Runner bean	*Phaseolus coccineus*	Leguminosae
Rye	*Secale cereale*	Gramineae
Safflower	*Carthamnus tinctorius*	Compositae
Saltbush	*Atriplex* sp.	Chenopodiaceae
Sarson	*Brassica campestris* ssp. *trilocularis*	Cruciferae
Sesame	*Sesamum indicum*	Pedaliaceae
Snapdragon	*Antirrhinum majus*	Scropulariaceae
Soapwort	*Saponaria officinalis*	Caryophyllaceae
Sorghum	*Sorghum bicolor*	Gramineae
Sour cherry	*Prunus cerasus*	Rosaceae
Soybean	*Glycine max*	Leguminosae
Sprouting broccoli	*Brassica oleracea* var. *italica*	Cruciferae
Strawberry	*Fragaria* × *ananassa*	Rosaceae
Stylosanthes	*Stylosanthes humilis*	Leguminosae
Subterranean clover	*Trifolium subterraneum*	Leguminosae
Sugar beat	*Beta vulgaris*	Chenopodiaceae
Sugar cane	*Saccharum officinarum*	Gramineae
Summer squash	*Cucurbita pepo*	Cucurbitaceae
Sunflower	*Helianthus annuus*	Compositae
Swede	*Brassica napus* (biennial types)	Cruciferae
Sweet cherry	*Prunus avium*	Rosaceae

Table 1. Continued

Common name	Latin name	Family
Sweet orange	*Citrus sinensis*	Rutaceae
Sweet pepper	*Capsicum annum*	Solanaceae
Sweet potato	*Ipomoea batatas*	Convolvulaceae
Taro	*Colocasia esculenta*	Araceae
Tea	*Camellia sinensis*	Camelliaceae
Tepary bean	*Phaseolus acutifolius*	Leguminosae
Thale cress	*Arabidopsis thaliana*	Cruciferae
Thorn apple	*Datura stramonium*	Solanaceae
Tobacco (ornamental)	*Nicotiana plumbaginifolia*	Solanaceae
Tobacco (ornamental)	*Nicotiana sylvestris*	Solanaceae
Tobacco	*Nicotiana tabacum*	Solanaceae
Tomato	*Lycopersicon esculentum*	Solanaceae
Toria	*Brassica campestris* ssp. *dichotoma*	Cruciferae
Triticale	*Triticosecale* spp.	Gramineae
Turnip	*Brassica campestris* ssp. *rapifera*	Cruciferae
Turnip rape	*Brassica campestris* ssp. *oleifera*	Cruciferae
Walnut	*Juglans regia*	Juglandaceae
Watercress	*Rorippa nasturtium-aquaticum*	Cruciferae
Watermelon	*Citrullus lanatus*	Cucurbitaceae
Wheat	*Triticum aestivum*	Gramineae
White clover	*Trifolium repens*	Leguminosae
White mustard	*Sinapis alba*	Cruciferae
Wild soybean	*Glycine argyrea*	Leguminosae
Wild soybean	*Glycine canescens*	Leguminosae
Winged bean	*Psophocarpus tetragonolobus*	Leguminosae
Yam	*Dioscorea* spp.	Dioscoreaceae
Zucchini	*Cucurbita pepo*	Cucurbitaceae

Table 2. Nomenclature of major crop and experimental plants: Latin name to common name and family

Latin name	Common name	Family
Abelmoschus esculentus	Okra	Malvaceae
Arbus precatorius	Crab's eyes Rosary pea	Leguminosae
Allium ampeloprasum	Leek	Liliaceae
Allium cepa	Onion	Liliaceae
Allium sativum	Garlic	Liliaceae
Allocasuarina verticillata	—	Casurinaceae
Alocasia macrorrhiza	Giant Taro	Araceae
Amaranthus spp.	Grain amaranth	Amaranthaceae
Ananas comosus	Pineapple	Bromeliaceae
Antirrhinum majus	Snapdragon	Scrophulariaceae

Table 2. Continued

Latin name	Common name	Family
Apium graveolens	Celery	Umbelliferae
Arabidopsis thaliana	Thale cress	Cruciferae
Arachis hypogaea	Groundnut	Leguminosae
	Peanut	
Armoracia rusticana	Horseradish	Cruciferae
Artocarpus altilis	Breadfruit	Moraceae
Asparagus officinalis	Asparagus	Liliaceae
Atriplex sp.	Saltbush	Chenopodiaceae
Avena sativa	Oats	Gramineae
Bambusa sp.	Bamboo	Gramineae
Beta vulgaris	Beetroot	Chenopodiaceae
Bidens alba	—	Compositae
Brassica campestris		Cruciferae
ssp. *chinensis*	Pak-choi	
	Chinese mustard	
ssp. *dichotoma*	Toria	
	Indian rape	
ssp. *oleifera*	Turnip rape	
ssp. *pekinensis*	Pe-tsai	
	Chinese cabbage	
ssp. *rapifera*	Turnip	
ssp. *trilocularis*	Sarson	
Brassica carinata	Ethiopian mustard	Cruciferae
Brassica juncea	Brown mustard	Cruciferae
	Indian mustard	
Brassica napus		Cruciferae
annual types	Oilseed rape	
biennial types	Forage rape	
	Swede	
Brassica nigra	Black mustard	Cruciferae
Brassica oleracea		Cruciferae
var. *acephala*	Kale	
var. *botrytis*	Cauliflower	
var. *capitata*	Cabbage	
var. *gemmifera*	Brussels sprout	
var. *italica*	Sprouting broccoli	
	Calabrese	
Cajanus cajan	Pigeon pea	Leguminosae
Camellia sinensis	Tea	Camelliaceae
Canavalia ensiformis	Jackbean	Leguminosae
Cannabis sativa	Hemp	Moraceae
Capsicum annuum	Sweet pepper	Solanaceae
	Chilli pepper	
	Paprika	
Carica papaya	Papaya	Caricaceae
Carthamnus tinctorius	Safflower	Compositae
Catharanthus roseus	Madagascan periwinkle	Apocyanaceae
	Rose periwinkle	

Table 2. Continued

Latin name	Common name	Family
Chenopodium spp.	Quinoa	Chenopodiaceae
Chichona spp.	Quinine	Rubiaceae
Chichorium intybus	Chicory	Compositae
Chrysanthemum cinerariifolium	Pyrethrum	Compositae
Cicer arietinum	Chickpea	Leguminosae
Citrullus lanatus	Watermelon	Cucurbitaceae
Citrus limon	Lemon	Rutaceae
Citrus × paradisi	Grapefruit	Rutaceae
Citrus sinensis	Sweet orange	Rutaceae
Cocos nucifera	Coconut	Palmae
Coffea arabica	Arabica coffee	Rubiaceae
Coffea canephora	Robusta coffee	Rubiaceae
Coix lachryma-jobi	Adlay Job's tears	Gramineae
Colocasia esculenta	Eddo Taro	Araceae
Cucumis melo	Cantaloup Muskmelon	Cucurbitaceae
Cucumis sativus	Cucumber	Cucurbitaceae
Cucurbita pepo	Courgette Pumpkin Summer squash Zucchini	Cucurbitaceae
Dactylis glomerata	Cock's foot (UK) Orchard grass (US)	Gramineae
Datura stramonium	Thorn apple	Solanaceae
Daucus carota	Carrot	Umbelliferae
Dendranthema indicum	Chrysanthemum	Compositae
Dianthus caryophyllus	Carnation	Caryophyllaceae
Digitalis sp.	Foxglove	Scrophulariaceae
Dioscorea spp.	Yam	Dioscoreaceae
Elaeis guinensis	Oil palm	Palmae
Eleusine coracana	Finger millet Ragi	Gramineae
Eugenia caryophyllus	Clove	Myrtaceae
Eustoma grandiflorum	Lisianthus	Gentianaceae
Fagopyrum spp.	Buckwheat	Polygonaceae
Ficus carica	Fig	Moraceae
Foeniculum vulgare	Fennel	Umbelliferae
Fragaria × ananassa	Strawberry	Rosaceae
Glycine argyrea	Wild soybean	Leguminosae
Glycine canescens	Wild soybean	Leguminosae
Glycine max	Soybean	Leguminosae
Gossypium hirsutum	Cotton	Malvaceae
Helianthus annuus	Sunflower	Compositae
Helianthus tuberosus	Jerusalem artichoke	Compositae
Hevea brasiliensis	Rubber	Euphorbiaceae
Hordeum vulgare	Barley	Gramineae

Table 2. Continued

Latin name	Common name	Family
Humulus lupulus	Hop	Moraceae
Ipomoea batatas	Sweet potato	Convolvulaceae
Ipomoea purpurea	Morning glory	Convolvulaceae
Juglans regia	Walnut	Juglandaceae
Kalanchoe blossfeldiana	—	Crassulaceae
Lactuca sativa	Lettuce	Compositae
Lens culinaris	Lentil	Leguminosae
Lilium auratum	Lily	Liliaceae
Linum usitatissimum	Flax Linseed	Linaceae
Lolium multiflorum	Italian ryegrass	Gramineae
Lolium perenne	Perennial ryegrass	Gramineae
Lotononis bainsii	—	Leguminosae
Lotus corniculatus	Bird's foot trefoil	Leguminosae
Lupinus albus	Lupin	Leguminosae
Lycopersicon esculentum	Tomato	Solanaceae
Malus pumila	Apple	Rosaceae
Mangifera indica	Mango	Anacardiaceae
Manihot esculenta	Cassava	Euphorbiaceae
Medicago sativa	Alfalfa Lucerne	Leguminosae
Mentha × *spicata*	Mint	Labiatae
Mimulus guttatus	Monkey flower	Scrophulariaceae
Musa spp.	Banana	Musaceae
Nicotiana plumbaginifolia	Ornamental tobacco	Solanaceae
Nicotiana sylvestris	Ornamental tobacco	Solanaceae
Nicotiana tabacum	Tobacco	Solanaceae
Olea europaea	Olive	Oleaceae
Oryza glaberrima	African rice	Gramineae
Oryza sativa	Asian rice	Gramineae
Parthenium argentatum	Quayule	Compositae
Pennisetum typhoides	Bulrush millet Pearl millet	Gramineae
Persea americana	Avocado	Lauraceae
Petunia hybrida	Petunia	Solanaceae
Pharbitis nil	—	Convolvulaceae
Phaseolus acutifolius	Tepary bean	Leguminosae
Phaseolus coccineus	Runner bean	Leguminosae
Phaseolus lunatus	Lima bean	Leguminosae
Phaseolus vulgaris	Common bean Field bean (USA) French bean Haricot bean Kidney bean Navy bean	Leguminosae
Phoenix dactylifera	Date palm	Palmae
Piper nigrum	Black pepper	Piperaceae
Pisum sativum	Pea	Leguminosae

Table 2. Continued

Latin name	Common name	Family
Populus spp.	Aspen Poplar	Salicaceae
Prunus amygdalus	Almond	Rosaceae
Prunus armeniaca	Apricot	Rosaceae
Prunus avium	Sweet cherry	Rosaceae
Prunus cerasus	Sour cherry	Rosaceae
Prunus domestica	Plum	Rosaceae
Prunus persica	Peach	Rosaceae
Psophocarpus tetragonolobus	Winged bean	Leguminosae
Pyrus communis	Pear	Rosaceae
Raphanus sativus	Radish	Cruciferae
Ribes nigrum	Blackcurrent	Grossulariaceae
Ricinus communis	Castor bean	Euphorbiaceae
Rorippa nasturtium-aquaticum	Watercress	Cruciferae
Rubus idaeus	Raspberry	Rosaceae
Rubus spp.	Blackberry	Rosaceae
Saccharum officinarum	Sugar cane	Gramineae
Saponaria officinalis	Soapwort	Caryophyllaceae
Secale cereale	Rye	Gramineae
Sesamum indicum	Sesame	Pedaliaceae
Sinapis alba	White mustard	Cruciferae
Solanum melongena	Eggplant Aubergine	Solanaceae
Solanum tuberosum	Potato	Solanaceae
Sorghum bicolor	Sorghum	Gramineae
Stylosanthes humilis	—	Leguminosae
Theobroma cacao	Cocoa	Sterculiaceae
Trifolium pratense	Red clover	Leguminosae
Trifolium repens	White clover	Leguminosae
Trifolium subterraneum	Subterranean clover	Leguminosae
Triticosecale spp.	Triticale	Gramineae
Triticum aestivum	Wheat	Gramineae
Triticum durum	Durum wheat Pasta wheat	Gramineae
Vaccinium corymbosum	Blueberry	Ericaceae
Vaccinium macrocarpon	Cranberry	Ericaceae
Vicia faba	Broad bean Faba bean Field bean (Europe)	Leguminosae
Vicia narbonensis	Narbonne vetch	Leguminosae
Vicia villosa	Hairy vetch	Leguminosae
Vigna aconitifolia	Moth bean	Leguminosae
Vigna radiata	Mung bean	Leguminosae
Vigna unguiculata	Cowpea	Leguminosae
Viscum album	Mistletoe	Viscaceae
Vitis vinifera	Grape	Vitaceae
Xanthosoma spp.	Cocoyam	Araceae
Zea mays	Corn Maize	Gramineae

6. REFERENCES

1. Hayward, H.E. (1938) *The Structure of Economic Plants.* MacMillan, New York.

2. Cutter, E.G. (1978) *Plant Anatomy. Part* 1. *Cells and Tissues.* Edward Arnold, London.

3. Johnson, G.D. and Aroujo Nogueira, G.M. (1981) *J. Immunol. Meth.*, **43,** 349.

4. Gahan, P.B. (1984) *Plant Histochemistry and Cytochemistry: An Introduction.* Academic Press, London.

5. Horobin, R.W. (1982) *Histochemistry.* Butterworths, London.

6. Jensen, W.A. (1962) *Botanical Histochemistry.* W.H. Freeman, San Francisco.

7. O'Brien, T.P. and McCully, M.E. (1981) *The Study of Plant Structure: Principles and Selected Methods.* Termacarphi Pty., Melbourne, Australia.

8. Pearce, A.G.E. (1980) *Histochemistry: Theoretical and Applied* (*4th edn*). Churchill-Livingstone, London, Vol. 1.

9. Hughes, J. and McCully, M.E. (1975) *Stain Technol.*, **50,** 319.

10. Harris, N. and Gates, P.J. (1984) *Plant Cell Environ.*, **7,** 699.

11. Jeffrey, C. (1973) *Biological Nomenclature.* Edward Arnold, London.

7. FURTHER READING

1. Bell, A.D. (1991) *Plant Form: An Illustrated Guide to Flowering Plant Morphology.* Oxford University Press, Oxford.

2. Fahn, A. (1974) *Plant Anatomy.* Pergamon Press, Oxford.

3. Gates, P.J. and Oparka, K.J. (1982) *Plant Cell Environ.*, **5,** 251.

4. Mabberley, D.J. (1987) *The Plant Book.* Cambridge University Press, Cambridge.

5. Mauseth, J.D. (1988) *Plant Anatomy.* Benjamin/Cummings, California.

6. Nicholas, J.R., Gates, P.J. and Grierson, D. (1986) *J. Hort. Sci.*, **61,** 417.

7. Simmonds, N.W. (ed.) (1976) *Evolution of Crop Plants.* Longman, London.

8. Toothill, E. (1984) *The Penguin Dictionary of Botany.* Penguin, London.

CHAPTER 2

PLANT NUCLEIC ACIDS

E.J. Croy, T. Ikemura, A. Shirsat and R.R.D. Croy

1. DNA ISOLATION FROM PLANTS

There are many difficulties associated with the isolation of undegraded plant nucleic acids which are free from contaminating proteins and polysaccharides. Different extraction methods are necessary since different plant groups contain diverse secondary compounds which may interfere with the isolation. Many of the papers referred to in *Table 1* tackle specific problems of plant DNA isolation and, generally, provide information which may be used elsewhere. The main experimental points on which most references agree are as follows: most plant cells have very tough cell walls, therefore vigorous methods are required to break cells open. The efficiency of the physical force exerted on a tissue is critical for a good recovery, but excessive force degrades very high molecular weight molecules through shearing. Plant tissues contain active nucleic acid degrading enzymes and the time between cell rupture and inactivation of these nucleases is a critical factor in the isolation of intact DNA and RNA (see *Chapter 3*). Tissues with a high carbohydrate content may benefit from extraction with larger volumes of buffer. Polysaccharides may be removed by binding to ethidium bromide (EtBr) or by precipitation with cetyl trimethylammonium bromide (CTAB). RNA can be removed by cesium chloride (CsCl)/EtBr density gradient centrifugation. Plant DNA accumulating a brown coloration may indicate the action of potent polyphenoloxidase activity which may be inhibited by inclusion of disodium diethyldithiocarbamate (SDDTC) in the extraction buffer. The methods given in *Table 1* do not give detailed procedures on how to obtain DNA from specific plant tissues, nor is this the result of an exhaustive search for all plant DNA isolation methods, but rather, it represents a compilation of those methods and background information which are referred to most frequently in molecular biology papers. *Table 2* is a key to the species listed in *Table 1*. *Table 3* lists the range of tissue types referred to in *Table 1*, from which DNA has been isolated.

2. ISOLATION OF *AGROBACTERIUM* Ti PLASMID DNA

Ti plasmid constructs are transferred to *Agrobacterium tumefaciens* host cells by direct transformation (e.g. by the freeze/thaw method (243)), tri-parental mating or by electroporation (see *Chapter 7, Table 20*), and it is important to ensure that the correct construct is present and intact, before use in plant transformation. Therefore, Ti plasmid gene DNA is isolated from selected *A. tumefaciens* clones and checked by restriction analysis and Southern blotting to confirm the integrity of the inserted gene construct. Due to the low copy number and the large size of Ti plasmids, particularly the co-integrative types, it is difficult to isolate high molecular weight, covalently closed circular plasmid DNA in good yields from *Agrobacterium* by normal plasmid mini-prep methods. *Table 4* lists the most commonly used methods employed for this task. Although numerous nutrient broths are recommended for the growth of *A. tumefaciens*, YEB (see *Table 4*) appears to give the most consistent results. Binary vectors such as pBIN19 (see *Chapter 4*) which are propagated in *E. coli* hosts can be prepared in quantity by standard plasmid preparative and purification methods.

► p. 30

Table 1. A compilation of plant DNA isolation methods

Ref.	Name and initial physical treatment	Extraction buffer	Purification	Tissue[a]	Species[b]	Yield ($\mu g\, g^{-1}$)	Size (kb)	Comments
1	**Bendich *et al.*** 1 Homogenize fresh tissue at 4°C 2 Grind frozen tissue in solid CO_2 3 Grind dry seeds	EDTA, buffer at osmoticum, sarkosyl	CsCl, EtBr	A, D, E	17		> 150	Total DNA, no organelle DNA isolation Use 0.1 M SDDTC if there is a lot of polyphenoloxidase activity
2	**Blin and Stafford** Grind in liquid N_2 in a Waring blender	EDTA, SAR, Prot.K at pH 8.0	Phenol, cresol, HQ	Eukaryotic cells				Background reading for many methods
3	**Boffey** Grind in liquid N_2 Isolation of organelles	Tris, EDTA, CTAB, NaCl, MERC	IAC	A	Many			Essential reading; good description of methods for extracting DNA from organelles
4	**Darby *et al.*** Shake with Ballotini beads in a rotary ball mill	Bentonite, CTAB, NaCl, acetate buffer at pH 6.0	Ethanol pptn	Bacterial cells				Background reading for many of the methods; good DNA isolation flow-chart
5	**Dellaporta *et al.*** Grind in liquid N_2 Polytron	Tris, EDTA, NaCl, SDS, MERC at pH 8.0	KAc Isopropanol pptn	A	1, 2, 11, 22, 23, 24, 25	50–100	> 50	Minipreparation method; for quantities of tissue < 1.0 g; difficult tissue can be lyophilized, e.g. 11 Further purification, precipitation of DNA by NaOAc and CTAB
6	**Domoney *et al.*** Grind in liquid N_2	SSC, EDTA, SDDTC, SDS, IAC at pH 8.0	CsCl/EtBr gradient Phenol, ethanol pptn	A	15			

7	**Doyle and Doyle** Grind fresh at 60°C	CTAB, Tris, NaCl, EDTA, MERC at pH 8.0	IAC, isopropanol pptn	A	44, 45, 46[c]	10–1000	> 30	Modification of Saghai-Maroof method (ref. 20); 0.5–1.5 g of tissue; RNase treatment may be necessary; EtBr and PVP used if phenolic substances present; rapid method; not high molecular weight DNA
8	**Ellis *et al.*** Grind in liquid N_2	NaCl, SDDTC, EDTA, SDS, TSC, IAC at pH 8.9	Phenol/cresol, ethanol pptn	A	15	100–500	> 30	May have RNA contamination; quantities of tissue from 1.0 g
9	**Ellis *et al.*** Grind in liquid N_2	TSC, EDTA, mannitol, BSA, TRIT at pH 6.0	LMP agarose (BRL), liquid paraffin SDS, Prot.K	A	15, 148		50–100 and 3000	Bimodal size distribution of isolated DNA; majority about 3 Mb
10	**Ganal and Tanksley** Protoplast preparation	Mannitol, MES at pH 5.6	Agarose, EDTA, Prot.K, sarkosyl	A	3, 6, 24	4 μg/240 μg	2000	Suitable for PFGE analysis; Pre-electrophoresis necessary to get rid of low molecular weight contaminants; excess restriction enzymes necessary for digestion
11	**Graham** Grind in liquid N_2	EDTA, SDS, PER, phenol, IAC	IAC, ethanol pptn	Whole organisms and tissues	[d]		4	Modified and used extensively for most plant tissues;[d] inclusion of pronase treatment and CsCl/EtBr gradient to remove protein and RNA contamination
12	**Guzman and Ecker** Protoplast preparation	Tris, EDTA, SAR at pH 9.0	Sucrose gradients	C	8, 40		> 12	Crude extraction but potentially very high molecular weight DNA
13	**Jokufu and Goldberg** Grind in liquid N_2 Isolation of nuclei	Tris, EDTA, SAR	CsCl, EtBr	A	7, 11, 22, 24, and many others		> 150	Informative paper; produces DNA suitable for RFLP analyses

Table 1. Continued

Ref.	Name and physical treatment	Extraction buffer	Purification	Tissue[a]	Species[b]	Yield ($\mu g\ g^{-1}$)	Size (kb)	Comments
14	**Landry *et al.*** Freeze tissue, blend in Waring blender	Sp, Sd, Tris, EDTA, PMSF, S, MERC, SAR, TRIT, KCl at pH 9.5	CsCl/EtBr gradient	A	43			
15	**Marmur** Various methods of cell distribution	EDTA, SDS, IAC	Isopropanol	50 different organisms				Essential background reading; reference quoted in many of the other methods in this table
16	**Murray and Thompson** 1 Lyophilize 2 Grind in liquid N_2	Tris, EDTA, NaCl, CTAB, MERC at pH 8.0	OCTC, SAR, CsCl/EtBr	A, B, J	1, 3, 6, 8, 15, 20, 22, 26, 28, 29, 30, 39, 40, 41, 42	20–700	> 50	RNA can be removed by RNase treatment or CsCl gradient centrifugation; a later Murray modification included the addition of NBIS, PT and proteinase K
17	**Nagao *et al.*** Grind in a polytron with buffer at 4°C	MES, $MgCl_2$, $CaCl_2$, KCl, NaCl, EtBr, MERC, TRIT, S, Gly	Percoll gradient CsCl/EtBr	L	11		> 60	
18	**Oono and Sugiura** Preparation of nuclei from embryos	MES, $CaCl_2$, $MgCl_2$, S at pH 6.0	Sucrose gradient. See ref. 15	F	32		40	RNase and proteinase K treatment necessary
19	**Rogers and Bendich** Grind in dry ice	At 65°C; CTAB, Tris, EDTA, NaCl, PVP, IAC at pH 8.0	Ethanol pptn	A	1, 2, 3, 4, 5, 7	0.3–200	0.4–30	Age and storage of tissue affected the yield; DNA can be extracted from milligram amounts of plant tissue; detailed discussion of results with reference to different species
				B	8			
				C	1			
				D, E	3, 7, 9, 10, 11, 12, 13, 14, 15			
				F	3, 4, 7, 10, 11, 12, 14, 15, 17, 18, 19			

				G	4, 10, 15			
				H	3, 7, 12			
				I	4, 7, 16			
				M	3, 4, 7, 44, 45, 46			
				N	55, 56, 57, 58, 59, 60			
20	**Saghai-Maroof *et al.*** Lyophilize and grind	Tris, NaCl, EDTA, CTAB, MERC, OCTC at pH 8.0	Isopropanol, AAc, ethanol pptn	B	12	200–300		Modification of Murray and Thompson (16)
21	**Sun *et al.*** See ref. 3	See ref. 3	Phenol, CsCl/EtBr	1	14		> 60	Modification of Blin and Stafford (2)
22	**Sing and Slightom** Dry grind at room temperature	GHCl, EDTA, Tris, SAR, MERC at pH 8.0	Prot.K, SDS, phenol, isopropanol, CsCl/EtBr gradient, NaOAc, HQ	A, F	11, 15, 21, 28	200	20–50	Two methods which have been used successfully for many other species, including 27
23	**Takaiwa *et al.*** See ref. 18 Preparation of nuclei from dry seeds	EDTA, iso-osmotic buffer See ref. 1	CsCl/EtBr	F	32			
24	**Taylor and Powell** 1 Lyophilize fresh tissue	Tris, CTAB, EDTA, NaCl, MERC at pH 8.0	IAC, CsCl, EtBr, ethanol pptn	A	1	25–40		Very informative paper; RNA can be extracted simultaneously; modified method of Murray and Thompson (16); sodium perchlorate is added if there is considerable carbohydrate; has been used for many other species
	2 Grind in dry ice	Tris, EDTA, SDS, phenol, SSC, HQ at pH 8.0	Phenol, IAC, ethanol pptn, CsCl, EtBr	J A	1 1	5–40		
25	**Thompson *et al.*** 1 Grind in liquid N_2	Tris, EDTA, NaCl, DTT, SDS, HQ, at pH 8.0	Phenol, HQ, IAC, ethanol pptn, CsCl	O	3, 33			Good description of methods using small quantities of tissue
	2 Isolation of nuclei	Tris, EDTA, NaCl, SDS, DPC, S at pH 8.0	Phenol, IAC, ethanol pptn	O	3, 33			
26	**Ward and Jen** Grind in liquid N_2	Tris, EDTA, SAR, Prot.K at pH 8.0	CsCl; concentration by ultrafiltration	A	8	10	> 50	Not appropriate for *Brassica* species

Table 1. Continued

Ref.	Name and initial physical treatment	Extraction buffer	Purification	Tissue[a]	Species[b]	Yield ($\mu g\ g^{-1}$)	Size (kb)	Comments
27	**Weber *et al.*** Grind lyophilized tissue with AlO_2	Tris, EDTA, NaCl, CTAB, MER, OCTC at pH 8.0	Ethanol pptn	A, J, K, P	1			
28	**Weising *et al.*** Grind fresh or lyophilized tissue in liquid N_2	Tris, EDTA, NaCl, MER, CTAB at pH 8.0	1 CsCl gradient 2 Quiagen tips 3 CL-6B sepharose columns 4 Ammonium acetate	Most tissues	Various	100–500		Modified Saghai-Maroof (20) and Doyle (7) methods; good descriptions of different methods of purification, species comparison and DNA yields
29	**Willmitzer and Wagner** Nuclei isolation at 4°C	Prot.K, phenol	CsCl	C	1			Large amounts of tissue; good study of isolation of nuclei
30	**Zimmer and Newton** Isolation of nuclei from seedlings	Tris, EDTS, SAR, Prot.K, phenol, IAC at pH 8.0	AAC; ethanol pptn	B	7	10–100		Rapid method; seedlings can be screened without sacrificing the plants; small quantities <0.1 g

[a]See *Table 3* for a key to tissue types.
[b]See *Table 2* for a key to species.
[c]See *Table 2*, note 1.
[d]See *Table 2*, note 2.

AAc, ammonium acetate; BSA, bovine serum albumin; CTAB, cetyltrimethylammonium bromide; DPC, diethyl pyrocarbonate; DTT, dithiothreitol; EDTA, ethylenediaminetetraacetic acid; EtBr, ethidium bromide; GHCl, guanidine hydrochloride; Gly, glycerol; HQ, hydroxyquinoline; IAC, isoamyl alcohol/chloroform 1:24; Isp, isopropanol; KAc, potassium acetate; MERC, mercaptoethanol; MES, 2-morpholinoethanesulfonic acid monohydrate; NaOAc, sodium acetate; NBIS, sodium bisulfite; OCTC, octanol/chloroform 1:24; PER, sodium perchlorate; PFGE, pulsed field gel electrophoresis; PMSF, phenylmethyl sulfonyl fluoride; pptn, precipitation; Prot.K, proteinase K; PT, phenanthroline; PVP, polyvinylpyrrolidone; RFLP, restriction fragment length polymorphism; S, sucrose; SAR, sarkosyl *N*-laurylsarcosine; Sd, Spermidine; SDDTC, disodium diethyldithiocarbamate; SDS, sodium dodecyl(lauryl)sulfate; Sp, spermine; SSC, sodium chloride/sodium citrate buffer; Tris, tris(hydroxymethyl)aminomethane; TRIT, triton X-100; TSC, trisodium citrate.

Table 2. Key to species listed in *Table 1*

No.	Species name	Common name	Refs
1	*Nicotiana tabacum*	Tobacco	5, 11, 16, 19, 24, 27, 29; see note 1
2	*Petunia hybrida*	Petunia	5, 19
3	*Triticum aestivum*	Wheat	10, 16, 19, 25
4	*Vicia faba*	Broad bean	19
5	*Vitis vinifora*	Grape	19
6	*Solanum tuberosum*	Potato	19, 16; see note 1
7	*Zea mays*	Maize	13, 19, 30
8	*Arabidopsis thaliana*	Thale cress	12, 16, 19, 26
9	*Allium cepa*	Onion	19
10	*Citrullus vulgaris*	Watermelon	19
11	*Glycine max*	Soybean	5, 13, 17, 19, 22
12	*Hordeum vulgare*	Barley	19, 20
13	*Linum usitatissimum*	Flax	19
14	*Phaseolus vulgaris*	French bean	19, 21
15	*Pisum sativum*	Pea	6, 8, 11, 16, 19, 22
16	*Cedrus deodera*	Cedar	19
17	*Cucumis melo*	Musk melon	1, 19
18	*Cucurbita pepo*	Courgette	19
19	*Vicia villosa*	Hairy vetch	19
20	*Lotus corniculatus*	Lotus	11, 16
21	*Canavalia ensiformis*	Jack bean	22
22	*Nicotiana plumbaginifolium*	Tobacco	5, 13, 16
23	*Nicotiana sylvestris*	Tobacco	5
24	*Lycopersicon esculentum*	Tomato	5, 10, 13
25	*Amaranthus* sp.	Amaranthus	5
26	*Avena sativa*	Oats	16
27	*Ricinus communis*	Castor bean	22
28	*Brassica napus*	Rapeseed	16, 22, 28
29	*Vigna unguiculata*	Cowpea	16
30	*Vigna radiata*	Mung bean	16
31	*Antirrhinum majus*	Snapdragon	16
32	*Oryza sativa*	Rice	7, 18, 23
33	*Secale cereale*	Rye	7, 25
34	*Sorghum vulgare*	Sorghum	7, 19
35	*Helianthus annuus*	Sunflower	19
36	*Gossypium hirsutum*	Cottonseed	19
37	*Eleusine coracana*	Millet	19
38	*Cocus nucifera*	Coconut	19
39	*Bambusa* sp.	Bamboo	19
40	*Daucus carota sativa*	Carrot	12, 16
41	*Mimulus guttatus*	Monkey flowers	16
42	*Atriplex* sp.	Saltbush	16
43	*Lactuca sativa*	Lettuce	14
44	*Capsicum annuum*	Pepper	7, 19; see note 2
45	*Agropyron junceum*	Herbarium species	7, 19; see note 2
46	*Poa juncifolia*	Herbarium species	7, 19; see note 2

Table 2. Continued

No.	Species name	Common name	Refs
47	*Poa palustris*	Herbarium species	7, 19; see note 1
48	*Raphanus sativus*	Radish	9
49		Miscellaneous mummified species	19
50	*Cannabis sativus*	Hempseed	28
51	*Psophocarpus tetragonorobus*	Winged bean	28
52	*Lupinus albus*	Lupin	28
53	*Lens culinaris*	Lentil	28
54	*Arachus hypogaea*	Peanut	28
55	*Encelia virginensis*	Mummified seeds	19
56	*Eschscholizia minutiflora*	Mummified seeds	19
57	*Lycium shockleyi*	Mummified seeds	19
58	*Juniperus osteosperma*	Mummified seeds	19
59	*Opuntia ramosissima*	Mummified seeds	19
60	*Symphoricarpos* sp.	Mummified seeds	19
61	*Coix lachryma-jobi*	Job's tears	16

Note 1: Extensively used for most species and tissues.
Note 2: Including large numbers of different angiosperm groups: monocots – palms, grasses, sedges, orchids; dicots – walnuts, hickories, oaks, beeches, legumes, apples and relatives, saxifrages, lobelias, brassica, portulacas. Also conifers and ferns and dried plant specimens.

Table 3. Tissue types cited in *Table 1*

A	Leaves
B	Whole seedlings
C	Suspension cultures
D	Grains
E	Seeds
F	Embryos
G	Cotyledons
H	Endosperm
I	Pollen
J	Callus
K	Roots
L	Hypocotyl
M	Herbarium leaves
N	Mummified seeds and embryos
O	Shoots

Table 4. Methods for isolation of plasmid DNA from *Agrobacterium*

Reference	Cell growth	Lysis and shearing	Removal of protein	Concentration	Purification	Remarks
65	Room temp., 12–24 h, large volume (1 l) cultures	TE, sarkosyl, pronase; mechanical shearing, pH adjustment. Modification by freezing to reduce mechanical treatment	1 × phenol extraction	Ethanol precipitation	CsCl–EtBr gradient centrifugation	Long laborious method; may take longer than 1 day. Mechanical treatment may lead to low plasmid yield. Good informative paper
66	28°C, 24–48 h, nutrient broth, small volume (1.5 ml) cultures	TE, sarkosyl, pronase; mechanical shearing by pipette	4 × phenol/chloroform extractions	Ethanol precipitation	None	Most commonly used method. Numerous modifications. Difficult to separate phases
67	28°C 24 h, YEB, small volume (1.5 ml) cultures	TE or TEA, SDS, NaOH heated to 65°C	1 × phenol/chloroform extraction	None	None	Very rapid, small volumes only but needs extended restriction time
68	28°C, 24 h, small volume (1.5 ml) cultures	TE, sarkosyl, pronase	Ammonium acetate at 0 °C	Equal volume of isopropanol	None	No phenol extraction; always yields easily restrictable DNA
69	28°C, 24 h, small volume (1 ml) cultures	Lysozyme, TE, SDS, NaOH – shake to mix	1 × phenol extraction (equilibrated with NaOH/SDS)	2 × ethanol precipitation	None	Rapid quick-screen method. Chromosomal DNA contamination high if neutral phenol used or cells mixed harshly

TE, Tris–HCl–EDTA buffer; TEA, Tris–acetate–EDTA buffer. For other abbreviations see *Table 1*.

3. CALCULATION OF GENE COPY NUMBER EQUIVALENTS

The estimation of the number of copies of a gene sequence in the genome of an experimental plant is an essential parameter for describing a gene family, and is useful for the isolation and characterization of the genes from a genomic library. Gene copy numbers are estimated by performing genomic Southern blotting with known amounts of pure genomic DNA and standard amounts of a cloned gene or cDNA sequence. The blot is then hybridized with a suitable labeled gene probe, usually the standard sequence or a sequence contained within the standard sequence under conditions of probe excess. The amount of labeled probe hybridizing to the gene fragments within the genomic DNA is compared with the amount hybridizing to the standard sequence. This is known as a gene copy reconstruction experiment. The best results are obtained with restriction enzymes which separate the members of a gene family on differently sized restriction fragments of intermediate sizes (2–10 kb) which will blot efficiently. In practice, it is usual to try several enzymes which do not cut within the gene sequence used for probing. Labels for probes can either be radioactive (^{32}P, with detection by enhanced autoradiography) or non-radioactive (either by digoxygenin labeling with antibody–conjugate detection (Boehringer Mannheim), by biotin labeling with streptavidin–conjugate detection (e.g. Gibco-BRL) or by horseradish peroxidase labeling with chemiluminescence detection (31) (Amersham International)). The relative amounts of probe hybridizing can be estimated by visual examination of exposed X-ray film or stained filters, or by densitometric scanning of either of these. Alternatively, the amount of radioactivity bound to DNA on a filter can be quantitatively estimated directly using a one-dimensional or two-dimensional scanning radioactivity counter, or by phosphorimaging (Molecular Dynamics, Bio-Rad), or by excision of the bands from the filter and counting by liquid scintillation counter.

A prerequisite for the estimation of gene copy reconstructions is a knowledge of the size of the haploid plant genome. *Table 5* lists the haploid genome sizes for a number of selected plant species. A large number of other plant genome sizes has been published, particularly in the extensive works of Bennet, Smith and Heslop-Harrison (32–34). Three methods are given here for the estimation of gene copy equivalents.

3.1. Formula

The following formula is commonly used to calculate gene copy equivalents of a cloned gene fragment in a genomic Southern blot hybridization. The result is expressed in picograms of the cloned fragment per gene copy per μg of genomic DNA:

For 1 μg genomic DNA:

$$1 \text{ gene copy} = \frac{\text{size of standard gene fragment (bp)}}{\text{haploid size of genome (bp)}} \times 10^{6} \text{ pg}$$

(size of standard = size of vector (bp) + size of cloned gene insert (bp)). For the equivalent of multiple gene copies or for larger amounts of genomic DNA analyzed, multiply this amount by the appropriate factor.

3.2. Nomogram

Figure 1 is a nomogram for the estimation of gene copy equivalents for plant genomes of different sizes. The left-hand bar represents the plant haploid genome sizes in base pairs covering the range from smallest plant genome sizes (e.g. *Arabidopsis*) to the largest plant genome sizes (e.g. *Lilium* (32)). The central bar represents the size, in base pairs, of the cloned gene fragment (including vector, if appropriate) to be used as the standard. The right-hand bar shows the amount of the cloned gene fragment, in picograms (10^{-12} g), representing

Table 5. Plant genome sizes (see *Chapter 1* for common names of species)

Species	Monocot. (M) Dicot. (D)	Ploidy level	Genome size 1 C value (bp)
Allium cepa L.	M	2	1.72×10^{10}
Anemone blanda	D	2	1.31×10^{10}
Antirrhinum majus L.	D	2	1.54×10^{9}
Arabidopsis thaliana	D	2	1.9×10^{8}
Arachis hypogeae L.	D	4	1.7×10^{9}
Avena sativa L.	M	6	1.3×10^{10}
Brassica campestris	D	2	7.7×10^{8}
Brassica napus L.	D	2	1.5×10^{9}
Brassica oleracea	D	2	8.7×10^{8}
Capsicum annuum L.	D	2	5.2×10^{9}
Coix lachryma-jobi L.	M	4	1.5×10^{9}
Datura innoxia Miller	D	2	2.2×10^{9}
Daucus carota L. ssp. *carota*	D	2	1.2×10^{9}
Lathyrus sativus L.	D	2	8.2×10^{9}
Lotus corniculatus	D	4	0.96×10^{9}
Lupinus albus	D	2	5.8×10^{8}
Lycopersicon esculentum Miller	D	2	2.2×10^{9}
Nicotiana tabacum L. var. *purpurea*	D	4	3.7×10^{9}
Oryza sativa L.	M	2	5.8×10^{8}
Petunia hybrida cv. blue dandy	D	2	1.5×10^{9}
Phaseolus vulgaris L.	D	2	1.7×10^{9}
Pisum sativum L.	D	2	5.0×10^{9}
Secale cereale L. cv. Petkus spring	M	2	8.5×10^{9}
Solanum tuberosum L.	D	4	2.0×10^{9}
Sorghum caudatum	M	4	4.7×10^{9}
Triticum aestivum L. cv. Chinese spring	M	6	1.7×10^{10}
Triticum monococcum L.	M	2	5.9×10^{9}
Vicia angustifolia L.	D	2	2.2×10^{9}
Vicia faba L.	D	2	1.2×10^{10}
Zea mays L. cv. Golden bantam	M	2	3.2×10^{9}

one gene copy per μg of genomic DNA of the selected haploid genome size. To calculate the required amount of a standard fragment for a gene reconstruction, use a straight edge to align the haploid genome size with the known size of standard gene fragment and read off the amount of probe DNA equivalent to one gene copy in one μg genomic DNA, where the line intersects the right-hand bar. For the analysis of larger amounts of genomic DNA, simply multiply the result by the number of μg analyzed; similarly for multiple gene copies multiply the result by 2, 5 or 10 as appropriate for the required number of gene copies for the reconstruction.

3.3. Computer program

The computer program GENE written by Croy and Gilroy, enables gene copy calculations to be performed rapidly and conveniently for any known plant genome and for any size of

Figure 1. Nomogram for the estimation of gene copy equivalents.

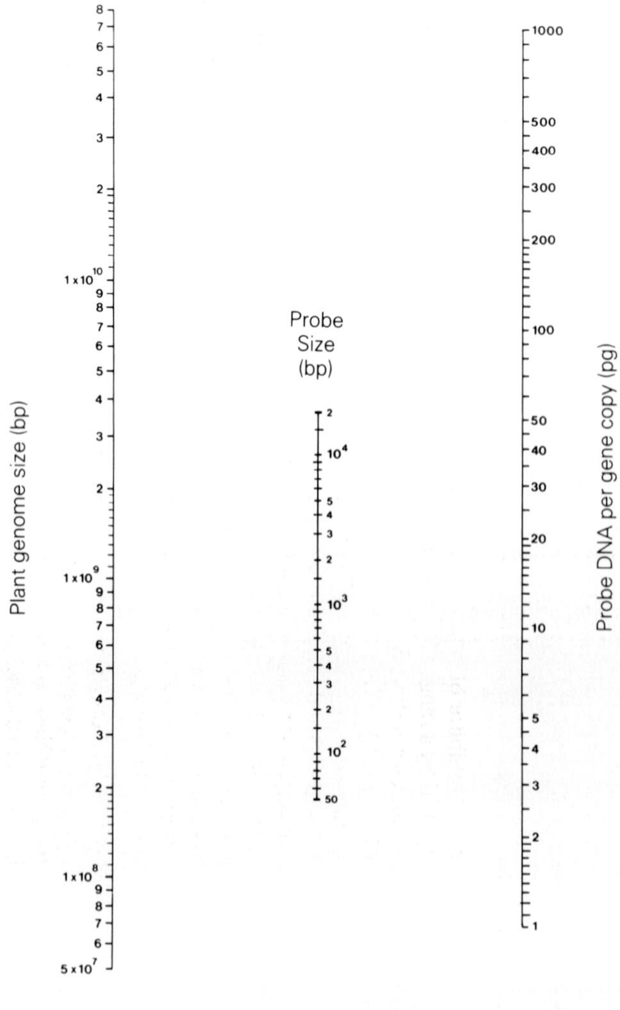

cloned gene fragment. It also allows the user to maintain a 'personal' list of plant genome sizes. GENE also includes the option for the estimation of gene library sizes for any plant species, a calculation which also depends on the genome size. (See also *Chapter 7, Figure 1*, of *Molecular Biology Labfax* for a graph to facilitate the calculation of genomic library sizes.) A disk copy of the GENE program can be obtained from the editor on request.

4. RNA ISOLATION FROM PLANTS

Just as there are many difficulties in isolating DNA from plant material, so there are also extreme difficulties with the isolation of intact RNA from plants. Major problems arise through the ubiquitous, highly active plant ribonucleases which add to the difficulties in handling plant tissues. The inclusion of RNase inhibitors in extraction media is often recommended to minimize the degradation by RNases; *Table 6* lists the main ones used. Unlike DNase enzymes, RNase enzymes are extremely resilient and will survive harsh treatment such as boiling and extremes of pH. The use of RNase inhibitors may therefore be essential for the isolation of intact RNA from certain types of tissues. Furthermore, although RNases

► p. 36

Table 6. RNase inhibitors

Inhibitor	Molecular weight	Example sources	Description (refs)
RNasin	49 847	Promega USB Gibco-BRL Boehringer	A protein inhibitor of RNase A-type enzymes, isolated from human placenta and supplied by a range of companies. Also available as a cloned (recombinant) version-rRNasin from Promega. The protein has an isoelectric point of 4.7 and is active over a pH range of 5–8 (optimum 7–8) and requires > 1 mM dithiothreitol. It inactivates by binding non-covalently to RNase and is a non-competitive inhibitor. It will inhibit RNases A, B and C but *not* RNase T1 or H, nuclease S1 nor the RNase from *Aspergillus*. RNasin is compatible with most enzyme systems used for transcription, translation and reverse transcription of RNA, and can be included in these reactions. It is not compatible with protein denaturants such as urea, SDS, phenol. One unit of RNasin is defined as the amount required to inhibit by 50% 5 ng of RNase A. RNasin is normally used at a concentration of 1 μg RNasin μl^{-1} of extraction or assay buffer (data from Promega) (35)
Diethylpyro-carbonate (DEPC, diethyl oxydiformate)	162	Sigma	0.2–1% (w/v) diethylpyrocarbonate can be used to pretreat most buffers and other solutions and glassware coming into contact with the isolated RNA, to remove traces of RNase. It is not compatible with Tris buffers. It is a potent non-specific inhibitor of RNAses and other nucleases through covalent modification by ethoxyformylation of amino and imidazole groups, (histidine) residues in the proteins and by cross-linking. DEPC must be removed from solutions by autoclaving prior to their use. DEPC can also be removed by ether extractions. DEPC destroys certain biological activities of RNA. **Care!** DEPC is a volatile liquid and a suspected carcinogen and should be handled only in a fume hood (36)

Table 6. Continued

Inhibitor	Molecular weight	Example sources	Description (refs)
Vanadyl ribonucleoside complexes (VRCs)	—	Gibco-BRL NEB	Vanadyl ribonucleoside complexes (VCRs) are transition-state analog competitive inhibitors of RNase, comprising complexes between oxovanadium (IV) ion and the four ribonucleosides. Supplied as a 200 mM, blue-green aqueous equimolar mixture which should be stored as aliquots at −20°C under nitrogen to prevent oxidation. VRCs are used at a concentration of 10 mM in buffers and other solutions but are incompatible with EDTA-containing buffers which will dissociate the complexes. VRCs absorb UV light at 260 nm. They are incompatible with certain enzyme systems, such as *in vitro* translation systems, and must be removed by multiple phenol extractions. One exception to this is reverse transcriptase, and VRCs are useful to protect RNA prior to cDNA synthesis. **Care!** VRCs are volatile toxic compounds harmful through skin absorption and by inhalation and should only be handled in a fume hood (37–39)
Aurintricarboxylic acid (ATA)	422	Sigma	Aurintricarboxylic acid is a triphenylmethane dye used at a concentration of 1 mM in RNA extraction buffers. RNA isolated in the presence of ATA can be translated *in vitro*. RNase A is inhibited by ATA levels as low as 10 μM. ATA also inhibits other nucleases including DNase, exonuclease III, nuclease S1 (40)
Macaloid clay	—	National Lead Company, Houston Texas	This is a neutral, mineral clay which adsorbs RNases. The clay is prepared as a slurry and used at 0.015% (w/v) in RNA extraction buffers. RNase is removed by continuous mixing with the Macaloid suspension. The clay is removed by centrifugation after phenol extraction of the extract. Macaloid can be used to specifically remove RNase from a DNase preparation. Its preparation and use is described in (41)
RNase Block II	Undisclosed	Stratagene	A potent protein RNase inhibitor isolated from an undisclosed source and reported to be 10–25 times more effective than RNasin in protecting RNA against RNase A. Similar in use to RNasin. Sensitive to freezing, heat and protein denaturants (42). 1 unit is defined as the amount of RNase Block II required to inhibit 1 ng of RNase A by 90% for 1 h

Heparin	~ 16 000	Sigma	A sulfated heteropolysaccharide of glucosamine-*N*-sulfate units, the sodium salt is soluble in aqueous buffers. Supplied as a white powder which may be autoclaved. Used at a concentration of 50 $\mu g\ \mu l^{-1}$, heparin will decrease RNase activity by more than 95%. Heparin is a potent inhibitor of protein synthesis initiation ($< 1\ \mu g\ ml^{-1}$) so must be removed from RNA by centrifugation through sucrose or cesium chloride or by lithium chloride precipitation (43, 44)
Bentonite	—	BDH/Merck	A clay, like Macaloid, containing many charged sites capable of binding metal ions and basic proteins such as RNase. A 5% (w/v) stock suspension in 0.01 M sodium acetate, pH 6.0 is added to the extraction medium and is subsequently removed with bound RNase by centrifugation. 35 $\mu g\ ml^{-1}$ is effective against plant RNases at pH 6.0 but 3.5 mg ml^{-1} is required at pH 7.4. Stock solutions are pre-treated with a 2% (w/v) bentonite suspension. Bentonite may also absorb RNA as well as protein (45)
Other inhibitors	—	Various	A large number of other RNase inhibitors have been reported, which exhibit variable effectiveness depending on the extraction conditions and the source of material to be extracted. These include polyamines (spermidine); polyvinylsulfate; low molecular weight RNA or ribonucleotides; copolymers of L-tyrosine and L-glutamic acid; pyrophosphate; proteases – pronase and proteinase K; tri-isopropylnaphthalene 1,5 disulfonate (TINS) (see refs 46 and 47 for more details)

Abbreviations: EDTA, ethylenediaminetetraacetic acid; SDS, sodium dodecyl sulfate; UV, ultraviolet.

can be inactivated by strong protein denaturants, removal of these chemicals may allow the enzyme to renature and the RNase activity to reappear. Inhibitors are selected depending on the purpose for which the RNA has been isolated, e.g. for *in vitro* translation, cDNA synthesis or further purification. Such inhibitors may be incompatible with each other, or with other components in the extraction media or with enzyme systems to be used subsequently with the RNA. Certain RNase inhibitors can also be inhibitors of other nucleases and may therefore be appropriate also for DNA isolations. The list in *Table 6* does not include the wide range of protein dissociation agents, such as detergents, chaotropic reagents or protein solvents, such as phenol, which find general use for the isolation of most nucleic acids. Reference 46 is a good review of the range and use of inhibitors for RNA isolation.

RNA isolation methods have been developed for a wide range of plant tissues and species; some of the major ones are listed in *Table 7*.

Table 7. A compilation of plant RNA isolation methods

Refs	Method	Plant	Tissues	Comments
48	**Chirgwin *et al.***			
	Guanidium thiocyanate	Tobacco Tomato Mustard	Callus Leaf, fruit Seeds	Long method, gives excellent quality RNA; 20 mg RNA per gram of fresh tissue
	Rice	Shoots, leaf		
		Maize	Leaf	
49	**Chomczynski *et al.***			
	Guanidium thiocyanate, phenol/ chloroform	Many plant species	Whole plants, leaf, root, shoot	Quick method, high yield
50	**Cox and Goldberg**			
	Guanidium thiocyanate	Many species	Whole plants, leaf, root, shoot	Pellet from some species may be difficult to resuspend
51	**Logeman *et al.***			
	Guanidine hydrochloride	Rape Tobacco Potato Alfalfa	Leaf, root Leaf, root Leaf Leaf	Widely applicable; contaminating poly-saccharides can be removed by precipitation of RNA with 4 M sodium acetate
52	**Schmidt *et al.***			
	Phenol/ chloroform	Pea Maize Many species	Leaf, seed Leaf Leaf, seed	Minimal contamination with phenolics and polysaccharides
53	**Prescott and Martin**			
	Phenol/ chloroform	Tobacco Pea Many species	Most tissues	DNA removed by LiCl pptn of RNA
54	**Bustos *et al.***			
	Phenol/ chloroform	Many species	Seedlings, immature seeds, leaf	Carbohydrate removed by pptn with 2 M sodium acetate

Table 7. Continued

Refs	Method	Plant	Tissues	Comments
55	**Rerie *et al.***			
	Phenol/ chloroform	Many species	Most tissues	Can be done using 100 mg of tissue
56	**Ausubel *et al.***			
	Phenol/ chloroform	Rice Barley	Seeds Seeds	—
57	**Martin and Northcote**			
	Phenol/ chloroform	Castor bean	Seeds	Remove polysaccharides by washing in 3 M sodium acetate
58	**Jones *et al.***			
	Phenol/ chloroform	Crucifers Petunia	Most tissues	200 μg RNA per gram fresh weight tissue
59	**Hall *et al.***			
	SDS	Pea, oat Tobacco Many species	Seeds, leaf, pods, cotyledons	Extraction at 100°C good RNA
60	**Loening**			
	TINS	Many monocot. species	Most tissues	
61	**Rochester *et al.***			
	TINS phenol/ chloroform	Maize Tomato	Leaf, seed Fruit	Good yield
62	**Hensgens and van-os-Ruygrok**			
	TINS	Rice	Cell cultures	

Abbreviations: pptn, precipitation; SDS, sodium dodecyl sulfate; TINS, tri-isopropylnaphthalene 1,5 disulfonate.

5. CODON USAGE

T. Ikemura

The data presented in *Tables 8* and *9* were computed from the gene sequences contained in GenBank Release 72.0 (1992). To reveal the characteristics of the codon use of individual plants, the frequency (per thousand) of codons used in each organism for which more than 20 genes are available was calculated by summing up the numbers of each codon used. The number of genes summed for each organism is given in the row designated 'No. of genes', and the total codon number thus summed is given in the bottom row (Total). Names of the organisms used are listed in *Table 10.* Amino acids are added simply according to the universal codon table. A more comprehensive review of codon usage in individual genes has been published by EMBL CD-ROM (63; refer to 64 for details).

This work was supported by a Grant-in-Aid of Scientific Research on Priority Areas, 'Genome Informatics', from the Ministry of Education, Science and Culture of Japan.

Table 8a. Codon usage in plant nuclear genes

		ALF[a]	ASN	ATH	BLY	BNA	COT	CRE	EME	MZE	NEU	PEA	PET
No. of genes		27	43	175	103	45	37	56	62	169	88	74	33
ARG	CGA	4.0	2.4	4.5	1.4	6.1	8.8	1.5	9.9	3.1	3.8	4.2	3.8
	CGC	3.6	12.0	3.5	16.7	5.4	7.7	38.8	15.8	17.0	21.3	4.4	2.6
	CGG	0.2	4.7	3.3	11.2	5.0	5.0	5.0	9.7	7.9	5.7	1.5	2.3
	CGU	10.8	8.7	10.9	6.1	8.7	9.5	6.6	11.0	5.6	13.0	8.1	7.8
	AGA	6.5	1.6	15.3	2.7	14.8	17.8	0.4	5.6	5.0	4.6	15.5	10.1
	AGG	7.2	2.9	12.1	11.2	12.0	25.7	2.0	5.7	14.7	7.5	10.8	9.8
LEU	CUA	7.2	4.0	8.3	3.6	8.8	6.0	1.5	9.4	8.2	3.6	7.3	6.7
	CUC	15.8	24.7	18.3	32.5	20.5	12.4	8.8	23.6	26.1	36.9	14.7	10.5
	CUG	5.3	21.9	8.1	23.2	11.8	8.2	66.1	20.7	28.5	15.7	8.0	6.7
	CUU	20.1	12.4	24.2	8.4	24.8	15.0	4.0	17.8	15.4	16.0	24.0	26.9
	UUA	8.6	0.8	8.3	0.7	9.1	5.4	0.7	6.4	3.6	1.4	9.0	8.2
	UUG	22.3	10.7	19.8	5.5	16.7	12.9	1.9	13.0	12.1	12.7	22.6	23.3
SER	UCA	21.3	3.5	12.9	4.2	11.9	12.3	2.0	11.8	8.2	4.6	15.9	13.7
	UCC	5.2	28.0	10.6	20.6	14.5	10.5	18.8	19.0	17.2	22.3	13.0	8.1
	UCG	2.1	12.7	6.9	10.4	8.2	7.8	19.2	13.0	9.8	12.3	5.3	3.2
	UCU	16.3	17.6	19.7	4.7	15.7	14.0	4.0	16.1	9.2	11.4	21.4	17.3
	AGC	9.8	23.6	11.2	17.1	12.3	14.3	14.9	14.4	17.1	15.1	10.2	10.8
	AGU	12.2	8.9	10.8	3.8	12.2	8.2	2.0	8.4	4.3	4.8	12.3	12.7
THR	ACA	14.4	7.4	14.7	5.3	17.4	14.1	2.7	14.0	8.4	6.6	16.4	16.1
	ACC	12.7	43.6	15.4	27.9	17.3	15.2	40.1	20.2	21.3	31.1	14.2	11.7
	ACG	1.5	10.3	6.8	14.0	8.8	7.4	9.0	11.4	12.7	8.8	3.8	3.2
	ACU	21.0	22.4	19.3	5.2	16.4	15.7	6.4	15.0	9.0	11.2	23.3	27.5
PRO	CCA	27.8	3.3	16.6	14.0	16.8	13.2	3.1	12.2	14.8	6.7	20.5	19.5
	CCC	6.7	20.3	6.9	18.6	8.9	7.7	32.7	15.5	17.8	26.5	6.0	7.7
	CCG	3.4	7.9	7.6	18.9	8.7	5.8	11.1	12.3	21.5	10.4	5.6	2.1
	CCU	26.3	9.8	17.6	8.7	16.7	16.2	5.5	15.1	13.2	13.4	19.3	18.6
ALA	GCA	26.3	11.8	16.2	11.1	17.0	23.5	5.6	15.4	15.0	6.9	23.8	20.1
	GCC	11.9	39.1	15.8	45.8	15.8	16.8	60.2	28.5	35.8	48.3	13.2	18.4
	GCG	4.0	13.6	8.9	27.6	10.9	11.2	20.7	16.4	26.4	12.3	5.9	4.7
	GCU	46.4	29.7	37.4	12.7	28.9	32.6	20.0	23.6	26.1	24.7	36.3	39.6

GLY	GGA	28.7	14.0	31.2	10.6	24.1	26.0	2.3	14.5	10.5	7.7	28.6	39.4
	GGC	8.8	38.4	10.8	50.9	10.2	13.3	59.2	23.8	34.5	34.7	11.2	13.5
	GGG	5.3	5.5	9.5	18.7	13.1	20.1	3.7	10.5	14.5	5.3	7.4	12.9
	GGU	27.8	30.4	30.0	13.3	24.4	26.1	12.9	20.7	15.6	28.1	27.1	40.8
VAL	GUA	11.0	3.0	6.5	3.0	8.2	5.4	1.4	6.2	4.7	3.0	9.9	8.6
	GUC	11.2	29.8	17.2	28.4	18.6	11.8	22.0	23.9	22.6	36.7	11.5	15.3
	GUG	20.1	19.0	18.6	33.6	18.5	18.4	45.8	13.7	29.9	10.2	18.4	16.3
	GUU	36.1	14.7	28.3	8.4	24.8	19.1	4.8	18.1	12.2	16.2	34.1	30.9
LYS	AAA	37.5	7.9	23.5	6.3	23.2	21.2	0.8	14.1	8.5	5.3	30.7	26.3
	AAG	46.6	35.3	39.4	44.3	27.7	32.0	53.9	35.1	38.3	48.8	39.2	39.5
ASN	AAC	20.8	36.6	25.4	32.2	29.0	26.9	34.5	28.3	26.4	32.6	22.1	17.5
	AAU	21.1	9.3	16.2	5.8	17.5	17.5	1.4	13.8	8.3	7.6	22.4	20.5
GLN	CAA	21.0	7.8	17.1	19.1	24.8	37.9	2.9	15.4	19.9	10.2	18.8	19.0
	CAG	11.2	25.3	18.3	32.0	23.5	27.2	34.1	26.2	31.4	29.8	12.6	8.3
HIS	CAC	10.0	13.8	9.5	17.2	9.9	12.4	15.6	13.8	14.8	17.9	8.0	7.6
	CAU	10.8	5.1	9.3	4.5	7.4	13.0	2.0	10.6	6.5	6.6	13.6	11.8
GLU	GAA	37.8	12.9	27.5	7.7	23.6	44.0	1.4	23.5	13.1	11.2	35.0	31.1
	GAG	33.3	30.2	33.8	42.8	32.0	41.5	64.7	36.3	42.8	48.1	29.4	34.4
ASP	GAC	15.5	38.3	22.4	43.4	21.7	19.1	45.1	29.0	32.0	34.6	19.0	17.6
	GAU	37.3	26.6	31.7	9.8	27.2	28.5	9.1	25.2	13.6	20.7	32.6	32.4
TYR	UAC	18.9	31.4	17.9	24.4	18.4	16.2	26.7	18.5	22.9	23.6	13.2	13.3
	UAU	13.7	10.2	10.5	3.8	10.3	10.9	2.2	10.3	5.3	6.4	14.9	9.9
CYS	UGC	3.4	9.7	7.5	15.7	8.1	7.4	14.5	8.0	12.7	9.2	4.5	7.3
	UGU	5.2	5.1	8.6	3.8	11.4	6.2	2.0	4.6	3.8	2.3	7.8	9.1
PHE	UUC	21.0	26.0	24.6	31.7	26.5	23.7	34.9	23.8	29.4	28.6	19.6	20.8
	UUU	22.7	7.2	16.6	6.7	20.2	13.4	3.1	11.1	9.0	8.0	23.0	28.6
ILE	AUA	7.9	1.5	8.6	4.0	10.3	8.8	1.0	6.1	5.0	2.6	10.9	8.1
	AUC	17.4	33.4	25.1	32.9	21.3	14.7	33.2	27.7	25.9	31.3	18.0	16.7
	AUU	23.7	16.0	22.9	6.9	18.4	16.2	8.9	18.9	11.3	14.8	26.4	27.2
MET	AUG	19.1	13.8	26.0	26.8	26.9	25.8	25.4	22.9	24.9	23.1	19.2	25.2
TRP	UGG	9.8	18.8	10.8	14.1	14.0	10.2	11.5	12.6	11.6	13.1	11.3	12.3

Table 8a. Continued

		ALF[a]	ASN	ATH	BLY	BNA	COT	CRE	EME	MZE	NEU	PEA	PET
TER	UAA	2.2	0.7	1.1	1.0	0.9	1.8	2.3	0.6	0.6	1.3	1.8	0.8
	UAG	0.7	1.0	0.6	1.0	1.3	1.4	0.7	0.6	1.1	0.4	0.5	0.8
	UGA	1.7	0.8	1.2	1.4	0.7	1.2	0.4	0.4	1.3	0.5	0.9	1.7
TOTAL		5820	17143	59725	30063	15835	8554	16191	37050	56690	39355	22625	9977

[a]These abbreviations are explained in *Table 10.*

Table 8b.

		PHV[a]	POT	RIC	SOY	SPI	TOB	TOM	USM	WHT	YSA	YSC	YSK	YSP
No. of genes		40	55	75	98	53	111	96	21	95	65	1416	46	119
ARG	CGA	2.9	6.1	3.9	4.5	4.3	5.2	4.2	12.5	2.6	1.2	2.5	1.0	6.9
	CGC	4.9	2.8	14.0	7.1	3.6	4.1	3.0	14.8	8.8	0.4	2.3	0.6	6.4
	CGG	2.6	1.9	7.9	2.1	3.0	2.4	2.0	6.6	6.6	0.5	1.5	0.8	2.7
	CGU	7.8	9.8	7.8	7.5	7.8	9.9	8.3	16.4	3.6	5.1	7.1	3.6	19.9
	AGA	11.0	12.1	7.7	14.1	12.5	12.4	16.4	6.7	4.9	31.6	22.4	25.4	10.9
	AGG	11.1	10.3	14.0	10.7	14.0	12.4	11.0	7.3	9.6	1.7	8.3	2.7	4.2
LEU	CUA	7.9	9.6	5.5	6.9	7.7	7.2	7.1	9.6	8.8	1.9	12.5	11.3	7.2
	CUC	17.5	10.7	27.9	14.9	13.4	13.0	10.8	31.7	20.6	3.1	4.8	2.5	7.5
	CUG	12.0	7.8	16.9	10.4	8.4	7.0	5.9	13.9	17.2	1.7	9.7	3.7	6.0
	CUU	23.9	26.2	13.7	22.5	22.0	24.3	24.4	17.1	9.4	7.7	10.9	8.4	26.2
	UUA	7.7	15.6	4.1	7.9	8.8	9.8	9.9	1.0	2.7	26.7	24.7	30.6	24.2
	UUG	18.3	22.3	10.8	19.4	24.3	21.6	24.6	11.1	12.5	46.5	30.0	26.4	24.2

SER	UCA	12.2	17.1	7.1	14.8	16.7	14.4	18.7	17.2	12.7	13.5	16.7	13.7	16.9
	UCC	18.4	8.0	13.1	10.2	15.9	10.0	9.1	18.0	13.2	16.3	14.9	10.5	12.1
	UCG	4.4	3.8	9.1	4.3	4.9	3.9	5.0	23.9	8.6	3.5	7.6	3.9	6.7
	UCU	22.2	18.1	9.3	16.8	20.6	21.4	17.5	19.2	15.4	23.7	24.1	30.7	32.0
	AGC	14.7	7.1	16.9	14.9	11.4	9.7	9.3	24.2	14.0	3.4	8.4	3.4	8.9
	AGU	10.9	14.0	8.6	13.1	10.9	12.1	11.2	9.5	4.5	9.1	12.4	14.8	13.3
THR	ACA	12.7	18.0	8.9	14.0	18.1	17.1	16.7	17.7	6.9	10.3	16.4	14.8	13.0
	ACC	18.3	9.4	20.6	15.4	18.8	11.1	11.4	16.7	18.1	20.9	13.3	11.8	12.2
	ACG	6.1	3.4	8.7	3.8	4.1	4.3	4.4	11.9	7.2	1.5	7.3	2.8	5.5
	ACU	15.3	27.4	9.7	17.8	17.3	20.5	20.2	12.9	13.0	31.5	21.3	22.9	24.5
PRO	CCA	16.9	18.1	11.9	33.2	19.2	26.6	28.2	14.8	48.6	31.5	19.9	19.6	10.9
	CCC	11.2	6.9	11.5	12.2	11.0	8.3	7.7	15.9	13.9	1.8	6.3	2.0	8.5
	CCG	3.1	3.5	14.2	4.9	5.7	4.2	3.4	15.9	19.4	1.4	4.7	2.0	3.9
	CCU	18.6	18.1	12.0	22.9	20.9	23.1	21.0	22.7	11.1	8.7	13.3	13.5	23.1
ALA	GCA	21.4	24.7	15.1	20.4	24.1	21.8	22.2	19.2	12.4	10.3	15.4	15.8	14.4
	GCC	21.7	11.5	28.1	16.1	18.2	16.7	13.2	24.2	26.5	18.4	14.5	9.4	13.5
	GCG	4.1	4.0	18.6	5.3	7.0	4.2	4.0	16.4	16.3	1.1	5.9	3.3	4.4
	GCU	29.1	36.2	20.2	25.1	36.9	38.1	34.1	20.3	13.9	43.1	25.1	28.7	32.8
GLY	GGA	37.3	29.3	17.1	22.1	24.5	29.3	30.3	10.9	26.9	8.2	9.6	15.6	16.2
	GGC	14.8	11.7	34.9	12.3	10.2	14.1	10.2	18.3	28.4	4.7	9.5	4.2	9.3
	GGG	13.9	9.1	15.0	12.0	13.6	9.7	9.1	4.4	29.6	3.6	5.5	2.5	3.9
	GGU	37.9	24.3	19.7	22.0	25.5	31.1	29.9	12.6	12.3	53.3	29.9	41.9	27.8
VAL	GUA	6.1	10.9	7.1	7.4	11.7	10.3	9.7	7.8	4.6	4.0	10.6	13.5	12.0
	GUC	11.7	9.8	27.6	9.4	14.0	14.1	11.1	23.9	17.1	18.2	13.4	12.4	13.3
	GUG	22.5	15.8	22.1	24.9	19.7	14.5	17.2	7.9	19.3	5.6	10.4	6.6	7.5
	GUU	26.1	30.8	15.2	24.5	32.9	27.7	30.7	19.4	11.9	38.4	24.3	29.3	31.2
LYS	AAA	26.0	32.4	8.6	24.7	27.4	22.7	31.0	23.0	6.7	39.6	39.7	51.5	35.7
	AAG	28.1	34.8	42.5	37.8	43.1	29.7	36.8	35.9	31.2	29.3	33.1	30.6	26.5
ASN	AAC	32.9	14.0	29.6	27.4	20.9	23.9	21.4	25.6	17.2	24.7	25.5	23.1	19.2
	AAU	19.9	30.6	12.8	20.1	19.3	24.9	26.5	14.4	7.1	25.3	33.0	40.8	31.9

Table 8b. Continued

		PHV[a]	POT	RIC	SOY	SPI	TOB	TOM	USM	WHT	YSA	YSC	YSK	YSP
GLN	CAA	19.4	21.4	22.9	26.0	18.4	22.9	21.2	14.4	108.6	34.7	28.5	23.2	27.3
	CAG	14.1	12.0	26.9	19.5	11.4	13.4	11.7	25.6	61.4	4.4	11.7	5.4	9.7
HIS	CAC	12.7	7.0	15.9	9.4	6.8	8.9	7.0	20.3	10.2	9.5	8.5	6.6	6.3
	CAU	11.2	15.1	8.3	13.1	9.2	10.5	10.8	13.0	6.6	9.5	13.0	11.7	15.0
GLU	GAA	28.1	36.2	14.5	34.8	29.1	28.6	32.7	21.4	11.8	51.3	47.2	51.5	42.9
	GAG	30.2	23.8	42.2	37.9	30.2	27.6	30.1	29.9	32.4	6.9	18.1	11.1	21.0
ASP	GAC	22.1	14.7	34.1	19.0	18.7	18.3	16.6	33.1	19.3	21.2	21.5	16.1	16.7
	GAU	25.9	37.2	18.7	29.9	30.8	32.4	34.0	19.6	10.5	37.5	37.4	45.5	39.1
TYR	UAC	19.3	15.5	24.4	18.5	15.5	16.0	17.2	12.3	23.8	18.6	15.7	15.8	12.1
	UAU	16.6	20.1	10.0	17.0	9.6	18.3	15.2	8.5	8.7	19.5	17.3	31.3	21.6
CYS	UGC	6.4	8.4	12.0	8.5	6.4	8.3	7.7	12.0	13.6	2.7	4.1	3.0	5.5
	UGU	4.5	11.8	4.4	5.3	8.7	9.2	9.1	5.4	4.6	10.5	8.0	13.1	8.6
PHE	UUC	25.3	15.9	29.5	21.7	18.4	19.0	17.8	23.7	16.7	21.8	19.4	20.9	13.3
	UUU	17.8	28.9	10.9	18.6	19.6	25.5	22.3	12.7	13.0	26.0	24.2	26.3	28.8
ILE	AUA	9.4	10.1	6.1	11.3	7.9	9.4	10.5	3.1	5.3	4.2	14.9	30.1	10.4
	AUC	21.4	14.3	31.2	14.3	19.4	14.8	16.6	23.0	19.8	18.4	17.9	14.3	14.3
	AUU	19.1	31.0	15.4	23.8	28.1	28.1	29.9	13.6	9.1	38.4	30.3	27.2	37.1
MET	AUG	15.5	25.5	25.0	20.4	23.5	22.5	25.5	20.8	18.1	18.9	21.1	19.7	22.0
TRP	UGG	12.1	10.3	14.9	10.3	10.3	14.2	12.3	11.5	8.0	11.0	10.3	12.5	8.6
TER	UAA	0.8	1.6	1.1	1.3	2.0	1.3	2.0	0.5	0.6	1.3	1.1	1.3	1.2
	UAG	0.9	0.5	0.9	0.9	0.6	0.5	0.5	0.4	0.6	0.7	0.4	0.2	0.4
	UGA	1.5	1.1	1.2	0.9	0.9	1.3	0.8	1.7	1.7	0.3	0.6	0.4	0.3
TOTAL		12868	17538	24283	31919	15245	35694	29293	8318	32822	28510	694410	23919	61600

[a]These abbreviations are explained in *Table 10.*

Table 9. Plant chloroplast codon usage and mitochondrial genes

		CRE[a] CP	EGR CP	MPO CP	MZE CP	PEA CP	RIC CP	SPI CP	TOB CP	WHT CP	MPO MT	MZE MT	YSC MT
No. of genes		26	45	42	46	30	78	42	105	21	66	24	33
ARG	CGA	1.8	7.2	10.3	15.1	9.2	11.4	12.8	13.7	10.6	11.8	9.9	0.0
	CGC	2.3	3.8	1.9	6.2	3.8	6.7	5.1	4.1	6.7	7.6	7.9	0.3
	CGG	0.0	1.1	1.0	4.4	3.7	4.9	4.8	4.8	3.4	4.8	5.4	0.7
	CGU	38.5	17.5	15.3	14.3	15.6	13.7	14.5	13.9	14.2	12.3	12.8	2.5
	AGA	5.1	19.5	14.6	20.8	13.5	16.4	15.4	17.3	16.0	14.9	12.2	24.9
	AGG	0.6	5.0	0.9	6.0	4.8	6.3	7.0	6.3	6.5	9.6	6.6	1.6
LEU	CUA	8.3	6.0	5.4	13.0	13.1	14.7	14.6	14.0	10.6	13.6	13.4	7.5
	CUC	0.0	0.5	0.9	6.8	4.6	8.0	5.3	7.2	10.3	8.4	14.9	0.8
	CUG	0.8	0.8	0.8	5.7	5.8	6.3	5.8	7.5	6.9	10.8	12.2	1.4
	CUU	20.1	17.4	22.1	19.9	21.9	24.0	22.3	21.8	19.1	20.2	17.6	5.3
	UUA	58.2	52.9	74.2	36.1	34.4	33.7	34.6	30.6	30.4	33.7	28.1	103.5
	UUG	1.9	16.9	6.1	19.8	20.8	20.1	21.1	20.7	17.7	22.2	19.4	5.1
SER	UCA	19.8	18.1	13.8	11.8	10.0	11.8	12.3	13.8	8.3	13.4	16.9	22.6
	UCC	1.2	3.9	3.1	14.0	10.0	14.0	10.5	12.7	13.8	11.8	12.1	2.8
	UCG	1.7	4.9	1.8	6.3	5.3	6.7	6.4	6.9	5.7	7.9	8.9	1.1
	UCU	19.4	24.9	25.3	16.6	21.2	18.6	20.4	22.0	14.0	16.3	14.9	16.7
	AGC	4.0	2.1	1.8	4.9	4.8	4.8	4.1	5.1	4.9	10.9	8.3	1.2
	AGU	10.7	12.9	16.5	14.5	14.1	14.6	14.6	14.3	14.2	14.4	16.0	14.3
THR	ACA	29.5	23.8	20.9	16.4	18.0	13.9	15.3	14.2	16.6	13.0	14.0	21.7
	ACC	1.1	1.9	2.6	9.8	10.9	11.1	10.9	11.2	10.1	12.6	11.7	3.1
	ACG	1.4	4.3	1.7	5.7	5.7	5.4	5.6	5.8	7.1	8.4	7.1	1.6
	ACU	29.5	24.4	26.8	21.2	24.3	23.5	21.3	20.5	23.9	17.1	17.5	17.8

Table 9. Continued

		CRE[a] CP	EGR CP	MPO CP	MZE CP	PEA CP	RIC CP	SPI CP	TOB CP	WHT CP	MPO MT	MZE MT	YSC MT
PRO	CCA	24.3	11.1	16.2	10.8	11.7	11.5	12.2	11.8	12.2	11.0	19.4	10.8
	CCC	0.6	2.8	1.7	9.9	7.2	9.5	8.3	7.7	10.4	9.3	8.3	2.2
	CCG	2.5	2.6	1.9	5.8	5.7	5.6	5.8	6.3	7.9	6.8	7.0	1.2
	CCU	19.1	22.8	21.4	15.9	19.9	18.3	19.3	17.6	19.1	14.7	16.0	19.9
ALA	GCA	24.0	21.4	22.8	19.5	21.7	19.0	17.9	15.8	20.5	14.9	17.9	15.4
	GCC	2.5	3.6	3.2	10.2	9.5	8.7	10.8	9.9	12.0	13.5	16.6	3.0
	GCG	2.1	4.9	2.6	7.7	9.1	7.5	8.3	5.7	8.5	9.6	9.3	2.1
	GCU	57.7	33.1	37.6	29.8	35.1	27.7	30.2	27.2	31.5	22.2	33.5	21.4
GLY	GGA	9.1	27.9	32.5	28.9	25.5	26.9	29.3	27.0	26.0	22.0	27.8	12.0
	GGC	3.6	5.0	4.1	8.6	8.2	7.7	9.1	8.0	14.6	11.1	11.8	1.2
	GGG	1.2	4.8	4.3	13.2	12.3	15.1	13.7	12.2	16.4	10.8	14.7	4.0
	GGU	69.5	36.0	32.2	23.4	32.4	24.2	27.7	25.9	27.2	19.9	25.7	33.5
VAL	GUA	31.0	19.8	20.7	21.3	24.0	22.5	23.2	22.7	25.0	16.6	17.6	24.9
	GUC	0.7	2.1	2.2	7.0	6.4	7.0	7.1	6.7	10.4	8.9	13.6	2.1
	GUG	1.8	3.3	2.2	7.9	7.7	8.8	8.3	8.0	10.8	13.6	20.8	2.1
	GUU	30.0	33.9	29.1	22.8	20.6	23.2	21.9	20.7	23.1	20.0	21.9	19.8
LYS	AAA	48.7	64.3	61.2	41.9	33.0	35.9	35.5	36.2	31.7	46.6	24.8	65.2
	AAG	2.4	10.1	2.6	14.9	12.3	14.5	10.9	13.4	17.5	19.6	16.0	7.9
ASN	AAC	28.7	8.2	6.7	9.3	10.6	11.2	11.6	12.7	9.9	14.6	12.0	7.1
	AAU	20.5	41.6	43.3	27.5	32.7	27.6	32.0	35.9	25.8	26.0	22.3	93.0
GLN	CAA	27.2	25.7	36.6	30.7	28.7	24.9	28.2	25.9	22.5	26.0	24.3	18.6
	CAG	1.3	3.6	2.1	9.1	7.6	8.2	8.1	8.6	11.2	9.6	8.7	2.0
HIS	CAC	21.3	3.3	2.9	6.1	5.6	6.2	6.7	6.0	5.3	7.6	5.4	1.9
	CAU	7.1	25.6	19.2	17.4	20.2	17.3	20.0	16.7	11.6	15.1	18.1	18.1

GLU	GAA	50.1	34.8	46.4	43.5	39.9	39.3	41.3	41.8	43.2	26.7	33.4	27.0
	GAG	4.1	7.3	3.5	14.3	13.9	15.8	12.6	14.1	16.4	13.7	15.3	3.9
ASP	GAC	17.8	6.7	3.5	9.0	8.3	8.8	9.1	9.6	14.4	13.4	12.2	2.7
	GAU	22.2	29.0	30.4	30.6	34.9	28.8	30.9	32.3	31.7	22.3	27.7	31.7
TYR	UAC	16.9	5.6	4.9	6.6	5.6	8.6	6.3	8.8	9.5	10.7	8.6	6.3
	UAU	14.7	30.7	31.1	29.3	25.0	26.7	25.2	26.9	23.5	24.8	22.6	53.3
CYS	UGC	2.0	3.0	2.0	2.8	3.1	3.8	2.4	3.2	3.5	6.9	3.4	0.7
	UGU	9.8	6.3	8.0	7.4	9.0	8.4	7.6	7.4	9.7	11.5	8.3	8.6
PHE	UUC	31.5	7.0	5.2	16.2	18.8	21.4	19.4	21.6	17.9	17.1	21.1	14.4
	UUU	20.7	52.4	55.2	29.6	32.9	35.7	36.7	33.5	29.0	46.5	31.0	36.9
ILE	AUA	4.9	32.3	26.2	24.3	21.5	21.5	22.6	22.4	18.3	27.2	17.8	32.7
	AUC	15.0	5.2	4.5	14.0	14.1	17.3	14.8	17.8	15.8	14.3	19.8	10.7
	AUU	46.8	47.5	60.0	42.0	46.3	38.0	37.5	38.1	41.0	36.3	38.2	80.3
MET	AUG	26.7	23.5	22.1	22.7	22.2	24.2	21.7	25.0	24.1	22.9	23.9	19.3
TRP	UGG	21.1	17.5	16.9	15.3	18.1	18.6	19.0	17.2	15.6	16.1	11.7	1.2
TER	UAA	3.1	3.0	2.6	1.8	1.7	1.4	1.6	1.6	1.2	1.6	1.5	8.5
	UAG	0.0	0.5	0.2	0.7	0.8	1.0	0.5	0.7	1.6	1.0	1.0	1.2
	UGA	0.0	0.5	0.1	0.9	0.9	1.3	0.6	1.1	1.4	1.0	1.0	14.2
TOTAL		8442	11123	14360	13564	8994	21424	15996	30695	5072	18277	6861	12186

[a]These abbreviations are explained in *Table 10.*

Table 10. Key to organisms cited in *Tables 8* and *9*

Nuclear genes	
ALF	Alfalfa (*Medicago sativa*)
ASN	*Aspergillus oryzae, A. niger, A. awamori*
ATH	*Arabidopsis thaliana*
BLY	Barley
BNA	*Brassica napus*
COT	Cotton (*Gossypium hirsutum*)
CRE	*Chlamydomonas reinhardtii*
EME	*Aspergillus nidulans*
MZE	Maize
NEU	*Neurospora*
PEA	Pea
PET	Petunia
PHV	Bean (*Phaseolus vulgaris*)
POT	Potato
RIC	Rice
SOY	Soybean
SPI	Spinach
TOB	Tobacco
TOM	Tomato
USM	*Ustilago maydis*
WHT	Wheat
YSA	Yeast, *Candida*
YSC	Yeast, *Saccharomyces cerevisiae*
YSK	Yeast, *Kluyveromyces*
YSP	Yeast, *Schizosaccharomyces pombe*
Chloroplast genes	
CRE CP	*Chlamydomonas reinhardtii* chloroplast
EGR CP	*Euglena gracilis* chloroplast
MPO CP	*Marchantia polymorpha* chloroplast
MZE CP	Maize chloroplast
PEA CP	Pea chloroplast
RIC CP	Rice chloroplast
SPI CP	Spinach chloroplast
TOB CP	Tobacco chloroplast
WHT CP	Wheat chloroplast
Mitochondrial genes	
MPO MT	*Marchantia polymorpha* mitochondria
MZE MT	Maize mitochondria
YSC MT	Yeast, *Saccharomyces cerevisiae* mitochondria

6. REFERENCES

1. Bendich, A.J., Anderson, R.S. and Ward, B.L. (1980) in *Genome Organisation and Expression in Plants* (C.J. Leaver ed.). Plenum, New York, p. 31.

2. Blin, N. and Stafford, D. (1976) *Nucleic Acids Res.*, **3,** 2303.

3. Boffey, S.A. (1991) *Methods in Plant Biochemistry* (L.J. Rogers, ed.). Academic Press, London, Vol. 5, p. 147.

4. Darby, G.K., Jones, A.S., Kennedy, J.F. and Walker, R.T. (1970) *J. Bacteriol.,* **103,** 159.

5. Dellaporta, S.L., Wood, J. and Hicks, J.B. (1983) *Plant Mol. Biol. Rep.*, **1,** 19.

6. Domoney, C. and Casey, R. (1985) *Nucleic Acids Res.*, **13,** 687.

7. Doyle, J.J. and Doyle, J.L. (1987) *Phytochem. Bull.*, **19,** 11.

8. Ellis, T.H.N., Davis, J.R., Castleton, J.A. and Bedford, I.D. (1984) *Chromosome*, **91,** 74.

9. Ellis, T.H.N., Delseny, M., Lee, D. and Burcham, K.W.G. (1989) *Plant Mol. Biol.,* **14,** 73.

10. Ganal, M.W. and Tanksley, S.D. (1989) *Plant Mol. Biol. Rep.*, **7,** 17.

11. Graham, D.E. (1978) *Anal. Biochem.*, **85,** 609.

12. Guzman, P. and Ecker, J.R. (1988) *Nucleic Acids Res.*, **16,** 11091.

13. Jofuku, K.D. and Goldberg, R.B. (1988) in *Plant Molecular Biology – A Practical Approach* (C.H. Shaw ed.). IRL Press, Oxford, p. 37.

14. Landry, B.S., Kesseli, R., Leung, H. and Michelmore, R.W. (1987) *Theor. Appl. Genet.*, **74,** 646.

15. Marmur, J. (1961) *J. Mol. Biol.*, **3,** 208.

16. Murray, M.G. and Thompson, W.I. (1980) *Nucleic Acids Res.*, **8,** 4321.

17. Nagao, R.T., Shah, D.M., Eckenrode, V.K. and Meagher, R.B. (1981) *DNA*, **2,** 1.

18. Oono, K. and Sugiura, M. (1980) *Chromosoma*, **76,** 85.

19. Rogers, S.O. and Bendich, A.J. (1985) *Plant Mol. Biol.*, **5,** 69.

20. Saghai-Maroof, M.A., Soliman, K.M., Jorgensen, R.A. and Allard, R.W. (1984) *Proc. Natl. Acad. Sci. USA*, **81,** 8014.

21. Sun, S.M., Slightom, J.L. and Hall, T.C. (1981) *Nature*, **289,** 37.

22. Sung, M.T. and Slightom, J.L. (1981) in *Genetic Engineering in the Plant Sciences* (N.J. Panopoulos ed.). Praeger, p. 39.

23. Takaiwa, F., Ebinuma, H., Kikuchi, S. and Oono, K. (1987) *FEBS Lett.*, **221,** 43.

24. Taylor, B. and Powell, A. (1982) *Focus*, **4,** 4.

25. Thompson, R.D., Bartels, D., Harberd, N.P. and Flavell, R.B. (1983) *Theor. Appl. Genet.*, **67,** 87.

26. Ward, E.R. and Jen, G.C. (1990) *Plant Mol. Biol.*, **14,** 561.

27. Weber, H., Ziechmann, C. and Graessmann, A. (1990) *EMBO J.*, **9,** 4409.

28. Weising, K., Beyermann, B., Ramser, J. and Kahl, G. (1991) *Electrophoresis*, **12,** 159.

29. Willmitzer, L. and Wagner, K.G. (1981) *Exp. Cell Res.*, **135,** 69.

30. Zimmer, E.A. and Newton, K.J. (1982) in *Maize for Biological Research*, University Press, Grand Forks, ND.

31. Renz, M. and Kurz, C. (1984) *Nucleic Acids Res.*, **12,** 3435.

32. Bennet, M.D. and Smith, J.B. (1976) *Phil. Trans. R. Soc. (Lond.), series B*, **274**, 227.

33. Bennet, M.D., Smith, J.B. and Heslop-Harrison, J.S. (1982) *Proc. R. Soc. (Lond.), series B*, **216**, 179.

34. Bennet, M.D. and Smith, J.B. (1991) *Phil. Trans. R. Soc. (Lond.), series B,* **334**, 309.

35. Blackburn, P., Wilson, G. and Moore, S. (1977) *J. Biol. Chem.*, **252,** 5904.

36. Summers, W.C. (1970) *Anal. Biochem.*, **33,** 459.

37. Berger, S.L. and Birkenmeier, C.S. (1979) *Biochemistry*, **18,** 5143.

38. Gray, J.C. (1974) *Arch. Biochem. Biophys.*, **163,** 343.

39. Berger, S.L. (1987) *Methods Enzymol.*, **152,** 227.

40. Hallick, R.B., Chelm, B.K., Gray, P.W. and Orozco, E.M. (1977) *Nucleic Acids Res.*, **4,** 3055.

41. Maniatis, T., Fritsch, E.F. and Sambrook, J. (1982) *Molecular Cloning: A Laboratory Manual.* Cold Spring Harbor Laboratory, New York, Appendix A, p. 452.

42. Bergseid, M. and Ho Sing Loy, P. (1989) *Strategies*, **2,** 48.

43. Palmiter, R.D., Christensen, A.K. and Schimke, R.T. (1970) *J. Biol. Chem.,* **245,** 833.

44. Schimke, R.T., Palacios, R., Sullivan, D., Kiely, M.L., Gonzales, C. and Taylor, J.M. (1974) *Methods Enzymol.*, **30,** 631.

45. Singer, B. and Fraenkel-Comrat, H. (1961) *Virology*, **14,** 59.

46. Poulson, R. (1977) in *The Ribonucleic Acids* (P.R. Stewart and D.S. Letham eds). Springer-Verlag, Berlin, Chapter 10, p. 333.

47. Kirby, K.S. (1968) *Methods Enzymol.,* **12,** 87.

48. Chirgwin, J.M., Przybyla, A.E., MacDonald, R.J. and Rutter, W.J. (1979) *Biochemistry,* **18,** 5294.

49. Chomczynski, P. and Sacchi, N. (1987) *Anal. Biochem.,* **162,** 156.

50. Cox, K.H. and Goldberg, R.B. (1988) in *Plant Molecular Biology – A Practical Approach* (C.H. Shaw ed.) IRL Press, Oxford.

51. Logemann, J., Schell, J. and Willmitzer, L. (1987) *Anal. Biochem.,* **163,** 16.

52. Schmidt, G.W., Bartlett, S.G., Grosmann, A.R., Cashmore, A.R. and Chua, N.H. (1981) *J. Cell. Biol.,* **91,** 468.

53. Prescott, A. and Martin, C. (1987) *Plant Mol. Biol. Rep.,* **4,** 219.

54. Bustos, M.M., Luckow, V.A., Griffing, L.R., Summers, M.D. and Hall, T.C. (1988) *Plant Mol. Biol.,* **10,** 475.

55. Rerie, W.G., Whitecross, M. and Higgins, T.G.V. (1991) *Mol. Gen. Genet.,* **225,** 148.

56. Ausubel, F.M., Brent, R., Kingston, R.E., Moore, D.D., Siedman, J.G., Smith, J.A. and Struhl, K. (1987) in *Current Protocols in Molecular Biology,* 4.3.1.

57. Martin, C. and Northcote, D.H. (1981) *Planta,* **151,** 189.

58. Jones, J.D.G., Dunsmuir, P. and Bedbrook, J. (1985) *EMBO J.,* **4,** 2411.

59. Hall, T.C., Ma, Y., Buchbinder, B.U., Payne, J.W., Sun, S.N. and Bliss, F.A. (1978) *Proc. Natl. Acad. Sci. USA,* **75,** 3196.

60. Loening, U. (1969) *Biochem. J.,* **113,** 131.

61. Rochester, D.E., Winer, J.A. and Shah, D.M. (1986) *EMBO J.,* **5,** 451.

62. Hensgens, L.A.M. and van-os-Ruygrok, P.E. (1989) *Rice Genetics Newsletter,* **6,** 163.

63. Wada, K., Wada, Y., Ishibashi, F., Gojobori, T. and Ikemura, T. (1991) EMBL CD-ROM (Dec. 1991), NAR Sequence Supplement.

64. Wada, K., Wada, Y., Ishibashi, F., Gojobori, T. and Ikemura, T. (1992) *Nucleic Acids Res.,* **20,** 2111.

65. Courrier, T.C. and Nester, E.W. (1976) *Anal. Biochem.,* **76,** 431.

66. Dhaese, P., De Grieve, H., Decraemer, H., Schell, J. and Von Montagu, M. (1979) *Nucleic Acids Res.,* **7,** 1837.

67. Nishiguchi, R., Takanami, M. and Oka, A. (1987) *Mol. Gen. Genet.,* **206,** 1.

68. Slusarenko, A.J.. *Plant Mol. Biol. Rep.,* **8,** 249.

69. An, G., Erbert, P.R., Mitra, A. and Ha, S.B. (1988) *Plant Mol. Biol. Manual,* **A3,** 1.

CHAPTER 3

PLANT GENE INDEX

M. Latterich and R.R.D. Croy

Gene Index

1. INTRODUCTION

Ever since the first successful efforts were undertaken to develop methods to determine the nucleic acid sequence of a gene, there has been an increasing number of published nucleic acid sequences in the literature. Current efforts to determine the complete genomic sequences of yeast (*Arabidopsis*) and man endeavour to understand life at its basic level and to establish the basis to determine what gene functions in which biochemical pathway. This flood of information requires the availability of computer resources which, for example, can assemble and store sequence data for homology searches and predictions of what sequence stretch potentially encodes proteins.

Simultaneously, many genes in a variety of organisms are sequenced because they feature certain properties that are desirable for study at the genetic level. This has especially been true for certain plant genes because of their potential importance in pest resistance and nutritional value of crop plants (1). Unfortunately, with the exception of *Arabidopsis*, there is no current project solely dedicated to determining the complete genomic sequence of a plant species, particularly economically important species, so a lot of effort must be made in cloning and sequencing plant genes, thus making them amenable to manipulation.

A number of identical genes of the same species have been sequenced by independent research groups. While this effort serves to confirm sequence data obtained by others, for large sequences it is expensive and time-consuming. The objective in producing this chapter was to compile a list of all currently available plant genes which have been published and submitted to sequence databases by June 1992. This list was obtained by searching the following databases for all entries under *Embryobionta* (higher plants) with the help of STRINGSEARCH of the UWGCG package (2) to selectively compile a list of all higher plant genes, omitting genes of algae, fungi and organelles:

GenBank	Release	72 (6/92)
EMBL (Mod.)	Release	28 (9/91)
PIR-Prot.	Release	29 (6/91)
PIR-NA	Release	36 (3/90)

This list of plant genes should aid the plant molecular biologist in rapidly determining whether a certain gene has been sequenced before, without the need for computer database searching. The list is intended to serve as an orientation only and does not pretend to be complete nor entirely up-to-date. Interested readers can obtain a more recent list by searching frequently updated sequence databases. The most important databases and suppliers of sequence analysis packages are listed in *Table 1*.

This collection of plant gene sequences is inevitably a one-dimensional system as opposed to the multi-dimensional computer database from which it was derived. The ordering and categorization of the entries in such a system is difficult and undoubtedly some sequences will not be found in classes where all users might expect to find them. Thus 'heat stress-induced transcription factor HSF24' is filed in a section containing heat shock and stress-induced genes rather than in a section containing transcription factors. Similarly 'a metal-dependent DNA-binding protein interacting with a constitutive element of a light-responsive promoter' is classified as DNA-binding *not* metal responsive *nor* light responsive. There is, therefore, a hierarchical system for ordering entries which depends on the information provided by the databases. Furthermore, several entries are poorly described, with reference only to the source of the sequence but no clue as to its identity. Most are rather more clear cut. Lack of space prevents cross-referencing of entries.

Entries in *Tables 3–21* have been compiled in the following way:

(i) Level 1: each table presents data for a specific type of molecule encoded in the nucleic acid sequences, e.g. *Table 4*, Antigens; *Table 8*, Enzymes; *Table 14*, Photosynthesis. To keep the task as simple as possible, these categories have been broadened to include a number of (loosely) associated types, e.g. photosynthesis also includes photoreception, plastocyanins, phytochromes, light-influenced.

(ii) Level 2: species' Latin names sorted alphabetically by genus. Those organisms with undefined species name as indicated by sp. (e.g. *Petunia* sp.) are sorted as *Petunia* S.

(iii) Level 3: locus identifiers sorted alphabetically. In this sort any numerical values have been ignored initially and the sort has been performed on the character order only. In cases where all characters are identical, numbers are then used to sort in increasing order — 2 followed by 23, followed by 221, e.g. BLYCAB2 comes before BLYLHBC, BLYRCAA2 comes before BLYRCAB, BLYLP58 comes before BLYLP60, POTU6SNRN comes before POTU211UR.

(iv) Level 4: gene description as entered in databases as *Definition*; *Keywords*; *Source*; *Reference*; or *Comment*. The most appropriate description abstracted from these categories and including: a. The identity of the encoded product (e.g. ribulose bisphosphate — small subunit) — common abbreviations are used; b. Gene designation (e.g. RbcS); c. Tissue specificity or site of synthesis or source of tissue used for the isolation of genomic DNA or RNA where indicated (e.g. leaves or seeds or suspension culture, etc.); d. Size in base-pairs or bases; e. DNA (ds-DNA) or mRNA (ss-RNA).

The classification procedure is demonstrated by the sample entries given in *Table 2*.

Permission to use these databases or to obtain sequence analysis packages can be obtained by writing to the addresses given in *Table 1* and requesting an identifier and password. These institutions also supply sufficient documentation of how to retrieve information from the database and also provide helpful tools for sequence analysis and comparison. References to sequence data compiled in *Tables 3–21* can be retrieved from the database by extracting the full sequence documentation with the help of the accession number.

Table 1. Sequence database and sequence analysis program suppliers

USA	Applied Genetic Technology, 3910 West Valley Drive, Fairview Park, OH 44126, USA
	GenBank Genetic Sequence Data Bank, Los Alamos National Laboratories, Los Alamos, NM 87545, USA
	GenBank On-line Service and GenBank Genetics Sequence Data Bank, IntelliGenetics, 700 East El Camino Real, Mountain View, CA 94040, USA
	Protein Data Bank, Chemistry Department, Brookhaven National Laboratory, Upton, NY 11973, USA
	Protein Identification Resource (PIR) and NBRF Protein Sequence Databases, National Biomedical Research Foundation, Georgetown University Medical Center, 3900 Reservoir Road N.W., Washington, DC 20007, USA
	University of Wisconsin Biotechnology Center, Madison, WI 53705, USA
Europe	EMBL Databases, EMBL Nucleotide Sequence Data Library, European Molecular Biology Laboratory, Meierhoffe Strasse 1, Postfach 10 2209, 6900 Heidelberg, Germany
	SEQNET SERC Database, Daresbury, UK
Japan	DNA Data Bank of Japan, Laboratory of Genetic information Research, Center for Genetic Information Research, National Institute of Genetics, Misuina, Shizuoda 411, Japan

Table 2. Example plant gene database, e.g. *Table 14*, Photosynthesis

Species	Identifier	Sequence specification	Tissue/site	Size (no. of base-pairs or bases)	Type	S[a]
Larix laricina	LAXRBC1C	Rubisco, SSU,	Needles	117	ss-mRNA	E
Lycopersicon esculentum	TOMCAB1AUP	cab 1a, 5′ region	Leaves	240	ds-DNA	E
Lycopersicon esculentum	TOMPSBXMR	Photosystem II 32 kDa protein, psbX	—	997	ss-mRNA	E
Silene pratensis	SIPCAB	cab, pSIAB1; Rubisco, SSU	—	642	ss-mRNA	—
Solanum tuberosum	POTPPOEC	33 kDa precursor of oxygen-evolving complex	Thylakoid	1116	ss-mRNA	E

[a]S = database source (E = EMBL, G = GenBank).

Abbrevations:

Rubisco ribulose bisphosphate carboxylase/oxygenase;
SSU small subunit;
LSU large subunit;
cab chlorophyll *a*/*b*-binding protein;
5′ upstream sequences;
3′ downstream sequences.

Table 3. ATP synthesis and electron transport (ATP synthases, ATPases, cytochromes, electron transport proteins, ferredoxins, thioredoxins)

Species	Identifier	Sequence specification	Tissue/site	Size (bp or b)	Type	S
A. sativa	ASTATPASEH	Vacuolar H^+-ATPase 16 kDa proteolipid subunit (vatp-P1)	Root tip	821	ds-DNA	—
B. napus	BNAATP6A	ATPase subunit 6 (*atp6*) gene	—	2498	ds-DNA	—
B. napus	BNAATP6B	ATPase subunit 6 gene	—	1358	ds-DNA	—
B. oleracea	BNACYTB5A	Cytochrome b-5 mRNA	—	738	ss-mRNA	—
D. carota	DARATPASE	V-type H^+-ATPase gene	—	1829	ds-DNA	—
D. carota	DARATPSBS	ATP synthase b subunit		3988	ds-DNA	E
D. carota	DARCOXII	Cytochrome c oxidase subunit II	Mitochondria	3532	ds-DNA	E
D. carota	DARVHATPA	70 kDa subunit vacuolar H^+-ATPase	—	2397	ss-mRNA	—
E. arvense	EARVATP1	Catalytic 70 kDa V-ATPase subunit	—	252	ds-DNA	E
E. arvense	EARVATP2	Catalytic 70 kDa V-ATPase subunit	—	276	ds-DNA	E
G. max	SOYCFOATP	*atpH* gene encoding CFO–ATPase subunit III	Chloroplast	440	ds-DNA	—
H. brasiliensis	HEVATPB	Mitochondrial ATP synthase β-subunit	—	2021	ss-mRNA	E
I. batatas	IPBCYXVC	Cytochrome c oxidase subunit Vc	—	455	ss-mRNA	E
I. batatas	IPBF1ATP	Mitcohondrial F1-ATPase δ-subunit	—	936	ss-mRNA	—
I. batatas	IPBFATPD	F1-ATPase (mitochondrial type) δ-subunit	—	1035	ss-mRNA	—
L. esculentum	TOMLHA1	Plasma membrane protein	—	3229	ss-mRNA	—
N. plumbaginifolia	TOBATP21	β-subunit of mitochondrial ATP synthase	—	5351	ds-DNA	E
N. plumbaginifolia	TOBATPASA	H^+-translocating ATPase	—	3178	ds-DNA	—
N. plumbaginifolia	TOBPMA1A	Plasma membrane H^+-ATPase (*pma1*)	—	7249	ds-DNA	—
N. plumbaginifolia	TOBPMA2A	Plasma membrane H^+-ATPase (*pma2*)	—	3872	ds-DNA	—
N. plumbaginifolia	TOBPMA3A	Plasma membrane H^+-ATPase (*pma3*)	—	3224	ss-mRNA	—
N. plumbaginifolia	TOBPMA3B	Plasma membrane H^+-ATPase (*pma3*)	—	3233	ds-DNA	—
N. tabacum	TOBRFESP	Mitochondrial Rieske Fe/S protein	Leaves	956	ss-mRNA	—
N. tabacum	TOBRFSCBFC	Rieske Fe/S protein of cytochrome b6/f complex	—	833	ss-mRNA	E
N. tabacum	TOBTRNA	Thioredoxin	—	698	ss-mRNA	E
O. sativa	RICCYTC	Cytochrome c (*OsCc-1*)	—	2408	ds-DNA	—
O. sativa	RICOSPSAB	Pseudo apocytochrome b	Shoots	2214	ds-DNA	E
P. vulgaris	PHVATPA	Adenosine triphosphatase (*atpa*)	—	5986	ds-DNA	—
P. sativum	PEAFDXI	Ferredoxin I	—	302	ss-mRNA	—
P. sativum	PEAFED1X	Ferredoxin I (*Fed-1*)	—	1995	ds-DNA	—
P. sativum	PEAFNADRA	Ferredoxin–$NADP^+$ reductase	—	1385	ss-mRNA	—
P. sativum	PEAFNR	Ferredoxin–$NADP^+$ reductase	—	1385	ss-mRNA	E
S. pratensis	SIPFDX	Ferredoxin precursor	—	1058	ss-mRNA	E
S. tuberosum	POTANTG	ADP/ATP translocator	Green leaves	1531	ss-mRNA	E
S. oleracea	SPIATP	ATP synthase δ-subunit (*atpD*)	—	1034	ds-DNA	—
S. oleracea	SPIATPC	Chloroplast ATP synthase γ-subunit	Chloroplast	3537	ds-DNA	E
S. oleracea	SPIATPE	ATP synthase δ-subunit (*atpD*)	Chloroplast	2760	ds-DNA	E
S. oleracea	SPIATPG	ATP synthase γ-subunit (*atpC*)	Chloroplasts	1140	ss-mRNA	—
S. oleracea	SPIFERRI	Ferredoxin I	—	632	ss-mRNA	—
S. oleracea	SPIFNADPO	Ferredoxin–$NADP^+$ oxidoreductase	—	1298	ss-mRNA	—
S. oleracea	SPIFNR	$NADP^+$ oxidoreductase	—	1492	ss-mRNA	E
S. oleracea	SPITHIORF	Thioredoxin f	Chloroplast	848	ss-mRNA	E
S. oleracea	SPITHM1	Thioredoxin m	—	927	ss-mRNA	E

Table 3. Continued

Species	Identifier	Sequence specification	Tissue/site	Size (bp or b)	Type	S
S. oleracea	SPITHM4	Thioredoxin m	–	893	ss-mRNA	E
Z. mays	MZEATP2MT	ATP2 mRNA for mitochondrial ATP synthase β-subunit	Coleoptile	2029	ss-mRNA	E
Z. mays	MZEFD1	Ferredoxin I	–	684	ds-DNA	–
Z. mays	MZEFD5	Ferredoxin	–	616	ds-DNA	–
Z. mays	MZEFD3	Ferredoxin III	Leaves	705	ds-DNA	–
Z. mays	MZEFD1P	Ferredoxin I	Leaves	757	ds-DNA	–
Z. mays	MZERFESPA	Mitochondrial Rieske Fe/S protein	–	1197	ss-mRNA	–

Table 4. Antigens (allergens, protein antigens, pollen allergens)

Species	Identifier	Sequence specification	Tissue/site	Size (bp or b)	Type	S
A. artemisii-folia	AMBAI1	Amb a I.1 (antigen E)	Pollen	1202	ss-mRNA	G
A. artemisii-folia	AMBAI2	Amb a I.2 (antigen E)	Pollen	1331	ss-mRNA	G
A. artemisii-folia	AMBAI3	Amb a I.3 (antigen E)	Pollen	1333	ss-mRNA	G
A. artemisii-folia	AMBAI1A	Amb a I.1 precursor protein	Pollen	1224	ss-mRNA	G
A. artemisii-folia	AMBAI2A	Amb a I.2 precursor protein	Pollen	1331	ss-mRNA	G
A. artemisii-folia	AMBAI3A	Amb a I.3 precursor protein	Pollen	1333	ss-mRNA	G
A. artemisii-folia	AMBAMBAII	Amb a II protein	Pollen	1395	ss-mRNA	G
A. artemisii-folia	AMBRAV	Amb a V	Pollen	314	ss-mRNA	G
A. trifida	AMBTVA	AmbtV= allergen	Pollen	290	ss-mRNA	–
G. max	SOYPCNA	Proliferating cell nuclear antigen	–	868	ss-mRNA	E
L. perenne	LOLLOLPI	RNA for *Lol pI* gene	Pollen	1132	ss-mRNA	E
L. perenne	LOLPISO1A	Pollen allergen (*Lol pI*)	Pollen	1035	ss-mRNA	–
L. perenne	LOLPISO5A	Pollen allergen (*Lol pI*)	Pollen	1047	ss-mRNA	–
P. pratensis	PPTKBG31	Pollen allergen (KBG 31)	Pollen	1322	ss-mRNA	–
P. pratensis	PPTKBG41	Pollen allergen (KBG 41)	Pollen	1363	ss-mRNA	–
P. pratensis	PPTKBG60	Pollen allergen (KBG 60)	Pollen	1222	ss-mRNA	–

Table 5. Development- and tissue-specific expression (tissue/organ-specific expression, developmental stage-specific expression — not seed or storage proteins (see *Table 18*) or lectins (see *Table 9*))

Species	Identifier	Sequence specification	Tissue/site	Size (bp or b)	Type	S
A. majus	AMATAP1G	*tap1* gene	Stamen	5280	ds-DNA	E
A. thaliana	ATHA9	*A9* gene	Tapetum	1952	ds-DNA	E
B. campestris	BNANS1	NS1 glycoprotein	Stigma	1422	ss-mRNA	E
B. campestris	BNASLG8	S-locus glycoprotein	Stigma	1584	ds-DNA	E
B. campestris	BNASLSGRA	S-locus-specific glycoprotein	—	1286	ss-mRNA	—
B. napus	BNAA9A9	A9 mRNA	Anther	504	ss-mRNA	E
B. napus	BNAA9A30	A9 mRNA	Anther	257	ss-mRNA	E
B. napus	BNABP4	*Bp4A* and *Bp4C* genes	Pollen	8585	ds-DNA	E
B. napus	BNACEXMR	Proline-rich protein	Anther	1383	ss-mRNA	E
B. napus	BNAI3	Microspore-specific mRNA	Anther	724	ss-mRNA	E
B. napus	BNALEA76	Late embryogenesis-abundant protein	Seed	1036	ss-mRNA	—
B. napus	BNARSLGLP	S-locus glycoprotein	Stigma	1477	ss-mRNA	E
B. napus	BNAWSLGLP	S-locus glycoprotein	—	1424	ss-mRNA	E
B. oleracea	BNAS291	S-locus-related glycoprotein (BS29-1)	Stigma	1416	ss-mRNA	E
B. oleracea	BNAS292	S-locus-specific glycoprotein (BS29-2)	Stigma	1465	ss-mRNA	E
B. oleracea	BNASKR6A	Receptor protein kinase (SRK6)	Stigma	2832	ss-mRNA	—
B. oleracea	BNASLG13	*SLG-13* gene for S-locus glycoprotein	Stigma	1308	ds-DNA	E
B. oleracea	BNASLR2RNA	S-locus-related glycoprotein SLR2	Stigma	1543	ss-mRNA	E
B. oleracea	BNASLRG	S-locus-related glycoprotein	Stigma	2139	ds-DNA	E
B. oleracea	BNASLSG	S-locus-specific glycoprotein	Stigma	1342	ss-mRNA	E
B. oleracea	BNASLSGR	S-locus-specific glycoprotein (SLSG)	Stigma	1286	ss-mRNA	E
D. carota	DARDC8	Embryonic-specific 66 kDa protein	—	3761	ds-DNA	E
D. carota	DAREMB1	EMB-1 mRNA	—	513	ss-mRNA	E
D. carota	DAR215G	2.15 mRNA for 14 kDa protein	—	581	ss-mRNA	E
D. caryophyllus	DINCARSR	Senescence-related protein	Flower	1250	ss-mRNA	—
D. caryophyllus	DINCARSR12	*CARSR12* gene, senescence-related	Flower	10971	ds-DNA	E
G. max	SOYLEAA	18 kDa late embryogenesis-abundant (Lea) protein		836	ds-DNA	—
G. max	SOYLEAB	Late embryogenesis-abundant (Lea) protein	—	1741	ss-mRNA	—
G. max	SOYMATSGA	Maturation-associated protein (MAT1)	Seed	971	ss-mRNA	—
G. max	SOYMATSGB	Maturation-associated protein (MAT9)	—	947	ss-mRNA	—
G. max	SOYSMATP	Seed maturation polypeptide	Cotyledon	724	ss-mRNA	E
G. hirsutum	COTLEA2	Embryogenesis-abundant protein gene	—	3784	ds-DNA	—
G. hirsutum	COTLEA11	Seed protein D-11	—	970	ds-DNA	E
G. hirsutum	COTLEA19	Seed protein D-19	—	1692	ds-DNA	E
G. hirsutum	COTLEA113	Seed protein D-113	—	2388	ds-DNA	E
G. hirsutum	COTLEA14A	Group 4 late embryogenesis-abundant protein (Lea14-A)	—	4635	ds-DNA	—
G. hirsutum	COTLEA14B	Group 4 late embryogenesis-abundant protein (Lea14-A)	—	595	ss-mRNA	—
G. hirsutum	GHLEA2	Embryogenesis-abundant protein	—	3784	DNA	—
H. annuus	HNN10KDLEA	10 kDa Lea (Late embryogenesis-abundant) protein	Cotyledon	494	ss-mRNA	E
H. annuus	HNN11KDLEA	11 kDa Lea (Late embryogenesis-abundant) protein	Cotyledon	891	ss-mRNA	E

Table 5. Continued

Species	Identifier	Sequence specification	Tissue/site	Size (bp or b)	Type	S
H. annuus	HNNSF2ASP	Anther-specific protein	–	3734	ds-DNA	E
H. vulgare	BLYALSPE	Aleurone-specific protein	–	485	ss-mRNA	–
H. vulgare	BLYBEG1	Embryo globulin (BEG1)	–	2122	ss-mRNA	–
H. vulgare	BLYSIP1A	Seed imbibition protein (Sip1) gene	Seed	4788	ds-DNA	–
L. esculentum	TOM9612	9612 mRNA	–	1401	ss-mRNA	–
L. esculentum	TOM2A11	Fruit-specific protein mRNA	–	526	ss-mRNA	–
L. esculentum	TOMETHYBR	mRNA expressed during fruit ripening potentially involved in ethylene biosynthesis	–	1332	ss-mRNA	E
L. esculentum	TOMFSPG	Fruit-specific protein	–	8028	ds-DNA	–
L. esculentum	TOM3H11	Fruit-specific protein	–	774	ss-mRNA	–
L. esculentum	TOMLAT52	Anther-specific *LAT52* gene	–	1810	ds-DNA	E
L. esculentum	TOMLAT56	*LAT56* gene	Pollen	1347	ds-DNA	E
L. esculentum	TOMLAT59	*LAT59* gene	Pollen	1544	ds-DNA	E
L. esculentum	TOMLAT56G	Protein P56	Pollen	1434	ds-DNA	E
L. esculentum	TOMLAT59G	*LAT59* gene	Pollen	1708	ds-DNA	E
L. esculentum	TOMRIPE	Fruit ripening-specific mRNA	–	1614	ss-mRNA	E
L. esculentum	TOMSTDR5	TDR5	Flowers	910	ss-mRNA	E
L. esculentum	TOMTDR3	TDR3	Flower	573	ss-mRNA	E
L. esculentum	TOMTDR4	TDR4	Flower	962	ss-mRNA	E
L. esculentum	TOMTDR5	TDR5	Flower	910	ss-mRNA	E
L. esculentum	TOMTDR6	TDR6	Flower	895	ss-mRNA	E
L. esculentum	TOMTDR8	TDR8	Flower	761	ss-mRNA	E
M. sylvestris	MAURRP	Ripening-related protein	Ripe fruit	1199	ss-mRNA	–
N. alata	TOBGPS	Stylar glycoprotein 2	–	804	ss-mRNA	–
N. alata	TOBSINCS2	S2 self-incompatibility glycoprotein	Style	754	ss-mRNA	E
N. tabacum	TOBROOTSP	Root-specific gene	Root	938	ss-mRNA	E
N. tabacum	TOBTA29	Anther-specific gene *TA-29* and stem-specific gene *TSJT1*	–	6254	ds-DNA	E
O. sativa	RICALEUROA	Aleurone layer tissue mRNA sequence	Seed	351	ss-mRNA	–
O. sativa	RICALEUROB	Aleurone layer tissue mRNA sequence	Seed	369	ss-mRNA	–
O. sativa	RICEMP1G	*Emp1* gene	–	1925	ds-DNA	E
O. sativa	RICGOS9G	Shoot-specific *GOS9* gene	–	3350	ds-DNA	E
P. sativum	PEAP4OSG	Organ-specific gene	–	2814	ds-DNA	E
P. sativum	PEAS2OSG	S2 organ-specific gene	Pod	3045	ds-DNA	E
P. sativum	PEASSP	Shoot-specific protein mRNA	–	743	ss-mRNA	–
R. sativus	RADDIN1AA	Senescence-associated (din1) mRNA	–	680	ss-mRNA	–
R. sativus	RADLEAPR	Late embryogenesis-abundant (Lea) protein	Mature seed	744	ss-mRNA	E
R. sativus	RADLMGP	Late maturation protein	Mature seed	447	ss-mRNA	–
T. aestivum	WHT3LEA	Group 3 late embryogenesis-abundant protein	Dormant seed	934	ss-mRNA	E
T. aestivum	WHTG3LEAA	Group 3 late embryogenesis-abundant protein	Mature seed	991	ss-mRNA	–
Z. mays	MZEEMBIV	Late embryogenesis-abundant protein	–	430	ss-mRNA	–
Z. mays	ZMEACI	Embryogenic abscisic acid-inducible gene	–	797	RNA	E

Table 6. DNA binding and associated proteins (polymerases, histones, DNA binding, transcription, translation, replication, cell cycle, homeogenes)

Species	Identifier	Sequence specification	Tissue/site	Size (bp or b)	Type	S
A. glauca	ABGTEF1G	Elongation factor 1-α	Vegetative mycelium	1936	ds-DNA	E
A. majus	AMADEFA	DEF A gene	Flower	4269	ds-DNA	E
A. majus	AMADEL	DEL	Flower bud corolla	2075	ss-mRNA	G
A. majus	AMAFLO	Floricaula	Flower	1545	ss-mRNA	G
A. majus	AMASQUAMOS	Squamosa gene	—	6420	ds-DNA	E
A. thaliana	ATHAGAMSG	Agamous (*AG*) gene	—	1458	ss-mRNA	E
A. thaliana	ATHAGL1A	Transcription factor (AGL1)	—	895	ss-mRNA	—
A. thaliana	ATHAGL2A	Transcription factor (AGL2)	—	1294	ss-mRNA	—
A. thaliana	ATHAGL4A	Transcription factor (AGL4)	—	1348	ss-mRNA	—
A. thaliana	ATH5DBINDN	(5′ region) for DNA binding protein homologous to nucleolin	—	872	ss-mRNA	E
D. carota	DARCAM1G	*Cam-1* gene for calmodulin	Suspension culture	704	ss-mRNA	E
D. carota	DAREF1A	Elongation factor 1A	Callus	1785	ds-DNA	E
D. carota	DARLPCNAM	Larger proliferating cell nuclear antigen	—	1312	ss-mRNA	E
D. carota	DARMCYC	Mitotic cyclin	Somatic embryo	1243	ss-mRNA	E
D. carota	DARPCNAM	Proliferating cell nuclear antigen (PCNA)	—	865	ss-mRNA	E
D. caryophyllus	DINCARSR12	CARSR12 gene	Petal	10971	ds-DNA	E
G. max	SOYHMG1	HMG-1 like protein	Embryo	761	ss-mRNA	E
G. max	SOYHMGYA	HMB-Y related protein	Embryo	832	ss-mRNA	E
G. max	SOYHMGYB	HMG-Y related protein	Embryo	585	ss-mRNA	E
G. max	SOYRNAPOLA	RNA polymerase II fifth largest subunit	—	884	ss-mRNA	—
G. max	SOYRPB1B1	Polymerase II (RPB1) largest subunit (B1)	—	3371	ds-DNA	E
G. max	SOYRPB1B2	B2 largest subunit of RNA polymerase II	—	2822	ds-DNA	E
G. max	SOYTEFS1	*tefS1* gene for elongation factor EF-1a	Seedling leaves	1979	ds-DNA	E
H. vulgare	BLYHISH3PA	Histone H3	Seed scutella	505	ss-mRNA	—
L. esculentum	TOMEF1A	Elongation factor 1-α	—	4565	ds-DNA	E
L. esculentum	TOMH1HSTL	H1 histone-like protein	Leaves	551	ss-mRNA	E
L. esculentum	TOMH4PSD	H4-histone pseudogene	Leaves	592	ds-DNA	E
L. esculentum	TOMLEEF1	Elongation factor 1-α	Ripe fruit	1692	ss-mRNA	E
M. sativa	ALFCAL1	Calmodulin	Seedling	807	ss-mRNA	E
M. sativa	ALFEIF4DMR	eIF-4D	—	741	ss-mRNA	E
M. sativa	ALFHIS311	H3 histone gene *ALH3-1. 1*	—	1089	ds-DNA	E
M. sativa	ALFHIS3C1	Histone H3	Callus	618	ss-mRNA	E
M. sativa	ALFHIS3C2	Histone H3	Callus	513	ss-mRNA	E
M. sativa	ALFHIS3C3	Histone H3	Callus	579	ss-mRNA	E
M. sativa	ALFHIS3C4	Histone H3	Callus	322	ss-mRNA	E
M. sativa	ALFHISH3A	Histone H3	—	1089	ds-DNA	—
M. sativa	ALFTCTPMR	Tumor protein	—	725	ss-mRNA	E
N. plumbaginifolia	TOBNEIF4A2	*Nicotiana* eukaryotic initiation factor 4A	Root	1368	ss-mRNA	E
N. plumbaginifolia	TOBNEIF4A3	*Nicotiana* eukaryotic initiation factor 4A	Root	1573	ss-mRNA	E
N. plumbaginifolia	TOBNEIF5A1	Initiation factor 5A(1)	Leaves	702	ss-mRNA	E

Gene Index

Table 6. Continued

Species	Identifier	Sequence specification	Tissue/site	Size (bp or b)	Type	S
N. plumbaginifolia	TOBNEIF5A2	Initiation factor 5A(2)	Leaves	1224	ss-mRNA	E
Nicotiana sp.	TOBCTGA1A	*TGA1a* DNA-binding protein	—	1617	ss-mRNA	E
Nicotiana sp.	TOBCTGA1B	*TGA1b* DNA-binding protein	—	726	ss-mRNA	E
N. tabacum	TOB3AF1	*3AF1* gene	Leaves	683	ss-mRNA	E
N. tabacum	TOBARS	Autonomous replicating sequence (t3-ars)	—	1187	ds-DNA	—
N. tabacum	TOBGT1A	DNA-binding protein (GT-1a)	—	1440	ss-mRNA	—
N. tabacum	TOBNEIF5A3	Initiation factor 5A(3)	Leaves	2443	ds-DNA	E
N. tabacum	TOBPG13A	PG13 (*ASF-1/G13*) gene	—	5837	ds-DNA	—
N. tabacum	TOBP20PR	P20 promoter region with a leader peptide gene	—	401	ds-DNA	—
N. tabacum	TOBPRMMG	Promoter activating a promoterless *nptII* marker gene	—	200	ds-DNA	—
N. tabacum	TOBTAF1	Truncated TAF-1 protein, binding to G-box motif	Leaves	1338	ss-mRNA	E
O. sativa	RICHISH2AA	Histone *H2A* gene	—	321	ds-DNA	—
O. sativa	RICHIS3	Histone 3 gene	—	568	ds-DNA	—
O. sativa	RICHIS311	H3 histone gene *H3R-11*	—	1200	ds-DNA	E
O. sativa	RICHIS321	H3 histone *H3R-21*	—	1200	ds-DNA	E
O. sativa	RICHIS3PS	H3 histone pseudogene *H3R-12*	—	1920	ds-DNA	E
O. sativa	RICPCNA	Cyclin (*PCNA*) gene	—	177	ds-DNA	—
O. sativa	RICPCNAGEN	Proliferating cell nuclear antigen (*PCNA*)	—	4556	ds-DNA	E
P. crispum	PUMHISH2A	Histone H2A	Tissue culture	556	ss-mRNA	E
P. crispum	PUMHISH3A	Histone H3 (*PcH3-7*) gene	—	1098	ds-DNA	—
P. crispum	PUMHISH3B	Histone H3 (*PcH3-20*) gene	—	1098	ds-DNA	—
P. crispum	PUMHISH3C	Histone H3 (*PcH3-16*) gene	—	1098	ds-DNA	—
P. hybrida	PETEPF1	*EPF1* gene for DNA-binding protein	Flower	3151	ds-DNA	E
P. sativum	PEAFED1GEN	5′ flanking region of *fed-1* gene	—	671	ds-DNA	E
P. sativum	PEAH2A	Histone H2A	—	609	ss-mRNA	—
P. sativum	PEAHISH1R	H1 histone	—	910	ss-mRNA	E
S. tuberosum	POTTATABP	TATA-binding protein	Tuber	1148	ss-mRNA	E
S. oleracea	SPIRBP28	28 kDa RNA binding protein	—	1009	ss-mRNA	E
T. aestivum	WHTARS	ORF flanked by ARS sequence	Embryo	1241	ds-DNA	E
T. aestivum	WHTARSREL	ARS-related dispersed repeat DNA	Embryo	371	ds-DNA	E
T. aestivum	WHTEMGC19	DNA binding protein (EmBP-1)	Embryo	567	ss-mRNA	—
T. aestivum	WHTEMGCF13	DNA binding protein (EmBP-1)	Embryo	1400	ss-mRNA	—
T. aestivum	WHTH3	Histone H3 gene	—	811	ds-DNA	E
T. aestivum	WHTH4	Histone 4 H4 gene	—	1200	ds-DNA	E
T. aestivum	WHTH4091	Histone H4 *TH091* gene	—	727	ds-DNA	—
T. aestivum	WHTHBP1A	Sequence-specific DNA binding protein	Seedling	1569	ss-mRNA	—

Table 6. Continued

Species	Identifier	Sequence specification	Tissue/site	Size (bp or b)	Type	S
T. aestivum	WHTHISTH1	Histone H1	Seedling	1028	ss-mRNA	E
T. aestivum	WHTHISTH2B	Histone H2B	Seedling	712	ss-mRNA	E
T. aestivum	WHTH3X	Histone H3 gene	Seed	107	ds-DNA	—
T. aestivum	WHT1A17	HBP-1a (leucine zipper type transcription factor)	—	1578	ss-mRNA	E
T. aestivum	WHT1B38	HBP-1b (leucine zipper type transcription factor)	—	2204	ss-mRNA	E
T. aestivum	WHTTFIID	TATA-binding protein (TFIID)	Seedling	1145	ss-mRNA	E
Z. mays	MZEARS1	*ARS1*, autonomous replicating sequence	—	536	ds-DNA	—
Z. mays	MZEARS2	*ARS2*, autonomous replicating sequence	—	537	ds-DNA	—
Z. mays	MZEARS3	*ARS3*, nuclear autonomously replicating sequence	—	669	ds-DNA	—
Z. mays	MZEBPERU	Regulatory protein	Husk	1970	ss-mRNA	E
Z. mays	MZEC1	*c1* locus	—	4060	ds-DNA	E
Z. mays	MZEH3A	Histone H3 (*H3C3*) gene	—	1177	ds-DNA	—
Z. mays	MZEH4A	Histone H4 (*H4C13*) gene	—	949	ds-DNA	—
Z. mays	MZEH3C2	Histone H3 gene (*H3C2*)	—	1001	ds-DNA	—
Z. mays	MZEH3C4	Histone H3 gene (*H3C4*)	—	1264	ds-DNA	—
Z. mays	MZEH4C7	Histone H4 gene (*H4C7*)	—	868	ds-DNA	—
Z. mays	MZEH4C14	Histone H4 gene (*H4C14*)	—	1135	ds-DNA	—
Z. mays	MZEHISH3A	Histone H3 gene	—	411	ds-DNA	—
Z. mays	MZEHMGPMR	High mobility group protein	Leaves	885	ss-mRNA	E
Z. mays	MZEKN1	Knotted-1 (*Kn-1*) gene	Juvenile chromosome	1627	ss-mRNA	E
Z. mays	MZEMYBAA	*c1* locus *myb* homolog	—	4059	ds-DNA	—
Z. mays	MZENBPAA	Nucleic acid-binding protein	Seedling leaf	1251	ss-mRNA	—
Z. mays	MZENBPG	Nucleic acid-binding protein	Leaves	3176	ds-DNA	E
Z. mays	MZENFYB	CAAT-box DNA-binding protein subunit B	—	870	ss-mRNA	E
Z. mays	MZEO2	Opaque (O2)	Endosperm	1751	ss-mRNA	E
Z. mays	MZEOPA2	Opaque-2 (*Os*) gene	—	4761	ds-DNA	E
Z. mays	MZEOPAQ2A	DNA-binding protein opaque-2 (*O2*)	—	1709	ss-mRNA	—
Z. mays	MZEPPR	*myb*-like transcription factor	—	1802	ss-mRNA	—
Z. mays	MZEPPRA	*myb*-like transcription factor (P) protein	—	945	ss-mRNA	—
Z. mays	MZEREGG	*Lc* regulatory protein	Female flower	2605	ss-mRNA	—
Z. mays	MZEREGPRO	Viviparous-1 mRNA	—	2456	ss-mRNA	—
Z. mays	MZERS	Anthocyanin regulatory R-S mRNA	Immature kernel	2309	ss-mRNA	E
Z. mays	MZE7SL	7SL RNA of signal recognition particle	Endosperm	309	ss-mRNA	E
Z. mays	MZESN	SN mRNA	Scutellar node	2572	ss-mRNA	E
Z. mays	ZMH1HIS	H1 histone	—	1075	RNA	E

Gene Index

Table 7. Structural DNA (repetitive DNA, satellite DNA, unidentified DNA)

Species	Identifier	Sequence specification	Tissue/site	Size (bp or b)	Type	S
A. cepa	ALCSAT1	Satellite DNA repeat	–	380	ds-DNA	E
A. cepa	ALCSAT2	Satellite DNA repeat	–	372	ds-DNA	E
A. cepa	ALCSAT3	Satellite DNA repeat	–	383	ds-DNA	E
B. campestris	BNAREPA	Tandemly repeated DNA	–	182	ds-DNA	–
B. campestris	BNAREPC	Tandem repeat	–	182	ds-DNA	E
B. juncea	BNAREPB	Tandemly repeated DNA	–	182	ds-DNA	–
B. juncea	BNASATAG	Satellite DNA	–	174	ds-DNA	E
B. nigra	BNREP34	Tandem repeat DNA	–	348	DNA	E
B. nigra	BNREP35	Repeat DNA	–	496	DNA	E
B. nigra	BNREPRUS	Tandem repeat DNA	–	177	DNA	E
B. nigra	BNTAREP	Tandem repeat DNA, subfamily P	–	347	DNA	E
B. nigra	BNTAREQ	Tandem repeat DNA, subfamily Q	–	348	DNA	E
B. oleracea	BNASAT	Satellite DNA	–	179	ds-DNA	E
B. oleracea	BNASATAA	Satellite DNA inverted direct repeat	–	176	ds-DNA	–
B. oleracea	BNASATAB	DNA inverted direct repeat	–	174	ds-DNA	–
B. oleracea	BNASATAC	Satellite DNA inverted direct repeat	–	175	ds-DNA	–
B. oleracea	BNASATAD	Satellite DNA inverted direct repeat	–	175	ds-DNA	–
B. oleracea	BNASATAE	Satellite DNA inverted direct repeat	–	176	ds-DNA	–
B. oleracea	BNASATAF	Satellite DNA inverted direct repeat	–	176	ds-DNA	–
C. ichangensis	CLISAT	Satellite DNA	–	181	ds-DNA	E
C. limon	CLIRSSAT	Satellite DNA	–	90	ds-DNA	–
D. erucoides	DILREPTAN	Tandem repeat	–	176	ds-DNA	E
H. vulgare	BLYREL1	Tandem repeat	–	208	ds-DNA	E
H. vulgare	BLYREL2	Tandem repeat	–	238	ds-DNA	E
H. vulgare	BLYREL6	Tandem repeat	–	147	ds-DNA	E
L. albus	LUPHA1	Mitochondrial plasmid-like DNA	–	1139	ds-DNA	E
L. luteus	LUPRSECOA	*Eco*RI-family repetitive DNA	–	1071	ds-DNA	–
L. luteus	LUPRSECOB	*Eco*RI-family repetitive DNA	–	1079	ds-DNA	–
L. esculentum	TOMPLDN2B2	Plastid DNA in the nuclear genome	–	199	ds-DNA	E
L. esculentum	TOMPLDN2D3	Plastid DNA in the nuclear genome	–	265	ds-DNA	E
N. tabacum	TOBHRS60D	*Bam*HI repeat element	–	366	ds-DNA	E
N. tabacum	TOBRPBAM1	*Bam*H1 tandem repeat element DNA (HRS60.1)	Leaves	184	ds-DNA	E
N. tabacum	TOBRPBAM2	*Bam*H1 tandem repeat element DNA (HRS60.3)	Leaves	182	ds-DNA	E
N. tabacum	TOBRPBAM3	Tandem repeat element DNA (HRS60.5)	Leaves	183	ds-DNA	E
O. sativa	OSRSOSC19	RSOsC193 insert DNA	–	353	DNA	E
O. sativa	OSRSR1R6	RSOs81 × 19 insert DNA	–	2124	DNA	E
O. sativa	OSRSR6R8	RSOs81 × 19 insert DNA	–	471	DNA	E
O. sativa	OSRSR8R9	RSOs81 × 19 insert DNA	–	404	DNA	E
O. sativa	RICB4P401	Plasmid-like sequence (clone 401)	–	1290	ds-DNA	E
O. sativa	RICB4P406	Plasmid-like sequence (clone 406)	–	610	ds-DNA	E
O. sativa	RICB4P411	Plasmid-like sequence (clone 411)	–	971	ds-DNA	E
O. sativa	RICNAT1	Taichung native I DNA	–	659	ds-DNA	E
O. sativa	RICREP1	Repetitive sequence RG1	–	354	ds-DNA	E
O. sativa	RICREP2	Repetitive sequence RG2	–	354	ds-DNA	E
O. sativa	RICREP5	Repetitive sequence RG5	–	354	ds-DNA	E
O. sativa	RICREP6	Repetitive sequence RG6	–	352	ds-DNA	E
O. sativa	RICREP7	Repetitive sequence RG7	–	355	ds-DNA	E
O. sativa	RICREP9	Repetitive sequence RG9	–	354	ds-DNA	E
O. sativa	RICREP11A	Repetitive sequence RG11A	–	354	ds-DNA	E
O. sativa	RICREP11B	Repetitive sequence RG11B	–	354	ds-DNA	E
O. sativa	RICREP48R	Repetitive sequence RC48	–	352	ss-mRNA	E
O. sativa	RICRSOSC15	RSOsC154 insert DNA	Callus	352	ds-DNA	E
O. sativa	RICVCN1	Variable copy number DNA (pRB301)	–	1653	ds-DNA	E

Table 7. Continued

Species	Identifier	Sequence specification	Tissue/site	Size (bp or b)	Type	S
P. hybrida	PETTBS	Pattern DNA	—	2054	ds-DNA	—
P. nil	PHNSNH39	Nil DNA for Snh39 repeat	—	76	ds-DNA	E
P. nil	PHNSNH40	Nil Snh40 repeat	—	60	ds-DNA	E
P. nil	PHNSNH43	Nil DNA for Snh43 repeat	—	120	ds-DNA	E
P. nil	PHNSNH52	Nil SNH 52 repeat	—	173	ds-DNA	E
P. nil	PHNSNH54	Nil DNA for Snh54 repeat	—	105	ds-DNA	E
P. juncea	PSJREPA	N8 family repetitive sequence	—	897	ds-DNA	—
P. juncea	PSJREPB	N8 family repetitive sequence DNA	—	1138	ds-DNA	—
R. sativus	RADSAT1	Alpoid-like satellite DNA	—	177	ds-DNA	E
R. sativus	RADSAT2	Alpoid-like satellite DNA	—	177	ds-DNA	E
S. siberica	SSIREP5	Satellite DNA fragment	—	34	ds-DNA	E
S. siberica	SSIRSSAT1	Satellite DNA fragment 1	—	33	ds-DNA	—
S. siberica	SSIRSSAT2	Satellite DNA fragment 2	—	48	ds-DNA	—
S. siberica	SSIRSSAT3	Satellite DNA fragment 3	—	57	ds-DNA	—
S. siberica	SSIRSSAT4	Satellite DNA fragment 4	—	66	ds-DNA	—
S. cereale	RYEHCRS	Heterochromatin repetitive sequences	—	649	ds-DNA	—
S. cereale	RYER1731	Dispersed repeat sequence (R173-1)	—	6676	ds-DNA	E
S. cereale	RYER1732	Dispersed repeat sequence (R173-2)	—	4247	ds-DNA	E
S. cereale	RYER1733	Dispersed repeat sequence (R173-3)	—	4659	ds-DNA	E
S. cereale	RYER1734	Dispersed repeat sequence (R173-4)	—	442	ds-DNA	E
S. alba	SALRSCRSB	Cryptic satellite sequence, *Hin*fI	Unopened buds	172	ds-DNA	—
S. alba	SALRSCRSC	Cryptic satellite sequence	Unopened buds	172	ds-DNA	—
S. irio	SSYRS6	219 bp repetitive sequence	—	225	ds-DNA	—
S. brevidens	POTREPDNA	Chromosome specific repetitive DNA sequence	*In vitro* callus	1728	ds-DNA	E
S. halepense	SRGXSR3	XSR3 DNA repetitive sequence	—	203	ds-DNA	E
S. halepense	SRGXSR6	XSR6 DNA repetitive sequence	—	184	ds-DNA	E
T. elongatum	THIREP	E genome repetitive sequence DNA	—	1375	ds-DNA	—
T. repens	TFRTRPT1	Tandem repeat region	—	347	ds-DNA	E
T. repens	TFRTRPT2	Tandem repeat region	—	347	ds-DNA	E
T. repens	TFRTRPT3	Tandem repeat region	—	693	ds-DNA	E
T. repens	TFRTRPT4	Tandem repeat region	—	694	ds-DNA	E
T. dactyloides	TRD2AA	Knob-specific repeat region	—	177	ds-DNA	—
T. dactyloides	TRD2AB	Knob-specific repeat region	—	180	ds-DNA	—
T. dactyloides	TRD2AC	Knob-specific repeat region	—	180	ds-DNA	—
T. dactyloides	TRD2AD	Knob-specific repeat region	—	179	ds-DNA	—
T. aestivum	TAARSREL	ARS-related dispersed repeat DNA	Embryo	371	DNA	E
T. aestivum	WHTREPTA1	Repetitive DNA pTa1	—	150	ds-DNA	E
T. aestivum	WHTREPTA2	Repetitive DNA pTa2	—	150	ds-DNA	E
T. aestivum	WHTREPTA7	Repetitive DNA pTa7	—	133	ds-DNA	E
T. aestivum	WHTREPTA8	Repetitive DNA	—	143	ds-DNA	E
T. aestivum	WHTWIS1	Middle repetitive insertion sequence WIS1	—	1601	ds-DNA	E
T. monococcum	WHTREPTA	Repetitive DNA sequence	Leaves	295	ds-DNA	—
T. monococcum	WHTREPTB	Repetitive DNA sequence	Leaves	273	ds-DNA	—
T. monococcum	WHTREPTC	Repetitive DNA sequence	Leaves	229	ds-DNA	—
V. faba	VFABAMRAA	*Bam*HI repetitive element, 850 bp family	Embryonic axes	901	ds-DNA	—
V. faba	VFABAMRBB	*Bam*HI repetitive element, 900 bp family	Embryonic axes	855	ds-DNA	—
V. faba	VFABAMRCC	*Bam*HI repetitive element, 990 bp family	Embryonic axes	993	ds-DNA	—
V. faba	VFABAMRDD	*Bam*HI repetitive element, 1150 bp family	Embryonic axes	1155	ds-DNA	—

Table 7. Continued

Species	Identifier	Sequence specification	Tissue/site	Size (bp or b)	Type	S
V. faba	VFABAMREE	*Bam*HI repetitive element, 1500 bp family	Embryonic axes	1426	ds-DNA	—
V. faba	VFABAMRFF	*Bam*HI repetitive element, 1750 bp family	Embryonic axes	1749	ds-DNA	—
V. faba	VFABAMRPA	*Bam*HI repetitive element	Embryonic axes	255	ds-DNA	—
V. faba	VFABAMRPB	*Bam*HI repetitive element	Embryonic axes	254	ds-DNA	—
V. faba	VFABAMRPC	*Bam*HI repetitive element	Embryonic axes	254	ds-DNA	—
V. faba	VFABAMRPD	*Bam*HI repetitive element	Embryonic axes	250	ds-DNA	—
V. faba	VFABAMRPE	*Bam*HI repetitive element	Embryonic axes	246	ds-DNA	—
V. faba	VFABAMRPF	*Bam*HI repetitive element	Embryonic axes	250	ds-DNA	—
V. faba	VFAFOKI	Fok I repeat element	Seed	59	ds-DNA	—
V. radiata	VIRRPT	Repetitive sequence DNA	Germinating seed	361	ds-DNA	—
V. radiata	VIRSAT01	174 bp satellite repeat homologous to rDNA subrepeats	—	174	ds-DNA	E
Z. diploperennis	MZEZEAH231	Repetitive DNA	Leaves	231	ds-DNA	E
Z. diploperennis	MZEZEAR266	Repetitive DNA	Leaves	266	ds-DNA	E
Z. diploperennis	MZEZEAR270	Repetitive DNA	Leaves	270	ds-DNA	E
Z. mays	MZECIN1	Cin1-001 allele middle repetitive sequence	—	691	ds-DNA	E
Z. mays	MZEHETRO	Heterochromatin repetitive DNA	—	184	ds-DNA	—
Z. mays	MZEZMA	180 bp knob-specific repeat region	—	180	ds-DNA	—
Z. mays	MZEZMB	180 bp knob-specific repeat region	—	180	ds-DNA	—
Z. mays	MZEZMC	180 bp knob-specific repeat region	—	180	ds-DNA	—
Z. mays	MZEZMD	180 bp knob-specific repeat region	—	180	ds-DNA	—
Z. mays	MZEZME	180 bp knob-specific repeat region	—	180	ds-DNA	—
Z. mays	MZEZMF	180 bp knob-specific repeat region	—	180	ds-DNA	—
Z. mays	MZEZMG	180 bp knob-specific repeat region	—	179	ds-DNA	—
Z. mays	MZEZMH	180 bp knob-specific repeat region	—	180	ds-DNA	—
Z. mays	MZEZMI	180 bp knob-specific repeat region	—	180	ds-DNA	—
Z. mays	MZEZMJ	180 bp knob-specific repeat region	—	180	ds-DNA	—
Z. mays	MZEZMK	180 bp knob-specific repeat region	—	179	ds-DNA	—
Z. mays	MZEZML	180 bp knob-specific repeat region	—	180	ds-DNA	—
Z. mexicana	MZEEMA	180 bp knob-specific repeat region	—	180	ds-DNA	—
Z. mexicana	MZEEMB	180 bp knob-specific repeat region	—	180	ds-DNA	—

Table 8. Enzymes (all identified enzyme sequences) (see also *Table 14*, photosynthetic and related enzymes, and *Table 15*, protease, kinase and phosphatase enzymes)

Species	Identifier	Sequence specification	Tissue/site	Size (bp or b)	Type	S
A. majus	AMACHS	Chalcone synthase	—	3574	ds-DNA	E
A. majus	AMACHALIS	Chalcone isomerase	Flowers	667	ss-mRNA	G
A. majus	AMAGADPH	Glyceraldehyde-3-phosphate dehydrogenase	—	1301	ss-mRNA	E
A. majus	AMAPALLID	Dihydroflavonol-4-reductase Pallida gene	Flower buds	1608	ss-mRNA	E
A. thaliana	ATHACP	Acyl carrier protein	—	1438	ds-DNA	—
A. thaliana	ATHACPA1	Acyl carrier protein (ACP) *A 1*	—	1927	ds-DNA	E
A. thaliana	ATHACPA2	Acyl carrier protein (ACP) *A 2*	—	1876	ds-DNA	E
A. thaliana	ATHADH	Alcohol dehydrogenase	—	3195	ds-DNA	—
A. hypogaea	ARCGSC2	Resveratrol synthase mRNA	—	1130	ss-mRNA	—
A. hypogaea	ARCGSG11	Resveratrol synthase	—	951	ds-DNA	E
A. hypogaea	ARCPNC1	Cationic peroxidase	—	1116	ss-mRNA	E
A. hypogaea	ARCPNC2	Cationic peroxidase	—	1330	ss-mRNA	E
A. hypogaea	ARCRSGSG10	Resveratrol synthase	—	1228	ds-DNA	E
A. rusticana	HRAHRPCA	Peroxidase isoenzyme C (*HPR C*)	—	2912	ds-DNA	—
A. rusticana	HRAHRPCB	Peroxidase isoenzyme C (*HPR C*)	—	3289	ds-DNA	—
A. rusticana	HRAHRPCC	Peroxidase isoenzyme C (*HPR C*)	—	1229	ds-DNA	—
A. rusticana	HRANEUPERO	Neutral peroxidase	Fresh sideroots	1397	ss-mRNA	E
A. rusticana	HRAPRXC2	Peroxidase isozyme (*prxC2*)	—	3095	ds-DNA	—
A. rusticana	HRAPRXC3	Peroxidase isozyme (*prxC3*) gene	—	2733	ds-DNA	—
A. sativa	ASTSPE1	Arginine decarboxylase (EC 4.1.1.19)	—	2124	ss-mRNA	E
B. vulgaris	BEUBADHD	Salt-inducible betaine aldehyde dehydrogenase	Leaves	1750	ss-mRNA	E
B. vulgaris	BEAUF16BPP	Cytosolic fructose-1,6-bisphosphatase	—	1278	ss-mRNA	—
B. campestris	BNAACPBC	Acyl carrier protein	Seeds	557	ss-mRNA	E
B. juncea	BNAEFEMR	Ethylene-forming enzyme	Developing leaves	1275	ss-mRNA	E
B. napus	BNAACCP05	Acyl carrier protein	Embryo	2227	ds-DNA	E
B. napus	BNAACP05	Acyl carrier protein	Embryo	1095	ds-DNA	E
B. napus	BNAACP09	Acyl carrier protein	Embryo	1083	ds-DNA	E
B. napus	BNAACP10C	Seed acyl-carrier protein	Embryo	337	ss-mRNA	E
B. napus	BNAACP22C	Seed acyl-carrier protein	Embryo	291	ss-mRNA	E
B. napus	BNAACP29C	Seed acyl-carrier protein	Embryo	405	ss-mRNA	E
B. napus	BNAACP34C	Seed acyl-carrier protein	Embryo	405	ss-mRNA	E
B. napus	BNAACP04F	Seed acyl-carrier protein	Embryo	319	ss-mRNA	E
B. napus	BNAACP28F	Seed acyl-carrier protein	Embryo	671	ss-mRNA	E
B. napus	BNAACP34F	Seed acyl-carrier protein	Embryo	368	ss-mRNA	E
B. napus	BNAACP10H	Seed acyl-carrier protein	Embryo	405	ss-mRNA	E
B. napus	BNAAHSYI	Actohydroxyacid synthase I	—	4724	ds-DNA	E
B. napus	BNAAHSYII	Acetohydroxyacid synthase II	—	3326	ds-DNA	E
B. napus	BNAAHSYIII	Acetohydroxyacid synthase III	—	3386	ds-DNA	E
B. napus	BNAALS	Acetolactate synthase (*ALS*) gene	Leaves	3315	ds-DNA	—
B. napus	BNAALS2	Acetolactate synthase	—	2019	ss-mRNA	—
B. napus	BNABP19G	*Bp19* gene	Pollen	4977	ds-DNA	E
B. napus	BNACYP	Cyclophilin (CyP)	Immature flower bud	661	ss-mRNA	—
B. napus	BNAEPSPG	5-enolpyruvylshikimate-3-phosphate synthase	—	3831	ds-DNA	E
B. napus	BNAIPMDH	3-isopropylmalate dehydrogenase	Leaves	1428	ss-mRNA	—
B. napus	BNAMALS	Malate synthase	—	1932	ss-mRNA	—
B. oleracea	BNAHDHQ	Histidinol dehydrogenase	—	1606	ss-mRNA	—

Table 8. Continued

Species	Identifier	Sequence specification	Tissue/site	Size (bp or b)	Type	S
C. ensiformis	CENUREA	Urease	–	2719	ss-mRNA	–
C. tinctorius	CAHSACPD	Stearoyl-acyl carrier protein desaturase	Embryo	1533	ss-mRNA	–
C. arietinum	CIENADPHIO	NADPH–isoflavone oxidoreductase	–	1183	ss-mRNA	E
C. vulgaris	CNAMD	Mitochondrial malate dehydrogenase (EC 1.1.1.37)	Cotyledon	1414	ss-mRNA	E
C. vulgaris	CNAGMDHA	Malate dehydrogenase	–	1334	ss-mRNA	–
C. unguiculata	CLRPGIA	Phosphoglucose isomerase (EC 5.3.1.9)	Leaves	2040	ds-DNA	–
D. carota	DARBFRUC	Cell wall β-fructosidase	–	1972	ss-mRNA	–
D. carota	DAREP2LTP	Lipid transfer protein	–	737	ss-mRNA	–
D. carophyllus	DINCARAA	S-adenosylmethionine synthetase	–	1632	ss-mRNA	–
D. caryophyllus	DINCARACC	1-aminocyclopropane-1-carboxylate synthase	Flower petals	1942	ss-mRNA	–
D. caryophyllus	DINCARSR8	Glutathione S-transferase	Flower petals	957	ss-mRNA	–
D. caryophyllus	DINCARSR9	Glutathione S-transferase	Flower petals	957	ss-mRNA	E
F. ananassa	FRRADH	Alcohol dehydrogenase (EC 1.1.1.1)	–	3208	ds-DNA	E
F. chloraefolia	FTRFLA3STA	Flavonol 3-sulfotransferase	Leaves	1158	ss-mRNA	–
F. chloraefolia	FTRFLA3STB	Flavonol 4′-sulfotransferase	Leaves	1236	ss-mRNA	–
F. linearis	FTRMRSEQ	NADP–malic enzyme	Leaves	2122	ss-mRNA	–
F. trinervia	FTRMRSEQB	NADP–malic enzyme	Leaves	762	ss-mRNA	–
F. trinervia	FTRNADPME	NADP-dependent malic enzyme	Leaves	2258	ss-mRNA	E
F. trinervia	FTRPDK	Pyruvate, orthophosphate dikinase	Leaves	3105	ss-mRNA	E
F. trinervia	FTRPEPC	Phosphoenolpyruvate carboxylase	Leaves	3174	ss-mRNA	E
G. max	GMCHS1	Chalcone synthase	–	2073	DNA	E
G. max	GMCHS13	Intergenic region (chalcone synthase)	–	1876	DNA	E
G. max	SOYB13ENDG	β-1,3-endoglucanase	–	1260	ss-mRNA	–
G. max	SOYCHS15	Chalcone synthetase gene *CHS1*	–	378	ds-DNA	E
G. max	SOYCHS25	Chalcone synthetase gene *CHS2*	–	405	ds-DNA	E
G. max	SOYCHS35	Chalcone synthetase *CHS3*	–	422	ds-DNA	E
G. max	SOYFESOD	Iron superoxide dismutase	–	1067	ss-mRNA	–
G. max	SOYGENE3	Chalcone synthetase protein	–	4047	ds-DNA	E
G. max	SOYLOX	Lipoxygenase-1	–	2805	ss-mRNA	–
G. max	SOYLOXB	Lipoxygenase-2	Embryo	2766	ss-mRNA	–
G. max	SOYLOXC	Lipoxygenase-1	Embryo	2046	ss-mRNA	–
G. max	SOYLOXD	Lipoxygenase-1	Embryo	978	ss-RNA	–
G. max	SOYLOX3G	Seed lipoxygenase-3 (EC 1.13.11.12)	Embryo	5709	ds-DNA	E
G. max	SOYLOX3R	Lipoxygenase-3 (EC 1.13.11.12)	–	2779	ss-mRNA	E
G. max	SOYPAL1	Phenylalanine ammonia lyase	Leaves	4085	ds-DNA	E
G. max	SOYP5CR	Pyrroline-5-carboxylate reductase	–	1109	ss-mRNA	E
G. max	SOYPHYDES	Phytoene desaturase	Embryo	2293	ss-mRNA	–
G. max	SOYREDUC	NAD(P)H dependent 6′-deoxychalcone synthase	Suspension culture	1108	ss-mRNA	E
G. max	SOYRPB1C	Largest subunit of RNA polymerase II RPB1 (EC 2.7.7.6)	–	4153	ds-DNA	E
G. max	SOYSC514	Lipoxygenease (sc514)	–	5663	ds-DNA	E
G. max	SOYSOYCHS	Chalcone synthase (EC 2.3.1.74)	–	2323	ds-DNA	E
G. max	SOYURA	Urease	–	1303	ds-DNA	–
G. hirsutum	COTCAT1	Catalase subunit 1 (EC 1.11.1.6)	24 h dark-grown cotyledon	1666	ss-mRNA	E
G. hirsutum	COTICL	Isocitrate lyase (ICL) (EC 4.1.3.1)	24 h dark-grown cotyledon	1872	ss-mRNA	E
G. hirsutum	COTLCCSU2	Catalase subunit 2	–	1763	ss-mRNA	E

Table 8. Continued

Species	Identifier	Sequence specification	Tissue/site	Size (bp or b)	Type	S
H. brasiliensis	HBHMGR1G	3-hydroxy-3-methylglutaryl coenzyme A reductase 1	Latex, leaves	3993	DNA	E
H. brasiliensis	HBHMGR1M	3-hydroxy-3-methylglutaryl coenzyme A reductase 1	Latex, leaves	2185	RNA	E
H. brasiliensis	HBHMGR2M	3-hydroxy-3-methylgutaryl coenzyme A reductase 2	Leaves	908	RNA	E
H. vulgare	BLY8339	Chalcone synthase	Leaves	5230	ds-DNA	E
H. vulgare	BLYACL1	Acyl carrier protein I (Acl1)	Leaves	3822	ds-DNA	—
H. vulgare	BLYACL2	Acyl carrier protein II (Acl2)	Leaves	706	ss-mRNA	—
H. vulgare	BLYACL3	Acyl carrier protein III (Acl3)	Leaves	4830	ds-DNA	—
H. vulgare	BLYACPI	Acyl carrier protein I (ACP I)	Leaves	768	ss-mRNA	—
H. vulgare	BLYACPIII	Acyl carrier protein III (ACP III)	Leaves	520	ss-mRNA	—
H. vulgare	BLYADH1	Alcohol dehydrogenase 1 (ADH1, EC 1.1.1.1)	Root	1754	ss-mRNA	—
H. vulgare	BLYADH2	Alcohol dehydrogenase gene Adh2 (EC 1.1.1.1)	—	2310	ds-DNA	E
H. vulgare	BLYADH3	Alcohol dehydrogenase gene Adh3 (EC 1.1.1.1)	—	2270	ds-DNA	E
H. vulgare	BLYADH1A	Alcohol dehydrogenase gene Adh1 (EC 1.1.1.1)	—	1506	ds-DNA	E
H. vulgare	BLYALREDR	Aldose reductase-related protein	Embryo	1209	ds-DNA	E
H. vulgare	BLYAMY1	α-amylase 1	—	1461	ss-mRNA	—
H. vulgare	BLYAMY2	α-amylase 2	—	1588	ss-mRNA	—
H. vulgare	BLYAMY46	α-amylase type B (Amy46)	Aleurone	2159	ds-DNA	—
H. vulgare	BLYAMY56	α-amylase (clone gRAmy56) (EC 3.2.1.1)	Aleurone	2203	ds-DNA	E
H. vulgare	BLYAMY152	α-amylase (clone gRAmy152) (EC 3.2.1.1)	Aleurone	2625	ds-DNA	E
H. vulgare	BLYAMY1A	α-amylase 1	Aleurone	1992	ds-DNA	—
H. vulgare	BLYAMY2A	α-amylase 2	Aleurone	2337	ds-DNA	—
H. vulgare	BLYAMYAA	α-amylase type A isozyme	Aleurone	1588	ss-mRNA	—
H. vulgare	BLYAMYABA	α-amylase type B isozyme	Aleurone	595	ss-mRNA	—
H. vulgare	BLYAMYABB	α-amylase type B isozyme	Aleurone	512	ss-mRNA	—
H. vulgare	BLYAMYABC	α-amylase type B isozyme	Aleurone	158	ss-mRNA	—
H. vulgare	BLYAMYABD	α-amylase type B isozyme (Amy6-4)	Aleurone	2910	ds-DNA	—
H. vulgare	BLYAMYABE	α-amylase type B isozyme	Aleurone	1119	ss-mRNA	—
H. vulgare	BLYAMYBA	β-amylase	—	1754	ss-mRNA	—
H. vulgare	BLYAMYG	High pI isozyme of α-amylase	Aleurone	2717	ds-DNA	—
H. vulgare	BLYAMY1G	α-amylase (EC 3.2.1.1)	Aleurone	950	ds-DNA	E
H. vulgare	BLYBAMYL	β-amylase	Endosperm Young shoots Leaves	1754	ss-mRNA	E
H. vulgare	BLYBDG	(1-3, 1-4)-β-D-glucanase (EC 3.2.1.73)	—	6260	ds-DNA	E
H. vulgare	BLYB13GLU	β 1-3 glucanase (EC 3.2.1.39)	Leaves	225	ss-mRNA	—
H. vulgare	BLYBRNZ1H	UDP–glucose flavonol 3-*O*-glucosyl transferase	Leaves	2326	ds-DNA	E
H. vulgare	BLYCBGL32	(1-3)-β-glucanase	Embryo	1234	ss-mRNA	—
H. vulgare	BLYCHI	26 kDa chitinase	Embryo	998	ss-mRNA	—
H. vulgare	BLYENDCHT	Endochitinase	Aleurone	556	ss-mRNA	—
H. vulgare	BLYFL3DOX	Flavonone-3-hydroxylase	Testa	1363	ss-mRNA	E
H. vulgare	BLYGADPH	Glycolytic glyceraldehyde-3-phosphate dehydrogenase	—	1365	ss-mRNA	
H. vulgare	BLYGAPDH	Glyceraldehyde-3-phosphate dehydrogenase	Scutellum	1141	ss-mRNA	—

Gene Index

Table 8. Continued

Species	Identifier	Sequence specification	Tissue/site	Size (bp or b)	Type	S
H. vulgare	BLYGEH	(1→3)-β-glucan endohydrolase	Scutellum	1250	ss-mRNA	–
H. vulgare	BLYGLB2	1,3-1,4-β-D glucan 4-glucanohydrolase (*Glb 2*)	–	5159	ds-DNA	–
H. vulgare	BLYGLN2	Glutamine synthetase precursor (EC 6.3.1.2)	Leaves	1549	ss-mRNA	E
H. vulgare	BLYGLN2R	Glutamine synthetase 2 (EC 6.3.1.2)	Leaves	1600	ss-mRNA	E
H. vulgare	BLYGSA	Glutamate 1-semialdehyde aminotransferase (GSA)	–	1621	ss-mRNA	–
H. vulgare	BLYGLUCB	β-glucanase	–	874	ss-mRNA	–
H. vulgare	BLYGLEUND	(1,3-1,4)-β-glucanase (EC 3.2.1.73)	–	4643	ds-DNA	E
H. vulgare	BLYKASI	β-ketoacyl-ACP synthase I (Kas12)	Leaves	1805	ss-mRNA	–
H. vulgare	BLYLDHB12	L-lactate dehydrogenase	–	1166	ss-mRNA	–
H. vulgare	BLYLDHA13	L-lactate dehydrogenase	–	1298	ss-mRNA	–
H. vulgare	BLYLTP1	Lipid transfer protein	Aleurone	1487	ds-DNA	E
H. vulgare	BLYLTP1G	LTP 1	Aleurone	2874	ds-DNA	E
H. vulgare	BLYLTPMR	Putative phospholipid transfer protein	Aleurone	517	ss-mRNA	E
H. vulgare	BLYMY32B	α-amylase type A (Amy32b)	Aleurone	2717	ds-DNA	–
H. vulgare	BLYNADPHP	NADPH-protochlorophyllide oxidoreductase (PCR)	Leaves	1490	ss-mRNA	E
H. vulgare	BLYPERO	Peroxidase	–	740	ss-mRNA	E
H. vulgare	BLYPEROX1	Peroxidase (BP1)	Endosperm	621	ss-mRNA	E
H. vulgare	BLYPEROXI	Peroxidase (EC 1.11.1.7)	–	621	ss-mRNA	E
H. vulgare	BLYPRX5A	Peroxidase BP 1 (Prx5)	Endosperm	1177	ss-mRNA	–
H. vulgare	BLYPRX6A	Peroxidase BP 2A (Prx6)	Embryo	2660	ds-DNA	–
H. vulgare	BLYWAXYG	*Waxy* locus encoding starch synthase (EC 2.4.1.11)	Leaves	5153	ds-DNA	E
H. vulgare	BLYWAXYR	*Waxy* locus	Leaves	2311	ss-mRNA	E
H. niger	HYSH6H	Hyoscyamine-6-β-hydroxylase	Root	1354	ss-mRNA	–
I. batatas	IPBCATR	Catalase (EC 1.11.1.6)	–	1804	ss-mRNA	E
I. batatas	IPBPAL	Phenylalanine ammonia-lyase	Root	2335	ss-mRNA	–
I. batatas	IPBSPP	Starch phosphorylase	Root	3292	ss-mRNA	–
L. sativa	LAUGLUSYN	Glutamine synthetase	Seed	1429	ss-mRNA	E
L. angustifolius	LUPAATP1A	Aspartate aminotransferase P1	Root	1454	ss-mRNA	–
L. angustifolius	LUPASNASE	Developing seed L-asparaginase	Seed	3113	ds-DNA	E
L. angustifolius	LUPASPAT	Aspartate aminotransferase	Root	1645	ss-mRNA	E
L. angustifolius	LUPGLUSYN	Glutamine synthetase (EC 6.3.1.2)	Root nodule	1043	ss-mRNA	E
L. polyphyllus	LUPPOD	Peroxidase	Undifferentiated suspension	701	ss-mRNA	E
L. esculentum	LEEFEMR	Ethylene-forming enzyme	–	1035	RNA	E
L. esculentum	TOMACC2G	1-aminocyclopropane-1-carboxylic acid synthase	Etiolated hypocotyls	7244	ds-DNA	E
L. esculentum	TOMACC2MR	1-aminocyclopropane-1-carboxylic acid synthase	Wounded ripe fruit	1775	ss-mRNA	E
L. esculentum	TOMACC4MR	1-aminocyclopropane-1-carboxylic acid synthase	Wounded ripe fruit	1616	ss-mRNA	E
L. esculentum	TOMACCS	1-aminocyclopropane-1-carboxylate synthase	Pericarp	1635	ss-mRNA	–
L. esculentum	TOMACS	1-aminocyclopropane-1-carboxylate synthase	Pericarp	1819	ss-mRNA	–
L. esculentum	TOMACT	1-aminocyclopropane-1-carboxylate synthase	Pericarp	420	ss-mRNA	–
L. esculentum	TOMADHMR	Alcohol dehydrogenase	–	594	ss-mRNA	E
L. esculentum	TOMAPP1A	Acid phosphate-1	–	3834	ds-DNA	–

Table 8. Continued

Species	Identifier	Sequence specification	Tissue/site	Size (bp or b)	Type	S
L. esculentum	TOMAPS1A	Acid phosphatase (aps1)	—	948	ss-mRNA	—
L. esculentum	TOMAROA	5-enolpyruvylshikimate-3-phosphate synthase	—	2045	ss-mRNA	—
L. esculentum	TOMCAT1A	Esculentum catalase	—	1804	ss-mRNA	—
L. esculentum	TOMCBPE	Phytoene synthetase	Fruit	1786	ss-mRNA	—
L. esculentum	TOMCHSOD	Chloroplast Cu–Zn superoxide dismutase	Leaves	1033	ss-mRNA	E
L. esculentum	TOMCYTSOD	Cytoplasmic Cu–Zn superoxide dismutase (EC 1.15.1.1)	Leaves	777	ss-mRNA	E
L. esculentum	TOMENOLASE	Enolase	—	3005	ds-DNA	E
L. esculentum	TOMETH1	ETH.1 gene, involved in ethylene biosynthesis	—	1620	ds-DNA	E
L. esculentum	TOMACIN	Invertase (TIV1)	Pericarp	2176	ss-mRNA	—
L. esculentum	TOMPGA	Polygalacturonase (PG)	Fruit	414	ss-mRNA	—
L. esculentum	TOMPGAAA	Polygalacturonase (PG)	Fruit	7456	ds-DNA	—
L. esculentum	TOMPG2AR	Polygalacturonase-2a	Fruit	1621	ss-mRNA	E
L. esculentum	TOMB13GLUA	β-1,3-glucanase	Leaves	1189	ss-mRNA	—
L. esculentum	TOMB13GLUB	β-1,3-glucanase	Leaves	1331	ss-mRNA	—
L. esculentum	TOMCYP	Cyclophilin (CyP)	Shoot tip	720	ss-mRNA	—
L. esculentum	TOMGTOM5	Phytoene synthase	—	3707	ds-DNA	E
L. esculentum	TOMHMG2A	3-hydroxy-3-methyl glutaryl coenzyme A reductase	—	2884	ds-DNA	—
L. esculentum	TOMILV1A	Threonine deaminase	Flower	1980	ss-mRNA	E
L. esculentum	TOMILV1B	Threonine deaminase (Td)	Flower	4240	ds-DNA	—
L. esculentum	TOMMRNAEN	Enolase	—	1553	ss-mRNA	E
L. esculentum	TOMPAL5A	Phenylalanine ammonia lyase	Leaves	3416	ds-DNA	—
L. esculentum	TOMPECES	Pectin esterase	—	1665	ss-mRNA	—
L. esculentum	TOMPGR	Polygalacturonase (PG, EC 3.2.1.15)	Fruit	1617	ss-mRNA	E
L. esculentum	TOMPGRA	Polygalacturonase (PG)	—	2207	ds-DNA	—
L. esculentum	TOMPHYTDES	Phytoene desaturase	Fruit	2373	ss-mRNA	—
L. esculentum	TOMPOLYGA	Cell wall degrading polygalcturonase	Fruit	7456	ds-DNA	E
L. esculentum	TOMPSPS	Phytoene synthase pseudogene	Leaves	4073	ds-DNA	E
L. esculentum	TOMRNA	1-aminocyclopropane-1-carboxylate synthase	Fruit pericarp	268	ss-mRNA	—
L. esculentum	TOMRNAB	1-aminocyclopropane-1-carboxylate synthase	—	271	ss-mRNA	—
L. esculentum	TOMRNAC	1-aminocyclopropane-1-carboxylate synthase	—	271	ss-mRNA	—
L. esculentum	TOMRNAD	1-aminocyclopropane-1-carboxylate synthase	Ripe pericarp	219	ss-mRNA	—
L. esculentum	TOMSODA	Superoxide dismutase	—	776	ss-mRNA	—
L. esculentum	TOMSODB	Superoxide dismutase	—	1034	ss-mRNA	—
L. esculentum	TOMTAP1	Anionic peroxidase	—	2907	ds-DNA	E
L. esculentum	TOMTAP2	Anionic peroxide	—	2690	ds-DNA	E
L. esculentum	TOMTSW12	TSW12	Stem	675	ss-mRNA	E
M. liliiflora	MAIGADPH	Glycolytic glyceraldehyde-3-phosphate dehydrogenase	—	1397	ss-mRNA	E
M. incana	MINCHSY	Chalcone synthase (EC 2.3.1.74)	Petals	1378	ss-mRNA	E
M. sativa	ALFAAT1MR	Aspartate aminotransferase	—	1514	ss-mRNA	E
M. sativa	ALFISREDMR	Isoflavone reductase	Callus	1181	ss-mRNA	E
M. sativa	ALFGS	Glutamine synthetase	—	276	ss-mRNA	—
M. sativa	ALFGLNAG	Glutamine synthetase	—	4464	ds-DNA	E
M. sativa	ALFMSCHI1	Chalcone isomerase (MsCHI-1)	*Rhizobium meliloti* induced root nodule	950	ss-mRNA	—

Table 8. Continued

Species	Identifier	Sequence specification	Tissue/site	Size (bp or b)	Type	S
M. sativa	ALFMSCHI2	Chalcone isomerase (MsCHI-2)	*Rhizobium meliloti* induced root nodule	845	ss-mRNA	—
M. sativa	ALFPAL	Phenylalanine ammonia lyase	—	2409	ss-mRNA	E
M. sativa	ALFSALMCAT	Caffeic acid 3-*O*-methyltransferase	—	1317	ss-mRNA	G
M. crystallinum	CIPFNRA	Ferredoxin–NADP$^+$ reductase	—	1419	ss-mRNA	—
M. crystallinum	CIPGPDNAD	NAD–glyceraldehyde-3-phosphate dehydrogenase	Leaves	1354	ss-mRNA	—
M. crystallinum	CIPIMT1A	Methyl transferase (Imt1) mRNA	Leaves	1494	ss-mRNA	—
M. crystallinum	CIPNADGAPD	Glyceraldehyde-3-phosphate dehydrogenase (NAD–GAPDH)	Leaves	1354	ss-mRNA	—
M. crystallinum	CIPPPCA	Phosphoenolpyruvate carboxylase (ppc1)	—	7694	ds-DNA	E
M. crystallinum	CIPPPCB	Phosphoenolpyruvate carboxylase (ppc2)	—	8050	ds-DNA	E
M. crystallinum	CIPPPCR	Phosphoenolpyruvate carboxylase (EC 4.1.1.31)	—	3348	ss-mRNA	—
M. crystallinum	CIPPRK	Phosphoribulokinase	Leaves	1520	ss-mRNA	—
N. plumbaginifolia	TOBB13GG	*β*-(1,3)-glucanase, vacuolar isoform	—	3022	ds-DNA	E
N. plumbaginifolia	TOBCPGS	Chloroplast glutamine synthetase	Leaves	1340	ss-mRNA	—
N. plumbaginifolia	TOBFESOD	Iron superoxide dismutase (FeSOD)	—	1012	ss-mRNA	—
N. plumbaginifolia	TOBGCBREG	*β*(1,3)glucanase regulator genes	—	5007	ds-DNA	—
N. plumbaginifolia	TOBGLUCB1	*β*-glucanase	—	127	ds-DNA	—
N. plumbaginifolia	TOBGLUCB2	*β*-glucanase	—	1232	ss-mRNA	—
N. plumbaginifolia	TOBMNSOD	Manganese superoxide dismutase (MnSOD) EC 1.15.1.1)	—	966	ss-mRNA	E
N. plumbaginifolia	TOBSODM	Superoxide dismutase	—	837	ss-mRNA	E
N. sylvestris	TOB6P229	6P229 polypeptide homologous to animal glutathione peroxidase	—	816	ss-mRNA	E
N. tabacum	TOBACHIPRP	Acidic chitinase PR-P	Leaves	997	ss-mRNA	E
N. tabacum	TOBACHIPRQ	Acidic chitinase PR-Q	Leaves	975	ss-mRNA	E
N. tabacum	TOBACIDCL3	Acidic chitinase III	—	1018	ss-mRNA	E
N. tabacum	TOBALSURA	Acetolactate synthase (EC 4.1.3.18)	—	2468	ds-DNA	E
N. tabacum	TOBALSURB	Acetolactate synthase (EC 4.1.3.18)	—	2461	ds-DNA	E
N. tabacum	TOBBASICL3	Basic chitinase III	—	1064	ss-mRNA	E
N. tabacum	TOBCHN50	Endochitinase (CHN50)	Suspension culture	2949	ds-DNA	E
N. tabacum	TOBDAHPS1	3-deoxy-D-arabino-heptulosonate 7-phosphate synthase	—	1928	ss-mRNA	—
N. tabacum	TOBECH	Endochitinase	—	1070	ss-mRNA	—
N. tabacum	TOBECHITG	Endochitinase (EC 3.2.1.14)	Leaves	3850	ds-DNA	E
N. tabacum	TOBECHITR	Endochitinase (EC 3.2.1.14)	Pith parenchyma	1132	ss-mRNA	E
N. tabacum	TOBEPSPS1	5-enolpyruvylshikimate-3-phosphate synthase	—	2011	ss-mRNA	E
N. tabacum	TOBEPSPS2	5-enolpyruvylshikimate-3-phosphate synthase	—	1322	ss-mRNA	—
N. tabacum	TOBGAPA	A-subunit of chloroplast glyceraldehyde-3-phosphate dehydrogenase	Leaves	1327	ss-mRNA	—

Table 8. Continued

Species	Identifier	Sequence specification	Tissue/site	Size (bp or b)	Type	S
N. tabacum	TOBGAPB	B-subunit of chloroplast glyceraldehyde-3-phosphate dehydrogenase	Leaves	1370	ss-mRNA	—
N. tabacum	TOBGAPC	Glyceraldehyde-3-phosphate dehydrogenase	Leaves	1174	ss-mRNA	—
N. tabacum	TOBGLA13B	Glucan β-1,3-glucanase	—	4482	ds-DNA	—
N. tabacum	TOBGLB13B	Glucan β-1,3-glucosidase	—	4699	ds-DNA	—
N. tabacum	TOBGLN2G	β-1,3-glucanase	—	4443	ds-DNA	E
N. tabacum	TOBGLUBA	β-1,3-glucanase	—	1104	ss-mRNA	—
N. tabacum	TOBGLUBB	β-1,3-glucanase	—	582	ss-mRNA	—
N. tabacum	TOBGLUBC	β-1,3-glucanase	—	1054	ss-mRNA	—
N. tabacum	TOBGLUCA	Basic-1,3-glucanase	—	4766	ds-DNA	—
N. tabacum	TOBGLUCB	β-1,3-glucanase	—	2259	ds-DNA	—
N. tabacum	TOBLTP1	Lipid transferase	Leaves	2075	ds-DNA	E
N. tabacum	TOBPAL	Phenylalanine ammonia lyase	—	372	ss-mRNA	E
N. tabacum	TOBPPC	Phosphoenolpyruvate carboxylase	—	3189	ss-mRNA	E
N. tabacum	TOBPTLG	Pectate lyase	Pollen	1389	ds-DNA	E
N. tabacum	TOBPTLMR	Pectate lyase	Pollen	967	ss-mRNA	E
N. tabacum	TOBPXDL5	Lignin-forming peroxidase	—	—	—	—
N. tabacum	TOBSP41A	(1-3)-β-glucanase	—	1375	ss-mRNA	E
N. tabacum	TOBSP41B	(1-3)-β-glucanase	Style	1297	ss-mRNA	E
N. tabacum	TOBTPA1A	Phenylalanine ammonia lyase (tpa1)	—	6976	ds-DNA	—
O. sativa	RICAAMYA	α-amylase	—	1553	ss-mRNA	—
O. sativa	RICAAMYB	α-amylase	—	1682	ss-mRNA	—
O. sativa	RICACLDE1	Alcohol dehydrogenase	—	1273	ss-mRNA	E
O. sativa	RICADH2A	Alcohol dehydrogenase	—	3779	ds-DNA	—
O. sativa	RICADP	ADP–glucose pyrophosphorylase 51 kDa subunit	Endosperm	1664	ss-mRNA	—
O. sativa	RICADPA	ADP–glucose pyrophosphorylase	Endosperm	5934	ds-DNA	—
O. sativa	RICALAM	α-amylase	—	4276	ds-DNA	E
O. sativa	RICAMAA	α-amylase gene	—	76	ds-DNA	—
O. sativa	RICAMYA	α-amylase (α-*Amy3*)	—	1317	ds-DNA	—
O. sativa	RICAMY2A	α-amylase (*amy2A*)	—	4128	ds-DNA	—
O. sativa	RICAMY1B	α-amylase (*RAmy1B*) gene	—	537	ds-DNA	—
O. sativa	RICAMYC	α-amylase (*OSamy-c*)	Leaves	2685	ds-DNA	E
O. sativa	RICAMYC2M	α-amylase (*Amyc2*)	Seed	1659	ss-mRNA	E
O. sativa	RICAMY3D	α-amylase (*RAmy3D*)	—	2938	ds-DNA	—
O. sativa	RICAMY3E	α-amylase (*RAmy3E*)	—	2683	ds-DNA	—
O. sativa	RICALCDE2	Alcohol dehydrogenase 2	—	1575	ss-mRNA	E
O. sativa	RICCATAL	Catalase	Immature seed	1865	ss-mRNA	E
O. sativa	RICCHIT	Endochitinase	Seedlings	1237	ds-DNA	E
O. sativa	RIC4CL	4-coumarate–CoA ligase (4-CL)	—	5225	ds-DNA	E
O. sativa	RICFBPA	Fructose-diphosphate aldolase (EC 4.1.2.13)	—	1305	ss-mRNA	E
O. sativa	RICOSAMYBG	α-amylase	Leaves	1558	ds-DNA	E
O. sativa	RICPAL	Phenylalanine ammonia lyase (EC 4.3.1.5)	—	4412	ds-DNA	E
O. sativa	RICRAMY3A	α-amylase	Etiolated leaf	2625	ds-DNA	E
O. sativa	RICRAMY3B	α-amylase	Etiolated leaf	1996	ds-DNA	E
O. sativa	RICRAMY3C	α-amylase	Etiolated leaf	2090	ds-DNA	E
O. sativa	RICRIC	Triosephosphate isomerase	—	1050	ss-mRNA	—
O. sativa	RICRIGS8	Cytosolic glutamine synthetase	Root	1169	ss-mRNA	E
O. sativa	RICRSS2	Sucrose–UDP glucosyltransferase (isozyme 2)	Leaves & root	9225	ds-DNA	E
O. sativa	RICSIGS28	Cytosolic glutamine synthetase (EC 6.3.1.2)	Shoot	1553	ss-mRNA	E

Table 8. Continued

Species	Identifier	Sequence specification	Tissue/site	Size (bp or b)	Type	S
O. sativa	RICSIGS31	Chloroplastic glutamine synthetase (EC 6.3.1.2)	Shoot	1649	ss-mRNA	E
O. sativa	RICWAXY	UDP–glucose starch glycosyl transferase	—	1074	ds-DNA	—
O. sativa	RICWAXYL	ADP (UDP)–glucose starch glycosyl transferase	Seedling	4100	ds-DNA	E
O. sativa	RICWAXZ	Glycogen (starch) synthetase (EC 2.4.1.11)	—	5499	ds-DNA	E
P. miliaceum	PNIAAT1	Aspartate aminotransferase	Seedling	1677	ss-mRNA	E
P. miliaceum	PNIAAT2	Aspartate aminotransferase	Seedling	1514	ss-mRNA	E
P. miliaceum	PNIAAT3	Aspartate aminotransferase	Seedling	1659	ss-mRNA	E
P. americanum	PENADH1S	Alcohol dehydrogenase	Roots	1469	ss-mRNA	E
P. glaucum	PENADH1	Alcohol dehydrogenase	—	3508	ds-DNA	—
P. crispum	PUMCHSA2	Chalcone synthase	—	732	ds-DNA	—
P. crispum	PUM4CL1A	4-coumarate–CoA ligase (*4CL-1*)	—	1190	ds-DNA	E
P. crispum	PUM4CL1AA	4-coumarate–CoA ligase (4CL; EC 6.2.1.12)	—	1914	ss-mRNA	E
P. crispum	PUM4CL1B	4-coumarate–CoA ligase (*4CL-1*)	—	295	ds-DNA	E
P. crispum	PUM4CL2A	4-coumarate–CoA ligase (*4CL-2*)	—	958	ds-DNA	E
P. crispum	PUM4CL2AA	4-coumarate–CoA ligase (4CL: EC 6.2.1.12)	—	1927	ss-mRNA	E
P. crispum	PUM4CL2B	4-coumarate–CoA ligase	—	300	ds-DNA	E
P. crispum	PUMCCOAMT	Caffeoyl-CoA 3-*O*-methyltransferase	—	1258	ss-mRNA	—
P. crispum	PUMCHSA1	Chalcone synthase	—	90	ds-DNA	G
P. crispum	PUMPAL4	Phenylalanine ammonia lyase	—	2375	ss-mRNA	E
P. crispum	PUMPAL1EX2	Phenylalanine ammonia lyase	—	1964	ds-DNA	E
P. crispum	PUMPAL1GN	Phenylalanine ammonia lyase	—	1036	ds-DNA	E
P. hortense	PUMCHL	Chalcone synthase	—	1431	ss-mRNA	—
P. hortense	PUMGADPH	Glycolytic glyceraldehyde-3-phosphate dehydrogenase	—	1266	ss-mRNA	E
P. hybrida	PETACHITIN	Acidic chitinase	Leaves	944	ss-mRNA	E
P. hybrida	PETADH1	Alcohol dehydrogenase-1	—	4672	ds-DNA	E
P. hybrida	PETAN3	Flavonone 3-β-hydroxylase	—	1227	ss-mRNA	E
P. hybrida	PETAROA	5-enolpyruvylshikimate 3-phosphate synthase	—	1944	ss-mRNA	—
P. hybrida	PETCFI	Chalcone flavanone isomerase	—	759	ss-mRNA	—
P. hybrida	PETCHIA	Chalcone flavanone isomerase	Leaves, flowers	2399	ds-DNA	E
P. hybrida	PETCHIB	Chalcone flavanone isomerase	Leaves, flowers	2170	ds-DNA	E
P. hybrida	PETCHSA	Chalcone synthase	Leaves, flowers	4966	ds-DNA	E
P. hybrida	PETCHSB	Chalcone synthase	Leaves	5973	ds-DNA	E
P. hybrida	PETCHSD	Chalcone synthase	Leaves	2948	ds-DNA	E
P. hybrida	PETCHSF	Chalcone synthase	Leaves	2542	ds-DNA	E
P. hybrida	PETCHSG	Chalcone synthase	Leaves	4680	ds-DNA	E
P. hybrida	PETCHSR	Chalcone synthase	Floral tissue	1391	ss-mRNA	E
P. hybrida	PETDFRA	Dihydroflavonol-4-reductase	—	1471	ss-mRNA	E
P. hybrida	PETEFE	Ethylene forming enzyme	Corolla	1250	ss-mRNA	—
P. hybrida	PETEPSP	5-enolpyruvylshikimate-3-phosphate synthase	—	1788	ds-DNA	—
P. hybrida	PETGADPHR	Glycolytic glyceraldehyde-3-phosphate dehydrogenase	—	1476	ss-mRNA	E
P. hybrida	PETSODI	Superoxide dismutase I	Mature leaf mesophyll	945	ss-mRNA	—
P. inflata	PETRNX2A	Ribonuclease	Leaves	1569	ds-DNA	—
P. vulgaris	PHV13BDGL	1,3-β-D-glucanase	Cell suspension	1158	ss-mRNA	E

Table 8. Continued

Species	Identifier	Sequence specification	Tissue/site	Size (bp or b)	Type	S
P. vulgaris	PHVCAD	Cinnamyl-alcohol dehydrogenase	—	1950	ss-mRNA	—
P. vulgaris	PHVCEL	Cellulase	Leaves	1687	ss-mRNA	—
P. vulgaris	PHVCHALCPO	Silencer region of chalcone synthase promoter	—	510	ds-DNA	E
P. vulgaris	PHVCHALCSN	Chalcone synthase	—	1342	ss-mRNA	E
P. vulgaris	PHVCHALIS	Chalcone isomerase	—	791	ss-mRNA	E
P. vulgaris	PHVCHITIN	Chitinase	—	976	ss-mRNA	E
P. vulgaris	PHVCHM	Chitinase	—	1132	ss-mRNA	—
P. vulgaris	PHVCHMA	Chitinase	—	72	ss-mRNA	—
P. vulgaris	PHVCHS	Chalcone synthase	—	226	ss-mRNA	—
P. vulgaris	PHVGLNB	Glutamine synthetase β subunit 5′-flank (EC 6.3.1.2)	—	1146	ds-DNA	E
P. vulgaris	PHVGLNGAM	Glutamine synthetase γ subunit	—	1310	ds-DNA	—
P. vulgaris	PHVGS3E	Glutamine synthetase	Root nodules	712	ss-mRNA	—
P. vulgaris	PHVGSCH	Glutamine synthetase (EC 6.3.1.2)	Leaves	1510	ss-mRNA	—
P. vulgaris	PHVGSG1	Glutamine synthetase γ-subunit (EC 6.3.1.2)	Nodule	1448	ss-mRNA	E
P. vulgaris	PHVGSR1	Glutamine synthetase (GS, EC 6.3.1.2)	Nodules, roots, leaves	1452	ss-mRNA	E
P. vulgaris	PHVGSR2	Glutamine synthetase (GS, EC 6.3.1.2)	Nodules, roots, leaves	1353	ss-mRNA	E
P. vulgaris	PHVLIPOXY	Lipoxygenase	Leaves	2226	ss-mRNA	E
P. vulgaris	PHVLOXA	Lipoxygenase	—	8117	ds-DNA	E
P. vulgaris	PHVPAL	Phenylalanine ammonia lyase	—	1707	ss-mRNA	—
P. sylvestris	PINBBS	Bibenzyl synthase	—	1997	ds-DNA	E
P. sylvestris	PINCHS	Chalcone synthase	—	1401	ds-DNA	E
P. sativum	PSSODR	Manganese superoxide dismutase	Leaves	984	RNA	E
P. sativum	PSCATAL	Catalase	Leaves	1738	RNA	E
P. sylvestris	PINCUZNP	Cu–Zn superoxide dismutase	—	676	ss-mRNA	E
P. sylvestris	PINCUZNPS3	Cu–Zn superoxide dismutase	—	821	ss-mRNA	E
P. sativum	PEAADH1	Alcohol dehydrogenase	—	4011	ds-DNA	E
P. sativum	PEAALAD	Aminolevulinic acid dehydratase	—	1458	ss-mRNA	—
P. sativum	PEAAS1	Asparagine synthase (EC 6.3.5.4)	Nodule	2200	ss-mRNA	E
P. sativum	PEAAS2	Asparagine synthase (EC 6.3.5.4)	Root	2002	ss-mRNA	E
P. sativum	PEACAMRA	Carbonic anhydrase	—	1239	ss-mRNA	—
P. sativum	PEACAP	Carbonic anhydrase (EC 4.2.1.1)	—	1300	ss-mRNA	E
P. sativum	PEACATAL	Catalase	—	1738	ss-mRNA	E
P. sativum	PEACHGS2A	Glutamine synthetase	Root	1540	ss-mRNA	—
P. sativum	PEACHITIN	Chitinase	Leaves	1143	ss-mRNA	E
P. sativum	PEACUZNSD	Cu–Zn superoxide dismutase	Leaves	738	ss-mRNA	—
P. sativum	PEAGDC	H-protein of glycine decarboxylase	Leaves	679	ss-mRNA	—
P. sativum	PEAGDCPG	P-protein of glycine decarboxylase enzyme complex	—	708	ss-mRNA	E
P. sativum	PEAGDPP	P protein, a part of glycine cleavage complex	Leaves	3404	ss-mRNA	E
P. sativum	PEAGPA1	Glyceraldehyde-3-phosphate dehydrogenase (GAPDH) subunit A	Green photosynthetic tissue	1480	ss-mRNA	E
P. sativum	PEAGPA2	Glyceraldehyde-3-phosphate dehydrogenase subunit A	Green photosynthetic tissue	3696	ds-DNA	E
P. sativum	PEAGPB1	Glyceraldehyde-3-phosphate dehydrogenase (GAPDH) subunit B (GapB) (EC 1.2.1.13)	—	1570	ss-mRNA	E
P. sativum	PEAGS132	Glutamine synthetase	—	52	ss-mRNA	—
P. sativum	PEAGS3AG	Glutamine synthase	—	1939	ds-DNA	E
P. sativum	PEAGSCY1A	Glutamine synthetase	Root	1434	ss-mRNA	—
P. sativum	PEAGSR1	Glutamine synthetase (EC 6.3.1.2)	Roots	1373	ss-mRNA	E

Gene Index

Table 8. Continued

Species	Identifier	Sequence specification	Tissue/site	Size (bp or b)	Type	S
P. sativum	PEAGSR2	Glutamine synthetase	Leaves	1304	ss-mRNA	E
P. sativum	PEAGSR3	Glutamine synthetase	Root nodules	1065	ss-mRNA	E
P. sativum	PEALIPDEHM	Lipoamide dehydrogenase	Leaves	1861	ss-mRNA	E
P. sativum	PEALIPOCY	Cytoplasmic lipoxygenase	Mature seed	2838	ss-mRNA	—
P. sativum	PEALIPOX	Lipoxygenase	Cotyledon	2805	ss-mRNA	E
P. sativum	PEAL2LLG	Lipoxygenase	Seed	958	ds-DNA	—
P. sativum	PEALPD	Dihydrolipoamide dehydrogenase	Leaves	1627	ss-mRNA	E
P. sativum	PEASHMTA	Serine hydroxymethyltransferase	—	1803	ss-mRNA	—
P. sativum	PEASOD	Chloroplastic Cu–Zn superoxide dismutase	Leaves	825	ss-mRNA	—
P. sativum	PEASODII	Cu–Zn superoxide dismutase II	—	901	ss-mRNA	E
P. sativum	PEASODR	Manganese superoxide dismutase	Leaves	984	ss-mRNA	E
P. deltoides	POPOME	*O*-methyltransferase	Leaves	1375	ss-mRNA	—
P. deltoides	POPSAMA	Deltoides S-adenosyl methionine synthetase	Leaves	1436	ss-mRNA	—
Populus sp.	POPCHIA	Chitinase	Leaves	726	ss-mRNA	—
Populus sp.	POPCHIB	Chitinase	Leaves	1073	ss-mRNA	—
Populus sp.	POPCHIC	Chitinase	Leaves	251	ss-mRNA	—
Populus sp.	POPCHIG	Chitinase	Leaves	193	ss-mRNA	—
P. tremuloides	POPLBCA	Lignin bispecific caffeic acid/5-hydroxyferulic acid *O*-methyl transferase	Xylem	1503	ss-mRNA	E
P. trichocarpa	POPGWIN62B	Chitinase gene *gwin6.2b*	—	7305	ds-DNA	E
P. trichocarpa	POPMAL	Malic enzyme	Leaves	2077	ss-mRNA	E
R. acris	RACGADPHR	Glycolytic glyceraldehyde-3-phosphate	—	1236	ss-mRNA	E
R. serpentina	RAUSTRICT	Strictosidine synthase	—	1215	ss-mRNA	E
R. communis	RCCATPPPA	ATP:pyruvate phosphotransferase	—	2115	ss-mRNA	—
R. communis	RCCATPPPB	ATP:pyruvate phosphotransferase (PK-p-β)	—	2022	ss-mRNA	—
R. communis	RCCCATAL	Catalase	—	456	ss-mRNA	E
R. communis	RCCICL4	Isocitrate lyase	—	2036	ss-mRNA	—
R. communis	RCCLTPA	Lipid transfer protein	Cotyledon	582	ss-mRNA	—
R. communis	RCCLTPB	Lipid transfer protein	Cotyledon	628	ss-mRNA	—
R. communis	RCCMSYNTH	Malate synthase (EC 4.1.3.2)	—	1826	ss-mRNA	E
R. communis	RCCSACPD	Stearoyl–ACP desaturase	Seed	1643	ss-mRNA	—
R. communis	RCCSTEA	Stearoyl–acyl carrier protein desaturase	Endosperm	1662	ss-mRNA	E
Saccharum hybrid	SCFSCPEPCD	Phosphoenolpyruvate carboxylase	—	8638	ds-DNA	—
S. officinarum	SCFSUS1MR	Sucrose synthase	—	654	ss-mRNA	E
S. guttatum	SURAOX1A	Alternative oxidase protein	—	1404	ss-mRNA	—
S. cereale	RYEBAMY1	β-amylase	Endosperm	762	ss-mRNA	E
S. alba	SALGAPDHA	Chloroplast glyceraldehyde-3-phosphate dehydrogenase (GAPDH, NADP-dependent)	—	829	ss-mRNA	E
S. alba	SALGAPDHR	Cytosolic glyceraldehyde-3-phosphate dehydrogenase (GAPDH, NAD-specific; EC 1.2.1.12)	—	1091	ss-mRNA	E
S. alba	SALMYRMA1	Thioglucoside glucohydrolase (myrosinase)	Seed	893	ss-mRNA	E
S. alba	SALMYRMB1	Thioglucoside glucohydrolase (myrosinase)	Seed	901	ss-mRNA	E
S. alba	SALMYRMB2	Thioglucoside glucohydrolase (myrosinase)	Seed	891	ss-mRNA	E

Table 8. Continued

Species	Identifier	Sequence specification	Tissue/site	Size (bp or b)	Type	S
S. alba	SALMYRMB3	Thioglucoside glucohydrolase (myrosinase)	Seed	1878	ss-mRNA	E
S. alba	SALSCHS1	Chalcone synthase	Cotyledon	986	ss-mRNA	E
S. alba	SALSCHS3	Chalcone synthase	Cotyledon	1331	ss-mRNA	E
S. alba	SALSCHSG	Chalcone synthase (EC 2.3.1.74)	Cotyledon	2873	ds-DNA	E
S. oleracea	SPIACPI	Acyl carrier protein I	Leaves	664	ss-mRNA	—
S. oleracea	SPIACPII	Acyl carrier protein II	Root, leaves, seeds	714	ss-mRNA	E
S. oleracea	SPIALA1	δ-aminolevulinic acid dehydratase	—	1727	ss-mRNA	E
S. oleracea	SPIBADH	Betaine-aldehyde dehydrogenase	Leaves	1797	ss-mRNA	—
S. oleracea	SPICA	Chloroplastic carbonic anhydrase	Leaves	965	ss-mRNA	—
S. oleracea	SPICPAHRI	Acetohydroxy acid reductoisomerase	Leaves	2050	ss-mRNA	E
S. oleracea	SPICPCCA	Chloroplast carbonic anhydrase	Leaves	1122	ss-mRNA	—
S. oleracea	SPIFBPY	Fructose-1,6-bisphosphatase	Leaves	1585	ss-mRNA	E
S. oleracea	SPIGLO	Glycolate oxidase	Leaves	1511	ss-mRNA	—
S. oleracea	SPIGPB1	Glyceraldehyde-3-phosphate dehydrogenase (GAPDH) subunit A (GapA) (EC 1.2.1.13)	Leaves	1549	ss-mRNA	E
S. oleracea	SPI37KDPM	Chloroplast 37 kDa inner envelope membrane protein	Leaves	1226	ss-mRNA	E
S. oleracea	SPIPHTRAN	Chloroplast phosphate translocator protein	Leaves	1432	ss-mRNA	E
S. oleracea	SPIPRK	Phosphoribulokinase (PRK)	Leaves	1411	ss-mRNA	—
S. oleracea	SPIPRKA	Phosphoribulokinase	—	1455	ss-mRNA	—
S. oleracea	SPISFS1A	10-formyltetrahydrofolate synthetase (sfs1)	Leaves	2221	ss-mRNA	—
S. oleracea	SPISODCY	Cytosolic copper/zinc-superoxide dismutase (EC 1.15.1.1)	Leaves	795	ss-mRNA	E
S. bicolor	SRGMDHCM3	Malate dehydrogenase (NADP)	Leaves	932	ss-mRNA	E
S. bicolor	SRGMDHCM7	Malate dehydrogenase (NADP)	Leaves	1441	ss-mRNA	E
S. vulgare	SRGMDH	NADP malate dehydrogenase	Leaf mesophyll	4638	ds-DNA	—
S. vulgare	SRGPEPC	Phosphoenolpyruvate involved in C4 photosynthesis (EC 4.1.1.31)	Leaf mesophyll	3147	ss-mRNA	E
S. vulgare	SRGPEPCG	Phosphoenolpyruvate carboxylase	—	6018	ds-DNA	E
S. vulgare	SRGPEPCGX	PEP carboxylase	—	7587	ds-DNA	E
S. vulgare	SVPEPCA	Phosphoenolpyruvate carboxylase (PEPC)	Root	3128	RNA	E
S. tuberosum	POTACC	Adenine nucleotide translocator	Tuber	1453	ss-mRNA	E
S. tuberosum	POTADH1	Alcohol dehydrogenase 1	Tuber	1448	ss-mRNA	—
S. tuberosum	POTADH2	Alcohol dehydrogenase 2	—	1351	ss-mRNA	—
S. tuberosum	POTADH3	Alcohol dehydrogenase 3	—	1431	ss-mRNA	—
S. tuberosum	POTADPGPP	ADP–glucose pyrophosphorylase subunit	Tuber	1627	ss-mRNA	E
S. tuberosum	POTAGPTHI	α-glucan phosphorylase type H isozyme	—	2655	ss-mRNA	—
S. tuberosum	POTARO1A	3-deoxy-D-arabino-heptulosonate-7-phosphate synthase (aro1)	—	1784	ss-mRNA	—
S. tuberosum	POTCHIT	Endochitinase	Leaves	1057	ss-mRNA	E
S. tuberosum	POTCHITIN	Endochitinase gene (EC 3.2.1.14)	—	3281	ds-DNA	E
S. tuberosum	POTCPKIN1	Pyruvate kinase	—	1386	ds-DNA	E
S. tuberosum	POTPAL1	Phenylalanine ammonia lyase	—	3888	ds-DNA	E
S. tuberosum	POTPAL2	Phenylalanine ammonia lyase	—	3106	ds-DNA	E
S. tuberosum	POTPFPA	Pyrophosphate-fructose 6-phosphate 1-phosphotransferase, (PFP) α-subunit	—	2041	ss-mRNA	—

Table 8. Continued

Species	Identifier	Sequence specification	Tissue/site	Size (bp or b)	Type	S
S. tuberosum	POTPFPB	Pyrophosphate-fructose 6-phosphate 1-phosphotransferase (PFP) β-subunit	–	1986	ss-mRNA	–
S. tuberosum	POTPXD	Anionic peroxidase	Periderm	1227	ss-mRNA	–
S. tuberosum	POTPYRK	Pyruvate kinase	–	2006	ss-mRNA	E
S. tuberosum	POTSACPD	Stearoyl-acyl carrier protein desaturase	–	1507	ss-mRNA	–
S. tuberosum	POTSSYN	Sucrose synthase	Tuber	2711	ss-mRNA	–
S. tuberosum	POTST4C11	4-coumarate-CoA ligase (*St4C1-1*)	–	4030	ds-DNA	–
S. tuberosum	POTUDPGP	UDP-glucose pyrophosphorylase	Tuber	1758	ss-mRNA	–
S. tuberosum	STSTARCH	Starch phosphorylase (EC 2.4.1.1)	–	4942	RNA	E
S. tuberosum	POTWAXYG1W	*Waxy* gene for starch synthase	–	1203	ds-DNA	E
S. tuberosum	POTWAXYG28	*Waxy* gene for starch synthase	–	1335	ds-DNA	E
T. repens	TFRBG104	Cyanogenic β-glucosidase	Leaves	1690	ss-mRNA	E
T. repens	TFRBG361	Non-cyanogenic β-glucosidase	Leaves	1859	ss-mRNA	E
T. repens	TFRADH1	Alcohol dehydrogenase 1	–	1268	ss-mRNA	–
T. aestivum	WHTAAM28	α-*Amy2/8*	–	940	ds-DNA	E
T. aestivum	WHTAAM234	α-*Amy2/34*	–	934	ds-DNA	E
T. aestivum	WHTAAM246	α-*Amy2/46*	–	787	ds-DNA	E
T. aestivum	WHTAAM253	α-*Amy2/53*	–	919	ds-DNA	E
T. aestivum	WHTAAM254	α-*Amy2/54*	–	847	ds-DNA	E
T. aestivum	WHTAGA1	ADP-glucose pyrophosphorylase (WL:AGA.1) (EC 2.7.7.27)	Leaves, endosperm	947	ss-mRNA	E
T. aestivum	WHTAGA3	ADP-glucose pyrophosphorylase (WE:AGA.3) (EC 2.7.7.27)	Endosperm	1207	ss-mRNA	E
T. aestivum	WHTAGA7	ADP-glucose pyrophosphorylase (WE:AGA.7) (EC 2.7.7.27)	Endosperm	1779	ss-mRNA	E
T. aestivum	WHTAMYA	α-amylase	–	2397	ds-DNA	–
T. aestivum	WHTDHDPD17	Dihydrodipicolinate synthase (DHDPS)	–	1458	ss-mRNA	–
T. aestivum	WHTDHDPD26	Dihydrodipicolinate synthase (DHDPS)	–	1361	ss-mRNA	–
T. aestivum	WHTFBP1	Fructose-1,6-bisphosphatase	Etiolated shoots, leaves	3600	ds-DNA	E
T. aestivum	WHTIVSS	Amylase gene	–	310	ds-DNA	–
T. aestivum	WHTNAD4G	NADH-ubiquinone oxidoreductase chain 4	Etiolated seedlings	7927	ds-DNA	E
T. aestivum	WHTPERO	Peroxidase	–	1247	ss-mRNA	E
T. aestivum	WHTPEROXIG	Peroxidase (EC 1.11.1.7)	–	3240	ss-mRNA	E
T. aestivum	WHTPGKCHL	Phosphoglycerate kinase (EC 2.7.2.3)	Leaves	1605	ss-mRNA	E
T. aestivum	WHTPGKCYT	Phosphoglycerate kinase (EC 2.7.2.3)	Leaves	1473	ss-mRNA	–
T. aestivum	WHTGSTA1	Glutathione-S-transferase	–	2178	ds-DNA	E
T. aestivum	WHTGSTAGST	Glutathione-S-transferase	–	3196	ds-DNA	E
T. aestivum	WHTPRK	Ribulose-5-phosphate kinase	Etiolated shoots	2501	ds-DNA	E
T. aestivum	WHTPRKGEN	Phosphoribulokinase	–	1533	ss-mRNA	–
T. aestivum	WHTTAFBP	Fructose 1,6-bisphosphatase (EC 3.1.3.11)	–	1399	ss-mRNA	E
T. aestivum	WHTWAXYSS	Granule-bound starch synthase	Endosperm	2186	ss-mRNA	E
V. faba	VFAPPORNA	Polyphenol oxidase	Leaves	2110	ss-mRNA	E
V. mungo	VIRALAMYL	α-amylase (EC 3.2.1.1)	Cotyledon	1525	ss-mRNA	E
V. radiata	VIRACCSY01	ACC synthase fragment gene	Hypocotyl	296	ds-DNA	–
V. radiata	VIRACCSY02	ACC synthase fragment gene	Hypocotyl	309	ds-DNA	–
V. radiata	VIRACCSY03	ACC synthase fragment gene	Hypocotyl	314	ds-DNA	–
Z. mays	MZEACC	Adenine nucleotide translocator	Coleoptile	1579	ss-mRNA	E

Table 8. Continued

Species	Identifier	Sequence specification	Tissue/site	Size (bp or b)	Type	S
Z. mays	MZEADH1	Alcohol dehydrogenase (adh1)	Root monosomes and polysomes	879	ss-mRNA	—
Z. mays	MZEADH1ALL	Adh1-Cm allele	—	6158	ds-DNA	E
Z. mays	MZEAHAS108	Acetohydroxyacid synthase	Seedling	2664	ds-DNA	E
Z. mays	MZEAHAS109	Acetohydroxyacid synthase	—	2544	ds-DNA	E
Z. mays	MZEADH1CM	Alcohol dehydrogenase	—	6167	ds-DNA	—
Z. mays	MZEADH1DS	Alcohol dehydrogenase	—	773	ds-DNA	—
Z. mays	MZEADH1F	Alcohol dehydrogenase	—	1590	ss-mRNA	—
Z. mays	MZEADH11F	Alcohol dehydrogenase 1	—	1611	ss-mRNA	E
Z. mays	MZEADH1FA	Alcohol dehydrogenase 1	—	5249	ds-DNA	—
Z. mays	MZEADH1FX	Alcohol dehydrogenase (*Adh1-1F* allele)	—	167	ds-DNA	—
Z. mays	MZEADH2N	Alcohol dehydrogenase 2	—	3535	ds-DNA	E
Z. mays	MZEADH2NR	Alcohol dehydrogenase 2	—	1600	ss-mRNA	E
Z. mays	MZEADH1P	Alcohol dehydrogenase (*Adh1-PrF* allele)	—	360	ds-DNA	—
Z. mays	MZEADH5P	Alcohol dehydrogenase (*ADH5*)	—	62	ds-DNA	—
Z. mays	MZEADH1RT	Alcohol dehydrogenase (*Adh1*)	—	368	ds-DNA	—
Z. mays	MZEADH1RV	Alcohol dehydrogenase (*Adh1*)	—	368	ds-DNA	—
Z. mays	MZEADH1S	Alcohol dehydrogenase 1 (*Adh1-1S*)	—	5904	ds-DNA	—
Z. mays	MZEA1G	40.1 kDa A1 protein	—	3001	ds-DNA	E
Z. mays	MZEALD	Aldolase	—	1388	ss-mRNA	—
Z. mays	MZEALDOAR	Anaerobically regulated gene for fructose bisphosphate aldolase (EC 4.1.2.13)	—	3324	ds-DNA	E
Z. mays	MZEANT	ATP/ADP translocator	—	1136	ss-mRNA	E
Z. mays	MZEANTG1	Adenine nucleotide translocator	—	2058	ds-DNA	E
Z. mays	MZEANTG2	Adenine nucleotide translocator	Coleoptile	2099	ds-DNA	E
Z. mays	MZEANT2MU	Adenine nucleotide translocator (ADP/ATP translocase)	Coleoptile	1511	ss-mRNA	E
Z. mays	MZEBRNZDS	UDPG-flavonal glucosyltransferase (mutant *bronze*) gene, (EC 2.4.1.91)	—	575	ds-DNA	—
Z. mays	MZEBRNZDSA	UDPG-flavonal glucosyltransferase (mutant *bronze*)	—	607	ds-DNA	—
Z. mays	MZEBRNZDSC	UDPG-flavonal glucosyltransferase (mutant *bronze*)	—	275	ds-DNA	—
Z. mays	MZEBRNZW	UDP-glucose flavonoid glucosyl transferase (*bronze* gene Bz-W22)	—	2908	ds-DNA	—
Z. mays	MZEBRNZY	*Bronze bz-R* pseudogene with a 340 bp deletion	—	3181	ds-DNA	—
Z. mays	MZEBRNZZA	UDP-glucose flavonoid glycosyl transferase (Bz-McC)	—	3485	ds-DNA	—
Z. mays	MZECAT1	Catalase isozyme 1 (CAT-1)	Scutella	2065	ss-mRNA	—
Z. mays	MZECAT2	Catalase (CAT-2)	Scutella	1798	ss-mRNA	—
Z. mays	MZECAT3	Catalase isozyme 3 (CAT-3)	Scutella	1790	ss-mRNA	—
Z. mays	MZECAT1I	Catalase-1 isoenzyme (CAT-1)	Endosperm	2065	ss-mRNA	—
Z. mays	MZECAT3I	Catalase-3 isoenzyme (CAT-3)	Epicotyl	1790	ss-mRNA	—
Z. mays	MZECAT2R	Catalase 2	Scutellum	1778	ss-mRNA	E
Z. mays	MZEC2CS	Chalcone synthase	Leaves	3829	ds-DNA	E
Z. mays	MZECHITA	Chitinase A	Seed	843	ds-DNA	—
Z. mays	MZECHITB	Chitinase B	Seed	810	ds-DNA	—
Z. mays	MZECIN4E	cin4 element	—	6822	ds-DNA	E
Z. mays	MZECYP	Cyclophilin	Mature embryo/ scutellum	792	ss-mRNA	—

Table 8. Continued

Species	Identifier	Sequence specification	Tissue/site	Size (bp or b)	Type	S
Z. mays	MZEDHPS	Dihydrodipicolinate synthase (EC 4.2.1.52)	—	1399	ss-mRNA	E
Z. mays	MZEENOLA	Enolase (2-phospho-D-glycerate hydrolase)	Root	1699	ss-mRNA	E
Z. mays	MZEFEGLU	Ferredoxin-dependent glutamate synthase	Leaves	5617	ss-mRNA	—
Z. mays	MZEGAPDH	Glyceraldehyde-3-phosphate dehydrogenase (GAPDH)	Leaves	1372	ss-mRNA	—
Z. mays	MZEGAPDHA	Glyceraldehyde-3-phosphate dehydrogenase (GAPDH)	Leaves	1382	ss-mRNA	—
Z. mays	MZEGGST3B	Glutathione-S-transferase GSTIII	—	859	ss-mRNA	E
Z. mays	MZEGPA1	Glyceraldehyde-3-phosphate dehydrogenase subunit A	—	6414	ds-DNA	E
Z. mays	MZEGPC1	Glyceraldehyde-3-phosphate dehydrogenase	—	6015	ds-DNA	E
Z. mays	MZEG3PD	Glyceraldehyde-3-phosphate dehydrogenase	Leaves	6414	ds-DNA	—
Z. mays	MZEGST3A	Glutathione-S-transferase GSTIII	—	840	ss-mRNA	E
Z. mays	MZEGST3AA	Glutathione-S-transferase III	—	913	ss-mRNA	—
Z. mays	MZEGSTI	Glutathione-S-transferase I	—	929	ss-mRNA	—
Z. mays	MZEGSTIA1	Glutathione-S-transferase (GST-I)	Coleoptile	770	ds-DNA	—
Z. mays	MZEGSTIA2	Glutathione-S-transferase (GST-I)	Coleoptile	1671	ds-DNA	—
Z. mays	MZEGSTIB	Glutathione-S-transferase (GST-I)	—	727	ss-mRNA	—
Z. mays	MZELACDEHG	Lactate dehydrogenase	Leaves	2999	ds-DNA	E
Z. mays	MZEMC2BZ1	UFGT (UDP–glucose:flavonol 3-*O*-glucosyltransferase)	—	2531	ds-DNA	E
Z. mays	MZEMCCBZ1	UFGT (UDP–glucose:flavonol 3-*O*-glucosyltransferase)	—	2919	ds-DNA	E
Z. mays	MZEMDH	Chloroplast NADP–malate dehydrogenase (NADP–MDH)	Leaves	1537	ss-mRNA	E
Z. mays	MZENAR1	Glyceraldehyde-3-phosphate dehydrogenase	Leaves	386	ss-mRNA	—
Z. mays	MZENAR2	Glyceraldehyde-3-phosphate dehydrogenase	Leaves	456	ss-mRNA	—
Z. mays	MZENDMEX	NADP-dependent malic enzyme	Leaves	2184	ss-mRNA	—
Z. mays	MZEOMTH	*O*-methyltransferase	Endosperm	2512	ss-mRNA	—
Z. mays	MZEORP1A	Tryptophan synthase β-subunit	Endosperm	1382	ss-mRNA	—
Z. mays	MZEORP2A	Tryptophan synthase β-subunit	Pericarp	1540	ss-mRNA	—
Z. mays	MZEPC21	Phosphoenolpyruvate carboxylase gene 5′ flanking region (EC 4.1.1.31)	Leaves	1182	ds-DNA	E
Z. mays	MZEPC91	Phosphoenolpyruvate carboxylase gene 5′ flanking region (EC 4.1.1.31)	Leaves	1185	ds-DNA	E
Z. mays	MZEPC324	Phosphoenolpyruvate carboxylase gene 5′ flanking region (EC 4.1.1.31)	Leaves	1108	ds-DNA	E
Z. mays	MZEPDCGENE	Pyruvate decarboxylase	—	2123	ss-mRNA	E
Z. mays	MZEPEP	Phosphoenolpyruvate carboxylase (EC 4.1.1.31)	—	6781	ds-DNA	E
Z. mays	MZEPEPC1	Phosphoenolpyruvate carboxylase	Leaves	168	ss-mRNA	—
Z. mays	MZEPEPC2	Phosphoenolpyruvate carboxylase 3′ flank	Leaves	80	ss-mRNA	E
Z. mays	MZEPEPC5A	Phosphoenolpyruvate carboxylase	Leaves	330	ss-mRNA	E

Table 8. Continued

Species	Identifier	Sequence specification	Tissue/site	Size (bp or b)	Type	S
Z. mays	MZEPEPCAA	Phosphoenolpyruvate carboxylase (EC 4.1.1.31)	Mesophyll	3265	ss-mRNA	—
Z. mays	MZEPEPCAB	Phosphoenolpyruvate carboxylase (EC 4.1.1.31)	Leaves	6151	ds-DNA	E
Z. mays	MZEPEPCR	Phosphoenolpyruvate carboxylase (PEPCase EC 4.1.1.31)	—	3027	ss-mRNA	E
Z. mays	MZEPGPA1	Glyceraldehyde-3-phosphate dehydrogenase subunit A	Leaves	5922	ds-DNA	E
Z. mays	MZEPGPA2	Glyceraldehyde-3-phosphate dehydrogenase subunit A	Leaves	8839	ds-DNA	E
Z. mays	MZEPK1	Putative protein kinase	—	2693	ss-mRNA	E
Z. mays	MZEPLTP	Phospholipid transfer protein	—	813	ss-mRNA	—
Z. mays	MZEPLTPA	Phospholipid transfer protein	—	677	ss-mRNA	—
Z. mays	MZEPOD	Pyruvate, orthophosphate dikinase	Leaves	3171	ss-mRNA	—
Z. mays	MZEPPDKA1	Pyruvate, orthophosphate dikinase	—	1987	ds-DNA	—
Z. mays	MZEPPDKA2	Pyruvate, orthophosphate dikinase	—	5519	ds-DNA	—
Z. mays	MZEPPDK2F	Pyruvate, orthophosphate dikinase (PPDK2)	—	322	ds-DNA	—
Z. mays	MZEPPDK1F1	Pyruvate, orthophosphate dikinase (PPDK1)	—	1272	ds-DNA	—
Z. mays	MZEPPDK1F2	Pyruvate, orthophosphatae dikinase (PPDK1)	—	265	ds-DNA	—
Z. mays	MZEPPDKLR	Pyruvate, orthophosphate dikinase (PPDK1)	—	246	ss-mRNA	—
Z. mays	MZESOD2	Superoxide dismutase-2 (SOD-2)	—	694	ss-mRNA	—
Z. mays	MZESOD2A	Superoxide dismutase 2 (SOD2)	—	694	ss-mRNA	—
Z. mays	MZESOD4	Superoxide dismutase-4 (SOD-4)	—	722	ss-mRNA	E
Z. mays	MZESOD3A	Manganese superoxide dismutase-3 (SOD-3)	—	1037	ss-mRNA	—
Z. mays	MZESOD4A	Superoxide dismutase-4A (SOD4A)	—	691	ss-mRNA	E
Z. mays	MZESOD3I	Superoxide dismutase-3 isoenzyme	Scutellum	1038	ss-mRNA	E
Z. mays	MZESUCS1	Sucrose synthase (EC 2.4.1.13)	—	2746	ss-mRNA	E
Z. mays	MZESUSYCD	Sucrose synthetase	Endosperm	459	ss-mRNA	—
Z. mays	MZESUSYSG	Sucrose synthase gene (shrunken)	—	6922	ds-DNA	E
Z. mays	MZESUSYSH	Sucrose synthetase gene (shrunken)	—	818	ds-DNA	—
Z. mays	MZETPI	Triosephosphate isomerase gene	—	3770	ds-DNA	—
Z. mays	MZETPI1	Triosephosphate isomerase 1	—	707	ds-DNA	—
Z. mays	MZETPI2	Triosephosphate isomerase 1	—	2870	ds-DNA	—
Z. mays	MZEWAXZ	*Waxy* (*wx*$^+$) locus for UDP-glucose starch glycosyl transferase	—	4800	ds-DNA	E
Z. mays	MZEW22BZ1	UDP-glucose:flavonol 3-*O*-glucosyltransferase	—	1594	ds-DNA	E
Z. mays	MZEWHPCS	White pollen gene for chalcone synthase	Leaves	4069	ds-DNA	E

Table 9. Lectins (phytohemagglutinins, agglutinins, ribosome-inactivating proteins (RIPs), anti-viral proteins)

Species	Identifier	Sequence specification	Tissue/site	Size (bp or b)	Type	S
A. precatorius	APAB	Preproabrin	—	2743	DNA	G
A. precatorius	ABRAB	Preproabrin	Leaves	2743	ds-DNA	E
A. precatorius	ABRRICA72	Ricin A-chain	Leaves	756	ds-DNA	E
A. precatorius	ABRRICA73	Ricin A-chain	Leaves	756	ds-DNA	E
A. sativum	ALCLECIA	I lectin	Bulb	1066	ss-mRNA	—
A. sativum	ALCLECIB	I lectin	Bulb	1067	ss-mRNA	—
A. sativum	ALCLECIC	I lectin	Bulb	1029	ss-mRNA	—
A. sativum	ALCLECID	I lectin	Bulb	1033	ss-mRNA	—
A. sativum	ALCIILECA	II lectin	Bulb	599	ss-mRNA	—
A. sativum	ALCIILECB	II lectin	Bulb	594	ss-mRNA	—
A. sativum	ALCIILECC	II lectin	Bulb	599	ss-mRNA	
C. gladiata	CAGCONA	Concanavalin A	Seedling	2314	ds-DNA	E
C. gladiata	CAGCONA5UP	*conA*	Seedling	414	ds-DNA	E
C. ensiformis	CENCONA	Concanavalin A	Immature seeds	1027	ss-mRNA	—
D. biflorus	DBILECS	Lectin subunits I and II	Cotyledon	1005	ss-mRNA	—
D. biflorus	DBIDB58	Lectin DB58 gene	Leaves, stem	2149	ds-DNA	—
D. biflorus	DBISLEC	Seed lectin (*SL*) gene	Seed	2230	ds-DNA	—
G. nivalis	GAAL1A	Lectin (LECGNA1)	Ovary	610	ss-mRNA	—
G. nivalis	GAAL2A	Lectin (LECGNA 2)	Ovary	570	ss-mRNA	—
G. nivalis	GAAL3A	Lectin (LECGNA3)	Ovary	667	ss-mRNA	—
G. nivalis	GAAL5A	Lectin (LECGNA 5)	Ovary	650	ss-mRNA	—
G. nivalis	GAAL8A	Lectin (LECGNA 8)	Ovary	597	ss-mRNA	—
G. max	SOYLEA	Lectin (*Le1*)	Seed	2152	ds-DNA	—
G. max	SOYTGM1	Transposable element Tgm1	Seed	3556	ds-DNA	—
Hippeastrum sp.	HIELECHHAA	Tetrameric mannose specific lectin protein (LECHHA 1)	Ovary	696	ss-mRNA	—
Hippeastrum sp.	HIELECHHAB	Tetrameric mannose specific lectin protein (LECHHA 2)	Ovary	455	ss-mRNA	—
Hippeastrum sp.	HIELECHHAC	Tetrameric mannose specific lectin protein (LECHHA 3)	Ovary	645	ss-mRNA	—
Hippeastrum sp.	HIELECHHAD	Tetrameric mannose specific lectin protein (LECHHA 4)	Ovary	665	ss-mRNA	—
Hippeastrum sp.	HIELECHHAE	Tetrameric mannose specific lectin protein (LECHHA 5)	Ovary	570	ss-mRNA	—
Hippeastrum sp.	HIELECHHAG	Tetrameric mannose specific lectin protein (LECHHA 6)	Ovary	516	ss-mRNA	—
Hippeastrum sp.	HIELECHHAH	Tetrameric mannose specific lectin protein (LECHHA 7)	Ovary	541	ss-mRNA	—
Hippeastrum sp.	HIELECHHAI	Tetrameric mannose specific lectin protein (LECHHA 8)	Ovary	586	ss-mRNA	—
H. vulgare	BLYCRIP30	Ribosome-inactivating protein mRNA	Seed	1031	ss-mRNA	—
H. vulgare	BLYCRIP30A	Ribosome-inactivating protein (isoform II)	Seed	479	ss-mRNA	—
H. vulgare	BLYLEC	Root-specific lectin	Embryo, root	972	ss-mRNA	—
M. jalapa	MIJMAP	Antiviral protein	Root	1066	ss-mRNA	—
M. charantia	MOMAMC	α-momorcharin	Seed	1044	ss-mRNA	E
Narcissus sp.	NAILECNPAA	Dimeric mannose specific lectin protein	—	635	ss-mRNA	—
Narcissus sp.	NAILECNPAB	Dimeric mannose specific lectin protein	Ovary	676	ss-mRNA	—
Narcissus sp.	NAILECNPAC	Dimeric mannose specific lectin protein	Ovary	642	ss-mRNA	—

Table 9. Continued

Species	Identifier	Sequence specification	Tissue/site	Size (bp or b)	Type	S
Narcissus sp.	NAILECNPAD	Dimeric mannose specific lectin protein	Ovary	635	ss-mRNA	—
Narcissus sp.	NAILECNPAE	Dimeric mannose specific lectin protein	Ovary	651	ss-mRNA	—
Narcissus sp.	NAILECNPAF	Dimeric mannose specific lectin protein	Ovary	605	ss-mRNA	—
Narcissus sp.	NAILECNPAG	Dimeric mannose specific lectin protein	Ovary	612	ss-mRNA	—
P. vulgaris	PHVDLEC1	Phytohemagglutinin (PHA-E)	—	1395	ds-DNA	E
P. vulgaris	PHVDLEC2	Phytohemagglutinin (PHA-L)	—	1661	ds-DNA	E
P. vulgaris	PHVPDLEC1	Phytohemagglutinin	Young leaves	1768	ds-DNA	E
P. lunatus	PHVLECA	Lectin	Cotyledon	926	ss-mRNA	—
P. vulgaris	PHVLECT	Lectin gene	—	1689	ds-DNA	—
P. vulgaris	PHVPDLEC2	Phytohemagglutinin	Young leaves	1992	ds-DNA	E
P. vulgaris	PHVDLEC2A	Phytohemagglutinin-L	Seed	1188	ds-DNA	E
P. vulgaris	PHVDLECA	Erythroagglutinating phytohemagglutinin (PHA-E)	—	1441	ds-DNA	—
P. vulgaris	PHVDLECB	Leucoagglutinating phytohemagglutinin	—	1784	ds-DNA	—
P. americana	PTCPAP	PAPA gene for anti-viral protein	Leaves	1164	ss-mRNA	E
P. sativum	PEALEC	Lectin	Cotyledon	967	ss-mRNA	—
P. sativum	PEALECA	Lectin	Seed	1701	ds-DNA	E
P. sativum	PEALECB	Lectin	Seed	1659	ss-mRNA	—
P. sativum	PEALECPGA	Lectin pseudogene	Seed	355	ss-mRNA	—
P. sativum	PEALECPGB	Lectin pseudogene	Seed	739	ss-mRNA	—
P. sativum	PEALECPGC	Lectin pseudogene	Seed	751	ss-mRNA	—
P. sativum	PEAPELEC	Lectin	Bud	216	ss-mRNA	—
P. sativum	PEAVLP	Vegetative lectin precursor	Bud	939	ss-mRNA	—
R. communis	RCCAGG	Agglutinin	Seed	1791	ss-mRNA	—
R. communis	RCCPPRICR	Ricin precursor	—	1884	ss-mRNA	E
R. communis	RCCRICA	Ricin E beta chain	—	942	ss-mRNA	—
R. communis	RCCRICIN	Gene for ricin precursor	—	2344	ds-DNA	E
R. communis	RCCRICIN2	Ricin	—	2362	ds-DNA	E
S. officinalis	SAOSAP2	Saporin	—	1300	ds-DNA	E
S. officinalis	SAOSAP3	Saporin	—	1115	ds-DNA	E
S. officinalis	SAOSAP4	Saporin	—	631	ds-DNA	E
T. kirilowii	TRHTCSA	Trichosanthin	—	1010	ss-mRNA	—
T. kirilowii	TRHTCS	α-trichosanthin	Leaves	1227	ds-DNA	—
T. aestivum	WHTAGGTA	Germ agglutinin isolectin A	Developing grain	998	ss-mRNA	—
T. durum	WHTAGGTD	Germ agglutinin	Developing grain	898	ss-mRNA	—
T. aestivum	WHTAGGTDD	Germ agglutinin isolectin D	Developing grain	998	ss-mRNA	—
Z. mays	MZERIP	Ribosome-inactivating protein	Immature kernel	1076	ss-mRNA	—
Z. mays	MZECRIP3A	Ribosome-inactivating protein	Endosperm	1039	ss-mRNA	—
Z. mays	MZECRIP9A	Ribosome-inactivating protein	Endosperm	976	ss-mRNA	—

Table 10. Membrane-associated proteins (membrane structural proteins, oleosins, envelope proteins, tonoplast proteins, transit proteins)

Species	Identifier	Sequence specification	Tissue/site	Size (bp or b)	Type	S
A. majus	AMTAP2	TAP2	–	2148	DNA	–
B. napus	BNABNO	Oleosin	–	2628	ds-DNA	–
B. napus	BNAOLEO	Oleosin BN-III	–	1403	ds-DNA	E
B. napus	BNAOLEOSIN	Major oleosin, nap-II	Embryo	810	ss-mRNA	E
D. carota	DAR215G	14 kDa protein	Embryo	581	ss-mRNA	E
D. carota	DAR12RNA	1.2 mRNA membrane protein	–	846	ss-mRNA	E
G. max	GMOLEOB	24 kDa oleosin isoform	–	864	RNA	E
G. max	SOYOLEOA	24 kDa oleosin	–	732	ss-mRNA	E
G. max	SOYOLEOB	24 kDa oleosin isoform	Seed oil body	864	ss-mRNA	E
G. max	SOYP34A	Oil bodies-associated protein P34	Seed oil body	1287	ss-mRNA	–
M. sativa	ALFGRERP	Glucose-regulated endoplasmic reticular protein	Suspension culture callus	1478	ss-mRNA	–
M. sativa	ALFPUTEND	Endomembrane protein	Suspension culture callus	1770	ss-mRNA	–
O. sativa	RICGOS5G	GOS5 putative chloroplast transit peptide	Shoot	2600	ds-DNA	E
P. vulgaris	PHVTIP	Tonoplast intrinsic protein	Seed	1208	ss-mRNA	E
P. sativum	PEAIM30	IM30 protein	Leaves	1449	ss-mRNA	–
P. sativum	PEAOM14A	Outer membrane protein	Leaves	456	ss-mRNA	–
S. oleracea	SPIOMCA	6.7 kDa chloroplast outer envelope membrane protein	Mesophyll	463	ss-mRNA	–
S. oleracea	SPIOMCB	6.7 kDa chloroplast outer envelope membrane protein	Mesophyll	548	ss-mRNA	–
Z. mays	MZEKD18	KD18 (KD18; L2)	Seed oil body	3101	ds-DNA	–
Z. mays	MZEMPL3	Major protein (L3) from the surface of lipid bodies	Embryo	757	ss-mRNA	–
Z. mays	MZEWAXY	Amyloplast-specific transit protein (*waxy*; *wx*+locus)	Amyloplast	4800	ds-DNA	–

Table 11. Metal-associated proteins (metallothioneins, metalloproteins, metal-binding, cysteine/sulfur-rich proteins)

Species	Identifier	Sequence specification	Tissue/site	Size (bp or b)	Type	S
D. carota	DARCAM1G	Calmodulin	Suspension culture	704	ss-mRNA	E
G. max	SOYFERRIT	Ferritin	—	568	ss-mRNA	—
G. max	SOYFERRITB	Ferritin	—	1050	ss-mRNA	—
G. max	SOYSOFH2	Ferritin	—	965	ss-mRNA	—
H. vulgare	BLYCAMA	Calmodulin	Leaves	1086	ss-mRNA	—
H. vulgare	BLYTH1AR	Toxin α-hordothionin	—	569	ss-mRNA	E
H. vulgare	BLYTHIOR1	Leaf-specific thionin	Leaves	637	ss-mRNA	E
H. vulgare	BLYTHIOR2	Leaf-specific thionin	Leaves	658	ss-mRNA	E
H. vulgare	BLYTHIOR3	Leaf-specific thionin	Leaves	453	ss-mRNA	E
H. vulgare	BLYTHIOR4	Leaf-specific thionin	Leaves	544	ss-mRNA	E
H. vulgare	BLYTHIOR5	Leaf-specific thionin	Leaves	491	ss-mRNA	E
H. vulgare	BLYTHNA	Leaf-specific thionin	Leaves	540	ss-mRNA	—
H. vulgare	BLYTHNB	Leaf-specific thionin	Leaves	608	ss-mRNA	—
H. vulgare	BLYTHNC	Leaf-specific thionin	Leaves	663	ss-mRNA	—
H. vulgare	BLYIDS1	ids-1	Root	503	ss-mRNA	E
H. vulgare	HVIDS1	ids-1 gene	Root	503	RNA	—
L. esculentum	TOMCALM1LE	Calmodulin mRNA	—	704	ss-mRNA	—
O. sativa	RICCALM	Calmodulin	Leaves	453	ss-mRNA	E
P. hybrida	PETCALPRO	Calmodulin-related protein	Petal	767	ss-mRNA	—
P. hybrida	PETCAM81	Calmodulin	Petal	815	ss-mRNA	—
Petunia sp.	PETRCAM72A	Calmodulin (CAM72)	Petal	719	ss-mRNA	—
P. vulgaris	PHVPFE	Ferritin	Seed	1246	ss-mRNA	E
S. tuberosum	POTCAM	Calmodulin	Tuber	906	ss-mRNA	—
T. aestivum	WHTTTH20MR	Type V thionin	Endosperm	688	ss-mRNA	E
T. aestivum	WHTTTHVG	Type V thionin	Endosperm	1106	ds-DNA	E

Table 12. Miscellaneous (unknown sequences, unidentified proteins, unassigned proteins)

Species	Identifier	Sequence specification	Tissue/site	Size (bp or b)	Type	S
G. max	SOY28GLA	28 kDa protein	Stem	1074	ss-mRNA	—
G. max	SOY31GLA	31 kDa protein 3′	Stem	1034	ss-mRNA	—
G. max	SOYMCS137	Mitotic cyclin	Root nodules	954	ss-mRNA	E
G. max	SOYMCYC	Mitotic cyclin	Root nodules	1596	ss-mRNA	E
H. brasiliensis	HEVHEV1	Hevein	Laticifer	1008	ss-mRNA	—
H. brasiliensis	HEVREF	Rubber elongation factor	Laticifer	681	ss-mRNA	E
H. vulgare	HVPAZPSE	Pseudogene PAZ	—	1554	DNA	E
L. polyphyllus	LUPPLB02	pPLB02	Leaves	459	ss-mRNA	E
L. polyphyllus	LUPPLB07M	pPLB07	Leaves	377	ss-mRNA	E
L. polyphyllus	LUPPLB08M	pPLB08	Leaves	640	ss-mRNA	E
L. polyphyllus	LUPPLZ12	Lupin-specific pPLZ12	Suspension culture	790	ss-mRNA	E
L. polyphyllus	LUPPLZ20	pPLZ20	Undifferentiated	424	ss-mRNA	E
L. polyphyllus	LUPPLZ01M	pPLZ01 mRNA	Undifferentiated	475	ss-mRNA	E
L. polyphyllus	LUPPLZ16M	pPLZ16	Undifferentiated	744	ss-mRNA	—
L. polyphyllus	LUPPLZ25M	pPLZ25 mRNA	Undifferentiated	611	ss-mRNA	E
L. polyphyllus	LUPPLZ26M	pPLZ26 mRNA	Undifferentiated	380	ss-mRNA	E
L. polyphyllus	LUPPLZ02	Lupin-specific pPLZ02 mRNA	Undifferentiated suspension culture	875	ss-mRNA	E
L. polyphyllus	LUPPLZ03M	pPLZ03 mRNA	Undifferentiated	438	ss-mRNA	—
L. esculentum	TOMANNP34G	Annexin P34	—	92	ds-DNA	E
L. esculentum	TOMGPRO	G protein α-subunit (TGA1)	Flower pistil	1634	ss-mRNA	—
L. esculentum	TOMPRO	Prosystemin	—	4526	ds-DNA	—
L. esculentum	TOMPRORNA	Prosystemin	—	951	ss-mRNA	—
N. tabacum	TOBI700	cDNA I-700, poly(A), binding to human β-interferon 3′UTR	—	148	ss-mRNA	E
N. tabacum	TOBE2TLP	Thaumatin-like protein	—	2297	ds-DNA	E
N. tabacum	TOBE22TLP	Thaumatin-like protein	—	2356	ds-DNA	E
O. sativa	RICGOS2G	GOS2 gene	—	3192	ds-DNA	E
O. sativa	RICRAB25	rab25	—	920	ss-mRNA	E
P. inflata	PETS1ALLE	S1 protein	—	4256	ds-DNA	—
P. inflata	PETS3ALLE	S3 protein	—	3409	ds-DNA	—
Populus sp.	POPPOP3A	pop3 mRNA	Leaves	586	ss-mRNA	—
Populus sp.	POPUCIB	Fragment of unknown coding identity	Leaves	174	ss-mRNA	—
Populus sp.	PSCHIC	mRNA fragment of unknown coding identity	—	142	RNA	—
S. officinarum hybrid	SCFBCS1A	*S. officinarum* protein gene	—	2328	ds-DNA	—
T. daniellii	TDATHAU2	Preprothaumatin-2	—	931	ss-mRNA	—
T. aestivum	WHTEMG	Em gene	Embryo	2465	ds-DNA	E
T. aestivum	WHTEMR	Em protein	Embryo	590	ss-mRNA	E
T. aestivum	WHTHMGW	High mobility group protein (HMGW)	Endosperm	923	ss-mRNA	E
V. faba	VFACREBL	CREB-like protein mRNA	—	1735	ss-mRNA	—
V. radiata	VIRSTORED	Stored cotyledon mRNA	Cotyledon	459	ss-mRNA	E
Z. mays	MZEBT1A	Brittle-1 protein	Endosperm	1733	ss-mRNA	—
Z. mays	MZERAB17G	RAB-17 gene	—	1986	ds-DNA	E

Table 13. Nodulins (proteins associated with nitrogen fixation, nodulins, leghemoglobins, nitrogenases)

Species	Identifier	Sequence specification	Tissue/site	Size (bp or b)	Type	S
C. glauca	CAAHBIIG	Hemoglobin	—	2001	ds-DNA	E
G. max	SOYDRP90	drp90 soybean leghemoglobin	Nodule	288	ss-mRNA	E
G. max	SOYENOD2A	ENOD2A gene for Ngm-75	—	3072	ds-DNA	E
G. max	SOYENOD2B	ENOD2B gene for Ngm-75	—	2992	ds-DNA	E
G. max	SOYGL02	Leghemoglobin gene	—	1523	ds-DNA	E
G. max	SOYGL03	Leghemoglobin	—	1886	ds-DNA	E
G. max	SOYGL04	Leghemoglobin	—	1204	ds-DNA	E
G. max	SOYGL05	Leghemoglobin	—	1464	ds-DNA	E
G. max	SOYGL06	Leghemoglobin	—	294	ds-DNA	E
G. max	SOYLBA	Leghemoglobin lba	—	1886	ds-DNA	E
G. max	SOYLBA5	Leghemoglobin (lba)	—	604	ds-DNA	E
G. max	SOYLBC	Leghemoglobin lbc	—	1523	ds-DNA	E
G. max	SOYLBC2	Leghemoglobin c-2	—	1292	ds-DNA	—
G. max	SOYLBC3	Leghemoglobin c-3	—	1118	ds-DNA	—
G. max	SOYLBC15	Leghemoglobin (lbc1)	—	663	ds-DNA	E
G. max	SOYLBC35E	Leghemoglobin C-3	—	772	ds-DNA	E
G. max	SOYLBC3N	Leghemoglobin C3	—	2013	ds-DNA	E
G. max	SOYLBCS25	Leghemoglobin	—	631	ds-DNA	E
G. max	SOYLBCS35	Leghemoglobin	—	750	ds-DNA	E
G. max	SOYLBGI	Leghemoglobin gene I	Root nodule	1204	ds-DNA	—
G. max	SOYLBGII	Leghemoglobin gene II	Root nodule	1463	ds-DNA	—
G. max	SOYLBPS1	Leghemoglobin pseudogene	—	852	ds-DNA	E
G. max	SOYLBPS2	Leghemoglobin pseudogene	—	1698	ds-DNA	—
G. max	SOYLBPS25	Leghemoglobin pseudogene	—	277	ds-DNA	—
G. max	SOYLBPSA15	Leghemoglobin pseudogene	—	607	ds-DNA	—
G. max	SOYLBPSI2	Leghemoglobin pseudogene	—	1046	ds-DNA	—
G. max	SOYPLB1	Leghemoglobin	—	145	ss-mRNA	E
G. max	SOYLBTRUN	Leghemoglobin	—	294	ds-DNA	E
G. max	SOYLGLO	Leghemoglobin	—	145	ds-DNA	E
G. max	SOYNGM16	Nodulin 16	Nodule	1548	ds-DNA	E
G. max	SOYNODAA	Nodulin-35	Nodule	980	ss-mRNA	—
G. max	SOYNOD21	Nodulin-21	Nodule	819	ss-mRNA	E
G. max	SOYNOD24	Nodulin 24	Nodule	382	ss-mRNA	E
G. max	SOYNOD26	Nodulin 26	Nodule	256	ss-mRNA	E
G. max	SOYNOD44	Nodulin 44	Nodule	353	ds-DNA	E
G. max	SOYNOD75	Nodulin (N-75)	Nodule	976	ss-mRNA	—
G. max	SOYNOD26B	Nodulin-26b	Nodule	912	ss-mRNA	E
G. max	SOYNOD20G	Nodulin-20	Nodule	2041	ds-DNA	E
G. max	SOYNOD22G	Nodulin 22	Nodule	1934	ds-DNA	E
G. max	SOYNOD23G	Nodulin-23	Root nodule	2223	ds-DNA	E
G. max	SOYNOD24G	Nodulin-24 gene	Root nodule	737	ds-DNA	E
G. max	SOYNOD35G	Uricase II	Root nodule	4783	ds-DNA	—
G. max	SOYNOD24H	Peribacteroid membrane protein (nod24)	Root nodule	2013	ds-DNA	—
G. max	SOYNOD35M	Uricase II	Root nodule	1120	ss-mRNA	—
G. max	SOYNODR1	Nodulin (C51)	Root nodule	954	ss-mRNA	E
G. max	SOYNODR2	Nodulin (E27)	Root nodule	1303	ss-mRNA	E
G. max	SOYNOD26R	Nodulin-26	Root nodule	1107	ss-mRNA	E
G. max	SOYNOD27R	Nodulin-27	Root nodule	892	ss-mRNA	E
G. max	SOYOXIDORE	Oxygen oxidoreductase (nodulin 35)	Root nodule	1219	ss-mRNA	—
H. vulgare	BLYGNIRE	Nitrate reductase	Seedling	7299	ds-DNA	E
H. vulgare	BLYMNIRE	Nitrate reductase	Leaves	2940	ss-mRNA	E
H. vulgare	BLYNAR7	NAD(P)H-bispecific nitrate reductase	—	3538	ds-DNA	E
L. luteus	LUPLBIIA	Leghemoglobin (lupin Lb II)	—	595	ss-mRNA	—
L. luteus	LUPLBR	Leghemoglobin I	Root nodule	754	ss-mRNA	E
L. luteus	LUPNOSP	Nodule-specific protein	Root nodule	213	ss-mRNA	—
L. esculentum	TOMNIA	Nitrate reductase (EC 1.6.6.1)	Leaves	5309	ds-DNA	E

Gene Index

Table 13. Continued

Species	Identifier	Sequence specification	Tissue/site	Size (bp or b)	Type	S
M. truncatula	ALFLB1	Leghemoglobin	Root nodule	3200	ds-DNA	E
M. truncatula	ALFLB2	Leghemoglobin	Root nodule	2126	ds-DNA	E
M. sativa	ALFLBA	Leghemoglobin	Root nodule	578	ss-mRNA	–
M. sativa	ALFLBB	Leghemoglobin	Root nodule	521	ss-mRNA	–
M. sativa	ALFLB3R	Leghemoglobin III	Root nodule	581	ss-mRNA	–
M. sativa	ALFLEGH12	Leghemoglobin	Root nodule	312	ss-mRNA	E
M. sativa	ALFLEGH13	Leghemoglobin	Root nodule	470	ss-mRNA	E
M. sativa	ALFLEGHE1	Leghemoglobin	Root nodule	521	ss-mRNA	E
M. sativa	ALFLEGHE2	Leghemoglobin	Root nodule	578	ss-mRNA	E
M. sativa	ALFLEGHEM	Leghemoglobin	Root nodule	559	ss-mRNA	E
M. sativa	ALFLEGHEMA	Leghemoglobin	Root nodule	3228	ds-DNA	–
M. sativa	ALFLEGHEMB	Leghemoglobin	Root nodule	221	ss-mRNA	–
M. sativa	ALFLEGHEMO	Leghemoglobin (*MsLb3*) gene	Root nodule	584	ss-mRNA	–
M. sativa	ALFMSENODA	MsENOD12-1	Root nodule	509	ss-mRNA	–
M. sativa	ALFMSENODB	MsENOD12-2	Root nodule	414	ss-mRNA	–
M. sativa	ALFMSENODC	MsENOD12-3	Root nodule	1459	ss-mRNA	–
M. sativa	ALFMSPRP	*MsPRP* gene	Root nodule	938	ss-mRNA	–
M. sativa	ALFNOD25	Nodulin-25	Root nodule	10048	ds-DNA	E
M. sativa	ALFNOD25A	Nodulin-25	Root nodule	809	ss-mRNA	E
M. sativa	ALFNOD25R	Nodulin-25	Root nodule	961	ss-mRNA	E
M. sativa	ALFNODUL	Soybean N-75 homologous	Root nodule	292	ss-mRNA	–
N. tabacum	TOBNIA1	Nitrate reductase (EC 1.6.6.1)	Leaves	6165	ds-DNA	E
N. tabacum	TOBNIA2	Nitrate reductase (EC 1.6.6.1)	Leaves	7096	ds-DNA	E
N. tabacum	TOBNR1	Nitrate reductase	Leaves	1663	ss-mRNA	E
O. sativa	RICNIA112	Nitrate reductase apoenzyme	–	1700	ds-DNA	E
O. sativa	RICNIA134	Nitrate reductase apoenzyme	–	2000	ds-DNA	E
P. andersonii	RHPHBEM	Hemoglobin	–	1520	ds-DNA	–
P. rhizobium	RHPNIFDK	*nifD* and *nifK* genes	–	3500	ds-DNA	E
P. rhizobium	RHPNIFH	Nitrogenase (*pifH*) gene	–	2030	ds-DNA	–
P. vulgaris	PHVLBA	Leghemoglobin	–	933	ds-DNA	–
P. sativum	PEAENOD2	Early nodulin protein	Nodule parenchyma	540	ss-mRNA	E
P. sativum	PEAENOD12A	Early nodulin protein 12	Root nodule	564	ss-mRNA	–
P. sativum	PEAENOD12B	Early nodulin	Root nodule, stem, flower	2763	ds-DNA	E
P. sativum	PEANOD6G	Nodulin	Leaves	1757	ds-DNA	E
P. sativum	PEANOD6MR	Nodulin	Root	356	ss-mRNA	E
S. rostrata	SESENOD2	Nodulin	–	2600	ds-DNA	E
S. rostrata	SESENODA	Nodulin	Nodules	456	ss-mRNA	–
S. rostrata	SESENODB	Nodulin	Nodules	460	ss-mRNA	–
S. rostrata	SESGLB3	Leghemoglobin	Nodules	1873	ds-DNA	E
S. rostrata	SESLBDRLA	Leghemoglobin (lbSr1)	Nodules	618	ss-mRNA	–
S. rostrata	SESLBDRLB	Leghemoglobin (lbSr6) protein	Nodules	541	ss-mRNA	–
S. rostrata	SESSRGLB2	Leghemoglobin	Leaves	1836	ds-DNA	E
S. rostrata	SESSRGLB3	*Srglb3* gene upstream region, leghemoglobin, nodulin	Leaves	409	ds-DNA	E
S. oleracea	SPINIS	Nitrite reductase (NIR)	Leaves	4234	ds-DNA	E
S. oleracea	SPINIR	Nitrite reductase (NiR)	Leaves	2062	ss-mRNA	–
S. oleracea	SPINR	NADH-nitrate reductase	–	3284	ss-mRNA	–
T. tomentosa	TRTHB	Hemoglobin gene	–	1727	ds-DNA	E
V. aconitifolia	VIRNOD35A	Nodulin-35 mRNA	–	1356	ss-mRNA	–
V. faba	VFALBA	Leghemoglobin mRNA	Nodules	441	ss-mRNA	–
V. faba	VFALBB	Leghemoglobin	Nodules	375	ss-mRNA	–
V. faba	VFALBC	Leghemoglobin	Nodules	256	ss-mRNA	–
Z. mays	MZENADH	NAD(P)H-nitrate reductase	Roots	931	ss-mRNA	E
Z. mays	MZENADHNR	NADH-nitrate reductase	Seedling scutellum	1801	ss-mRNA	–
Z. mays	MZENAR	NADH-nitrate reductase	Seedling leaf	2126	ss-mRNA	–

Table 14. Photosynthesis (proteins associated with photosynthesis, photoreception, phytochromes, plastocyanins, light-influenced genes)

Species	Identifier	Sequence specification	Tissue/site	Size (bp or b)	Type	S
A. magnifica	AMRBCLG	Ribulose-1,5-bisphosphate carboxylase	—	1545	DNA	E
A. thaliana	ATH5FCAB2	5′ flanking region of *cab2* gene and *cab2* gene	Leaves	825	ds-DNA	E
A. thaliana	ATH5FCAB3	5′ flanking region of *cab3* gene and *cab2* gene	Leaves	1004	ds-DNA	E
A. longa	ALRBCL	*rbcL*, *rp15*, *rps8*, *rpl36*, *rps14*, *rps2*, *trnI*, *trnF*, *trnC* and *rpoB* (partial) genes	—	13133	DNA	E
A. sativa	ASTAP3R	Phytochrome (AP3)	—	3696	ss-mRNA	E
A. sativa	ASTAP4R	Phytochrome (AP4)	—	3705	ss-mRNA	E
A. sativa	ASTAP5R	Phytochrome (AP5)	—	1623	ss-mRNA	—
A. sativa	ASTPCR	Protochlorophyllide reductase	Etiolated leaves	1146	ss-mRNA	E
A. sativa	ASTPHT3A	Type 3 phytochrome gene (phy3)	Shoot	6746	ds-DNA	—
B. oerstedii	BAARBCSB1	Ribulose bisphosphate carboxylase small subunit, clone B1	—	758	ss-mRNA	E
B. oerstedii	BAARBCSB2	Ribulose bisphosphate carboxylase small subunit, clone B2	—	524	ss-mRNA	—
B. oleracea	BNARBCL	Ribulose bisphosphate carboxylase (rbcL)	—	1699	ds-DNA	—
B. napus	BNARBCS	Small subunit of ribulose-1, 5-bisphosphate carboxylase	Cotyledons	2714	ds-DNA	E
B. napus	BNARBCSF1	Ribulose-1,5-bisphosphate carboxylase/oxygenase small subunit	—	2006	ds-DNA	E
B. napus	BNARBCSSU	Ribulose-1,5-bisphosphate carboxylase/oxygenase small subunit	—	726	ss-mRNA	—
D. caryophyllus	DINRBCLA	Ribulose bisphosphate carboxylase	—	1407	ds-DNA	—
F. trinervia	FTRBCR	Ribulose-1,5-bisphosphate carboxylase	—	752	ss-mRNA	E
G. max	SOYCAB2	Minor PSII LHCII chlorophyll *a/b* binding protein	Leaves	1354	ds-DNA	—
G. max	SOYCAB3	PSII LHCII chlorophyll *a/b* binding protein	Leaves	1361	ds-DNA	E
G. max	SOYCAB4	Cab4	Leaves	1470	ds-DNA	E
G. max	SOYCAB5	Cab5	Leaves	1467	ds-DNA	E
G. max	SOYCAB1P	*Cab1* pseudogene for chlorophyll *a/b* binding protein	Leaves	872	ds-DNA	—
G. max	SOYCBPA	Light-harvesting chlorophyll *a/b* binding protein	Leaves	911	ss-mRNA	—
G. max	SOYRUBP	Ribulose-1,5-bisphosphate carboxylase small subunit	Leaves	1601	ds-DNA	E
G. max	SOYRUBPA	Ribulose-1,5-bisphosphate carboxylase small subunit	—	2185	ds-DNA	—
G. max	SOYRUBQ	Ribulose-1,5-bisphosphate carboxylase small subunit	—	1601	ds-DNA	E
G. max	SOYSRS1	Ribulose-1,5-bisphosphate carboxylase small subunit	—	241	ss-mRNA	E
G. max	SOYSRS4	Ribulose-1,5-bisphosphate carboxylase small subunit	—	248	ss-mRNA	E
G. hirsutum	COTCAB	Chlorophyll *a/b* binding protein	Leaves	1746	ds-DNA	E
H. annuus	HNNRBCS	(EC 4.1.1.39) Ribulose-1,5-bisphosphate carboxylase/oxygenase	—	1400	ds-DNA	E

Gene Index

Table 14. Continued

Species	Identifier	Sequence specification	Tissue/site	Size (bp or b)	Type	S
H. annuus	HNNRUBISC	Ribulose-1,5-bisphosphate carboxylase small subunit (EC 4.1.1.39)	–	674	ss-mRNA	–
H. vulgare	BLYCAB2	Major light-harvesting chlorophyll *a/b* binding protein (LHCP)	Leaves	1030	ds-DNA	E
H. vulgare	BLYLHBC	Type III LHCII CAB precursor protein	8-day-old seedlings	928	ss-mRNA	E
H. vulgare	BLYLP58	Early light-inducible protein	Leaves	860	ss-mRNA	E
H. vulgare	BLYLP60	Early light-inducible protein (ELIP)	Leaves	764	ss-mRNA	E
H. vulgare	BLYLP90	Early light-inducible protein	Leaves	660	ss-mRNA	E
H. vulgare	BLYPCR	Plastocyanin precursor	–	662	ss-mRNA	E
H. vulgare	BLYPSA2	Photosystem I protein psa2	Seedling	610	ss-mRNA	–
H. vulgare	BLYPSAH	10.2 kDa photosystem I polypeptide	Leaves	613	ss-mRNA	E
H. vulgare	BLYPSAL	Photosystem I protein (PSI-L)	5-day-old seedling	777	ss-mRNA	–
H. vulgare	BLYRCAA1	Rubisco activase (RcaA1)	7-day-old leaf	1394	ss-mRNA	–
H. vulgare	BLYRCAA2	Rubisco activase (RcaA2)	7-day-old leaf	1709	ss-mRNA	–
H. vulgare	BLYRCAB	Rubisco activase (RcaB)	7-day-old leaf	1656	ss-mRNA	–
H. vulgare	BLYRCABG	Rubisco activase	Leaves	8766	ds-DNA	–
L. laricina	LAXRBC1C	Ribulose bisphosphate carboxylase (LLRBC1C)	Needles	117	ss-mRNA	E
L. laricina	LAXRBC2CM	Ribulose bisphosphate carboxylase (LLRBC2C)	Needles	163	ss-mRNA	E
L. laricina	LAXRBC3CM	Ribulose bisphosphate carboxylase (LLRBC3C)	Needles	663	ss-mRNA	E
L. laricina	LAXRBC4CM	Ribulose bisphosphate carboxylase (LLRBC4C)	Needles	378	ss-mRNA	E
L. laricina	LAXRBC5CM	Ribulose bisphosphate carboxylase (LLRBC5C)	Needles	678	ss-mRNA	E
L. laricina	LAXRBC7CM	Ribulose bisphosphate carboxylase (LLRBC7C)	Needles	263	ss-mRNA	E
L. laricina	LAXRBCS	Ribulose-1,5-bisphosphate carboxylase/oxygenase (EC 4.1.1.39)	Needles	3679	ds-DNA	E
L. gibba	LG106	Negatively light-regulated mRNA (Lg106)	–	398	RNA	E
L. gibba	LGIAB19A	Chlorophyll *a/b* apoprotein	–	1913	ds-DNA	–
L. gibba	LGILG106NL	Negatively light-regulated protein (Lg106)	–	378	ss-mRNA	–
L. gibba	LGILHCPABP	Light-harvesting chlorophyll *a/b* protein	–	1633	ds-DNA	–
L. gibba	LGINPR1	NPR1	–	1573	ds-DNA	E
L. gibba	LGINPR2	NPR2	–	2118	ds-DNA	E
L. gibba	LGIRBCS1	Small subunit of ribulose bisphosphate carboxylase (EC 4.1.1.39)	–	764	ss-mRNA	E
L. gibba	LGIRBCS13	Small subunit of ribulose bisphosphate carboxylase (EC 4.1.1.39)	–	234	ds-DNA	E
L. gibba	LGIRBCS26	Small subunit of ribulose bisphosphate carboxylase (EC 4.1.1.39)	–	936	ds-DNA	E
L. gibba	LGIRBCS4A	Small subunit of ribulose bisphosphate carboxylase (EC 4.1.1.39)	–	791	ds-DNA	E
L. gibba	LGIRBCS5A	Small subunit of ribulose bisphosphate carboxylase (EC 4.1.1.39)	–	858	ds-DNA	E

Table 14. Continued

Species	Identifier	Sequence specification	Tissue/site	Size (bp or b)	Type	S
L. gibba	LGIRBCS4B	Small subunit of ribulose bisphosphate carboxylase (EC 4.1.1.39)	—	791	ds-DNA	E
L. gibba	LGIRBCS5B	Small subunit of ribulose bisphosphate carboxylase (EC 4.1.1.39)	—	800	ds-DNA	E
L. gibba	LGIR15BP1	Ribulose-1,5-bisphosphate carboxylase small subunit pseudogene 5A	—	159	ds-DNA	—
L. gibba	LGIR15BP2	Ribulose-1,5-bisphosphate carboxylase small subunit gene 5B	—	183	ds-DNA	—
L. gibba	LGIR15BPC	Ribulose-1,5-bisphosphate carboxylase small subunit gene	—	758	ss-mRNA	E
L. esculentum	TOMCAB	Type I chlorophyll *a/b* binding protein	—	1395	ds-DNA	—
L. esculentum	TOMCAB7	*Cab 7* type II chlorophyll *a/b* binding protein	—	4261	ds-DNA	E
L. esculentum	TOMCAB8	Type III chlorophyll *a/b* binding polypeptide of photosystem I, *Cab8*	—	2659	ds-DNA	E
L. esculentum	TOMCAB11	*Cab 11*	—	1393	ds-DNA	E
L. esculentum	TOMCAB12	*Cab 12*	—	1493	ds-DNA	E
L. esculentum	TOMCAB4A	Chlorophyll *a/b* binding protein	Leaves	895	ss-mRNA	—
L. esculentum	TOMCAB5A	*Cab 5* gene encoding chlorophyll *a/b* binding protein	—	810	ss-mRNA	—
L. esculentum	TOMCAB1AUP	*Cab 1a* gene 5′ upstream region	—	240	ds-DNA	E
L. esculentum	TOMCABIN	Chlorophyll a binding protein (*Cab 10A*)	—	1676	ds-DNA	—
L. esculentum	TOMCBBIN	Chlorophyll b binding protein (*Cab 10b*)	—	1586	ds-DNA	—
L. esculentum	TOMCBPA	Chlorophyll *a/b* binding protein gene *Cab 1B*	—	1253	ds-DNA	—
L. esculentum	TOMCBPB	Chlorophyll *a/b* binding protein gene *Cab 3C*	—	1135	ds-DNA	—
L. esculentum	TOMCBPC1	Chlorophyll *a/b* binding protein *Cab 1A*	—	184	ds-DNA	—
L. esculentum	TOMCBPC2	Chlorophyll *a/b* binding protein *Cab 1A*	—	351	ds-DNA	—
L. esculentum	TOMCBPD1	Chlorophyll *a/b* binding protein *Cab 1C*	—	187	ds-DNA	—
L. esculentum	TOMCBPD2	Chlorophyll *a/b* binding protein *Cab 1C*	—	351	ds-DNA	—
L. esculentum	TOMCBPE1	Chlorophyll *a/b* binding protein *Cab 3A*	—	192	ds-DNA	—
L. esculentum	TOMCBPE2	Chlorophyll *a/b* binding protein *Cab 3A*	—	351	ds-DNA	—
L. esculentum	TOMCBPF1	Chlorophyll *a/b* binding protein *Cab 3B*	—	178	ds-DNA	—
L. esculentum	TOMCBPF2	Chlorophyll *a/b* binding protein *Cab 3B*	—	351	ds-DNA	—
L. esculentum	TOMCBPG	Chlorophyll *a/b* binding protein *Cab 1D*	—	351	ds-DNA	—
L. esculentum	TOMCBPH	Chlorophyll *a/b* binding protein gene *Cab 3T*	—	339	ds-DNA	—
L. esculentum	TOMCBPI	Photosystem I chlorophyll *a/b* binding protein	Leaves	4261	ds-DNA	—
L. esculentum	TOMCRCAB	Dispersed repeat CR1 associated with *Cab 1*	—	1658	ds-DNA	—

Table 14. Continued

Species	Identifier	Sequence specification	Tissue/site	Size (bp or b)	Type	S
L. esculentum	TOMLHBC1G	LHCII type III	—	1802	ds-DNA	E
L. esculentum	TOMPLACY	Pre-plastocyanin	—	510	ss-mRNA	—
L. esculentum	TOMPSAD	(PS I) reaction center protein subunit II	—	802	ss-mRNA	—
L. esculentum	TOMPSBXMR	Photosystem II 23 kDa protein	—	997	ss-mRNA	E
L. esculentum	TOMPSI	Photosystem I protein	Leaves	896	ss-mRNA	—
L. esculentum	TOMRBCS1	Ribulose-1,5-bisphosphate carboxylase/oxygenase small subunit (*rbcS1*)	—	1338	ds-DNA	—
L. esculentum	TOMRBCS2	Ribulose-1,5-bisphosphate carboxylase/oxygenase small subunit (*rbcS2*)	—	2726	ds-DNA	—
L. esculentum	TOMRBCSA	Ribulose-1,5-bisphosphate carboxylase/oxygenase (RBCS) small subunit	—	1032	ds-DNA	—
L. esculentum	TOMRBCSB	Ribulose-1,5-bisphosphate carboxylase/oxygenase small subunit	—	599	ss-RNA	—
L. esculentum	TOMRBCSC	Ribulose-1,5-bisphosphate carboxylase/oxygenase (RBCS) small subunit	—	1027	ds-DNA	—
L. esculentum	TOMRBCS3A	Ribulose-1,5-bisphosphate carboxylase/oxygenase small subunit (*rbcS3A*)	—	1346	ds-DNA	—
L. esculentum	TOMRBCS3B	Ribulose-1,5-bisphosphate carboxylase/oxygenase small subunit (*rbcS3B*)	—	1054	ds-DNA	—
L. esculentum	TOMRBCS3C	Ribulose-1,5-bisphosphate carboxylase/oxygenase small subunit (*rbcS3C*)	—	1207	ds-DNA	—
L. esculentum	TOMRBCSD	Ribulose bisphosphate carboxylase small subunit	—	778	ss-mRNA	—
L. esculentum	TOMRBCSE	Ribulose bisphosphate carboxylase small subunit LESS5	—	742	ss-mRNA	—
M. domestica	MAUCHBP	Light-harvesting chlorophyll *a*/*b* binding polypeptide of photosystem II	Leaves	1306	ds-DNA	E
M. crystallinum	CIPRBC	Ribulose-1,5-bisphosphate carboxylase (*rbc5*)	—	776	ss-mRNA	—
M. crystallinum	CIPRBCS2	Ribulose-1,5-bisphosphate carboxylase/oxygenase small subunit (*rbcS-2*)	—	680	ss-mRNA	—
M. crystallinum	CIPRBCS3	Ribulose-1,5-bisphosphate carboxylase/oxygenase small subunit (*rbcS-3*)	—	765	ss-mRNA	—
M. crystallinum	CIPRBCS4	Ribulose-1,5-bisphosphate carboxylase/oxygenase small subunit (*rbcS-4*)	—	747	ss-mRNA	—
N. plumbaginifolia	TOBCABA	Chlorophyll *a*/*b* binding protein (*Cab-C*)	Leaves	1240	ds-DNA	—
N. plumbaginifolia	TOBCABB	Chlorophyll *a*/*b* binding protein	Leaves	1815	ds-DNA	—
N. plumbaginifolia	TOBCABC	Chlorophyll *a*/*b* binding protein (*Cab-F*)	Leaves	477	ds-DNA	—
N. plumbaginifolia	TOBCABEPR	*Cab-E* gene 5′-flanking region	Leaves	1613	ds-DNA	E
N. plumbaginifolia	TOBRBCS8B	Ribulose bisphosphate carboxylase (*rbcS-8B*)	—	2293	ds-DNA	—

Table 14. Continued

Species	Identifier	Sequence specification	Tissue/site	Size (bp or b)	Type	S
N. plumbaginifolia	TOBRBCS8C	Ribulose bisphosphate carboxylase small subunit	—	2293	ds-DNA	E
N. sylvestris	TOBPSADAG	*psaDa*	Whole plant	1260	ss-mRNA	E
N. sylvestris	TOBRBPC	Ribulose-1,5-bisphosphate carboxylase small subunit	—	453	ss-mRNA	—
N. sylvestris	TOBRBPCS	Ribulose bisphosphate carboxylase small subunit	—	796	ss-mRNA	E
N. sylvestris	TOBRUBSA	Ribulose-1,5-bisphosphate carboxylase small subunit gene	Leaves	73	ss-mRNA	—
N. sylvestris	TOBRUBSB	Ribulose-1,5-bisphosphate carboxylase small subunit gene	Leaves	60	ss-mRNA	—
N. sylvestris	TOBRUBSSU	NySS41 and partial NySS42 genes for Rubisco small subunit (SSU)	—	2795	ds-DNA	E
N. tabacum	NTPHII	Photosystem II 23 kDa polypeptide	Leaves	1035	RNA	E
N. tabacum	TOBCAB7	Chlorophyll *a*/*b* binding protein	—	1656	ds-DNA	E
N. tabacum	TOBCAB16	Major chlorophyll *a*/*b* binding protein	Leaves	1025	ss-mRNA	E
N. tabacum	TOBCAB21	Major chlorophyll *a*/*b* binding protein	Leaves	953	ss-mRNA	E
N. tabacum	TOBCAB36	Chlorophyll *a*/*b* binding protein	—	1986	ds-DNA	E
N. tabacum	TOBCAB40	Major chlorophyll *a*/*b* binding protein	Leaves	1080	ss-mRNA	E
N. tabacum	TOBCAB50	Major chlorophyll *a*/*b* binding protein	Leaves	964	ss-mRNA	E
N. tabacum	TOBOEE2AG	Photosystem II oxygen-evolving complex 23 kDa polypeptide	Leaves	2670	ds-DNA	E
N. tabacum	TOBOEE2AR	Photosystem II oxygen-evolving complex 23 kDa polypeptide	—	1012	ds-DNA	E
N. tabacum	TOBRBPCO	Ribulose bisphosphate carboxylase/oxygenase small subunit	—	2362	ds-DNA	E
N. tabacum	TOBRUBPA	Ribulose-1,5-bisphosphate carboxylase small subunit gene	—	979	ds-DNA	—
N. tabacum	TOBRUBPB	Ribulose-1,5-bisphosphate carboxylase small subunit pseudogene	—	1337	ds-DNA	—
O. sativa	RICCABR1	Light-harvesting chlorophyll *a*/*b* binding protein	—	1664	ds-DNA	E
O. sativa	RICCABR2	Light-harvesting chlorophyll *a*/*b* binding protein	—	1542	ds-DNA	E
O. sativa	RICLHCP1	Type I light-harvesting chlorophyll *a*/*b* binding protein of photosystem II (LHCPII)	—	1022	ss-mRNA	—
O. sativa	RICLHCP2	Type II light-harvesting chlorophyll *a*/*b* binding protein of photosystem II (LHCPII)	—	989	ss-mRNA	—
O. sativa	RICPHY18	*phy 18*, phytochrome	—	9979	ds-DNA	E
O. sativa	RICPHYB1GE	*phyB1* gene	Seedling shoots	4090	ds-DNA	—
O. sativa	RICRBCS	Ribulose-1,5-bisphosphate carboxylase/oxygenase small subunit	—	1110	ds-DNA	E
O. sativa	RICRUBPC1	Small subunit ribulose-1,5-bisphosphate carboxylase mRNA	—	830	ss-mRNA	—
O. sativa	RICRUBPC2	Small subunit of ribulose-1,5-bisphosphate carboxylase mRNA	—	895	ss-mRNA	—
P. crispum	PUMCPRF1	Light-inducible protein	—	1740	ss-mRNA	E
P. crispum	PUMCPRF2	Light-inducible protein	—	1435	ss-mRNA	E

Table 14. Continued

Species	Identifier	Sequence specification	Tissue/site	Size (bp or b)	Type	S
P. crispum	PUMCPRF3	Light-inducible protein	—	1302	ss-mRNA	E
P. hybrida	PETCAB37	Chlorophyll *a/b* binding protein	Leaves	1470	ds-DNA	E
P. hybrida	PETLCHI	Chlorophyll binding protein	Leaves	1013	ss-mRNA	—
P. hybrida	PETRBCA	Ribulose-1,5-bisphosphate carboxylase small subunit	Leaves	318	ss-mRNA	—
P. hybrida	PETRBCB	Ribulose-1,5-bisphosphate carboxylase small subunit	Leaves	212	ss-mRNA	—
P. hybrida	PETRBCC	Ribulose-1,5-bisphosphate carboxylase small subunit	Leaves	251	ss-mRNA	—
P. hybrida	PETRBCSA1	*SSU301* gene for rbcS, small subunit ribulose-1,5-bisphosphate carboxylase		412	ds-DNA	E
P. hybrida	PETRBCSA2	*SSU301* gene for rbcS, small subunit ribulose-1,5-bisphosphate carboxylase	—	264	ds-DNA	E
P. hybrida	PETRBCSB1	*SSU611* gene for rbcS, small subunit ribulose-1,5-bisphosphate carboxylase	—	413	ds-DNA	E
P. hybrida	PETRBCSB2	*SSU611* gene for rbcS, small subunit ribulose-1,5-bisphosphate carboxylase	—	264	ds-DNA	E
P. hybrida	PETRBCSC	*SSU231* gene for rbcS, small subunit ribulose-1,5-bisphosphate carboxylase	—	233	ds-DNA	E
P. hybrida	PETRBCSF2	*SSU112* gene for rbcS, small subunit ribulose-1,5-bisphosphate carboxylase	—	264	ds-DNA	E
P. hybrida	PETRBCSD1	*SSU491* gene for rbcS, small subunit ribulose-1,5-bisphosphate carboxylase	—	453	ds-DNA	E
P. hybrida	PETRBCSD2	*SSU491* gene for rbcS, small subunit ribulose-1,5-bisphosphate carboxylase	—	142	ds-DNA	E
P. hybrida	PETRBCSE1	*SSU911* gene for rbcS, small subunit ribulose-1,5-bisphosphate carboxylase	—	440	ds-DNA	E
P. hybrida	PETRBCSE2	*SSU911* gene for rbcS small subunit ribulose-1,5-bisphosphate carboxylase	—	259	ds-DNA	E
P. hybrida	PETRBCSF1	*SSU112* gene for rbcS small subunit ribulose-1,5-bisphosphate carboxylase	—	351	ds-DNA	E
P. hybrida	PETRBCSG	*SSU211* gene for rbcS small subunit ribulose-1,5-bisphosphate carboxylase	—	222	ds-DNA	E
Petunia sp.	PETCAB3	Chlorophyll *a/b* binding protein	Leaves	511	ss-mRNA	—
Petunia sp.	PETCAB4	Chlorophyll *a/b* binding protein (Cab)	Leaves	387	ss-mRNA	—
Petunia sp.	PETCAB10	Chlorophyll *a/b* binding protein	Leaves	367	ss-mRNA	—
Petunia sp.	PETCAB13	Chlorophyll *a/b* binding protein, Cab 13	—	1019	ds-DNA	E
Petunia sp.	PETCAB25	Chlorophyll *a/b* binding protein Cab 25	—	1184	ds-DNA	E
Petunia sp.	PETCAB102	Major chlorophyll *a/b* binding protein	Leaves	302	ss-mRNA	—

Table 14. Continued

Species	Identifier	Sequence specification	Tissue/site	Size (bp or b)	Type	S
Petunia sp.	PETCAB146	Chlorophyll *a*/*b* binding protein	Leaves	498	ss-mRNA	—
Petunia sp.	PETCABA	Chlorophyll *a*/*b* binding protein	Leaves	86	ss-mRNA	—
Petunia sp.	PETCABB	Chlorophyll *a*/*b* binding protein	Leaves	107	ss-mRNA	—
Petunia sp.	PETCAB22L	Chlorophyll *a*/*b* binding protein	—	1187	ds-DNA	E
Petunia sp.	PETCAB22R	Chlorophyll *a*/*b* binding protein	—	1187	ds-DNA	E
Petunia sp.	PETCAB91R	Chlorophyll *a*/*b* binding protein	—	1173	ds-DNA	E
Petunia sp.	PETRBCS	rbcS	—	66	ss-mRNA	—
Petunia sp.	PETRBCS08	Ribulose-1,5-bisphosphate carboxylase small subunit	Leaves	2242	ds-DNA	E
Petunia sp.	PETRBCSCA	Ribulose bisphosphate carboxylase small subunit	Leaves	124	ss-mRNA	—
Petunia sp.	PETRBCSA	Ribulose bisphosphate carboxylase small subunit	Leaves	135	ss-mRNA	—
Petunia sp.	PETRBCS1A	Ribulose-1,5-bisphosphate carboxylase small subunit	Leaves	1314	ds-DNA	E
Petunia sp.	PETRBCSB	Ribulose bisphosphate carboxylase small subunit	Leaves	75	ss-mRNA	—
Petunia sp.	PETRBCS1B	Ribulose-1,5-bisphosphate carboxylase small subunit	Leaves	192	ds-DNA	—
Petunia sp.	PETRBCS1C	Ribulose-1,5-bisphosphate carboxylase small subunit	Leaves	150	ds-DNA	—
Petunia sp.	PETRBCSD	rbcS protein	Multiple sites	87	ss-mRNA	—
Petunia sp.	PETRBCS1D	Ribulose-1,5-bisphosphate carboxylase small subunit	Leaves	189	ds-DNA	—
Petunia sp.	PETRBCS1E	Ribulose-1,5-bisphosphate carboxylase small subunit	Leaves	156	ds-DNA	—
Petunia sp.	PETRBCS1F	Ribulose-1,5-bisphosphate carboxylase small subunit	Leaves	164	ds-DNA	—
Petunia sp.	PETRBCS1G	Ribulose-1,5-bisphosphate carboxylase small subunit	Leaves	156	ds-DNA	—
P. vulgaris	PHVRBCOS	Ribulose-1,5-bisphosphate carboxylase/oxygenase	Leaves	849	ss-mRNA	E
P. vulgaris	PHVRBCS	Ribulose-1,5-bisphosphate carboxylase/oxygenase	Leaves	591	ss-mRNA	E
P. vulgaris	PHVSS15BCO	Small subunit of ribulose-1,5-bisphosphate carboxylase/oxygenase	—	1071	ss-mRNA	E
P. patens	PCPCBP	Major chlorophyll binding protein	—	2544	ds-DNA	—
P. sylvestris	PINCABII2	Chlorophyll *a*/*b* binding protein	Cotyledons	583	ss-mRNA	—
P. sylvestris	PINCABIIA	Chlorophyll *a*/*b*-binding protein	Cotyledons	1083	ss-mRNA	E
P. sylvestris	PINCABIIB	Chlorophyll *a*/*b* binding protein	Cotyledons	1006	ss-mRNA	E
P. sativum	PEAAB805	AB80 chlorophyll *a*/*b* binding protein	—	1040	ds-DNA	E
P. sativum	PEACAB15	Chlorophyll *a*/*b* binding thylakoid protein (polypeptide 15)	—	822	ss-mRNA	—
P. sativum	PEACAB66	*Cab* gene	Leaves	1666	ds-DNA	—
P. sativum	PEACAB80	Major light-harvesting chlorophyll *a*/*b* binding protein	Leaves	1993	ds-DNA	—
P. sativum	PEACABII	Chlorophyll *a*/*b* binding protein	Leaves	2368	ds-DNA	E
P. sativum	PEACABIIC	Chlorophyll *a*/*b* binding protein	Leaves	2236	ds-DNA	E
P. sativum	PEACH01	Chloroplast carboxylase (ribulose-1,5-bisphosphate carboxylase)	—	668	ss-mRNA	E
P. sativum	PEACHLROPH	*Cab II* gene	—	1919	ds-DNA	—
P. sativum	PEAELIP	Plastid early light-induced protein	—	2919	ds-DNA	E
P. sativum	PEA33KDA	33 kDa polypeptide of the water-oxidizing complex of photosystem II	Leaves	1054	ss-mRNA	—

Gene Index

Table 14. Continued

Species	Identifier	Sequence specification	Tissue/site	Size (bp or b)	Type	S
P. sativum	PEA23KDAP	23 kDa polypeptide of the oxygen-evolving complex of photosystem II	—	987	ss-mRNA	E
P. sativum	PEALHCP3R	LHCPII mRNA 3′ region	—	199	ss-mRNA	E
P. sativum	PEAPHTAP	Phytochrome	—	5690	ds-DNA	—
P. sativum	PEAPHY	Phytochrome apoprotein	—	6264	ds-DNA	E
P. sativum	PEAPHYPPG	*phyP* pseudogene	Leaves	608	ds-DNA	E
P. sativum	PEAPLAS	Plastocyanin	Shoots	1505	ds-DNA	E
P. sativum	PEAPOEC17P	Precursor of the oxygen-evolving complex 17 kDa protein	Leaves	810	ss-mRNA	—
P. sativum	PEARP36A	Chloroplast import receptor p36	Seedling	1490	ss-mRNA	E
P. sativum	PEARBPC	Ribulose-1,5-bisphosphate carboxylase small subunit precursor	—	669	ss-mRNA	—
P. sativum	PEARPC	Small subunit ribulose-1,5-bisphosphate carboxylase (*rbcS-3A*)	—	435	ds-DNA	—
P. sativum	PEARBCOSS	Ribulose-1,5-bisphosphate carboxylase/oxygenase small subunit (*rbcS*) gene	—	645	ds-DNA	—
P. sativum	PEARBCS3A	Ribulose-1,5-bisphosphate carboxylase	—	2061	ds-DNA	E
P. sativum	PEARBCS3C	Ribulose-1,5-bisphosphate carboxylase	—	1381	ds-DNA	E
P. sativum	PEARUBP1	Ribulose-1,5-bisphosphate carboxylase	—	206	ss-mRNA	—
P. sativum	PEARUBP2	Ribulose-1,5-bisphosphate carboxylase	—	674	ss-mRNA	—
P. sativum	PEARUBPA	Ribulose-1,5-bisphosphate carboxylase small subunit	—	619	ss-mRNA	—
P. sativum	PEARUBPS	Ribulose-1,5-bisphosphate carboxylase small subunit	—	2351	ds-DNA	E
P. thunbergii	PINCAB	Light-harvesting chlorophyll *a*/*b* binding protein	Cotyledon	978	ss-mRNA	E
P. thunbergii	PINRBCS	Ribulose-1,5-bisphosphate carboxylase small subunit	Cotyledon	911	ss-mRNA	E
P. munitum	POMCABF3	Chlorophyll *a*/*b* binding protein F3 gene	—	1846	ds-DNA	—
P. munitum	POMCABF6	Chlorophyll *a*/*b* binding protein F6, pseudogene	—	1020	ds-DNA	—
P. munitum	POMCABF7	Chlorophyll *a*/*b* binding protein F7 pseudogene	—	1280	ds-DNA	—
P. munitum	POMCABF8	Chlorophyll *a*/*b* binding protein F8 pseudogene	—	499	ds-DNA	—
P. aerugineum	PPUERUBISCO	Ribulose-1,5-bisphosphate carboxylase	—	2249	ds-DNA	E
P. capensis	PLGRBCLC	Ribulose-1,5-bisphosphate carboxylase	—	1428	ds-DNA	—
P. pyrifolia	PYPLHABBP	Light-harvesting *a*/*b* binding protein	Leaves	1055	ss-mRNA	—
P. pyrifolia	PYPRBCO	Ribulose-1,5-bisphosphate carboxylase/oxygenase subunit	Leaves	844	ss-mRNA	—
R. sativus	RADCHABBP	Chlorophyll *a*/*b* binding protein of photosystem II	—	518	ss-mRNA	—
R. sativus	RADRBCS	Ribulose-1,5-bisphosphate carboxylase SSU	—	704	ss-mRNA	—
Rheum × *cultorum*	RXCRBCLD	Ribulose-1,5-bisphosphate carboxylase	—	1513	ds-DNA	—
R. communis	RCCRSBA	Rubisco subunit binding protein α subunit	Seed endosperm	1488	ss-mRNA	—

Table 14. Continued

Species	Identifier	Sequence specification	Tissue/site	Size (bp or b)	Type	S
S. pratensis	SIPCAB	Chlorophyll *a/b* binding protein	—	642	ss-mRNA	—
S. pratensis	SIPPCY	Plastocyanin precursor	—	773	ss-mRNA	E
S. pratensis	SIPRUBPCS	Ribulose-1,5-bisphosphate carboxylase small subunit	—	692	ss-mRNA	—
S. alba	SALCAB	Chlorophyll *a/b* binding protein	Cotyledons	2774	ds-DNA	E
S. alba	SALCAB1	Chlorophyll *a/b* binding polypeptide	—	2775	ds-DNA	E
S. alba	SAL23KDP2	23 kDa polypeptide of the oxygen-evolving complex of photosystem II	Seedling	1936	ds-DNA	E
S. alba	SALOES23	23 kDa subunit of oxygen-evolving system of photosystem II	Seedling	867	ss-mRNA	E
S. alba	SALRUBISC	Small subunit of ribulose bisphosphate carboxylase	Cotyledons	322	ss-mRNA	E
S. tuberosum	POTLS1G	Light-inducible tissue-specific *ST-LS1* gene	Shoots	3699	ds-DNA	E
S. tuberosum	POTLS1R	Light inducible tissue-specific *ST-LS1*	Shoots	560	ss-mRNA	E
S. tuberosum	POTPPOEC	33 kDa precursor protein of the oxygen-evolving complex	Leaves	1116	ss-mRNA	E
S. tuberosum	POTRBCS	Ribulose-1,5-bisphosphate carboxylase	—	546	ss-mRNA	—
S. tuberosum	POTRBCSA	rbcS1 mRNA sequence	—	147	ss-mRNA	—
S. tuberosum	POTRBCSB	rbcS2a mRNA sequence	—	144	ss-mRNA	—
S. tuberosum	POTRBCSC	rbcS2b mRNA sequence	—	144	ss-mRNA	—
S. tuberosum	POTRBCSD	rbcS2c mRNA sequence	—	144	ss-mRNA	—
S. tuberosum	POTRUBISC	Ribulose-1,5-bisphosphate carboxylase (*rbcL*)	Mature leaves	1469	ds-DNA	—
S. oleracea	SPICABPA	Chlorophyll *a/b* binding protein	Leaves	980	ss-mRNA	—
S. oleracea	SPICAB1PRO	cab-1 promoter	Seedling	451	ds-DNA	E
S. oleracea	SPIOEC16	16 kDa protein of the photosynthetic oxygen-evolving protein	Green tissue	867	ss-mRNA	E
S. oleracea	SPIOEC23	23 kDa protein of the photosynthetic oxygen-evolving complex	Green tissue	1000	ss-mRNA	E
S. oleracea	SPIPCG	Plastocyanin	—	1000	ds-DNA	E
S. oleracea	SPIPCP	Promoter DNA for plastocyanin	Leaves	1831	ds-DNA	E
S. oleracea	SPIPSI2	Subunit II precursor of photosystem I	Seedling	863	ss-mRNA	E
S. oleracea	SPIPSI3	Subunit III precursor of photosystem I	Seedling	653	ss-mRNA	E
S. oleracea	SPIPSI4	Photosystem I subunit V	Green tissue	659	ss-mRNA	E
S. oleracea	SPIPSI5	Photosystem I subunit IV	Green tissue	923	ss-mRNA	E
S. oleracea	SPIPS33	33 kDa precursor protein	—	1153	ss-mRNA	E
S. oleracea	SPIPSACR	Photosystem I reaction centre subunit VII	—	509	ss-mRNA	E
S. oleracea	SPIPSAHR	Photosystem I reaction centre subunit VI	—	639	ss-mRNA	E
S. oleracea	SPIPSBIA	10 kDa polypeptide of photosystem II	Seedling leaf	590	ss-mRNA	—
S. oleracea	SPIPSI20	PSI reaction center 20 kDa polypeptide	Leaves	767	ss-mRNA	—
S. oleracea	SPIRBCO	Rubisco activase	Leaves	1832	ss-mRNA	—
S. oleracea	SPIRFES	FeS precursor protein	—	1024	ss-mRNA	—
T. repens	TFRRBPCX	Ribulose bisphosphate carboxylase (EC 4.1.1.49)	—	3974	ds-DNA	E
T. aestivum	WHTCAB	Chlorophyll *a/b* binding protein	Leaves	1191	ds-DNA	—
T. aestivum	WHTCAB1G5	Chlorophyll *a/b* binding protein	Leaves	929	ds-DNA	E

Table 14. Continued

Species	Identifier	Sequence specification	Tissue/site	Size (bp or b)	Type	S
T. aestivum	WHTPPSBP	23 kDa oxygen-evolving protein of photosystem II	Leaves	967	ss-mRNA	E
T. aestivum	WHTPSB0	33 kDa oxygen-evolving protein of photosystem II	Young leaves	1217	ss-mRNA	E
T. aestivum	WHTRBCA	Ribulose-1,5-bisphosphate carboxylase, small subunit	Seedling leaf	743	ss-mRNA	—
T. aestivum	WHTRBCB	Ribulose-1,5-bisphosphate carboxylase small subunit	Seedling leaf	611	ss-mRNA	—
T. aestivum	WHTRBCC	Ribulose-1,5-bisphosphate carboxylase small subunit	Seedling leaf	317	ss-mRNA	—
T. aestivum	WHTRSBA	Rubisco subunit binding-protein α subunit	Leaves	1834	ss-mRNA	—
T. aestivum	WHTRUBIAA	Ribulose-1,5-bisphosphate carboxylase/oxygenase gene	—	2205	ds-DNA	—
T. aestivum	WHTRUBIAB	Ribulose-1,5-bisphosphate carboxylase/oxygenase gene	—	779	ds-DNA	—
Z. mays	MZECAB1	Chlorophyll *a*/*b* binding protein	Leaves	2100	ds-DNA	E
Z. mays	MZECATM1A	Light-harvesting chlorophyll *a*/*b* binding protein	Leaves	1057	ds-DNA	—
Z. mays	MZELHCABB	Light-harvesting chlorophyll *a*/*b* binding protein	Leaves	869	ss-mRNA	E
Z. mays	MZELHCP	Light-harvesting chlorophyll *a*/*b* binding protein	Leaves	967	ss-mRNA	—
Z. mays	MZEPCSSU	Ribulose-1,5-bisphosphate carboxylase small subunit	Leaves	830	ss-mRNA	—
Z. mays	MZERBCL	Ribulose-1,5-bisphosphate carboxylase large subunit gene	Leaves	407	ds-DNA	—
Z. mays	MZERBCS	Ribulose-1,5-bisphosphate carboxylase/oxygenase small subunit (EC 4.1.1.39)	Leaves	1648	ds-DNA	E
Z. mays	MZERUBSSU	Ribulose-1,5-bisphosphate carboxylase small subunit (RuBPC; EC 4.1.1.39)	Leaves	830	ss-mRNA	E

Table 15. Protein modification (proteases, kinases, phosphatases, protease inhibitors, amylase inhibitors, enzyme inhibitors/activators, chaperonins, ubiquitins)

Species	Identifier	Sequence specification	Tissue/site	Size (bp or b)	Type	S
A. deliciosa	ACTACT	Actinidin	—	3399	ds-DNA	—
A. chinensis	ACTACT1	Actinidin	—	1138	ss-mRNA	E
A. chinensis	ACTACTA	Actinidin	—	1137	ss-mRNA	—
A. chinensis	ACTACTB	Actinidin	—	777	ss-mRNA	—
A. chinensis	ACTACT7	Actinidin	—	809	ss-mRNA	E
A. deliciosa	ACTACTINID	Actinidin	—	634	ds-DNA	—
A. deliciosa	ACTCPA	Actinidin	Leaves	634	ds-DNA	—
A. thaliana	ATH60CPNA	Plastid 60 kDa chaperonin-60 α polypeptide	—	735	ss-mRNA	—
A. thaliana	ATH60CPNB	Plastid 60 kDa chaperonin-60 β polypeptide	—	559	ss-mRNA	—
A. thaliana	ATH1PPHA	Phosphoprotein phosphatase type 1 catalytic subunit	—	1340	ss-mRNA	—
B. napus	BNA60CPNA	60 kDa chaperonin-60 α polypeptide	—	1783	ss-mRNA	—
B. napus	BNA60CPNB	60 kDa chaperonin-60 β polypeptide	—	1982	ss-mRNA	—
D. carota	DARPK431	Calcium-dependent protein kinase	Embryo	1380	ss-mRNA	E
G. max	SOYBBIM	Bowman–Birk protease inhibitor	Immature seeds	208	ss-mRNA	—
G. max	SOYCADPK	Calcium-dependent protein kinase	—	1768	ds-DNA	—
G. max	SOYCIIPI	CII protease inhibitor	—	645	ds-DNA	—
G. max	SOYCIIPIA	Protease inhibitor C-II	Endosperm	484	ss-mRNA	—
G. max	SOYCIIPIB	Protease inhibitor IV	Endosperm	449	ss-mRNA	—
G. max	SOYPROKIN	Protein kinase (PK6)	—	1724	ss-mRNA	—
G. max	SOYTIAR	Ti-a for Kunitz trypsin inhibitor subtype A	Seed	773	ss-mRNA	E
G. max	SOYTIBR	Ti-b for Kunitz trypsin inhibitor subtype B	Seed	784	ss-mRNA	E
G. max	SOYUBI	Ubiquitin	—	392	ss-mRNA	—
H. annuus	HNNUBB1M	Hexaubiquitin protein	—	1584	ss-mRNA	E
H. annuus	HNNUBIQ	Ubiquitin	—	1379	ss-mRNA	—
H. annuus	HNNGUBB1	Hexaubiquitin protein	—	3552	ds-DNA	E
H. annuus	HNNGUBB2	Polyubiquitin protein	—	3016	ds-DNA	E
H. vulgare	BLY656	α-amylase/trypsin inhibitor	Endosperm	672	ss-mRNA	—
H. vulgare	BLYASPROT	Aspartic proteinase	Developing embryo	1800	ss-mRNA	E
H. vulgare	BLYBASI	α-amylase/subtilisin inhibitor (BASI)	Seed	792	ss-mRNA	E
H. vulgare	BLYCBX	Carboxypeptidase I	Scutellum	1424	ss-mRNA	—
H. vulgare	BLYCI1A	Chymotrypsin inhibitor (CI-1A)	Endosperm	446	ss-mRNA	—
H. vulgare	BLYCI2A	Chymotrypsin inhibitor-2 (CI-2A)	Endosperm	451	ss-mRNA	E
H. vulgare	BLYCI1B	Chymotrypsin inhibitor (CI-1B)	Endosperm	452	ss-mRNA	—
H. vulgare	BLYCI2B	Chymotrypsin inhibitor-2 (CI-2B)	Endosperm	324	ss-mRNA	E
H. vulgare	BLYCMA	α-amylase inhibitor	—	609	ss-mRNA	E
H. vulgare	BLYCMD	Seed protein CMd	Endosperm	579	ss-mRNA	E
H. vulgare	BLYCME2	Trypsin inhibitor CMe	Endosperm	582	ss-mRNA	E
H. vulgare	BLYIAM1	Monomeric α-amylase inhibitor	Endosperm	579	ss-mRNA	E
H. vulgare	BLYICA2	Chymotrypsin inhibitor-2	—	1385	ds-DNA	E
H. vulgare	BLYLEU	Thiol protease aleurain	Aleurone cells	5372	ds-DNA	—
H. vulgare	BLYMUB1	Ubiquitin (*mub1*)	—	1815	ds-DNA	—
H. vulgare	BLYMUB2	Ubiquitin (*mub2*)	—	1730	ds-DNA	—
H. vulgare	BLYPAPI	Amylase/protease inhibitor (PAPI)	Aleurone cells	650	ss-mRNA	—
H. vulgare	BLYTHIOL	Region homologous to thiol protease	—	3333	ds-DNA	E
H. vulgare	BLYUBIQR	Ubiquitin polyprecursor	Leaves	703	ss-mRNA	E
L. usitatissimum	LINUBQ	Ubiquitin	Leaves	916	ds-DNA	—
L. esculentum	TOMCD4A	ATP-dependent protease	—	5429	ds-DNA	—
L. esculentum	TOMLTITP	Low-temperature-induced thiol protease	Mature green fruit pericarp	1353	ss-mRNA	—

Gene Index

Table 15. Continued

Species	Identifier	Sequence specification	Tissue/site	Size (bp or b)	Type	S
L. esculentum	TOMMCPI	Metallocarboxypeptidase inhibitor	—	454	ss-mRNA	E
L. esculentum	TOMPROI	Protease inhibitor I-like	Fruit	132	ss-mRNA	—
L. esculentum	TOMUBI3	Ubiquitin	Hydroponic root	2374	ds-DNA	E
L. esculentum	TOMWIPI	Leaf wound-induced protease inhibitor I	Leaves	511	ss-mRNA	—
L. esculentum	TOMWIPII	Leaf wound-induced protease inhibitor II	Leaves	660	ss-mRNA	—
L. esculentum	TOMWIPIG	Wound-inducible protease inhibitor I	Leaves	2506	ds-DNA	—
L. peruvianum	TOMPII	Protease inhibitor I	Fruit	480	ss-mRNA	—
L. peruvianum	TOMTRYINHI	Protease inhibitor I	—	2650	ds-DNA	—
L. polyphyllus	LUPUBIQ	Polyubiquitin	Leaves	550	ss-mRNA	E
M. sativa	ALFPDIISOM	Protein disulfide isomerase	—	1781	ss-mRNA	E
M. sativa	ALFPKSTA	CDC2MS serine/threonine protein kinase	—	998	ds-DNA	—
N. tabacum	TOBUBICEP	Ubiquitin carboxyl extension protein	—	219	ds-DNA	E
O. hookeri	OBEPKCIH	Protein kinase C inhibitor homolog	—	1017	ss-mRNA	E
O. sativa	RICCDC28	cdc2 + /CDC28-related protein kinase	Suspension culture	1764	ss-mRNA	E
O. sativa	RICLNOCI	Oryzacystatin-II	—	645	ss-mRNA	—
O. sativa	RICOCS	Oryzacystatin gene	—	1937	ds-DNA	—
O. sativa	RICOZA	Oryzain α (EC 3.4.22)	Seed	1929	ss-mRNA	—
O. sativa	RICOZB	Oryzain β (EC 3.4.22)	Seed	1675	ss-mRNA	—
O. sativa	RICOZC	Oryzain γ (EC 3.4.22)	Seed	1468	ss-mRNA	—
O. sativa	RICG11A	G11A protein	Dark-grown seedlings	1593	ss-mRNA	—
O. sativa	RICCPI	Protease inhibitor (oryzacystatin)	Seed	598	ss-mRNA	—
P. sativum	PEAP34KR	p34 kinase-related protein (*cdc2* homolog)	Etiolated epicotyl hooks	453	ss-mRNA	E
P. sativum	PEAPSPK1	Putative protein serine/threonine kinase	Seedlings	372	ss-mRNA	—
P. sativum	PEAPSPK2	Putative protein serine/threonine kinase	Seedlings	297	ss-mRNA	—
P. sativum	PEAPSPK3	Putative protein serine/threonine kinase	Seedlings	258	ss-mRNA	—
P. sativum	PEAPSPK4	Putative protein serine/threonine kinase	Seedling	243	ss-mRNA	—
P. sativum	PEAPSPK5	Putative protein serine/threonine kinase	Seedling	210	ss-mRNA	—
P. vulgaris	PHVENDO	Endopeptidase	Seedling	3652	ds-DNA	E
P. vulgaris	PHVEPC1	Endopeptidase (EP-C1)	Pod	1221	ss-mRNA	E
P. vulgaris	PHVPVPK	PVPK-1 protein	—	2224	ss-mRNA	—
P. sativum	PEASUMPUI	Polyubiquitin	—	1241	ss-mRNA	E
P. sativum	PSP34PK	p34 protein kinase	Shoot tip	444	RNA	E
Populus sp.	POPPIA	Protease inhibitor mRNA	Leaves	258	ss-mRNA	—
Populus sp.	POPPIB	Protease inhibitor mRNA	Leaves	142	ss-mRNA	—
Populus sp.	POPPIC	Protease inhibitor	Leaves	231	ss-mRNA	—
S. cereale	RYERKIN1	RKIN1	Endosperm	1809	ss-mRNA	—
S. tuberosum	POTAPI	Inhibitor of aspartic proteinase	Tuber	828	ss-mRNA	E
S. tuberosum	POTIIKA	Inhibitor II-chloramphenicol acetyltransferase gene (IIK)	—	1241	ds-DNA	—
S. tuberosum	POTPI2G	Protease inhibitor II	—	1914	ds-DNA	E
S. tuberosum	POTINHWI	Wound-inducible protease inhibitor I	—	2170	ds-DNA	—
S. tuberosum	POTPIIIR1	Protease inhibitor II	—	512	ss-mRNA	—

Table 15. Continued

Species	Identifier	Sequence specification	Tissue/site	Size (bp or b)	Type	S
S. tuberosum	POTPIIIR2	Protease inhibitor II	—	554	ss-mRNA	E
S. tuberosum	POTPPIIIK	Protease inhibitor IIK	—	945	ds-DNA	—
S. tuberosum	POT322R	Tuber protein (p322)	Tuber	470	ss-mRNA	E
S. tuberosum	POT340R	Tuber protein p340/p34021	Tuber	786	ss-mRNA	E
S. tuberosum	POTTRPI	Homology to Bowman–Birk trypsin inhibitor	Tuber	470	ss-mRNA	—
S. tuberosum	POTTRPIA1	Homology to Kunitz trypsin inhibitor	Tuber	276	ss-mRNA	—
S. tuberosum	POTTRPIA2	Homology to Kunitz trypsin inhibitor	Tuber	178	ss-mRNA	—
S. tuberosum	POTTRPIB	Kunitz trypsin inhibitor	Tuber	561	ss-mRNA	—
S. tuberosum	POTTRYPIN	Trypsin inhibitor	Tuber	796	ss-mRNA	E
S. tuberosum	POTUQ10	Ubiquitin monomer/ribosomal protein	Tuber disc	688	ss-mRNA	E
S. tuberosum	POTUQ23	Ubiquitin	Tuber disc	281	ss-mRNA	E
S. tuberosum	POTUQ25	Ubiquitin	Tuber disc	186	ss-mRNA	E
S. tuberosum	POTUQ93	Ubiquitin	Tuber disc	230	ss-mRNA	E
S. tuberosum	POTUQ95	Ubiquitin	Tuber disc	123	ss-mRNA	—
S. tuberosum	POTUQ163	Ubiquitin	Tuber disc	167	ss-mRNA	E
S. tuberosum	POTUQ165	Ubiquitin	Tuber disc	236	ss-mRNA	E
S. oleracea	SPIPKCIH	Protein kinase C inhibitor homolog	—	930	ss-mRNA	E
T. cacao	TC21KD	21 kDa seed protein, homologous to soybean trypsin inhibitor	Seed	930	RNA	E
T. aestivum	WHTCM1	Subunit CM1 of α-amylase tetrameric inhibitor	Endosperm	620	ss-mRNA	E
T. aestivum	WHTCM3	Subunit CM3 of α-amylase tetrameric inhibitor	Endosperm	696	ss-mRNA	E
T. aestivum	WHTCM16	Subunit CM16 of α-amylase tetrameric inhibitor	Endosperm	597	ss-mRNA	E
T. aestivum	WHTCM16MR	Chloroform/methanol-soluble (CM16) protein	Seed	617	ss-mRNA	E
T. aestivum	WHTE1A	Ubiquitin activating enzyme (E1)	Seedling	3695	ss-mRNA	—
T. aestivum	WHTE1B	Ubiquitin activating enzyme (E1)	Seedling	3588	ss-mRNA	—
T. aestivum	WHTMYR1	Putative amylase or protease inhibitor	Aleurone cells	650	ss-mRNA	E
T. aestivum	WHTUBA1	Ubiquitin activating enzyme E1 (UBA1)	—	3586	ss-mRNA	—
T. aestivum	WHTUBICP	Ubiquitin carrier protein	Seedling	802	ss-mRNA	—
T. aestivum	WHTUBIQU	Ubiquitin	—	436	ss-mRNA	E
T. aestivum	WHTUCP1A	Ubiquitin carrier protein (UBC1)	Etiolated seedling	773	ss-mRNA	—
T. durum	WHTCM17	Chloroform/methanol-soluble (CM16) protein α-amylase inhibitor	Seed	623	ss-mRNA	E
T. durum	WHTCM2MR	Chloroform/methanol-soluble (CM2) protein	Seed	617	ss-mRNA	E
T. durum	WHTCM3MR	CM3 protein	Seed	707	ss-mRNA	E
T. vulgare	WHTUBE2	Ubiquitin carrier protein	Seedling	965	ss-mRNA	—
V. mungo	VIRSHEP	Sulfhydryl-endopeptidase	Cotyledon	1314	ss-mRNA	—
V. mungo	VIRSHEPG	Sulfhydryl-endopeptidase	Leaves	4236	ds-DNA	E
V. unguiculata	VIRTRYINH	Trypsin inhibitor	—	377	ss-mRNA	E
V. unguiculata	VIRTRYPIFV	Trypsin inhibitor fIV	—	583	ss-mRNA	E
Z. mays	MZEACK2	Casein kinase II α subunit	—	1169	ss-mRNA	E
Z. mays	MZEB70A	Polypeptide chain-binding protein	Kernel	1628	ss-mRNA	—
Z. mays	MZECHHSP60	Mitochondrial chaperonin hsp60	—	2138	ss-mRNA	E
Z. mays	MZEKINAA	Protein cdc2 kinase	Seedling	1325	ss-mRNA	—
Z. mays	MZESMDPK	Putative protein kinase	—	1663	ss-mRNA	—
Z. mays	MZEUBFA	Ubiquitin fusion protein (UBF9)	—	1624	ds-DNA	—
Z. mays	MZEZMPP1	Protein phosphatase-1	—	1644	ss-mRNA	—

Table 16. Structural RNA (ribosomal RNAs, ribosomal proteins, transfer RNAs, small nuclear ribonucleoproteins (snRNPs), splicing proteins, RNA binding proteins)

Species	Identifier	Sequence specification	Tissue/site	Size (bp or b)	Type	S
A. bicornis	AEG5SDNAA	5S DNA	–	410	ds-DNA	E
A. caudata	AEG5SDNAK	5S DNA	–	411	ds-DNA	E
A. comosa	AEG5SDNAB	5S DNA	–	409	ds-DNA	E
A. longissima	AEG5SDNAL	5S DNA	–	411	ds-DNA	E
A. searsii	AEG5SDNAE	5S DNA	–	489	ds-DNA	E
A. searsii	AEG5SDNAF	5S DNA	–	410	ds-DNA	E
A. shoronensis	AEG5SDNAI	5S DNA	–	489	ds-DNA	E
A. shoronensi	AEG5SDNAJ	5S DNA	–	411	ds-DNA	E
A. speltoides	AEG5SDNAG	5S DNA	–	484	ds-DNA	E
A. tauschii	AEGRRN5	5S rDNA	–	421	ds-DNA	–
A. tauschii	AEG5SDNAC	5S DNA	–	491	ds-DNA	E
A. tauschii	AEG5SDNAD	5S DNA	–	420	ds-DNA	E
A. umbellutata	AEG5SDNAM	5S DNA	–	489	ds-DNA	E
A. umbellutata	AEG5SDNAN	5S DNA	–	411	ds-DNA	E
A. uniaristata	AEG5SDNAH	5S DNA	–	425	ds-DNA	E
A. cristatum	AGR5SDNAB	5S DNA	–	483	ds-DNA	E
A. glutinosa	AG18SRDN	18S rRNA	Leaves	1698	DNA	E
A. muticum	AMU5SDNAA	5S DNA	–	412	ds-DNA	E
A. vulgaris	AQI18SRNA	18S rRNA	Leaves	1115	ds-DNA	–
A. thaliana	ATH3RRNA	5.8S rRNA and 25S rRNA with 18S rRNA	Leaves	4310	ds-DNA	E
A. thaliana	ATH7SLRNA	7SL RNA	–	1881	ds-DNA	E
B. oleracea	BNAISR	25S and 18S ribosomal RNAs	–	4124	ds-DNA	E
B. disti-chophylla	BUC18SRRN	18S ribosomal RNA	–	1767	ss-mRNA	E
B. sempervirens	BUX18SRRN	18S ribosomal RNA	–	1747	ss-mRNA	E
C. floridus	CLY186RNA	18S rRNA	Leaves	1066	ds-DNA	E
C. limon	CLIRN26S3	26S ribosomal RNA gene	–	1475	ds-DNA	–
C. limon	CLIRRN26	26S ribosomal RNA	–	3393	ds-DNA	–
C. rehderiana	CMT18SRNA	18S rRNA	Leaves	1365	ds-DNA	E
C. americana	CONRRN16S	16S rRNA	Fruit	1978	ds-DNA	E
C. americana	CONRRN455	4.5S rRNA and 5S rRNA genes	Fruit	429	ds-DNA	E
C. costata	COS16S	16S rRNA	–	1799	ds-DNA	E
D. villosum	DAY5SDNAA	5S DNA	–	470	ds-DNA	E
D. carota	DARRDDER1	rDNA repeating unit	–	1787	ds-DNA	E
D. carota	DARRDDER2	rDNA repeating unit	–	1331	ds-DNA	E
D. domingensis	DED18SRRN	18S ribosomal RNA	–	1767	ss-mRNA	E
E. kokanica	EPK58RRNA	5.8S ribosomal RNA	–	161	ss-mRNA	E
E. alatus	EUO18SRRN	18S ribosomal RNA	–	1757	ss-mRNA	E
F. arundinacea	FES26SRNA	26S rRNA	Mycelial culture	307	ds-DNA	E
F. ananassa	FRRRNASPA	Intergenic spacer between 26S and 18S rRNA genes	–	4274	ds-DNA	E
F. ananassa	FRR18S	18S ribosomal rRNA	–	1804	ds-DNA	E
F. ananassa	FRR58S	5.8S ribosomal rRNA	–	620	ds-DNA	E
F. ananassa	FRR26SRN	26S rRNA gene	–	3377	ds-DNA	E
G. cymosium	GET18SRNA	18S rRNA	Leaves	1015	ds-DNA	E
G. max	SOYRGE	18S ribosomal RNA	–	1807	ds-DNA	–
G. max	SOYRN45SI	4.5–5S rRNA intergenic region	–	472	ds-DNA	–
G. max	SOYRPS11A	Protein S11 mRNA	Leaves	677	ss-mRNA	–
G. max	SOY5SRNAG	5S rRNA	–	556	ds-DNA	E
G. max	SOYTGDA	Putative *Asp*-tRNA gene	–	570	ds-DNA	–
G. max	SOYTGMAA	*Met*-tRNA gene	–	580	ds-DNA	–
G. max	SOYTGMI	*Met-i*-tRNA gene	–	372	ds-DNA	–
G. max	SOYU1A	U1a small nuclear	–	529	ds-DNA	E

Table 16. Continued

Species	Identifier	Sequence specification	Tissue/site	Size (bp or b)	Type	S
G. max	SOYU1B	U1B small nuclear RNA	—	532	ds-DNA	E
H. helix	HED18SRRN	18S ribosomal RNA	—	1759	ss-mRNA	E
H. bogdanii	BLY5SDNAC	5S DNA	—	489	ds-DNA	E
H. bulbosum	BLY5SDNAB	5S DNA	—	475	ds-DNA	E
H. bulbosum	BLY18SRP	18S ribosomal DNA	—	1230	ds-DNA	E
H. vulgare	BLYRG5SRN	5S rRNA gene	—	300	ds-DNA	—
H. vulgare	BLYRNPL171	Ribosomal protein L17-1	Leaves	785	ss-mRNA	E
H. vulgare	BLYRNPL172	Ribosomal protein L17-2	Leaves	854	ss-mRNA	E
H. vulgare	BLY5SDNAA	5S DNA	—	301	ds-DNA	E
H. vulgare	BLY58SRDNX	5.8S ribosomal RNA	—	598	ds-DNA	E
H. vulgare	BLYTRNPHE	Transfer RNA-*Phe* (GAA)	Embryo	76	ss-mRNA	E
H. vulgare	HVTRNPHE	Transfer RNA-*Phe* (GAA)	Embryo	76	RNA	—
H. sibthorpioides	HYC18SRRN	18S ribosomal RNA	—	1735	ss-mRNA	E
L. sativa	LAU5SRRNG	5S rRNA	Seed	119	ss-mRNA	E
L. rosea	LPA18SRNA	18S rRNA	Leaves	1016	ds-DNA	E
L. minor	LGI5SRRN	5S rRNA	—	119	ss-mRNA	—
L. usitatissimum	LINRGAD	5S ribosomal RNA gene short repeat unit D	—	341	ds-DNA	—
L. usitatissimum	LINRGAC	5S ribosomal RNA gene short repeat unit C	—	341	ds-DNA	—
L. usitatissimum	LINRGAB	5S ribosomal RNA gene short repeat unit B	—	341	ds-DNA	—
L. usitatissimum	LINRGAA	5S ribosomal RNA gene short repeat unit A	—	341	ds-DNA	—
L. usitatissimum	LINRGA	5S ribosomal RNA gene long repeat unit	—	362	ds-DNA	E
L. heterophylla	LHERRN5S	5S ribosomal RNA	—	119	ss-mRNA	E
L. albus	LUPLA5SRNA	5S rRNA	—	121	ss-mRNA	E
L. albus	LUPTRNTYR	tRNA-Tyr	—	76	ss-mRNA	—
L. angustifolius	LUP18SRNA	18S rRNA	Leaves	915	ds-DNA	E
L. luteus	LUPRGA1	5S rRNA gene	—	342	ds-DNA	—
L. luteus	LUPRGA2	5S rRNA gene	—	342	ds-DNA	—
L. luteus	LUPRGB	5.8S ribosomal RNA	—	346	ds-DNA	—
L. luteus	LUP58SRRNA	5.8S ribosomal RNA	Seeds	164	ss-mRNA	E
L. luteus	LUP5S18SRR	5S and 18S ribosomal RNA	Seedling	2896	ds-DNA	E
L. luteus	LUPTRNN	RNAs tRNA-*Asn(1)* and tRNA-*Asn(2)*	Seeds	76	ds-DNA	E
L. polyphyllus	LUPRPS16	rps16 ribosomal protein S16	Undifferentiated suspension	588	ss-mRNA	E
L. esculentum	LEPT62	Ribosomal RNA	Fruit pericarp	268	RNA	E
L. esculentum	TOMRDNAIS	Ribosomal DNA intergenic spacer	—	3253	ds-DNA	E
L. esculentum	TOMRPL2MR	Ribosomal protein L2	Shoot apex	816	ss-mRNA	E
L. esculentum	TOMSNRU11	U1 small nuclear RNA gene U1.1	—	397	ds-DNA	E
L. esculentum	TOMSNRU12	U1 small nuclear RNA gene U1.2	—	370	ds-DNA	E
L. esculentum	TOMSNRU13	U1 small nuclear RNA gene U1.3	—	368	ds-DNA	E
L. esculentum	TOMSNRU14	U1 small nuclear RNA gene U1.4	—	388	ds-DNA	E
L. esculentum	TOMSNRU15	U1 small nuclear RNA gene U1.5	—	385	ds-DNA	E
L. esculentum	TOMSNRU16	U1 small nuclear RNA gene U1.6	—	396	ds-DNA	E
L. esculentum	TOMSNRU17	U1 small nuclear RNA gene U1.7	—	394	ds-DNA	E
L. esculentum	TOMSNRU18	U1 small nuclear RNA gene U1.8	—	393	ds-DNA	E
L. esculentum	TOM25SRIB	25S ribosomal RNA gene	Leaves	3381	ds-DNA	—
L. esculentum	TOM7SRNA	7S RNA for signal recognition particle	Leaves	302	ss-mRNA	E
L. esculentum	TOM5SRRN	5S RNA	—	120	ss-mRNA	—

Table 16. Continued

Species	Identifier	Sequence specification	Tissue/site	Size (bp or b)	Type	S
L. esculentum	TOM17S25S	17S–25S ribosomal DNA spacer	—	1800	ds-DNA	—
L. esculentum	TOM17S25T	17S–25S ribosomal DNA spacer	—	1131	ds-DNA	E
L. esculentum	TOMU11PS	U1 small nuclear RNA pseudogene U1.1ps	—	350	ds-DNA	E
L. esculentum	TOMU6SNR	U6 small nuclear RNA gene	—	539	ds-DNA	E
L. esculentum	TOMU3SNRN	U3 small nuclear RNA gene	—	1526	ds-DNA	E
M. ternifolia	MCD18SRNA	18S rRNA	Leaves	1365	ds-DNA	E
M. cobus	MAIS5SRNA	5S rRNA	—	119	ss-mRNA	E
M. polymorpha	MPORRN5S	5S ribosomal RNA	—	119	ss-mRNA	E
M. quadrifolia	MAQRN58S	5.8S rRNA	—	155	ss-mRNA	E
M. incana	MINRGA	5S rRNA gene	—	509	ds-DNA	—
M. sativa	ALFC29	snRNP-related protein	—	376	ss-mRNA	E
M. sativa	ALFRRN26S	26S rRNA	—	1617	ss-mRNA	E
N. plumbaginifolia	TOBRBP30	RNA binding protein 30	Leaves	1246	ss-mRNA	E
N. plumbaginifolia	TOBRBP31	RNA binding protein 31	Leaves	1228	ss-mRNA	E
N. tabacum	TOBNTLEU	tRNA-Leu (UAG)	—	278	ds-DNA	E
N. tabacum	TOBRL2RNA	60S ribosomal protein L2	—	917	ss-mRNA	E
N. tabacum	TOBRNP33	Plastid 33 kDa ribonucleoprotein	Leaves	1356	ss-mRNA	E
N. tabacum	TOBRPL12	Ribosomal protein L12	Leaves	773	ss-mRNA	E
N. tabacum	TOBRPL121A	Ribosomal protein L12-1a	Leaves	825	ss-mRNA	E
N. plumbaginifolia	TOB16S	16S rRNA gene	Leaves	1490	ds-DNA	—
N. tabacum	TOBTRNYX	tRNA-Tyr-1	—	400	ds-DNA	E
N. rustica	TOBTGNYP	Transfer RNA-*Tyr* pseudogene	—	383	ds-DNA	E
N. rustica	TOBTRNY	Cytoplasmic tRNA-*Tyr* gene	—	210	ds-DNA	E
N. rustica	TOBTY1	Cytoplasmic tRNA-*Tyr* gene	Leaves	410	ds-DNA	E
N. rustica	TOBTY3	Cytoplasmic tRNA-*Tyr* gene	Leaves	445	ds-DNA	E
N. rustica	TOBTY4	Cytoplasmic tRNAC-*Tyr* gene	Leaves	488	ds-DNA	E
N. rustica	TOBTY5	Cytoplasmic tRNA-*Tyr* gene	Leaves	412	ds-DNA	E
N. rustica	TOBTY6	Cytoplasmic tRNA-*Tyr* gene	Leaves	393	ds-DNA	E
N. rustica	TOBTY7	Cytoplasmic tRNA-*Tyr* gene	Leaves	359	ds-DNA	E
N. rustica	TOBTY8	Cytoplasmic tRNA-*Tyr* gene	Leaves	435	ds-DNA	E
N. rustica	TOBTY9	Cytoplasmic tRNA-*Tyr* gene	Leaves	419	ds-DNA	E
N. rustica	TOBTY10	Cytoplasmic tRNA-*Tyr* gene	Leaves	491	ds-DNA	E
N. rustica	TOBTY11	Cytoplasmic tRNA-*Tyr* gene	Leaves	458	ds-DNA	E
N. rustica	TOBTY12	Cytoplasmic tRNA-*Tyr* gene	Leaves	404	ds-DNA	E
N. rustica	TOBTY13	Cytoplasmic tRNA-*Tyr* gene	Leaves	452	ds-DNA	E
N. rustica	TOBTY2P	Cytoplasmic tRNA-*Tyr* pseudogene	Leaves	386	ds-DNA	E
N. sylvatica	NYS18SRRN	18S ribosomal RNA	—	1751	ss-mRNA	E
N. sylvestris	TOBCHRIB	31 kDa chloroplast ribonucleoprotein	—	5040	ds-DNA	E
N. sylvestris	TOBCPRNP29	Chloroplast 29 kDa A ribonucleoprotein	Young leaves	1155	ss-mRNA	E
N. sylvestris	TOBRNP28	Plastid 28 kDa ribonucleoprotein	Leaves	1079	ss-mRNA	E
N. sylvestris	TOBRNP31	Plastid 31 kDa ribonucleoprotein	Leaves	1201	ss-mRNA	E
N. tabacum	TOBCPRNP33	Unknown chloroplast ribonucleoprotein	Young leaves	1250	ss-mRNA	E
O. australiensis	RIC5SDNAK	5S DNA	—	302	ds-DNA	E
O. brachyantha	RIC5SDNAA	5S DNA	—	490	ds-DNA	E
O. brachyantha	RIC5SDNAB	5S DNA	—	390	ds-DNA	E
O. latifolia	RIC5SDNAC	5S DNA	—	305	ds-DNA	E
O. minuta	RIC5SDNAD	5S DNA	—	240	ds-DNA	E
O. nivarra	RIC5SDNAE	5S DNA	—	336	ds-DNA	E
O. officinalis	RIC5SDNAF	5S DNA	—	333	ds-DNA	E

Table 16. Continued

Species	Identifier	Sequence specification	Tissue/site	Size (bp or b)	Type	S
O. punctata	RIC5SDNAG	5S DNA	—	405	ds-DNA	E
O. punctata	RIC5SDNAH	5S DNA	—	333	ds-DNA	E
O. rufipogon	RIC5SDNAI	5S DNA	—	322	ds-DNA	E
O. sativa	RICGICR	Gly-tRNA gene	—	524	ds-DNA	—
O. sativa	RICINTER	25S and 17S rDNA	—	2300	ds-DNA	E
O. sativa	RICRGAB	5S ribosomal RNA gene	Embryo	119	ds-DNA	—
O. sativa	RICRGAC	5S ribosomal RNA gene	Germinated embryo	303	ds-DNA	—
O. sativa	RICRGC1	Rice 17S rRNA gene	Germinated embryo	145	ds-DNA	—
O. sativa	RICRGC2	Rice 5.8S rRNA	Germinated embryo	260	ds-DNA	—
O. sativa	RICRGC3	Rice 25S rRNA gene	Germinated embryo	91	ds-DNA	—
O. sativa	RICRGC4	Rice 25S rRNA gene	Germinated embryo	161	ds-DNA	—
O. sativa	RICRGE	Rice 17S ribosomal RNA gene	—	1812	ds-DNA	—
O. sativa	RICRGH	Rice 25S ribosomal RNA gene	—	3377	ds-DNA	—
O. sativa	RICRGSBHA	Rice gene encoding three ribosomal RNAs: the 17S, 3′ end; 5.8S, complete; 25S, 5′ end	—	1044	ds-DNA	—
O. sativa	RIC5SDNAJ	5S DNA	—	302	ds-DNA	E
O. sativa	RIC5SRRNMR	5S ribosomal RNA	Aleurone layer	1114	ss-mRNA	E
O. sativa	RICTRNAG	tRNA-Gly(GCC) gene	Embryo	524	ds-DNA	E
O. sativa	RICVCN2	Variable copy number DNA	—	1456	ds-DNA	E
O. regalis	OSM58SRRN	5.8S rRNA	—	156	ss-mRNA	E
P. hybrida	PETRGA	5S ribosomal RNA	—	463	ds-DNA	—
P. hybrida	PETRN5S2	5S ribosomal RNA	Leaves	461	ds-DNA	E
P. hybrida	PETRN5S3	5S ribosomal RNA	Leaves	461	ds-DNA	E
Petunia sp.	PETTGN	*Asn*-tRNA-*aac*	—	249	ds-DNA	—
P. rubens	PIARDNA	18S ribosomal gene	Needle	404	ds-DNA	—
P. sativum	PEACL22	Ribosomal protein (CL22)	Leaves	2179	ds-DNA	—
P. sativum	PEACL22A	Ribosomal protein (CL22)	Seedling	861	ss-mRNA	—
P. sativum	PEACPRPS17	Ribosomal protein S17	Leaves	614	ss-mRNA	—
P. sativum	PEAGSNU2	U2 snRNA	—	528	ds-DNA	E
P. sativum	PEARDNAS	Large spacer of ribosomal DNA	Embryo	2981	ds-DNA	E
P. sativum	PEARPCL9	Plastid ribosomal protein CL9	Seedling	901	ss-mRNA	E
P. sativum	PEARPCL18	Plastid ribosomal protein	Seedling	792	ss-mRNA	E
P. sativum	PEARPCL24	Plastid ribosomal protein	Seedling	727	ss-mRNA	E
P. sativum	PEARPCL25	Plastid ribosomal protein	Seedling	427	ss-mRNA	E
P. sativum	PEARRNPSS	18S (partial) 25S (partial)	—	3876	ds-DNA	E
P. sativum	PEASNU1A	U1 snRNA	—	162	ss-mRNA	E
P. sativum	PEASNU4A	U4 snRNA	—	152	ss-mRNA	E
P. sativum	PEASNU5A	U5 snRNA	—	122	ss-mRNA	E
P. sativum	PEASNU1B	U1 snRNA	—	160	ss-mRNA	E
P. sativum	PEASNU4B	U4 snRNA	—	152	ss-mRNA	E
P. sativum	PEASNU1C	U1 snRNA	—	160	ss-mRNA	E
P. sativum	PEASNU4C	U4 snRNA	—	150	ss-mRNA	E
P. sativum	PEASNU2B	U2 snRNA	—	195	ss-mRNA	E
P. sativum	PEASNU5B	U5 snRNA	—	122	ss-mRNA	E
P. sativum	PEATRNF	Transfer RNA-*Phe* (G3 AA)	—	76	ss-mRNA	E
P. sativum	PEAU2SNRN	U2 snRNA	—	194	ss-mRNA	E
P. sativum	PSRRNPSS	18S (partial) and 25S (partial) rRNAs	—	3876	DNA	E
P. sativum	PSTRNF	Transfer RNA-*Phe*	—	76	RNA	—
P. vulgaris	PHVDSRNA	Enigmatic double stranded RNA	Leaves	630	ss-mRNA	E
P. vulgaris	PHVTGL	Transfer RNA-Leu	—	184	ds-DNA	—

Table 16. Continued

Species	Identifier	Sequence specification	Tissue/site	Size (bp or b)	Type	S
P. vulgaris	PHVTGP	*Pro*-tRNA (*TGG*)	—	172	ds-DNA	—
P. vulgaris	PHVTPA	*Pro*-tRNA pseudogene	—	134	ds-DNA	—
P. vulgaris	PHVTRNPR1	tRNA-*Pro* (UGG1) gene cluster	—	3505	ds-DNA	E
P. vulgaris	PHVTRNPR2	Transfer RNA-*Pro*	—	998	ds-DNA	E
P. vulgaris	PHVTRNPR3	tRNA-*Pro* (UGG3)	—	1040	ds-DNA	E
P. vulgaris	PHVUG1	U1 small nuclear RNA	—	605	ds-DNA	—
P. trichomanes	PTRRRN5S	5S ribosomal RNA	—	119	ss-mRNA	E
P. falcatus	PDO18SRNA	18S rRNA	Leaves	1215	ds-DNA	E
P. henkellii	PDO18SRNB	18S ribosomal RNA	Leaves	1215	ds-DNA	E
P. serotinum	PON18SRRN	18S ribosomal RNA	—	1769	ss-mRNA	E
P. deltoides	POP58SRDNA	18S rRNA	—	705	ds-DNA	E
P. repens	PRZ18SRNA	18S rRNA	Leaves	1013	ds-DNA	E
P. juncea	PSJ5SDNAA	5S DNA	—	448	ds-DNA	E
R. sativus	RADRRNAES	25S (partial) and 18S (partial) ribosomal RNA	—	2902	ds-DNA	E
R. sativus	RADRREQ	18S ribosomal RNA	—	236	ss-RNA	—
S. arenaria	SHO18SRRN	18S ribosomal RNA	—	1762	ss-mRNA	E
S. cereale	RYENORR1	26S rRNA 3′ end and 18S rRNA, 5′ end	—	4507	ds-DNA	—
S. cereale	RYE5SDNAA	5S DNA	—	485	ds-DNA	E
S. cereale	RYERRN5A	5S ribosomal RNA	—	118	ss-mRNA	E
S. cereale	RYE5SDNAB	5S DNA	—	462	ds-DNA	E
S. montanum	RYE5SDNAC	5S DNA	—	482	ds-DNA	E
S. montanum	RYE5SDNAD	5S DNA	—	461	ds-DNA	E
S. sylvestre	RYE5SDNAE	Host 5S DNA	—	483	ds-DNA	E
S. sylvestre	RYE5SDNAF	Host 5S DNA	—	459	ds-DNA	E
S. vavilovii	RYE5SDNAG	5S DNA	—	485	ds-DNA	E
S. vavilovii	RYE5SDNAH	5S DNA	—	461	ds-DNA	E
S. montanum	RYE58SRDNX	5.8S ribosomal RNA	—	601	ds-DNA	E
S. alba	SALNARRNA	18S and 25S rRNA genes	Bud	5500	ds-DNA	E
S. alba	SALRRN18	18S–25S rRNA spacer DNA	—	922	ds-DNA	E
S. alba	SAL25SR	25S rRNA	—	3381	ds-DNA	E
S. alba	SAL5SRRNA	5S rRNA gene	Bud	491	ds-DNA	E
S. irio	SSYRDREP	Ribosomal intergenic spacer, *Bam*HI repetitive seq.	—	433	ds-DNA	E
S. aviculare	POT18SRNA	18S rRNA	Leaves	1265	ds-DNA	E
S. melongena	POT18SRNB	18S rRNA	Leaves	1265	ds-DNA	E
S. petrophilum	POT18SRNC	18S rRNA	Leaves	1265	ds-DNA	E
S. tuberosum	POTCARSEQ	Cara sequence	Leaves	195	ds-DNA	—
S. tuberosum	POTSNRNAA	U2 small nuclear RNA	Leaves	195	ds-DNA	—
S. tuberosum	POTSNRNAB	U2 small nuclear RNA	Leaves	195	ds-DNA	—
S. tuberosum	POTSNRNAC	U2 small nuclear RNA	Leaves	196	ds-DNA	—
S. tuberosum	POTSNRNAD	U2 small nuclear RNA	Leaves	196	ds-DNA	—
S. tuberosum	POTSNRNAE	U2 small nuclear RNA	Leaves	196	ds-DNA	—
S. tuberosum	POTSTTRNK	Maturase, trnK for tRNA-Lys (UUU), rps16 for ribosomal protein S16	Leaves	3654	ds-DNA	E
S. tuberosum	POTU2B	Spliceosomal protein (U2B)	Tuber	921	ds-DNA	—
S. tuberosum	POTU211BG	*U2-11B* gene for U2 snRNA	—	871	ds-DNA	E
S. tuberosum	POTU24G	*U2-4* gene for U2 snRNA	—	1177	ds-DNA	E
S. tuberosum	POTU211G	*U2-11* gene for U2 snRNA	—	589	ds-DNA	E
S. tuberosum	POTU222G	*U2-22* gene for U2 snRNA	—	680	ds-DNA	E
S. tuberosum	POTU6SNRN	U6 small nuclear RNA	—	1288	ss-mRNA	E
S. tuberosum	POTU211UR	*U2-11* gene upstream region	—	432	ds-DNA	E
S. tuberosum	STU6SNRN	U6 small nuclear RNA	—	1288	RNA	E
S. bicolor	SRGTRNG	*tRNA-Gly* gene	—	160	ds-DNA	—

Table 16. Continued

Species	Identifier	Sequence specification	Tissue/site	Size (bp or b)	Type	S
S. oleracea	SPIL12A	Chloroplast ribosomal protein L12	Seedling leaf	886	ss-mRNA	—
S. oleracea	SPIRPAA	Ribosomal protein 30S subunit	—	1212	ss-mRNA	—
S. oleracea	SPIRPL11	Chloroplast ribosomal protein L11	Seedling	836	ss-mRNA	E
S. oleracea	SPIRPL13	Ribosomal protein L13	Seedling	982	ss-mRNA	—
S. oleracea	SPIRPL21	Ribosomal protein L21	—	1060	ss-mRNA	—
S. oleracea	SPIRPL22	Chloroplast rp121	—	1072	ss-mRNA	E
S. oleracea	SPIRPL24	Ribosomal protein rp124	—	795	ss-mRNA	—
S. oleracea	SPIRPL35A	Ribosomal protein L35	Seedling leaf	740	ss-mRNA	—
S. oleracea	SPIRPL40	Ribosomal protein rp140	—	688	ss-mRNA	—
S. oleracea	SPISORPS22	Chloroplast ribosomal protein S22	Leaves	1185	ss-mRNA	E
T. purpurescens	TAI18SRNA	18S rRNA	Leaves	1365	ds-DNA	E
T. aestivum	WHTRDN1	26S–18S rDNA non-transcribed spacer region 1	—	233	ds-DNA	E
T. aestivum	WHTRDN2	26S–18S rDNA non-transcribed spacer region 2	—	965	ds-DNA	E
T. aestivum	WHTRGEG	18S ribosomal RNA; 25S ribosomal RNA	—	4642	ds-DNA	—
T. aestivum	WHTRGNOR	Nor-D3 locus ribosomal RNA gene	—	3673	ds-DNA	—
T. aestivum	WHTRNA1	5S ribosomal RNA	—	120	ss-mRNA	E
T. aestivum	WHT5SDNAA	5S DNA	—	414	ds-DNA	E
T. aestivum	WHT5SDNAB	5S DNA	—	414	ds-DNA	E
T. aestivum	WHT5SDNAC	5S DNA	—	340	ds-DNA	E
T. aestivum	WHT5SDNAD	5S DNA	—	328	ds-DNA	E
T. aestivum	WHT5SDNAE	5S DNA	—	419	ds-DNA	E
T. aestivum	WHT5SDNAF	5S DNA	—	416	ds-DNA	E
T. aestivum	WHT5SDNAG	5S DNA	—	439	ds-DNA	E
T. aestivum	WHT5SDNAH	5S DNA	—	398	ds-DNA	E
T. aestivum	WHT5SDNAI	5S DNA	—	262	ds-DNA	E
T. aestivum	WHT5SDNAJ	5S DNA	—	488	ds-DNA	E
T. aestivum	WHT5SDNAK	5S DNA	—	490	ds-DNA	E
T. aestivum	WHT5SDNAL	5S DNA	—	485	ds-DNA	E
T. aestivum	WHT5SDNAM	5S DNA	—	480	ds-DNA	E
T. aestivum	WHT5SDNAN	5S DNA	—	485	ds-DNA	E
T. aestivum	WHT5SDNAO	5S DNA	—	409	ds-DNA	E
T. aestivum	WHT5SDNAP	5S DNA	—	340	ds-DNA	E
T. aestivum	WHT5SDNAQ	5S DNA	—	328	ds-DNA	E
T. aestivum	WHT5SDNAR	5S DNA	—	488	ds-DNA	E
T. aestivum	WHT5SDNAS	5S DNA	—	416	ds-DNA	E
T. aestivum	WHT5SDNAT	5S DNA	—	485	ds-DNA	E
T. aestivum	WHT5SDNAU	5S DNA	—	409	ds-DNA	E
T. aestivum	WHT58SRDNX	5.8S ribosomal RNA	—	602	ds-DNA	E
T. aestivum	WHT5SRRNL	5S DNA	—	500	ds-DNA	—
T. aestivum	WHT5SRRNS	5S DNA	Embryo	408	ds-DNA	—
T. aestivum	WHTTAR6B	Tar6B rRNA gene intergenic region	Seedling	393	ds-DNA	E
T. aestivum	WHTTGNS	tRNA-*Ser*(UGA)	Germ	501	ds-DNA	E
T. aestivum	WHTTRNA	Cytoplasmic initiator methionine tRNA	Germ	75	ss-mRNA	E
T. aestivum	WHTTRNTYR	tRNA-*Tyr* gene	Seedling	226	ds-DNA	E
T. aestivum	WHTTRTY1	tRNA-*Tyr*	Seedling	210	ds-DNA	E
T. aestivum	WHTTRTY2	tRNA-*Tyr*	Seedling	210	ds-DNA	E
T. aestivum	WHTTTARRNG	*rrn26* gene for rRNA large subunit (26S)	—	7242	ds-DNA	E
T. aestivum	WHTU3SNRA	U3 snRNA gene	—	1286	ds-DNA	E
T. aestivum	WHTU6SNRA	U6 snRNA gene	—	644	ds-DNA	E
T. monococcum	WHT5SDNAV	5S DNA	—	482	ds-DNA	E
T. monococcum	WHT5SDNAW	5S DNA	—	324	ds-DNA	E

Gene Index

Table 16. Continued

Species	Identifier	Sequence specification	Tissue/site	Size (bp or b)	Type	S
T. monococcum	WHT5SRRN	5S rRNA	—	327	ds-DNA	E
Triticum sp.	WHTRDNA	Ribosomal RNA spacer DNA	—	134	ds-DNA	—
Triticum sp.	WHTRDNAX	Ribosomal RNA spacer DNA	—	140	ds-DNA	—
T. urartu	WHTRDNASPA	External transcribed spacer in rDNA	—	1157	ds-DNA	E
T. vulgare	WHTTRNS	tRNA-*Ser*	Germ	2530	ds-DNA	E
V. angulari	VIRRDREP	174 bp subrepeats of rDNA spacer	—	174	ds-DNA	E
V. angustifolia	VFARRNINT	25S/18S rRNA intergenic spacer DNA	—	2721	ds-DNA	E
V. faba	VFARDNAA	Ribosomal spacer DNA repeat	—	1234	ds-DNA	—
V. faba	VFARDNAB	Ribosomal spacer DNA repeat	—	149	ds-DNA	—
V. faba	VFARDNAS	Large spacer of ribosomal DNA (rDNA)	Embryo	3490	ds-DNA	E
V. faba	VFARGHERP	25S–18S rDNA intergenic region subrepeating element	—	325	ds-DNA	—
V. faba	VFASN45LR	Small nuclear 4.5S (I)-like RNA	—	94	ss-mRNA	E
V. faba	VFASNU6R	U6 small nuclear RNA	—	98	ss-mRNA	E
V. faba	VFA5SRRN	5S rRNA	—	119	ss-mRNA	—
V. faba	VFASSDNA	Subrepeating sequence related to those of ribosomal intergenic spacer	—	336	ds-DNA	E
V. faba	VFAU4A	U4A small nuclear RNA	—	154	ss-mRNA	E
V. faba	VFAU4B	U4B small nuclear RNA	—	152	ss-mRNA	E
V. faba	VFAU2RNA	U2 small nuclear RNA	—	193	ss-mRNA	E
V. hirsuta	VFARDNA	rDNA intergenic spacer region	—	2910	ds-DNA	E
V. radiata	VIRI18S25	18S–25S rDNA spacer region	—	933	ds-DNA	E
V. radiata	VIREA	18S ribosomal RNA gene	—	2845	ds-DNA	—
V. radiata	VIRRGA	5S rRNA gene	—	215	ds-DNA	—
V. radiata	VIRRRN25	25S rRNA–18S rDNA spacer region	—	5270	ds-DNA	E
Z. pumila	ZAMRGEA	Small-subunit (18S) ribosomal RNA gene	—	1813	ds-DNA	—
Z. mays	MZEETS1	26S–17S rDNA spacer region	—	3131	ds-DNA	E
Z. mays	MZERG17S	17S ribosomal RNA gene	Leaves	2215	ds-DNA	—
Z. mays	MZERNETS1	Spacer DNA	—	3199	ds-DNA	E
Z. mays	MZERPS11C	Cytoplasmic ribosomal protein S11	Seedling	789	ss-mRNA	E
Z. mays	MZEU6RNA	U6 small nuclear RNA (snRNA)	—	105	ds-DNA	E
Z. mays	MZEU2SNRN	U2 small nuclear RNA (U2snRNA)	—	694	ds-DNA	E

Table 17. Specific ligand-binding proteins (co-factors, vitamins, growth regulators)

Species	Identifier	Sequence specification	Tissue/site	Size (bp or b)	Type	S
L. esculentum	TOMBIOBR	Biotin-binding protein	—	383	ss-mRNA	E
N. tabacum	TOBBLP1	Luminal binding protein (BiP)	—	1110	ss-mRNA	E
N. tabacum	TOBBLP2	Luminal binding protein (BiP)	—	1085	ss-mRNA	E
N. tabacum	TOBBLP3	Luminal binding protein (BiP)	—	807	ss-mRNA	E
N. tabacum	TOBBLP4	Luminal binding protein (BiP)	—	2345	ss-mRNA	E
N. tabacum	TOBBLP5	Luminal binding protein (BiP)	—	2273	ss-mRNA	E
N. tabacum	TOBBLP8	Luminal binding protein (BiP)	—	1067	ss-mRNA	E
O. sativa	RICRGP1	rgp 1 (ras-related GTP-binding) protein)	Seedling	1303	ss-mRNA	E
Z. mays	MZEABP	Putative auxin binding protein	Coleoptile	791	ss-mRNA	E
Z. mays	MZEABP1	Auxin-binding protein	Coleoptile	770	ss-mRNA	E
Z. mays	MZEAUX	Auxin-binding protein	Shoot	812	ss-mRNA	—
Z. mays	MZEAUX311	Auxin-binding protein	Shoot	5337	ds-DNA	E

Table 18. Storage proteins (seed storage proteins, tuber proteins, abundant plant proteins, other seed proteins. Not lectins (see *Table 9*); see also *Table 5*, development and tissue-specific development, and *Table 15*, enzyme inhibition)

Species	Identifier	Sequence specification	Tissue/site	Size (bp or b)	Type	S
A. sativa	ASTAVE	Avenin	Endosperm	758	ss-mRNA	—
A. sativa	ASTAVEA	Avenin	Seed	725	ss-mRNA	—
A. sativa	ASTAVEB	Avenin	Seed	889	ss-mRNA	—
A. sativa	ASTAVEC	Avenin	Seed	875	ss-mRNA	—
A. sativa	ASTGLOB	12S globulin	Seed	1042	ss-mRNA	—
A. sativa	AST12SSP	12S seed storage protein	Leaves	2387	ds-DNA	E
A. sativa	ASTSSPA	Seed storage protein	Seed	1548	ds-DNA	—
A. sativa	ASTSSP12S	12S globulin	Seed	3341	ds-DNA	—
A. sativa	ASTSPGLBA	Seed storage globulin	Seed	1805	ss-mRNA	—
B. excelsa	BRN2S1	2S albumin	Seed	991	ds-DNA	E
B. excelsa	BRN2SG1	2S seed storage protein	Seed	441	ds-DNA	E
B. excelsa	BRN2SG2	2S seed storage protein	Seed	530	ds-DNA	E
B. excelsa	BRNALB2S1	2S albumin	Seed	563	ss-mRNA	—
B. excelsa	BRNALB2S2	2S albumin	Seed	590	ss-mRNA	—
B. excelsa	BRNWSP	Sulfur-rich water-soluble seed protein (9 kDa subunit)	Nut	621	ss-mRNA	—
B. napus	BNACRUA	Cruciferin (*cruA*)	Seed	3113	ds-DNA	E
B. napus	BNACRUC	Cruciferin	Seed	1624	ss-mRNA	—
B. napus	BNACRUCA	Cruciferin	Seed	1686	ss-mRNA	—
B. napus	BNANAP	Napin mRNA	Seed	718	ss-mRNA	—
B. napus	BNANAP1	Napin	Leaves	1731	ds-DNA	E
B. napus	BNANAPA	1.7S seed storage protein napin	—	3289	ds-DNA	—
B. napus	BNANAPB	Napin	Seed	1257	ds-DNA	E
B. napus	BNANAPIN	Napin	Seed	1993	ds-DNA	E
B. napus	BNASSPA	Napin	Embryo	566	ss-mRNA	—
B. napus	BNASSPB	Napin	Embryo	713	ss-mRNA	—
B. napus	BNASSPC	(1.7S embryo-specific) storage protein gene	—	1026	ds-DNA	—
C. gladiata	CAGCANAV	Canavalin	Seed	1534	ss-mRNA	E
C. gladiata	CAGCANAVA	Canavalin	Seedling	3263	ds-DNA	E
C. gladiata	CAGCANVAP	Canavalin pseudogene	Seedling	1241	ds-DNA	E
C. ensiformis	CENCANAV	Canavalin	Cotyledon	1403	ss-mRNA	E
C. lachryma-jobi	CIXJACOIX	α-coixin 17 kDa	Seed	1019	ss-mRNA	E
G. max	SOYABAB	Glycinin A(1a) B(1b) and A(2) B(1a) boundary DNA	Seed	3210	ds-DNA	—
G. max	SOYASBC	α subunit of β-conglycinin	Immature cotyledons	1955	ss-mRNA	E
G. max	SOYBCONGL	α subunit of β-conglycinin	Coytledon	794	ss-mRNA	—
G. max	SOYBG7SG	7S globulin	Seed	1499	ss-mRNA	E
G. max	SOYBPSP	α' type β-conglycinin storage protein gene	Seed	3636	ds-DNA	—
G. max	SOYGA2B1A	Glycinin A2B1a subunit	Cotyledons	1708	ss-mRNA	—
G. max	SOYGLY	Glycinin A2B1a subunit	Leaves	1607	ds-DNA	—
G. max	SOYGLYA1A	Glycinin subunits A-1a and Bx precursor	Seed	1746	ss-mRNA	E
G. max	SOYGLYAAB	Glycinin A5-A4-B3 subunit	Seed	1899	ss-mRNA	E
G. max	SOYGLYAB	Glycinin seed storage protein	Cotyledons	1786	ss-mRNA	—
G. max	SOYGLYB4	Glycinin A3-B4 subunit	Seed	911	ss-mRNA	—
G. max	SOYGLYBSU	Glycinin A-1a-B-x subunit	Cotyledons	1743	ss-mRNA	—
G. max	SOYGLYCAB	Glycinin A2B1a subunit	Cotyledons	3176	ds-DNA	—
G. max	SOYGLYMAX	Max *vspA* gene	Leaves	3346	ds-DNA	—
G. max	SOYGLY4MU	Glycinin	—	3543	ds-DNA	E
G. max	SOYGLYR	Glycinin subunit A2B1a	Cotyledons	1709	ss-mRNA	—

Table 18. Continued

Species	Identifier	Sequence specification	Tissue/site	Size (bp or b)	Type	S
G. max	SOYGLY4R	Glycinin	—	622	ss-mRNA	E
G. max	SOYGY1	Glycinin subunit	Leaves	3527	ds-DNA	E
G. max	SOYGY2	Glycinin subunit G2	Leaves	4617	ds-DNA	E
G. max	SOYGY3	Glycinin subunit G3	Leaves	3573	ds-DNA	E
G. max	SOY7SAA	7S seed storage protein α and α'	Seed	1035	ss-mRNA	—
G. max	SOY7SAB	7S seed storage protein α subunit	Seed	530	ss-mRNA	—
G. max	SOY7SAC	7S seed storage protein α subunit	Seed	936	ss-mRNA	—
G. max	SOY7SAP	7S seed storage protein α'	Seed	1480	ds-DNA	—
G. max	SOY7SAPA	7S seed storage protein α'	Seed	1047	ss-mRNA	—
G. max	SOY7SP60	7S-like *in vitro* seed storage protein (60 kDa/53 kDa)	Seed	671	ss-mRNA	—
G. max	SOY7SP68	7S-like *in vitro* seed storage protein (68 kDa)	Seed	853	ss-mRNA	—
G. max	SOYVSPA	Vegetative storage protein	Leaves	942	ss-mRNA	—
G. max	SOYVSPB	Vegetative storage protein	Leaves	974	ss-mRNA	—
G. max	SOYVSPBA	Vegetative storage protein (*vspB*)	Leaves	3841	ds-DNA	—
G. hirsutum	COTASP	2S albumin storage protein	Cotyledon	570	ss-mRNA	—
G. hirsutum	COTDGALA	D-genome alloallele gene	Cotyledon	5795	ds-DNA	—
H. annuus	HNN5FHAG3A	Helianthinin	Seed	2376	ds-DNA	E
H. annuus	HNN5FHAG3D	Helianthinin	Seed	2457	ds-DNA	E
H. annuus	HNNG5ALB2	2S albumin storage protein	—	2299	ds-DNA	E
H. annuus	HNNSF8	Seed albumin	Seed	600	ss-mRNA	E
H. vulgare	BLYBHO31	β-hordein (pcr31)	Seed	1255	ds-DNA	E
H. vulgare	BLYBHO47	β-hordein (pcr47)	Seedling	1226	ds-DNA	E
H. vulgare	BLYB1HOR	β1-hordein	Endosperm	152	ss-mRNA	—
H. vulgare	BLYB1HOR1	β1-hordein	—	552	ss-mRNA	E
H. vulgare	BLYB1HORD	β-hordein	Endosperm	868	ss-mRNA	E
H. vulgare	BLYB3HORD	β3-hordein	Endosperm	954	ss-mRNA	E
H. vulgare	BLYCHORD1	C-hordein	Endosperm	475	ss-mRNA	E
H. vulgare	BLYCHORD2	C-hordein	Endosperm	310	ss-mRNA	—
H. vulgare	BLYCHORDA	C-hordein storage protein	Endosperm	391	ss-mRNA	—
H. vulgare	BLYCHORDB	C-hordein storage protein	Endosperm	255	ss-mRNA	—
H. vulgare	BLYG1HORDA	γ 1-hordein storage protein	Leaves	1614	ds-DNA	—
H. vulgare	BLYGHRDSP	γ-hordein storage protein	Leaves	1614	ds-DNA	E
H. vulgare	BLYHOR117	C-hordein	Endosperm	1420	ds-DNA	E
H. vulgare	BLYHORB	β 1-hordein	Endosperm	2900	ds-DNA	—
H. vulgare	BLYHORDCA	C-hordein	Immature endosperm	2065	ds-DNA	—
H. vulgare	BLYMRNA	Seed protein	Endosperm, aleurone, root, shoot, leaf	2384	ss-mRNA	E
H. vulgare	BLYPAZ1	Protein Z	—	3133	ds-DNA	E
H. vulgare	BLYPROTZ	Protein Z	—	703	ss-mRNA	E
H. vulgare	BLYPROZ	Protein Z	Endosperm	75	ss-mRNA	—
I. batatas	IPBAGSP	Sporamin	Parenchyma	840	ss-mRNA	E
I. batatas	IPBGSPOA1	Sporamin A	Tuber	2490	ds-DNA	E
I. batatas	IPBGSPOB1	Sporamin B	Tuber	2352	ds-DNA	E
I. batatas	IPBIMO355	Sporamin A	Tuber	904	ss-mRNA	E
I. batatas	IPBIMO535	Sporamin B	Tuber	854	ss-mRNA	E
I. batatas	IPBSPORB	Sporamin B	Tuber parenchyma	954	ss-mRNA	—
I. batatas	IPBSSPORA	Sporamin A	Tuber	889	ss-mRNA	—
L. angustifolius	LUPCONGLD	Conglutin δ	—	684	ss-mRNA	E
O. officinalis	RICCONS	Glutelin promoter	—	449	ds-DNA	—
O. sativa	RICGLU	Glutelin	—	913	ss-mRNA	E
O. sativa	RICGLUA3	Glutelin	Immature seeds	2936	ds-DNA	E

Table 18. Continued

Species	Identifier	Sequence specification	Tissue/site	Size (bp or b)	Type	S
O. sativa	RICGLUB1	Glutelin	Immature seeds	3373	ds-DNA	E
O. sativa	RICGLUB2	Glutelin	Immature seeds	3024	ds-DNA	E
O. sativa	RICGLUB3	GluB-3 gene for glutelin	Immature seeds	2142	ds-DNA	E
O. sativa	RICGLUI1	Glutelin type I	—	1617	ss-mRNA	E
O. sativa	RICGLUI2	Glutelin type I	—	1635	ss-mRNA	E
O. sativa	RICGLUII1	Glutelin type II	—	1654	ss-mRNA	E
O. sativa	RICGLUII2	Glutelin type II	—	1753	ss-mRNA	E
O. sativa	RICGLUT	Glutelin	Seed	1634	ss-mRNA	E
O. sativa	RICGLUT21	Preproglutelin	Seed	1632	ss-mRNA	E
O. sativa	RICGLUTA	Rice glutelin subunit	—	1640	ss-mRNA	—
O. sativa	RICGLUTE	Prepro-glutelin gene	—	5494	ds-DNA	—
O. sativa	RICGLUTEC	Glutelin gene	—	716	ds-DNA	—
O. sativa	RICGLUTEL	Glutelin	Seed	1637	ss-mRNA	E
O. sativa	RICGLUTG	Glutelin	Embryo	3435	ds-DNA	E
O. sativa	RICPRL	13 kDa prolamin	Seed	603	ss-mRNA	E
O. sativa	RICPRO10	10 kDa prolamin	—	1546	ds-DNA	E
O. sativa	RICPROL	Prolamine	Endosperm	769	ss-mRNA	E
O. sativa	RICPROLA1	Prolamin	Endosperm	609	ss-mRNA	E
O. sativa	RICPROLA2	Prolamin	Endosperm	596	ss-mRNA	E
O. rufipogon	RICPROLA3	Prolamin	—	332	ds-DNA	E
O. rufipogon	RICPROLA4	Prolamin	—	343	ds-DNA	E
O. rufipogon	RICPROLA5	Prolamin	—	339	ds-DNA	E
O. rufipogon	RICPROLA6	Prolamin	—	344	ds-DNA	E
O. rufipogon	RICPROLA7	Prolamin	—	328	ds-DNA	E
O. rufipogon	RICPROLA8	Prolamin	—	330	ds-DNA	E
O. rufipogon	RICPROLA9	Prolamin	—	343	ds-DNA	E
O. rufipogon	RICPROLAA	Prolamin	—	348	ds-DNA	E
O. longistaminata	RICPROLAB	Prolamin	—	349	ds-DNA	E
O. sativa	RICPROLAMI	10 kDa prolamin	Seed	562	ss-mRNA	E
O. sativa	RICR22CDNA	R22 mRNA	Seed	629	ss-mRNA	E
P. somniferum	PAPMLPG	Opium poppy major latex protein	Latex	2210	ds-DNA	—
P. somniferum	PAPMLPR	Major latex protein	Latex	774	ss-mRNA	E
P. somniferum	PAPMLPY	Major latex protein	Latex	401	ss-mRNA	—
P. somniferum	PAPMLPZ	Major latex protein	Latex	463	ss-mRNA	—
P. sativum	PEA5FRLEG1	Legumin A1 gene	—	583	ds-DNA	E
P. sativum	PEA5FRLEG2	Legumin A2 gene	—	476	ds-DNA	E
P. sativum	PEAABN1	Albumin 1	Seed	949	ds-DNA	—
P. sativum	PEAABN2	Albumin 2	Seed	957	ss-mRNA	—
P. sativum	PEAABN1M	Albumin 1	Seed	609	ss-mRNA	—
P. sativum	PEAALBMAJ	Major seed albumin	Seed	957	ss-mRNA	—
P. sativum	PEAALBUMI1	Albumin 1	Seed	2148	ds-DNA	—
P. sativum	PEACONVR1	Convicilin	—	591	ss-mRNA	E
P. sativum	PEACVCA	Convicilin	—	3736	ds-DNA	E
P. sativum	PEACVCB	Convicilin	—	2086	ds-DNA	—
P. sativum	PEAIVSS	Legumin J	—	350	ds-DNA	—
P. sativum	PEALEGA	Legumin	Seed	2310	ds-DNA	E
P. sativum	PEALEGA2	Legumin	Seed	4156	ds-DNA	E
P. sativum	PEALEGAA	Legumin	—	613	ss-mRNA	—
P. sativum	PEALEGAB	Legumin	—	137	ss-mRNA	—
P. sativum	PEALEGAG	Legumin	Seed	3347	ds-DNA	E
P. sativum	PEALEGB5	Legumin	Seed	1404	ds-DNA	E
P. sativum	PEALEGBB	Legumin	Seed	1274	ss-mRNA	—
P. sativum	PEALEGBC	Legumin	Seed	910	ss-mRNA	—

Table 18. Continued

Species	Identifier	Sequence specification	Tissue/site	Size (bp or b)	Type	S
P. sativum	PEALEGC5	Legumin	Seed	1671	ds-DNA	E
P. sativum	PEALEGJ	*legJ* gene for minor legumin	–	2976	ds-DNA	E
P. sativum	PEALEGK	*legK* gene for minor legumin	–	1543	ds-DNA	E
P. sativum	PEALEGPS	Legumin pseudogene	–	2506	ds-DNA	E
P. sativum	PEALGN	Legumin subunit	Cotyledon	787	ss-mRNA	–
P. sativum	PEAUSP43	Unknown seed protein	Seed	811	ds-DNA	E
P. sativum	PEAVIC	Vicilin	–	1433	ss-mRNA	–
P. sativum	PEAVIC2	Vicilin	–	907	ss-mRNA	–
P. sativum	PEAVIC4	Vicilin	–	212	ss-mRNA	–
P. sativum	PEAVIC7	Vicilin	–	1086	ss-mRNA	–
P. sativum	PEAVICA	Vicilin	–	81	ss-mRNA	–
P. sativum	PEAVICIL	Vicilin	–	190	ds-DNA	–
P. sativum	PEAVICIM	Vicilin	Seed	4239	ds-DNA	E
P. sativum	PSUSP20	Unknown seed protein	Embryo	664	RNA	E
P. vulgaris	PHVARC	Arcelin-1 seed protein	Seed	948	ss-mRNA	–
P. vulgaris	PHVARCA	Arcelin 2	Seed	902	ss-mRNA	–
P. vulgaris	PHVARC1A	Arcelin	Leaves, seed	4564	ds-DNA	–
P. vulgaris	PHVBCSP	β-type phaseolin storage protein	Seed	2921	ds-DNA	–
P. vulgaris	PHVPHAS	Phaseolin, g1-globin gene	–	895	ds-DNA	E
P. vulgaris	PHVPHASAR	α-type phaseolin	Seed	1535	ss-mRNA	E
P. vulgaris	PHVPHASBR	β-type phaseolin	Seed	1475	ss-mRNA	E
P. vulgaris	PHVAPHASE	α-phaseolin	Seed	4764	ds-DNA	E
P. vulgaris	PHVPHAT	Phaseolin	–	895	ds-DNA	E
P. glauca	PIALEGSP	Legumin-like storage protein	Embryo	1738	ss-mRNA	E
P. glauca	PIA2SSP	2S-like storage protein	Somatic embryo	712	ss-mRNA	E
P. glauca	PIAVICSP	Vicilin-like storage protein	Somatic embryo	1590	ss-mRNA	E
P. strobus	PINGLBN1	Pine globulin-1	Prothallium	1956	ss-mRNA	E
P. strobus	PINGLBN2	Pine globulin-2	Prothallium	1744	ss-mRNA	E
P. deltoides	POPBSP	Bark storage protein	Bark	1153	ss-mRNA	–
R. sativus	RADNAPA	Storage protein napin	Immature seed	623	ss-mRNA	–
R. sativus	RADNAPB	Napin	Immature seed	580	ss-mRNA	–
R. sativus	RADNAPC	Napin	Immature seed	653	ss-mRNA	–
R. sativus	RADPAAA	Seed protein	Seed	447	ss-mRNA	–
R. communis	RCC2SALBG	2S albumin	Ripening endosperm	1591	ds-DNA	E
S. cereale	RYESEC1A	*Sec1* gene for σ secalin	Endosperm	1342	ds-DNA	E
S. cereale	RYESEC1B	*Sec1* gene for σ secalin	Endosperm	1832	ds-DNA	E
S. cereale	RYESECGSR	S-rich γ-secalin-related storage protein	Seed	714	ss-mRNA	E
S. tuberosum	POTB33UPRG	*B33* gene upstream region	Tuber	1975	ds-DNA	E
S. tuberosum	POTPAT	Patatin	Tuber	1437	ss-mRNA	–
S. tuberosum	POTPAT1	Patatin	Tuber	544	ds-DNA	–
S. tuberosum	POTPAT2	Patatin	Tuber	279	ds-DNA	–
S. tuberosum	POTPATA	Patatin class I	Tuber	4029	ds-DNA	–
S. tuberosum	POTPATA1	Class II patatin	Tuber	297	ds-DNA	E
S. tuberosum	POTPATA2	Class II patatin	Tuber	474	ds-DNA	E
S. tuberosum	POTPATA3	Class II patatin	Tuber	456	ds-DNA	E
S. tuberosum	POTPATA4	Class II patatin	Tuber	465	ds-DNA	E
S. tuberosum	POTPATAA	Patatin	Tuber	464	ds-DNA	–
S. tuberosum	POTPATAB	Patatin gene	Tuber	454	ds-DNA	–
S. tuberosum	POTPATAC	Patatin	Tuber	409	ds-DNA	–
S. tuberosum	POTPATAD	Patatin	Tuber	287	ds-DNA	–
S. tuberosum	POTPATAE	Patatin	Tuber	461	ds-DNA	–
S. tuberosum	POTPATAF	Patatin class II	Tuber	89	ds-DNA	–

Table 18. Continued

Species	Identifier	Sequence specification	Tissue/site	Size (bp or b)	Type	S
S. tuberosum	POTPATAG	Patatin class II	Tuber	99	ds-DNA	—
S. tuberosum	POTPATAH	Patatin	Tuber	83	ds-DNA	—
S. tuberosum	POTPATATA	Patatin	Tuber	1378	ss-mRNA	—
S. tuberosum	POTPATATB	Patatin	Tuber	1404	ss-mRNA	—
S. tuberosum	POTPATB	Patatin class I	Tuber	1138	ds-DNA	—
S. tuberosum	POTPATB1	Patatin B1	Tuber	1354	ss-mRNA	E
S. tuberosum	POTPATB2	Patatin B2	Tuber	1378	ss-mRNA	E
S. tuberosum	POTPATC	Patatin class I	Tuber	1139	ds-DNA	—
S. tuberosum	POTPATD	Patatin class I	Tuber	708	ds-DNA	—
S. tuberosum	POTPATE	Patatin class II	Tuber	747	ds-DNA	—
S. tuberosum	POTPATF	Patatin class II	Tuber	738	ds-DNA	—
S. tuberosum	POTPATG	Patatin	Tuber	5313	ds-DNA	E
S. tuberosum	POTPATG1	Patatin	Tuber	6094	ds-DNA	E
S. tuberosum	POTPATGA	Patatin class II	Tuber	747	ds-DNA	—
S. tuberosum	POTPATGK	Patatin	Tuber	217	ss-mRNA	—
S. tuberosum	POTPATH	Patatin class II	—	433	ds-DNA	—
S. tuberosum	POTPATP1	Patatin	Tuber	6799	ds-DNA	E
S. tuberosum	POTPATP2	Patatin pseudogene	—	7373	ds-DNA	E
S. tuberosum	POTPB24	Patatin	Tuber	1407	ds-DNA	E
S. tuberosum	POTPGT2	5′ upstream region and first exon of PGT2 gene	—	1356	ds-DNA	E
S. tuberosum	POTPGT3TR	Patatin	Tuber	1458	ds-DNA	E
S. tuberosum	POTPPGT4	Patatin	Tuber	1611	ds-DNA	E
S. tuberosum	POTPPGT12	Patatin	Tuber	1180	ds-DNA	E
S. tuberosum	POTPPGT16	Patatin	Tuber	1391	ds-DNA	E
S. bicolor	SRGENSPMRN	Endosperm tissue mRNA	Endosperm	848	ss-mRNA	—
S. bicolor	SRGKAFGK1	Kafirin	Seed	2124	ds-DNA	E
S. bicolor	SRGKAFGK4	Kafirin	Seed	2203	ds-DNA	E
S. bicolor	SRGKAFSK2	Kafirin	Seed	955	ss-mRNA	E
S. bicolor	SRGKAFSK8	Kafirin	Seed	894	ss-mRNA	E
T. cacao	THBCSV	Seed vicilin	Leaves	3318	ds-DNA	E
T. cacao	THBCSVSV	Seed vicilin	Cotyledon	1380	ss-mRNA	E
T. aestivum	WHT1026	High mol. wt. glutenin subunit	Endosperm	2996	ds-DNA	E
T. aestivum	WHTABGLIA	α-/β-gliadin MM1	Endosperm	994	ss-mRNA	E
T. aestivum	WHTAGCNN16	α-gliadin pseudogene	—	3581	ds-DNA	E
T. aestivum	WHTAGCNN17	α-gliadin pseudogene	—	3566	ds-DNA	E
T. aestivum	WHTAGCNN35	α-gliadin pseudogene	—	3573	ds-DNA	E
T. aestivum	WHTCM18	CM 17 protein	—	654	ss-mRNA	E
T. aestivum	WHTGLA11A	High mol. wt glutenin subunit	Embryo	3077	ds-DNA	E
T. aestivum	WHTGL1AX1	High mol. wt glutenin subunit 1A× 1	Endosperm	2885	ds-DNA	E
T. aestivum	WHTGLB11B	High mol. wt glutenin subunit	Embryo	2999	ds-DNA	E
T. aestivum	WHTGL1D1B	High mol. wt glutenin subunit 5	—	3156	ds-DNA	E
T. aestivum	WHTGLIA	α-type gliadin gene	—	1152	ss-mRNA	—
T. aestivum	WHTGBL1A	Storage protein (Gbl1)	—	4747	ds-DNA	—
T. aestivum	WHTGLG	Storage protein (γ-gliadin)	Endosperm	818	ss-mRNA	—
T. aestivum	WHTGLGAP	γ-gliadin pseudogene	Seed	2450	ds-DNA	—
T. aestivum	WHTGLGB	γ-gliadin gene	Seed	2450	ds-DNA	—
T. aestivum	WHTGLIAA	Storage protein gliadin α-1Y	Seed	2347	ds-DNA	E
T. aestivum	WHTGGLN	γ-gliadin	—	2086	ds-DNA	—
T. aestivum	WHTGLD12B	High mol. wt glutenin subunit 10	Embryo	2556	ds-DNA	—
T. aestivum	WHTGLIABA	α-/β-gliadin class A-II	Seed	1102	ss-mRNA	—
T. aestivum	WHTGLIABB	α-/β-gliadin	—	3043	ds-DNA	—
T. aestivum	WHTGLIABC	α-/β-gliadin class A-V	Seed	1156	ss-mRNA	—
T. aestivum	WHTGLIABD	α-/β-gliadin	Seed	3310	ds-DNA	—
T. aestivum	WHTGLIABE	α-/β-gliadin class I gene	Seed	3022	ss-mRNA	—
T. aestivum	WHTGLIABF	α-/β-gliadin class A-I	Seed	950	ss-mRNA	—

Table 18. Continued

Species	Identifier	Sequence specification	Tissue/site	Size (bp or b)	Type	S
T. aestivum	WHTGLIABG	α-/β-gliadin	Seed	1126	ss-mRNA	—
T. aestivum	WHTGLIABH	α-/β-gliadin	Seed	1039	ss-mRNA	—
T. aestivum	WHTGLIAG1	α-/β-gliadin	—	1679	ds-DNA	E
T. aestivum	WHTGLIAG2	α-/β-gliadin storage protein	—	1672	ds-DNA	E
T. aestivum	WHTGLIAG3	α-/β-gliadin storage protein	—	1753	ds-DNA	E
T. aestivum	WHTGLIB	α-type gliadin gene	—	725	ss-mRNA	—
T. aestivum	WHTGLIC	Gliadin	—	3022	ss-mRNA	E
T. aestivum	WHTGLIGBA	γ-gliadin class B-I	Seed	1348	ss-mRNA	—
T. aestivum	WHTGLIGBB	γ-gliadin class B-I	Seed	1130	ss-mRNA	—
T. aestivum	WHTGLIGBC	γ-gliadin class B-III	Seed	927	ss-mRNA	—
T. aestivum	WHTGLIGP	γ-gliadin	Seed	798	ss-mRNA	—
T. aestivum	WHTGLIGR1	Seed storage protein	Seed	684	ss-mRNA	E
T. aestivum	WHTGLIGY	γ-gliadin	Seed	1397	ds-DNA	—
T. aestivum	WHTGLU1D1	Low mol. wt glutenin-1D1 gene	Seed	3165	ds-DNA	E
T. aestivum	WHTGLU1DG	High mol. wt glutenin subunit from chromosome 1D	Seed	3095	ds-DNA	E
T. aestivum	WHTGLUMPS	High mol. wt glutenin subunit pseudogene	Seed	2915	ds-DNA	E
T. aestivum	WHTGLUMRA	High-M_r gluten polypeptide	Endosperm	434	ss-mRNA	E
T. aestivum	WHTGLUMRB	High-M_r gluten polypeptide	Endosperm	204	ss-mRNA	E
T. aestivum	WHTGLUT	Low mol. wt glutenin	Endosperm	1827	ds-DNA	E
T. aestivum	WHTGLUT1	High mol. wt glutenin subunit	Seed	3193	ds-DNA	E
T. aestivum	WHTGLUTA	High mol. wt glutenin subunit	Seed	3077	ds-DNA	—
T. aestivum	WHTGLUTB	High mol. wt glutenin subunit	Seed	2999	ds-DNA	—
T. durum	WHTGLUU	Glutenin	Seed	1102	ss-mRNA	E
T. durum	WHTLMWG	Low mol. wt glutenin	Seed	856	ds-DNA	E
T. urartu	WHTGLNA	α-/β-type gliadin	—	3179	ds-DNA	—
T. urartu	WHTGLNB	α-/β-type gliadin pseudogene	—	3179	ds-DNA	—
V. faba	VFALEA1	Legumin A1 pre-pro-polypeptide	—	1585	ss-mRNA	E
V. faba	VFALEA2	Legumin A1 pre-pro-polypeptide	—	1666	ss-mRNA	E
V. faba	VFALEB1	*VfLEB1* pseudogene	Leaves	2009	ds-DNA	E
V. faba	VFALEB2	*VfLEB2* gene for legumin storage protein	Leaves	1984	ds-DNA	E
V. faba	VFALEB4	Legumin B	Seed	2844	ds-DNA	E
V. faba	VFALEB5	*VfLEB5* pseudogene	Leaves	1381	ds-DNA	E
V. faba	VFALEB6	*VfLEB6* gene	Leaves	1905	ds-DNA	E
V. faba	VFALEB7	*VfLEB7* gene	Leaves	1631	ds-DNA	E
V. faba	VFAUSP	Unknown seed protein	Leaves	3539	ds-DNA	E
V. faba	VFAUSP13	Unknown seed protein USP	Seed	163	ss-mRNA	E
V. faba	VFAUSP46	Unknown seed protein	Seed	172	ss-mRNA	E
V. faba	VFAUSP87	USP87 mRNA for unknown seed protein	Seed	1091	ss-mRNA	E
V. faba	VFAUSP92	USP92 mRNA for unknown seed protein	Seed	1071	ss-mRNA	E
V. faba	VFAUSPA	Unknown seed protein	Seed	1071	ss-mRNA	—
V. faba	VFAUSPB	Unknown seed protein	Seed	1091	ss-mRNA	—
V. faba	VFAUSPC	Unknown seed protein	Cotyledon	1193	ss-mRNA	—
V. faba	VFAVIC	Vicilin storage protein	Seed	1561	ss-mRNA	E
V. faba	VFAVICG	Vicilin gene	Seed	3217	ds-DNA	E
Z. mays	MZE27KZNA	27 kDa zein locus	Leaves	3704	ds-DNA	E
Z. mays	MZE27KZNB	27 kDa zein	Leaves	3695	ds-DNA	E
Z. mays	MZEALBB32	Albumin b-32	Endosperm	1033	ss-mRNA	—
Z. mays	MZEAZ22KD	22 kDa α-zein	—	7343	ds-DNA	E
Z. mays	MZEEG2R	Glutelin-2 protein	Endosperm	837	ss-mRNA	E
Z. mays	MZEGLB10	Vicilin-like storage protein	—	3604	ds-DNA	E
Z. mays	MZEGLB1LG	Vicilin-like embryo storage protein	Embryo	3583	ds-DNA	E

Table 18. Continued

Species	Identifier	Sequence specification	Tissue/site	Size (bp or b)	Type	S
Z. mays	MZEGLB1SA	Globulin S allele	Embryo	2003	ss-mRNA	—
Z. mays	MZEGLB1SG	Vicilin-like embryo storage protein	Embryo	3546	ds-DNA	E
Z. mays	MZEGLOB2	Globulin-2	—	1637	ss-mRNA	E
Z. mays	MZEGLUT2E	Glutelin-2	Endosperm	1857	ds-DNA	—
Z. mays	MZEI19	α-zein	—	3391	ds-DNA	E
Z. mays	MZEPML1	Zein	Leaves	3362	ds-DNA	E
Z. mays	MZEPML2	Zein	Leaves	3360	ds-DNA	E
Z. mays	MZEPMS1G	19 kDa zein protein	—	4461	ds-DNA	E
Z. mays	MZEPMS2G	19 kDa zein protein	—	7622	ds-DNA	E
Z. mays	MZEZC1	Zein Zc1, (14 kDa zein-2)	Endosperm	3864	ds-DNA	E
Z. mays	MZEZC2ZEI	Zein Zc2 (28 kDa glutelin-2)	Endosperm	2975	ds-DNA	E
Z. mays	MZEZE01	Zein (λ-ZG99)	—	2938	ds-DNA	E
Z. mays	MZEZE02	Zein (clone PZ19.1)	—	616	ss-mRNA	E
Z. mays	MZEZE09	Zein (pZ22.1)	—	897	ss-mRNA	E
Z. mays	MZEZE10	Zein (ZG99)	—	862	ss-mRNA	E
Z. mays	MZEZE11	Zein (pZ22.3)	—	893	ss-mRNA	E
Z. mays	MZEZE12	Zein (M6)	—	931	ss-mRNA	E
Z. mays	MZEZE15A3	15 kDa zein	Endosperm	851	ss-mRNA	—
Z. mays	MZEZE15G	Sulfur-rich zein protein of M_r 15 000	—	1240	ds-DNA	—
Z. mays	MZEZE16	16 kDa zein-2	Endosperm	800	ss-mRNA	—
Z. mays	MZEZE19A	19 kDa zein	Tissue culture	2938	ds-DNA	—
Z. mays	MZEZE19A2	19 kDa zein	Endosperm	773	ss-mRNA	—
Z. mays	MZEZE19B	19 kDa zein	Endosperm	614	ss-mRNA	—
Z. mays	MZEZE19B1	19 kDa zein	Endosperm	852	ss-mRNA	—
Z. mays	MZEZE19BA	Zein mutant (zE19)	—	2626	ds-DNA	—
Z. mays	MZEZE19C1	19 kDa zein	Endosperm	866	ss-mRNA	—
Z. mays	MZEZE19C2	19 kDa zein	Endosperm	849	ss-mRNA	—
Z. mays	MZEZE19D1	19 kDa zein	Endosperm	849	ss-mRNA	—
Z. mays	MZEZE22A	22 kDa (mol. wt = 26.53 kDa) zein	Endosperm	897	ss-mRNA	—
Z. mays	MZEZE22B	22 kDa (26.99 kDa) zein	Endosperm	891	ss-mRNA	—
Z. mays	MZEZE22C	2 kDa zein	Endosperm	939	ss-mRNA	—
Z. mays	MZEZEA20M	Zein	—	921	ss-mRNA	E
Z. mays	MZEZE25A	Zein (*zE25*) pseudogene	—	1305	ds-DNA	—
Z. mays	MZEZEA30M	Zein	—	794	ss-mRNA	E
Z. mays	MZEZEAZ124	Zein	Endosperm	853	ss-mRNA	—
Z. mays	MZEZEAZC7	Zein	Endosperm	950	ss-mRNA	—
Z. mays	MZEZEB49M	Zein	—	467	ss-mRNA	E
Z. mays	MZEZEG	Zein	Kernel	909	ss-mRNA	—
Z. mays	MZEZEIAA	Zein	Endosperm	711	ss-mRNA	—
Z. mays	MZEZEIAB	Zein	Endosperm	874	ss-mRNA	—
Z. mays	MZEZEIAC	Zein	Endosperm	868	ss-mRNA	—
Z. mays	MZEZEIN	10 kDa zein	Endosperm	570	ss-mRNA	—
Z. mays	MZEZEIN27	27 kDa storage protein	Endosperm	3108	ds-DNA	E
Z. mays	MZEZEINP	Zein	Germinating cotyledon stem and leaves	2085	ds-DNA	—
Z. mays	MZEZEINPR	Zein gene promoter	Whole immature cob	1452	ds-DNA	E
Z. mays	MZEZEISP	Zein 5′ flanking region	Endosperm	215	ds-DNA	—
Z. mays	MZEZEP	Zein	Endosperm	1303	ds-DNA	—
Z. mays	MZEZEPCM1	Zein heavy chain	Unfertilized ear	943	ss-mRNA	E
Z. mays	MZEZEPR	Zein promoter region	—	300	ds-DNA	E
Z. mays	MZEZEZ4G	Zein	Leaves	1364	ds-DNA	—
Z. mays	MZEZEZA1G	Zein heavy chain	—	1218	ds-DNA	E
Z. mays	MZEZEZ7	W22 zein pseudogene	—	1587	ds-DNA	—
Z. mays	MZEZEZG3A	Zein	—	730	ss-mRNA	—

Table 19. Stress and induction (heat shock, cold-induced, drought-induced, salt-induced, pathogen-related, wound-induced, growth regulator-induced)

Species	Identifier	Sequence specification	Tissue/site	Size (bp or b)	Type	S
C. rubrum	CHEHSP	Heat shock protein	—	742	ss-mRNA	E
C. rubrum	CHERNAHSP	Heat shock protein	Photoautotropic cell culture	738	ss-mRNA	E
D. carota	DARHSP70	Heat shock protein 70	Callus suspension culture	2372	ds-DNA	E
D. carota	DCHSP70	Heat shock protein 70	Callus suspension culture	2372	DNA	—
D. carota	DCHSP175	Heat shock protein	Callus derived from hypocotyl	394	DNA	E
D. carota	DCHSP177	Heat shock protein	Callus derived from hypocotyl	1250	DNA	E
D. carota	DCHSP179	Heat shock protein	Callus derived from hypocotyl	1152	DNA	E
G. max	SOYAUI	Auxin-regulated protein	—	908	ds-DNA	—
G. max	SOYAUX22	Auxin-regulated protein	—	1945	ss-mRNA	—
G. max	SOYAUX28	Auxin-regulated protein	—	1784	ss-mRNA	—
G. max	SOYGH3G	Auxin-regulated protein	—	3569	ds-DNA	E
G. max	SOYH4CDNA	Stress-induced gene (H4)	Root	742	ss-mRNA	E
G. max	SOYHSP	Heat shock protein	—	2763	ds-DNA	—
G. max	SOYHSP22	Heat shock protein	—	696	ss-mRNA	—
G. max	SOYHSP175	Low mol. wt heat shock protein gene	—	2484	ds-DNA	—
G. max	SOYHSP176	Low mol. wt heat shock protein gene	—	1711	ds-DNA	—
G. max	SOYHSP179	Heat shock protein	—	2427	ds-DNA	E
G. max	SOYHSP185	Heat shock protein	—	1833	ds-DNA	E
G. max	SOYHSPA1	Heat shock protein 6834	—	558	ss-mRNA	—
G. max	SOYHSPA2	Heat shock gene *hs6871*	—	1536	ds-DNA	E
G. max	SOYHSPGM	Heat shock protein	—	1683	ds-DNA	—
G. max	SOYHSP22L	Low mol. wt heat shock protein	Endomembrane system	3012	ds-DNA	E
G. max	SOYSAM22	Stress-induced gene (*SAM22*)	Cultured cells derived from hypocotyl	783	ss-mRNA	E
G. hirsutum	COTDPAA	Desiccation protectant (*Lea4-D9*)	Embryo	4107	ds-DNA	—
H. vulgare	BLYABA	Abscisic acid	Aleurone layer	998	ss-mRNA	—
H. vulgare	BLYABAIP	ABA-inducible protein	Aleurone layer	997	ss-mRNA	E
H. vulgare	BLYBLT4	BLT4 mRNA	Meristem	724	ss-mRNA	E
H. vulgare	BLYBLT14	BLT14	—	514	ss-mRNA	E
H. vulgare	BLYCLDAA	Cold-regulated	—	652	ss-mRNA	—
H. vulgare	BLYCLDAB	Cold-regulated mRNA	—	770	ss-mRNA	—
H. vulgare	BLYDHN3	Dehydrin (dhn3)	Seedling	852	ss-mRNA	—
H. vulgare	BLYDHN8	Dehydrin (dhn8)	Seedling	683	ss-mRNA	—
H. vulgare	BLYDHN9	Dehydrin (dhn9)	Seedling	683	ss-mRNA	—
H. vulgare	BLYDHN17	Dehydrin (dhn17)	Seedling	812	ss-mRNA	—
H. vulgare	BLYDHN18	Dehydrin (dhn18)	Seedling	1049	ss-mRNA	—
H. vulgare	BLYIDS2	Iron-deficiency specific clone No. 2	Root	1235	ss-mRNA	—
H. vulgare	BLYIDS3	Iron-deficiency specific clone No. 3	Root	1668	ss-mRNA	—
H. vulgare	BLY1433PH	14-3-3 protein homolog	Young leaves	1361	ss-mRNA	E
H. vulgare	BLYPR1	Fungal pathogen-induced mRNA for pathogen-related protein	Leaves	862	ss-mRNA	E
H. vulgare	BLYPRP1A	Pathogenesis-related protein (Hv-1a)	Primary leaf	809	ss-mRNA	E
H. vulgare	BLYPRP1B	Pathogenesis-related protein (Hv-1b)	Primary leaf	713	ss-mRNA	E
H. vulgare	BLYPRP1C	Pathogenesis-related protein (Hv-1c)	Primary leaf	748	ss-mRNA	E

Table 19. Continued

Species	Identifier	Sequence specification	Tissue/site	Size (bp or b)	Type	S
L. polyphyllus	LUPHSP70	hsp70	Undif-ferentiated suspension	1005	ss-mRNA	E
L. esculentum	LEPT52	Wound-induced gene	Fruit pericarp	306	RNA	E
L. esculentum	LEPT53	Wound-repressed gene	Fruit pericarp	461	RNA	E
L. esculentum	LEPT58	Wound-induced gene	Fruit pericarp	192	RNA	E
L. esculentum	TOME8	Ethylene-responsive fruit ripening gene	Fruit	3042	ds-DNA	E
L. esculentum	TOMGTOMA	Ethylene-related	—	3526	ds-DNA	E
L. esculentum	TOMHSC170	Heat shock protein 70 kDa	Pistil	2120	ss-mRNA	E
L. esculentum	TOMHSC270	Heat shock protein cognate 70 kDa	Pistil	2117	ss-mRNA	E
L. esculentum	TOMHSPIR	Small heat shock protein (class I)	Pericarp	652	ss-mRNA	E
L. esculentum	TOMLE25	*LE25* (le25)	Developing seeds, drought-stressed leaves	2232	ds-DNA	—
L. esculentum	TOMNP24	NP24 protein	Root	917	ss-mRNA	—
L. esculentum	TOMPRP2	Pathogenesis-related protein P2	Leaves	651	ss-mRNA	E
L. esculentum	TOMPRP4	(Pathogenesis related) protein (P4)	—	705	ss-mRNA	—
L. esculentum	TOMPRP6	(Pathogenesis related) protein (P6)	—	763	ss-mRNA	—
L. esculentum	TOMTAS14	TAS14 mRNA inducible by abscisic acid and environmental stress	Seedling	746	ss-mRNA	E
L. peruvianum	TOMHSF24	Heat stress transcription factor HSF24	—	1530	ss-mRNA	E
M. falcata	ALFSTRESS	Environmental stress and abscisic acid inducible protein	Young leaves	679	ss-mRNA	E
M. sativa	ALFESIPA	Environmental stress-induced protein	Seedling	569	ss-mRNA	—
M. sativa	ALFESIPB	Environmental stress-induced protein	Seedling	436	ss-mRNA	—
M. sativa	ALFESIPC	Environmental stress-inducible protein	Seedling	822	ss-mRNA	—
M. sativa	ALFHSP181	Heat shock protein MsHsp18	*In vitro* callus	553	ss-mRNA	E
M. sativa	ALFHSP182	Heat shock protein MsHsp18	*In vitro* callus	655	ss-mRNA	E
N. tabacum	TOBAUX1	Auxin-induced protein	Leaves	2820	ds-DNA	E
N. tabacum	TOBAUX35	Auxin-induced protein	Leaves	2820	ds-DNA	E
N. tabacum	TOBAUX103	Auxin-induced	Cell suspension	842	ss-mRNA	E
N. tabacum	TOBAUX107	Auxin-induced mRNA	Cell suspension	941	ss-mRNA	E
N. tabacum	TOBAUX110	Auxin-induced mRNA	Cell suspension	839	ss-mRNA	E
N. tabacum	TOBAUX111	Auxin-induced mRNA	Cell suspension	874	ss-mRNA	E
N. tabacum	TOBAUX115	Auxin-induced mRNA	Cell suspension	1309	ss-mRNA	E
N. tabacum	TOBC7	mRNA C-7	Mesophyll protoplasts	951	ss-mRNA	E
N. tabacum	TOBHSP70	Heat shock protein 70	Leaves	1894	ss-mRNA	E
N. tabacum	TOBHSP82	Heat shock protein 82	—	1747	ss-mRNA	E
N. tabacum	TOBOLP	Osmotin-like protein (OLP1)	Cultured cells	1004	ss-mRNA	—
N. tabacum	TOBOSM	Osmotin	—	732	ss-mRNA	—
N. tabacum	TOBOSMOTIN	Osmotin	Cell suspension	738	ss-mRNA	E
N. tabacum	TOBPAR	Auxin-regulated	Mesophyll protoplasts	1071	ss-mRNA	—
N. tabacum	TOBPARC	Auxin-regulated gene	Mesophyll protoplasts	989	ss-mRNA	E

Table 19. Continued

Species	Identifier	Sequence specification	Tissue/site	Size (bp or b)	Type	S
N. tabacum	TOBPARG	Auxin-regulated gene	Mesophyll protoplasts	1275	ds-DNA	—
N. tabacum	TOBPR1A	PR1a protein	—	783	ss-mRNA	—
N. tabacum	TOBPR1A1	Pathogenesis-related protein 1a	Leaves	2038	ds-DNA	E
N. tabacum	TOBPR4A	Pathogenesis-related protein 4A	Leaves	686	ss-mRNA	E
N. tabacum	TOBPR1AA	Pathogenesis-related protein PR1a	—	738	ss-mRNA	—
N. tabacum	TOBPR1AG	Pathogenesis-related proteins	—	2334	ds-DNA	E
N. tabacum	TOBPR1AG1	Pathogenesis-related protein PR1a	Leaves	1261	ds-DNA	E
N. tabacum	TOBPR1AG2	Pathogenesis-related protein	—	1363	ds-DNA	E
N. tabacum	TOBPR4AX	PR-4a protein	—	645	ss-mRNA	E
N. tabacum	TOBPR1B	Pathogenesis-related protein PR1b	—	693	ss-mRNA	E
N. tabacum	TOBPR1B1	PR1b protein	—	762	ss-mRNA	—
N. tabacum	TOBPR4B	Pathogenesis-related protein 4B	Leaves	599	ss-mRNA	E
N. tabacum	TOBPR1BA	Pathogenesis-related protein 1b	Leaves	1149	ds-DNA	E
N. tabacum	TOBPR4BX	PR-4b protein	—	641	ss-mRNA	E
N. tabacum	TOBPR1C	Pathogenesis-related protein PR1c	—	663	ss-mRNA	E
N. tabacum	TOBPR1CA	Pathogenesis-related protein 1c	Leaves	1160	ds-DNA	E
N. tabacum	TOBPR1CR	PR-1c protein	—	721	ss-mRNA	E
N. tabacum	TOBPRP	Pathogenesis-related protein P	—	968	ss-mRNA	—
N. tabacum	TOBPRP1	Pathogenesis-related protein (PRP) 1	—	2256	ds-DNA	—
N. tabacum	TOBPRPR	Pathogenesis-related (PR) protein 1b	—	768	ss-mRNA	E
N. tabacum	TOBPR1PS	PR-1 related pseudogene	Leaves	873	ds-DNA	E
N. tabacum	TOBPRRMAJ	Pathogenesis-related protein R major form	Leaves	900	ss-mRNA	—
N. tabacum	TOBPRQ	Pathogenesis-related protein Q (PR-Q)	—	1005	ss-mRNA	—
N. tabacum	TOBTHAUR	TMV-induced protein homologous to thaumatin	—	845	ss-mRNA	E
N. tabacum	TOBW381	PR-1 pathogenesis-related protein	—	1551	ds-DNA	E
N. tabacum	TOBW383	Pseudogene for PR-1 pathogenesis-related protein	—	1157	ds-DNA	E
O. sativa	OSLMWHSP	16.9 kDa low mol. wt heat shock protein	Seedling	806	RNA	—
O. sativa	RICRAB21	RAB21	Seed	2537	ds-DNA	E
O. sativa	RICLMWHSP	Low mol. wt heat shock protein	Seedling	801	ss-mRNA	—
O. sativa	RICLMWHSQ	16.9 kDa low mol. wt heat shock protein	Seedling	806	ss-mRNA	E
P. crispum	PUMPR2	Pathogenesis-related protein 2	Cultured cells	800	ss-mRNA	E
P. crispum	PUMPR12	Pathogenesis-related protein type A	—	807	ss-mRNA	—
P. crispum	PUMPR13	Pathogenesis-related protein type B	—	689	ss-mRNA	—
P. crispum	PUMPR113	Pathogenesis-related protein type A	—	279	ds-DNA	E
P. crispum	PUMPR115	Pathogenesis-related protein type A	—	930	ds-DNA	E
P. crispum	PUMPR2G	Pathogenesis-related protein 2	Cultured cells	2683	ds-DNA	E
P. hybrida	PETHSP70G	hsp70	Leaves	2958	ds-DNA	E
P. hybrida	PETHSP70R	Heat shock protein hsp70	Leaves	2079	ss-mRNA	—
P. vulgaris	PHVPRP1	Pathogenesis-related protein 1	Suspension culture	769	ss-mRNA	E
P. vulgaris	PHVPRP2	Pathogenesis-related protein 2	Suspension culture	693	ss-mRNA	E

Table 19. Continued

Species	Identifier	Sequence specification	Tissue/site	Size (bp or b)	Type	S
P. vulgaris	PHVPVPR3A	PVPR3 protein	—	871	ss-mRNA	—
P. sativum	PEADRRA	Disease resistance response protein	Pod	728	ss-mRNA	—
P. sativum	PEADRRB	Disease resistance response protein	Pod	594	ss-mRNA	—
P. sativum	PEADRRC	Disease resistance response protein	Pod	734	ss-mRNA	—
P. sativum	PEADRRG	Disease resistance response protein	—	2290	ds-DNA	—
P. sativum	PEAGAMRA	GA mRNA	Stem	574	ss-mRNA	E
P. sativum	PEAGAMRF	GA mRNA	Stem	901	ss-mRNA	E
P. sativum	PEAPHSP1	HSP70	—	2314	ss-mRNA	E
P. sativum	PEAHSP21	Heat shock protein 21	—	985	ss-mRNA	—
P. sativum	PEAHSP177A	17.7 kDa heat shock protein	Leaves	772	ss-mRNA	—
P. sativum	PEAHSP179A	17.9 kDa heat shock protein	Leaves	700	ss-mRNA	—
P. sativum	PEAHSP181A	18.1 kDa heat shock protein	Leaves	862	ss-mRNA	—
P. sativum	PEAHSP227A	22.7 kDa heat shock protein	Leaves	795	ss-mRNA	—
P. trichocarpa	POPGWIN3	Wound responsive gwin3	Leaves	1509	ds-DNA	E
P. trichocarpa	POPPD	Wound-responsive mRNA	Leaves	649	ss-mRNA	E
S. tuberosum	POTPR1A	Pathogenesis-related protein (prpl)	Leaves	1526	ds-DNA	—
S. tuberosum	POTPSTH2	pSTH-2 protein	Tuber	791	ss-mRNA	—
S. tuberosum	POTPSTH21	pSTH-21 protein	Tuber	767	ss-mRNA	—
S. tuberosum	POTSTHA	Pathogenesis-related protein	—	2896	ds-DNA	—
S. tuberosum	POTSTHB	Pathogenesis-related protein (STH-21)	—	2278	ds-DNA	—
S. tuberosum	POTWIN12G	WIN1 and WIN2	—	7200	ds-DNA	E
S. oleracea	SPISCE70	70 kDa heat shock protein	—	2535	ss-mRNA	E
T. aestivum	WHTCOAC	WCS120 protein	Seedling shoot	1501	ss-mRNA	—
T. aestivum	WHTGIR	Gibberellin responsive protein gene	Aleurone layer	5315	ds-DNA	—
T. aestivum	WHTTHAU	Thaumatin-like protein	—	674	ss-mRNA	E
T. aestivum	WHTHSP26	26.6 kDa heat shock protein	Young leaves	940	ss-mRNA	E
T. aestivum	WHTHSP173	Heat shock protein 17.3	Young leaves	722	ss-mRNA	E
T. aestivum	WHTHSPLW	Low mol. wt heat shock protein	Coleoptile	741	ss-mRNA	—
T. aestivum	WHTPUTPER	Putative peroxidase	Young leaves	240	ss-mRNA	E
T. aestivum	WHTRAB	ABA responsive gene	Root	781	ss-mRNA	E
T. aestivum	WHTRAB15B	rab15B	Root	1068	ss-mRNA	E
Z. mays	MZEHSP701	70 kDa heat shock protein	—	371	ds-DNA	E
Z. mays	MZEHSP702	70 kDa heat shock protein	—	1934	ds-DNA	E
Z. mays	MZEHSP70CT	hsp70cat	Protoplasts	125	ds-DNA	—
Z. mays	MZEHSP70I	70 kDa heat shock protein	—	858	ds-DNA	E
Z. mays	MZEHSP18K1	18 kDa heat shock protein	Radicle	982	ss-mRNA	E
Z. mays	MZEHSP18K2	18 kDa heat shock protein	Radicle	843	ss-mRNA	E
Z. mays	MZEIN21	In2-1	Seedlings	906	ss-mRNA	E
Z. mays	MZEPRMS	Pathogenesis-related protein	Germinating seed	784	ds-DNA	E
Z. mays	MZER28DNA	*rab28* gene	Leaves	1846	ds-DNA	E
Z. mays	ZMHSP18K1	18 kDa heat shock protein	Radicle	982	RNA	E
Z. mays	ZMHSP18K2	18 kDa heat shock protein	Radicle	843	RNA	E
Z. mays	ZMPRMS	Pathogenesis-related protein	Germinating seed	784	DNA	E

Table 20. Structural proteins (cell wall proteins, cytoskeletal proteins, proline-/hydroxyproline-/glycine-rich proteins)

Species	Identifier	Sequence specification	Tissue/site	Size (bp or b)	Type	S
A. thaliana	ATHACT1A	Actin-1	—	1830	ds-DNA	—
A. sativa	ASTTUB1MR	β-tubulin	Etiolated coleoptiles	1420	ss-mRNA	E
C. rubrum	CHEHC1	Glycine-rich protein	—	689	ss-mRNA	E
C. rubrum	CHETLP	Tubulin-like protein	—	434	ss-mRNA	E
D. carota	DAREXT	Extensin	—	2230	ds-DNA	E
D. carota	DAREXTA	Extensin	Wounded root	399	ss-mRNA	—
D. carota	DAREXTR	33 kDa extensin-related protein	Wounded root	832	ss-mRNA	—
D. carota	DARGLYR	Glycine-rich protein	Suspension culture	725	ss-mRNA	E
D. carota	DARGRP	Glycine-rich protein	Cell culture	563	ss-mRNA	E
D. carota	DARTLP	Tubulin-like protein	—	434	ss-mRNA	—
G. max	SOYAC6L	Actin *SAc6*	—	1669	ds-DNA	E
G. max	SOYAC7L	Actin *SAc7*	—	1320	ds-DNA	E
G. max	SOYACTI	Actin	—	1620	ds-DNA	E
G. max	SOYACT1G	Actin 1	—	1898	ds-DNA	—
G. max	SOYACT3G	Actin 3	—	1620	ds-DNA	—
G. max	SOYHPR	Hydroproline-rich protein	Seedling	364	ss-mRNA	—
G. max	SOYPRP1	Proline-rich protein	—	2835	ds-DNA	—
G. max	SOYRPRP3	Repetitive proline-rich protein 3	—	2821	ds-DNA	E
G. max	SOYSBRPR2	Proline-rich cell wall protein (*SbPRP2*)	—	2775	ds-DNA	—
G. max	SOYSBRPR3	Proline-rich cell wall protein (*SbPRP3*)	—	2400	ds-DNA	—
G. max	SOYSB1TUB	β-tubulin (*S-β-1*)	—	2929	ds-DNA	—
G. max	SOYSB2TUB	β-tubulin (*S-β-2*)	—	3070	ds-DNA	—
G. max	SOYTUBB	β-tubulin	Hypocotyl	1327	ss-mRNA	E
H. annuus	HNNHPRG	Hydroxyproline-rich protein	—	2204	ds-DNA	—
H. vulgare	BLYGRP	Glycine rich protein	Leaves, kernel	1877	ds-DNA	E
L. esculentum	TOMEXTEN	Extensin	All stem tissue	940	ss-mRNA	E
L. esculentum	TOMEXTEN1	Extensin	All stem tissue	459	ss-mRNA	E
L. esculentum	TOMEXTEN2	Extensin	All stem tissue	357	ss-mRNA	E
L. esculentum	TOMEXTEN3	Extensin	All stem tissue	288	ss-mRNA	E
L. esculentum	TOMEXTEN5	Extensin	All stem tissue	739	ss-mRNA	E
L. esculentum	TOMEXTEN6	Extensin	All stem tissue	309	ss-mRNA	E
L. esculentum	TOMEXTEN7	Extensin	All stem tissue	231	ss-mRNA	E
L. esculentum	TOMEXTEN8	Glycine-rich protein	All stem tissue	693	ss-mRNA	E
L. esculentum	TOMEXTEN9	Glycine-rich protein	All stem tissue	586	ss-mRNA	E
L. esculentum	TOMEXTEN10	Glycine-rich protein	All stem tissue	385	ss-mRNA	E
L. esculentum	TOMEXTEN11	Glycine-rich protein	All stem tissue	619	ss-mRNA	E
L. esculentum	TOMEXTEN12	Extensin	All stem tissue	619	ss-mRNA	E
L. esculentum	TOMEXTEN13	Glycine-rich protein	All stem tissue	402	ss-mRNA	E
L. esculentum	TOMEXTEN14	Extensin	All stem tissue	402	ss-mRNA	E
L. esculentum	TOMEXTEN15	Glycine-rich protein	All stem tissue	598	ss-mRNA	E
L. esculentum	TOMEXTEN16	Glycine-rich protein	All stem tissue	404	ss-mRNA	E
L. esculentum	TOMEXTENA	Extensin	Leaves	2383	ds-DNA	—
L. esculentum	TOMEXTENB	Extensin	Leaves	1900	ds-DNA	—
L. esculentum	TOMHRGP	Cell wall HRGP (hydroxproline-rich glycoprotein)	—	279	ds-DNA	—
L. esculentum	TOMTPRPF1	*TPRP-F1*	Fruit	1246	ss-mRNA	E
L. esculentum	TOMTPRPF2	*TPRP-F1*	—	1837	ds-DNA	E
N. plumbaginifolia	TOBEXT	Extensin	Leaves	3994	ds-DNA	—
N. sylvestris	TOBGRP2MR	Glycine-rich protein	—	991	ss-mRNA	E
N. tabacum	TOBEXU	*HRGPnt3* gene for extensin	Leaves	4343	ds-DNA	E

Table 20. Continued

Species	Identifier	Sequence specification	Tissue/site	Size (bp or b)	Type	S
N. tabacum	TOBGRP	Glycine-rich protein	Leaves	3063	ds-DNA	—
N. tabacum	TOBGRPA	Glycine-rich protein	—	3063	ds-DNA	—
O. sativa	RICGRP1	Glycine-rich cell wall protein	—	2208	ds-DNA	E
O. sativa	RICRAC1	*RAc1*	Etiolated shoot	1460	ss-mRNA	E
O. sativa	RICRAC1D	Actin (*RAc1*)	Shoot	1550	ds-DNA	E
O. sativa	RICRAC2	Actin	Shoot	1611	ds-DNA	E
O. sativa	RICRAC3	Actin (*RAc3*)	Shoot	1724	ds-DNA	E
O. sativa	RICRAC7	Actin (*RAc7*)	Shoot	1919	ds-DNA	E
O. sativa	RICTUBA1M	α-1-tubulin	Anther	1726	ss-mRNA	E
P. hybrida	PETGCR1	Glycine-rich protein, *grp-1*	—	1934	ds-DNA	E
P. vulgaris	PHVGRP10	Glycine-rich cell wall protein *grp-1.0*	Young leaves	1812	ds-DNA	E
P. vulgaris	PHVGRP18	Glycine-rich wall protein *grp 1.8*	Young leaves	1925	ds-DNA	E
P. vulgaris	PHVHRGPA	Hydroxyproline-rich glycoprotein	—	1113	ss-mRNA	—
P. vulgaris	PHVHRGPB	Hydroxyproline-rich glycoprotein	—	692	ss-mRNA	—
P. vulgaris	PHVHRGPC	Hydroxyproline-rich glycoprotein	—	1504	ss-mRNA	—
P. vulgaris	PHVPRP1R	Proline-rich protein	Suspension culture	1111	ss-mRNA	E
P. contorta	PINACT	Actin	Seed	745	ds-DNA	—
P. sativum	PEACHSAR	Chromosomal scaffold-associated region	—	1380	ds-DNA	—
S. tuberosum	POTACTP	Actin pseudogene	—	2082	ds-DNA	E
S. vulgare	SRGGLYRNA1	Glycine-rich RNA-binding protein	—	731	ss-mRNA	E
S. vulgare	SRGHRGPG	Hydroxyproline-rich glycoprotein	Leaves	4231	ds-DNA	E
T. aestivum	WHTGERMIN	*Germin 9f-2.8*	—	2822	ds-DNA	—
T. aestivum	WHTGERMINA	*Germin 9f-3.8*	—	3761	ds-DNA	—
T. aestivum	WHTPRP	Proline-rich protein	Seedling	1487	ss-mRNA	E
Triticum sp.	WHTGAMINA	Germin	Germinating embryo	1012	ss-mRNA	—
Z. diploperennis	MZEHRGP	Hydroxyproline-rich glycoprotein	Leaves	4478	ds-DNA	E
Z. mays	MZEACT1G	Actin-1	—	2424	ds-DNA	—
Z. mays	MZEALPTUB	α-1-tubulin and α-2-tubulin	Root endosperm	9242	ds-DNA	E
Z. mays	MZEB1TUB	β-1-tubulin	—	2461	ds-DNA	E
Z. mays	MZEB2TUBR	β-2-tubulin	Endosperm	1579	ss-mRNA	E
Z. mays	MZECWAA	Cell wall protein	Coleoptile and primary root	1192	ss-mRNA	—
Z. mays	MZECWAB	Cell wall protein	Coleoptile and primary root	710	ss-mRNA	—
Z. mays	MZECWAC	Cell wall protein	Coleoptile and primary root	664	ss-mRNA	—
Z. mays	MZEE41CWG	Cell wall glycoprotein	—	664	ss-mRNA	E
Z. mays	MZEGLYR	Glycine-rich protein	—	813	ss-mRNA	E
Z. mays	MZEGRP	Glycine-rich protein	Endosperm	1114	ds-DNA	E
Z. mays	MZEW64CWG	Cell wall glycoprotein	—	1193	ss-mRNA	E
Z. mays	ZMTUBA	α-3-tubulin	—	4397	DNA	—

Table 21. Transposons (transposable elements, insertional elements, T-DNA insertions, transposon-mutated genes)

Species	Identifier	Sequence specification	Tissue/site	Size (bp or b)	Type	S
A. majus	AMACHSTA1	Junction *Tam 1*-chalcone synthase gene	–	262	ds-DNA	E
A. majus	AMACHSTA2	Insertion element *Tam 1*	–	601	ds-DNA	E
A. majus	AMACHSY	Chalcone synthase (Am3) gene	–	301	ds-DNA	E
A. majus	AMAPALA	*Pallida* gene with *Tam 3* transposable element	–	471	ds-DNA	G
A. majus	AMATAM1TR	Transposable element *Tam 1*	–	15164	ds-DNA	E
A. majus	AMATAM2G	Chalcone synthase	–	5187	ds-DNA	E
A. majus	AMATAM2R	Chalcone synthase, *Tam 2*	–	1464	ss-mRNA	E
A. majus	AMATAM3TN	*Tam 3* — transposase	–	3629	ds-DNA	E
A. majus	AMATAM4	Transposable element *Tam 4*	–	4329	ds-DNA	E
C. arvensis	COARIINT	Ri plasmid insertion	–	570	ds-DNA	–
C. arvensis	COARIPJ	RI plasmid TL-DNA/plant junction	–	444	ds-DNA	–
C. arvensis	COARIPJ1	Ri TL-DNA/plant junction	–	1319	ds-DNA	E
C. arvensis	COARIPJ2	RI TL-DNA/plant junction	–	2216	ds-DNA	E
G. max	SOYTGM6A	Transposable element Tgm 6	Young leaves	348	ds-DNA	E
G. max	SOYTGM6B	Transposable element Tgm 6	Young leaves	341	ds-DNA	E
G. max	SOYTGM5OR	Transposable element Tgm 5	Young leaves	1002	ds-DNA	E
L. henryi	LILDELTX	Del transposon	–	9345	ds-DNA	–
N. glauca	TOBROLBC	*rolB–C* genes homologous to TL-DNA of *A. rhizogenes*	–	3141	ds-DNA	E
N. tabacum	TOBTAR7	Target DNA 7	–	600	ds-DNA	E
N. tabacum	TOBTINS2	Ti plasmid 3′ junction	Leaves	468	ds-DNA	–
N. tabacum	TOBTINS1	Ti plasmid 5′ junction	Leaves	466	ds-DNA	–
N. tabacum	TOBTINS	Ti plasmid DNA insertion site	Leaves	595	ds-DNA	–
O. sativa	RICMU	Mu-like transposable element	Seedling	1260	ds-DNA	E
P. americanum	PENACTIV	Ac-like (activator transposable element)	–	100	ds-DNA	E
P. sativum	PEAPIS1TP	*legC* gene 5′-flank with Pis1 insertion	–	594	ds-DNA	E
S. tuberosum	POTCOPT	Copia-like transposable element	–	5060	ds-DNA	E
S. tuberosum	STCOPT	Copia-like transposable element	–	5060	DNA	E
T. aestivum	WHTGLUIN1	*Glu–Aly* gene; Wis-2 insertion sequence; glutenin	–	794	ds-DNA	E
T. aestivum	WHTGLUIN2	*Glu–Aly* gene; Wis-2 insertion sequence; glutenin	–	1515	ds-DNA	E
T. dactyloides	TRDDS130	Ds130 element (Ds1-related)	–	405	ds-DNA	E
T. dactyloides	TRDDS132	Ds132 element (Ds1-related)	–	407	ds-DNA	E
Z. mays	MZEAC1	Ac element; transposable element	–	4565	ss-mRNA	E
Z. mays	MZEACCR11	Cryptic Ac-11 sequence	Leaves	4150	ds-DNA	E
Z. mays	MZEAC9WX	Transposable element Activator (Ac9) from the *waxy* locus	–	4810	ds-DNA	–
Z. mays	MZEBRNZX	Ins2 insertion element 5′ of *bronze* gene	–	447	ds-DNA	–
Z. mays	MZEBRNZZ	Ins1 insertion element located 5′ of *bronze* gene	–	279	ds-DNA	–
Z. mays	MZEBS1RET	Transposable element	Seedlings	3203	ds-DNA	E
Z. mays	MZEBS1TRA	Transposon Bs1	–	304	ds-DNA	E
Z. mays	MZEBZRA	Spontaneous *del*etion sequence from *waxy* (bz-R) gene	–	81	ds-DNA	–
Z. mays	MZEBZWMTN	bz1 gene 5′ region with DS1 element *Bz–wm* inserted	–	505	ds-DNA	–
Z. mays	MZEDS101	Ds101 controlling element	–	394	ds-DNA	E
Z. mays	MZEDS101A	Ds101 controlling element	–	405	ds-DNA	E
Z. mays	MZEDS103	Ds103 controlling element	–	406	ds-DNA	E
Z. mays	MZEDS105	Ds105 controlling element	–	406	ds-DNA	E

Table 21. Continued

Species	Identifier	Sequence specification	Tissue/site	Size (bp or b)	Type	S
Z. mays	MZEDS123A	Ds123 controlling element	—	406	ds-DNA	E
Z. mays	MZEDSBZWM	Ds(bz–wm) controlling element	—	406	ds-DNA	E
Z. mays	MZEDSWXM1	Wx–m1 dissociation (Ds) insertion sequence	—	441	ds-DNA	E
Z. mays	MZEDSEWXM5	Ds element in *Wx–m5* allele	Leaves	2013	ds-DNA	E
Z. mays	MZEDSWXM1A	Ds(*wx–m1*) controlling element (Ds1-related)	—	407	ds-DNA	E
Z. mays	MZEDSPMBT	Defective suppressor-mutator from the brittle-1-mutable allele	—	3326	ds-DNA	—
Z. mays	MZEISWAX	Insertion sequence 5′ of *waxy* gene	—	316	ds-DNA	—
Z. mays	MZEMU4	Transposable element Mu4	—	2233	ds-DNA	E
Z. mays	MZEMU5	Transposable element Mu5	—	1651	ds-DNA	E
Z. mays	MZEMU8	Mutator element Mu8	—	1428	ds-DNA	E
Z. mays	MZEMUA2	Mutator transposable element MuA2	—	4845	ds-DNA	E
Z. mays	MZERALONC	*Wx–m1* dissociation (Ds) insertion sequence	—	441	ds-DNA	E
Z. mays	MZERCYMU7	rcy: *Mu7* Cy transposable element system	—	2199	ds-DNA	E
Z. mays	MZERUQSTTR	*ruq-st* transposon	—	396	ds-DNA	E
Z. mays	MZERUQ31TR	*ruq31* transposon	—	405	ds-DNA	E
Z. mays	MZERUQ66TR	*ruq66* transposon	—	397	ds-DNA	E
Z. mays	MZES30341	Alcohol dehydrogenase gene with transposon *Mu-1* insertion	Dark-grown seedling	296	ds-DNA	—
Z. mays	MZES30342	Alcohol dehydrogenase gene with transposon *Mu-1* insertion	Dark-grown seedling	55	ds-DNA	—
Z. mays	MZES3034A1	Alcohol dehydrogenase gene with transposon *Mu-1* insertion	Dark-grown seedling	222	ds-DNA	—
Z. mays	MZES3034A2	Alcohol dehydrogenase gene with transposon *Mu-1* insertion	Dark-grown seedling	55	ds-DNA	—
Z. mays	MZESUTZ81	Sucrose synthetase gene mutant (Sh-5586 allele)	—	250	ds-DNA	—
Z. mays	MZESUTZ82	Sucrose synthetase gene mutant (Sh-5586 allele)	—	250	ds-DNA	—
Z. mays	MZETAC	Transposable element *Ac*	—	4565	ds-DNA	E
Z. mays	MZETACA1	Transposon *Ac* insertion sequence 1	—	74	ds-DNA	—
Z. mays	MZETACA2	Transposon *Ac* insertion sequence 2	—	58	ds-DNA	—
Z. mays	MZETACB1	Transposon *Ac* insertion sequence 1	—	74	ds-DNA	—
Z. mays	MZETACB2	Transposon *Ac* insertion sequences 2 and 3	—	92	ds-DNA	—
Z. mays	MZETNDS	Transposable element dissociation (Ds)	—	4384	ds-DNA	E
Z. mays	MZETNENSPM	Autonomous transposable element En-1 mosaic protein gene	—	8287	ds-DNA	—
Z. mays	MZETNMU1	Transposable element *Mu1*, inserted in *Adh1*	—	1504	ds-DNA	E
Z. mays	MZETNMU14	*Mu*-homologous element *Mu1.4.-B37*	—	1396	ds-DNA	E
Z. mays	MZETNMU17	Transposable element *Mu1.7*	—	1754	ds-DNA	E
Z. mays	MZETNMUT	*Mu* transposable element *MRSA*	—	1557	ds-DNA	—
Z. mays	MZETNSPM	Deletion derivative (Spm-I8) of *En-1* transposon	—	2242	ds-DNA	E
Z. mays	MZETRNBS1	Transposon *Bs1*	Seedling	3203	ds-DNA	—
Z. mays	MZETRNMU	Transposable element *Mu9* DNA	Immature cob	4942	ds-DNA	—
Z. mays	MZEWX	*Waxy* (*wx*) locus with target site for *Ac9* transposon	—	241	ds-DNA	—

Gene Index

Table 21. Continued

Species	Identifier	Sequence specification	Tissue/site	Size (bp or b)	Type	S
Z. mays	MZEWXBA	Spontaneous deletion sequence from *waxy* (*wx-B*) gene	—	187	ds-DNA	—
Z. mays	MZEWXB1A	Spontaneous deletion sequence from (*waxy* (*wx-B1*) gene	—	73	ds-DNA	—
Z. mays	MZEWXB6A	Spontaneous deletion sequence from *waxy* (*wx-B6*) gene	—	71	ds-DNA	—
Z. mays	MZEWXC4A	Deletion sequence from *waxy* (*wx-C4*) gene	—	74	ds-DNA	—
Z. mays	MZEWXX23A	Deletion sequence from *waxy* (*Wx-W23*) gene	—	79	ds-DNA	—

ACKNOWLEDGEMENTS

We are grateful to the SEQNET operators at the Daresbury SEQNET Computing Facility and the MENDEL network at the University of California at Berkeley for allowing access to their networks and making available excess CPU time. This work was in part sponsored by an SERC award to M.L.

2. REFERENCES

1. Croy, R.R.D. and Gatehouse, J.A. (1985) in *Plant Genetic Engineering* (J.H. Dodds, ed.). Cambridge University Press, Cambridge. p. 143.

2. Deveraux, J. *et al.* (1984) *Nucl. Acids Res.*, **12**, 387.

CHAPTER 4

PLANT TRANSFORMATION AND EXPRESSION VECTORS

F. Guerineau and P. Mullineaux

1. INTRODUCTION

Recent developments in plant molecular biology have created the need for versatile cassettes for the expression of foreign genes, binary vectors for gene transfer using *Agrobacterium*, marker genes or phenotypic markers for selection or detection of transformants, and reporter genes for assaying gene expression. We present here descriptions and restriction maps of commonly used plasmids. The chapter is divided into five parts: (i) expression cassettes; (ii) reporter genes; (iii) selectable markers; (iv) phenotypic markers; and (v) binary vectors. Nucleotide sequence information is provided at the end of the chapter. See also *Chapter 5* for more detailed information on reporter and selectable marker genes.

Requests for published plasmids should be addressed to the relevant authors. Unpublished constructs can be obtained from P. Mullineaux.

Table 1 lists the abbreviations used in this chapter.

2. EXPRESSION CASSETTES

Cloning a coding sequence into the polylinker of an expression cassette in the proper orientation will result in the expression of the gene after its transfer to plant cells. Transcription will occur from the promoter located upstream from the polylinker and the mRNA will be polyadenylated at the site(s) present downstream of the polylinker. The polylinker may contain a translation start codon, in which case the coding sequence must be positioned in frame with the cassette ATG codon for correct translation. *Table 2* lists nucleotide sequences at cloning junctions of expression cassettes; *Table 3* gives the codes and accession numbers of relevant sequences from EMBL and GenBank databases; and *Chapter 5* presents these sequences and the corresponding restriction maps.

2.1. pRT103

The cauliflower mosaic virus (CaMV) strain Cabb-B was used to construct this expression cassette (1) (*Figure 1*). In this strain, an *Eco*RV restriction site is located between the CaMV 35S promoter enhancer region and the TATA box. The cassette allows translational fusions with an ATG codon embedded in the consensus sequence CCACCATGG optimal for translation initiation in eukaryotes. Other constructs with various polylinkers are described (1).

2.2. pJIT74

The CaMV strain Cabb-JI was the source of the components of this cassette (*Figure 2*). The CaMV 35S promoter has been fused to the CaMV gene VI first codons. Translation fusion of the *gus* coding sequence with this construct resulted in a high level of expression, as compared to the expression given by a transcriptional fusion with the same 35S promoter.

Table 1. Abbreviated terms used in this chapter

*aad*A	streptomycin-resistance gene
adh1	alcohol dehydrogenase gene
amp	ampicillin-resistance gene
*aph*IV	hygromycin-resistance gene
bar	bialaphos-resistance gene
bom	mobilization sequence of conjugative plasmids
CaMV	cauliflower mosaic virus
cat	chloramphenicol-resistance gene
cos	packaging signal for lambda phage
Gm	gentamycin resistance
gus	β-glucuronidase gene
hph	hygromycin-resistance gene
Km	kanamycin resistance
*lac*Z	β-galactosidase gene
LB	left border of *Agrobacterium* T-DNA
luc	luciferase gene
mas	mannopine synthase gene
nos	nopaline synthase gene
*npt*II	kanamycin-resistance gene
ocs	octopine synthase gene
ori	ColE1 origin of replication
*ori*T	origin of transfer of a conjugative plasmid
*ori*V	broad host-range plasmid origin of replication
plac	promoter of the *lac* operon
polyA	sequence directing polyadenylation
RB	right border of *Agrobacterium* T-DNA
rbcS	Rubisco small subunit gene
Rubisco	ribulose bisphosphate carboxylase/oxygenase
spc:sm	spectinomycin/streptomycin-resistance gene
*sul, sul*I	sulfonamide-resistance gene
tet	tetracycline-resistance gene
TP	Rubisco transit peptide sequence

Table 2. Nucleotide sequences at cloning junctions of expression cassettes and at 5′ ends of coding sequences (see also *Chapter 5* for complete nucleotide sequences of reporter and selectable marker genes and promoters)

```
pRT 103
+1  XhoI      BalI       NcoI            SacI         KpnI  SmaI     BamHI
  ACCTCGAGTGGCCACC ATG GGC GAG CTC GGT ACC CGG GGA TCC
  XbaI
  TCT AGA

pJIT 74
+1                                     PstI       SalI        BamHI  SmaI
ACAGCCCAAGC ATG GAG AAC CGA CCT GCA GGT CGA CGG ATC CCC
                 [CaMV gene VI]→
     EcoRI
GGG AAT TCG

pDH51
+1      KpnI SmaI    BamHI   XbaI     SalI      PstI      SphI
  ACACGGTACCCGGGGATCCTCTAGAGTCGACCTGCAGGCATGC

pJIT 60
+1         HindIII      PstI       SalI       BamHI SmaI    EcoRI
ACAGCCCAAGCTTGGCTGCAGGTCGACGGATCCCCGGGAATTC

junction between                  TCCCAC-CC-CTACT
enhancer regions                         :    :
                                      7376   7040

pMA 406
+1
GTCAAACGAAGAAAAAAGTTACCTGTTTGCGATCTCATTACAATCTCCCTAGTTTCT
                                 HindIII    SphI   PstI     SalI     XbaI     BamHI
AATCTCAGCTAAGAAAAACCCCAAGCTTGCATGCCTGCAGGTCGACTCTAGAGGATC
SmaI KpnI    SacI       EcoRI
CCCGGGTACCGAGCTCGAATTC

pKYLX3
                +1
HindIII   SphI     MluI     XhoI   ApaI      PstI      SalI      XbaI     BamHI  SmaI
AAGCTTGCATGCCACGCGTCTCGAGGGCCCCTGCAGGTCGACTCTAGAGGATCCCCG
KpnI     SacI                   XbaI
GGTACCGAGCTCGAATTGATCCTCTAGA

pJIT 117
 SphI      PstI       SalI        BamHI  SmaI    EcoRI
TGC ATG CCT GCA GGT CGA CGG ATC CCC GGG AAT TCG
←[TP]
(transit peptide sequence)

pBI101.1
HindIII      SphI        PstI        SalI       XbaI       BamHI   SmaI
AAG  CTT  GCA  TGC  CTG  CAG  GTC  GAC  TCT  AGA  GGA  TCC  CCG
GGT  GGT  CAG  TCC  CTT  ATG  TTA
                         [gus]→
```

Table 2. Continued

---

pBI101.2

*Hin*dIII *Sph*I *Pst*I *Sal*I *Xba*I *Bam*HI *Sma*I
A AGC TTG CAT GCC TGC AGG TCG ACT CTA GAG GAT CCC CGG
GTA GGT CAG TCC CTT ATG TTA
[*gus*]→

pBI101.3

*Hin*dIII *Sph*I *Pst*I *Sal*I *Xba*I *Bam*HI *Sma*I
AA GCT TGC ATG CCT GCA GGT CGA CTC TAG AGG ATC CCC GGG
TAC GGT CAG TCC CTT ATG TTA
[*gus*]→

pJIT26

*Stu*I
AGGCCTAGCTTGCGAGATTTTCAGGAGCTAAGGAAGCTAAA ATG GAG
[*cat*]→

pJIT27

*Stu*I
AGGCCT ATG GAA
[*luc*]→

pJIT134

*Hin*dIII *Sph*I *Pst*I
AAGCTTGCATGCCTGCAGGC ATG ATT
[*npt*II]→

pJIT6

*Stu*I
AGGCCT ATG AAA
[*aph*IV]→

pIJ4104

*Eco*RI *Sac*I *Kpn*I *Sma*I
GAATTCGAGCTCGGTACCCGGGGATCTACC ATG AGC
[*bar*]→

pJIT92

*Hin*dIII *Sph*I *Pst*I
AAGCTTGCATGCCTGCAGGTCC ATG GTG
[*sul*]→

pPM19

*Bsp*HI
TC ATG AGG
[*aad*A]→

Table 3. Codes and accession numbers of relevant sequences from EMBL and GenBank databases (see also *Chapter 5* for complete sequences)

Sequence	EMBL code	Accession number	GenBank code
pRT100	ARPRT100	X05868 M31331	PRTCAMV
gus	ECUIDAA	M14641	ECOIDAA
cat	ISTN9X	V00622 J01841	TRN9CAT
luc	PPLUC	M15077	PPYLUC
*npt*II	CV8KMRCG	M17626	PK18KMRCG
*aph*IV	PJHPH	K01193	PJRHPH
bar	SHBARPA	X17220	X17220
*sul*I	ECR46SUL	X15024	ECOR46SUL
*aad*A	SFPCN1AT	X03886	SHFPCN1AT

Figure 1. Map for plasmid pRT103.

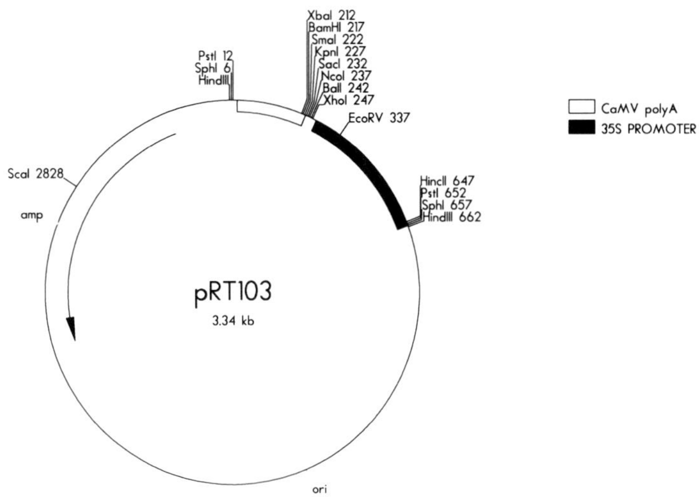

2.3. pDH51

In contrast with the two previous constructs, insertion of a coding sequence into the polylinker of pDH51 (2) (*Figure 3*) will result in a transcriptional fusion with the 35S promoter. Translation initiation will occur at the first initiation codon(s) present in the introduced sequence.

2.4. pJIT60

In this construct (3) (*Figure 4*), the enhancer region of the 35S promoter has been duplicated for a higher level of transcription initiation. The level of expression of the *cat* gene cloned in this cassette was 2.3 times higher than the expression of the same gene under the control of a single 35S promoter. As no ATG sequence is located in the polylinker, this construct generates transcriptional fusions.

Figure 2. Map for plasmid pJIT74.

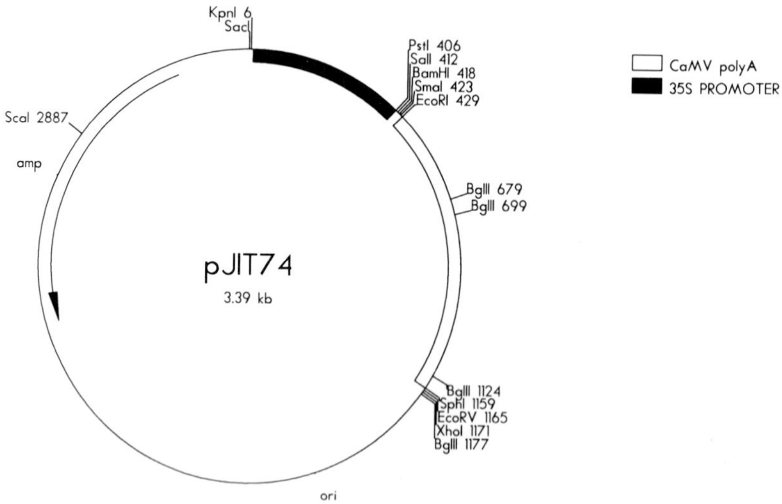

Figure 3. Map for plasmid pDH51.

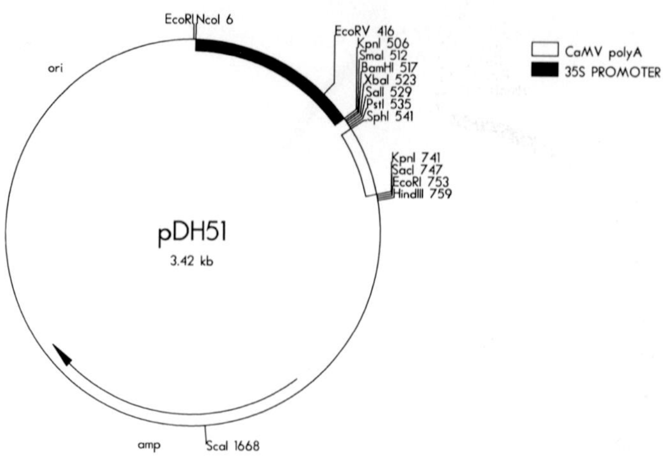

2.5. pMA406

The promoter present in pMA406 (4) (*Figure 5*) originates from the soybean *Gmhsp17.5-E* gene (*Glycine max* heat-shock protein also designated *2019E*), expression of which is heat inducible. The *gus* gene was cloned downstream of the *2019E* promoter of pMA406 and the chimeric construct was transferred to protoplasts by electroporation. Protoplasts were subjected to heat shock. The expression of the *gus* gene was 10 times higher after a heat shock at 40°C than after a shock at 29°C. Data also suggest that the *2019E* promoter directs transcription at a higher level than the CaMV 35S promoter (4).

Figure 4. Map for plasmid pJIT60.

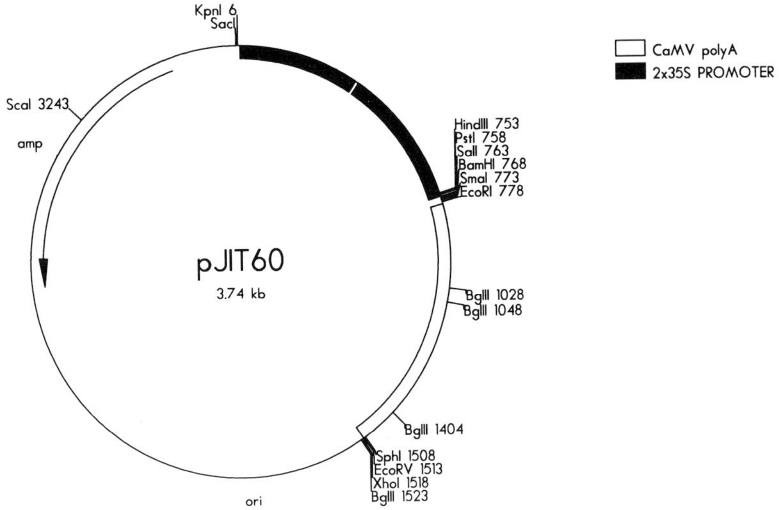

Figure 5. Map for plasmid pMA406.

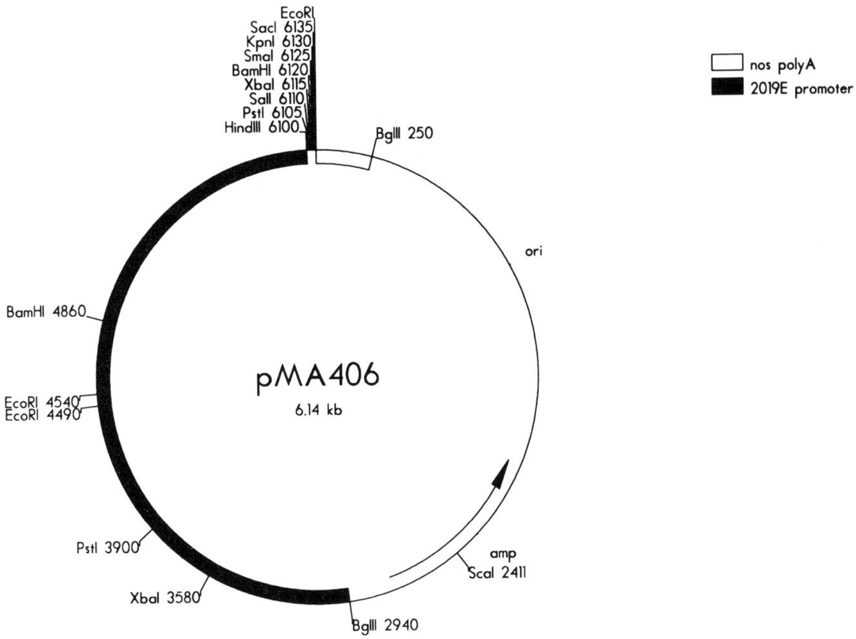

2.6. pKYLX3

The expression of the ribulose bisphosphate carboxylase/oxygenase (Rubisco) small subunit gene (*rbcS*) is regulated by light and is tissue specific. The pea *rbcS-E9* promoter and polyadenylation sequence are present in pKYLX3 (5) (*Figure 6*). Expression of the *cat* gene cloned in this expression cassette was detected in transgenic tobacco callus. The expression cassette from pKYLX3 has been cloned into the pGA472 binary vector. This construct and other expression vectors are described in ref. 5.

2.7. pJIT117

This expression cassette (6) (*Figure 7*) is based on the ability of the Rubisco small subunit transit peptide to direct, import and release foreign proteins into chloroplasts. A translational fusion with the transit peptide sequence is created by inserting a coding sequence into the polylinker of pJIT117. After transfer to plant cells, the fusion protein is transported and the mature protein is released in the chloroplast stroma. This cassette has been used to target bacterial β-glucuronidase and dihydropteroate synthase to plant chloroplasts (see *Chapter 5*).

Figure 6. Map for plasmid pKYLX3.

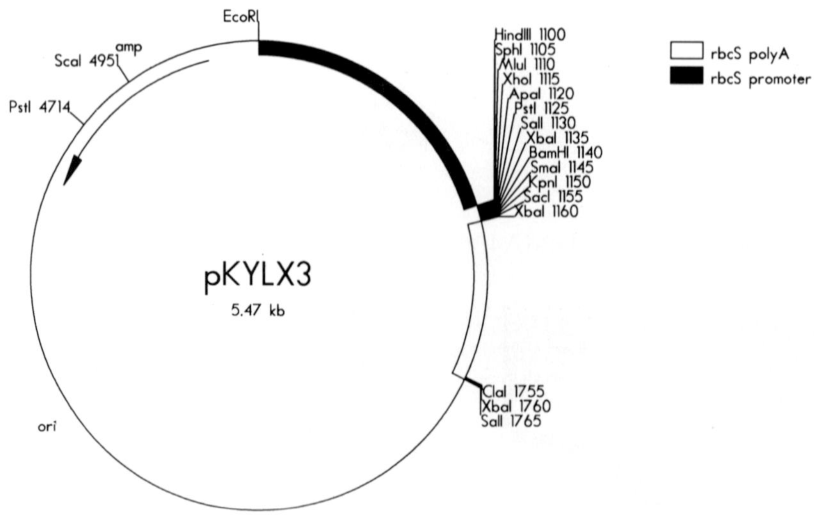

Figure 7. Map for plasmid pJIT117.

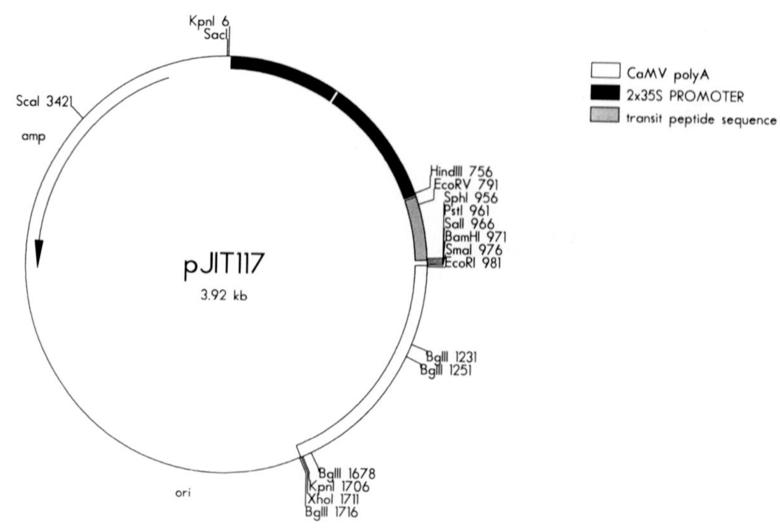

3. REPORTER GENES

A reporter gene codes for an enzyme or other protein which can be quantified using a biochemical assay. The most widely used reporter gene in plant transformation is the *E. coli* *gus* gene coding for β-glucuronidase (7). The *cat* and the *luc* genes coding for a bacterial chloramphenicol acetyltransferase and firefly luciferase, respectively, can also be used in plants for accurate assays of gene expression (8, 9) (see *Chapter 5*).

3.1. *gus* gene

One of the interesting properties of the β-glucuronidase enzyme is its ability to tolerate N-terminal fusions. Plasmids pBI101.1,.2,.3 (10) (*Figure 8*), supplied commercially by Clontech, provide the three frames for translational fusions. To construct pJIT75 (*Figure 9*), the *gus* coding sequence recovered from pBI101.2 was inserted into the polylinker of the expression cassette pJIT74 (*Figure 2*).

Figure 8. Map for plasmids pBI101.1,.2,.3.

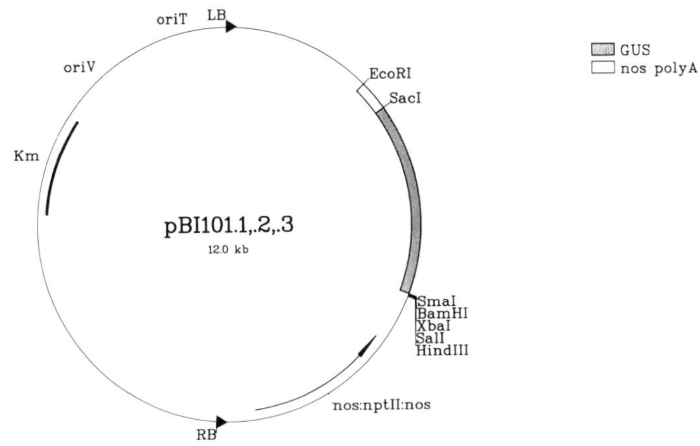

Figure 9. Map for plasmid pJIT75.

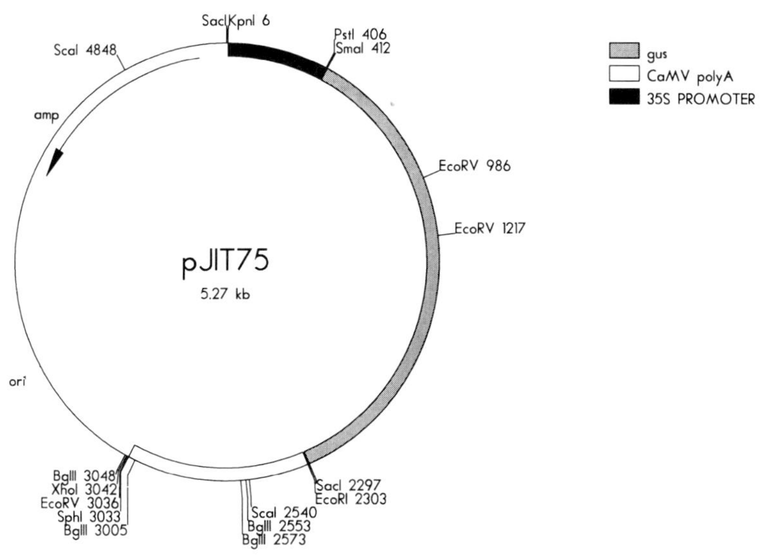

Assay for GUS activity
The activity of β-glucuronidase (GUS) can be assayed as shown in *Figure 10* (method taken from ref. 10, see also ref. 11).

Relative fluorimetric units are plotted on a graph and the activity in nmoles or pmoles of 4-methylumbelliferone per minute is determined. The specific activity can be calculated after determination of the amount of protein present in the extract. Alternatively, when no quantification is required, fluorescence can be detected on a UV transilluminator.

Figure 10. Enzymatic β-glucuronidase assay.

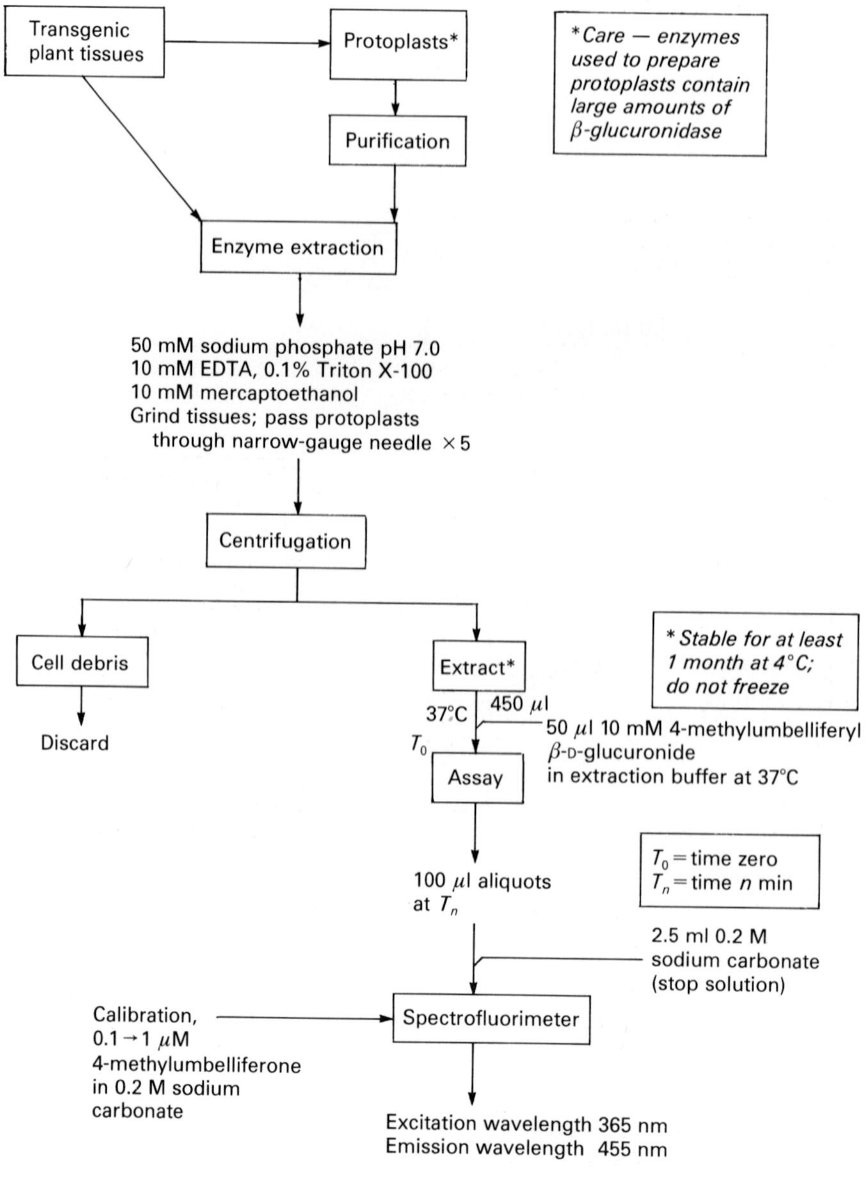

Histochemical assay for GUS activity

Histochemical determination of GUS activity is also possible using 5-bromo-4-chloro-3-indoyl β-D-glucuronide (X-Gluc) (10): it allows the expression of β-glucuronidase to be visualized in specific tissues or cell types. This method is represented by *Figure 11*.

Figure 11. Histochemical β-glucuronidase assay. (Reagents: X-Glu, 5-bromo-4-chloro-3-indolyl-β-D-glucuronide (Sigma Chemical Company); MUG, 4-methyl-umbelliferyl β-D-glucuronide (Sigma); naphthol AS-BI β-D-glucuronide (Sigma); GUS antiserum (Clontech Products).)

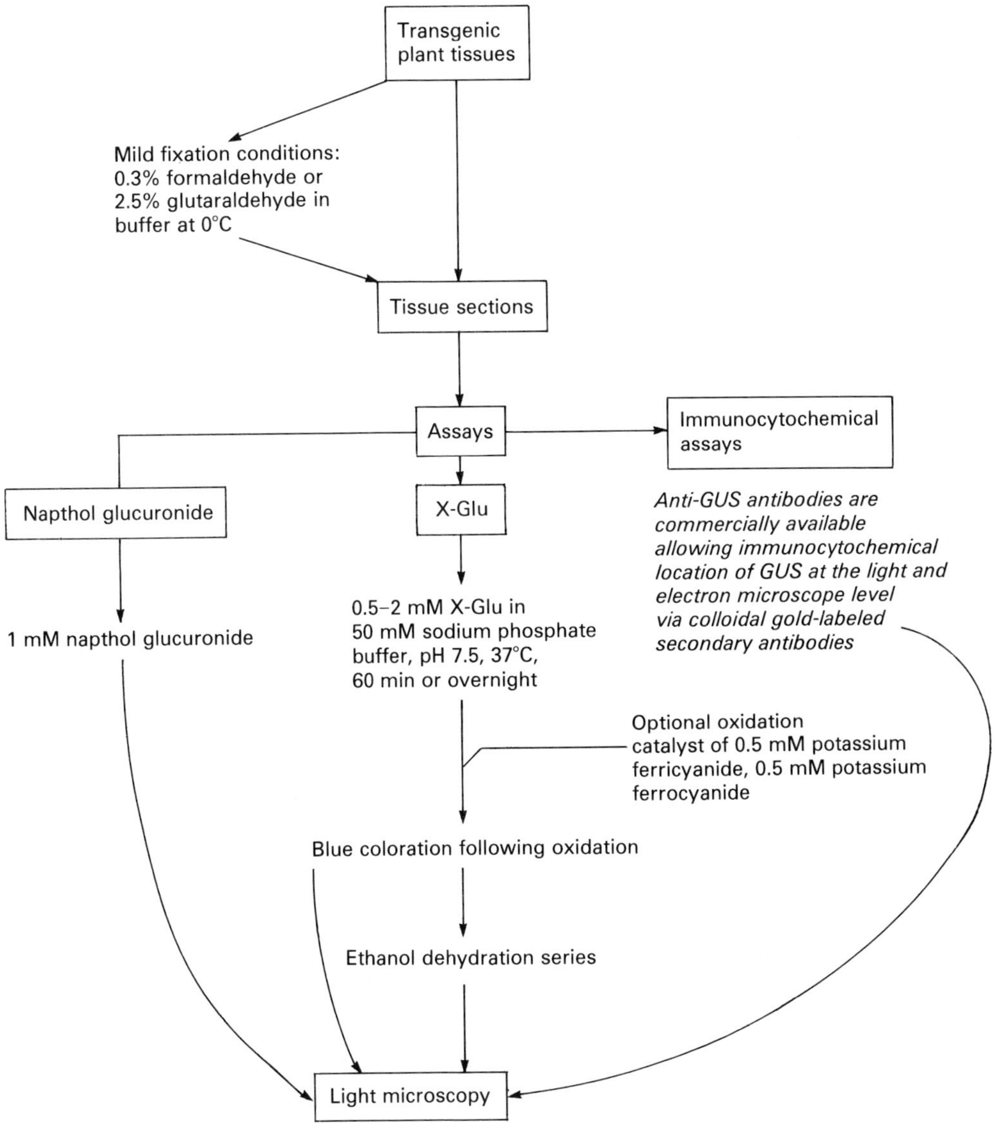

3.2. *cat* and *luc* genes

The *cat* and *luc* genes originate from Tn*9* (12) and pKW101 (13), respectively. Plasmids pJIT26 (*Figure 12*) and pJIT27 (14) (*Figure 13*) can be used for recovering the *cat* and *luc* coding sequences, respectively (see *Chapter 5*). They both contain a *Stu*I restriction site immediately upstream of their translation initiation codon, and a polylinker at the 3′ end of the coding sequence. Transcriptional fusions of the *cat* and *luc* genes with the CaMV 35S promoter were constructed in pJIT52 (*Figure 14*) and pJIT53 (14) (*Figure 15*), respectively. These two chimeric genes were expressed at a high level in transient expression and in transgenic tobacco plants.

Figure 12. Map for plasmid pJIT26.

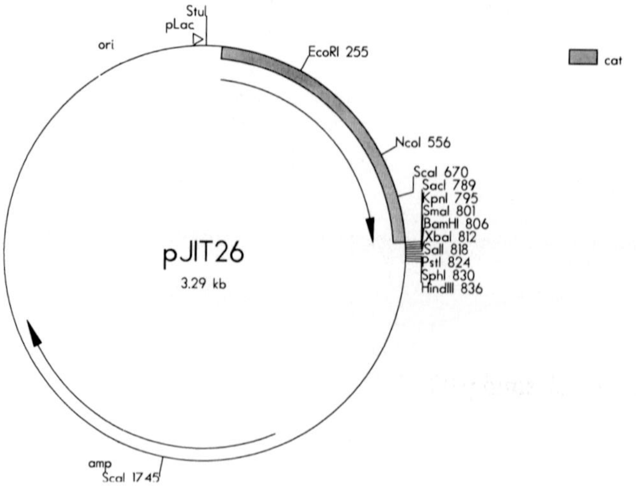

Figure 13. Map for plasmid pJIT27.

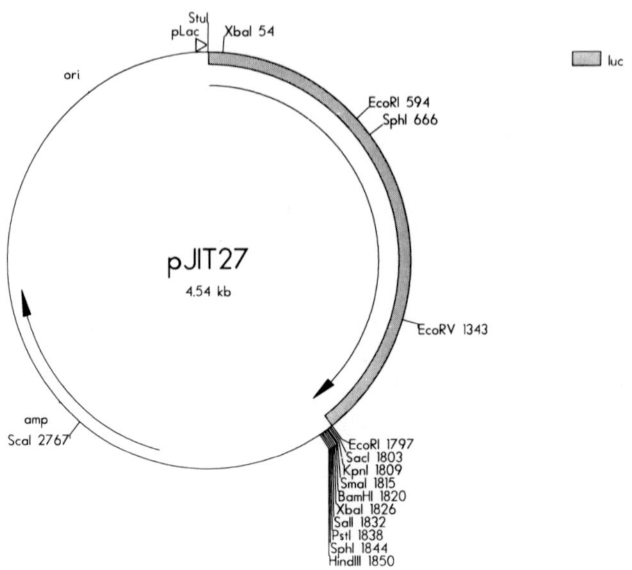

Figure 14. Map for plasmid pJIT52.

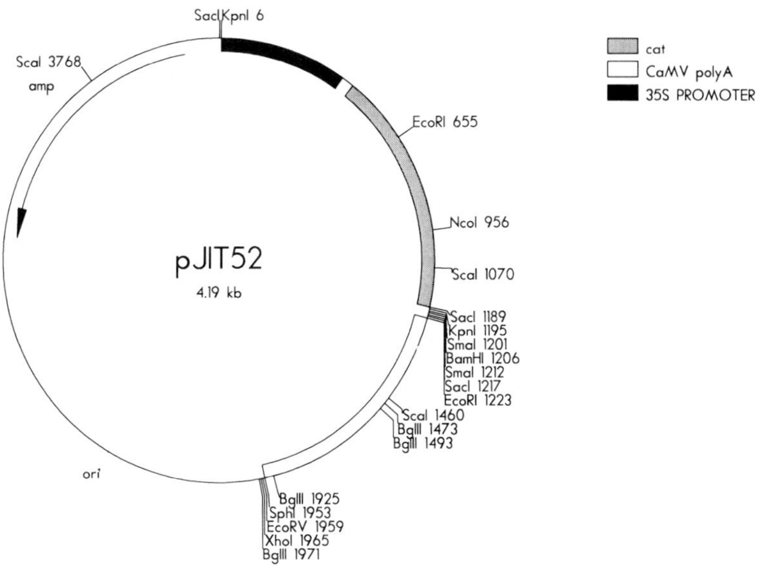

Figure 15. Map for plasmid pJIT53.

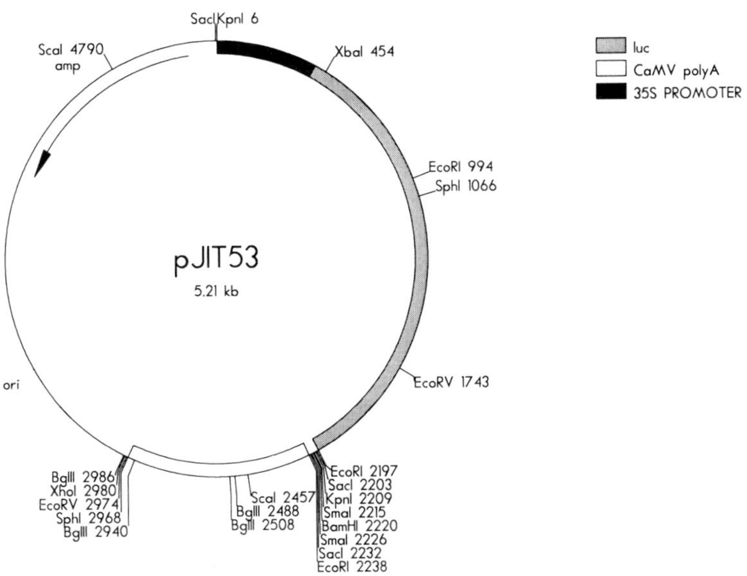

Transformation, Expression Vectors

Assay for CAT activity
Chloramphenicol acetyltransferase (CAT) activity can be assayed by the thin-layer chromatography (TLC) method (15) or the quantitative counting method (16) as shown in *Figure 16.*

Although the TLC method allows the comparison of the CAT activity present in various extracts by visualization of the spots given by acetylated chloramphenicol on the autoradio-

Figure 16. Chloramphenicol acetyltransferase assay.

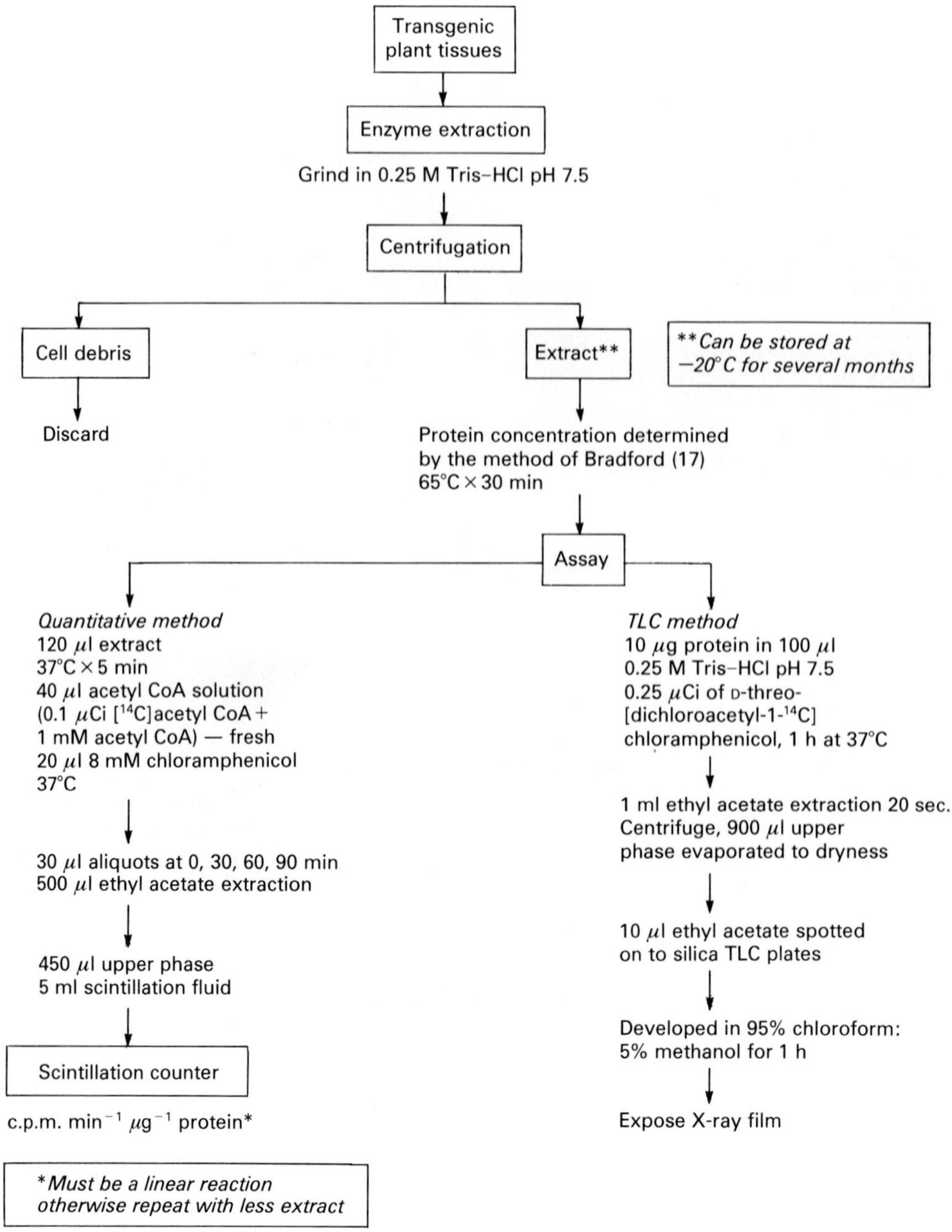

graph, the quantification is difficult and no evidence is given that the reaction occurred within the range of linearity, i.e. in excess of substrate. Accurate quantification is possible, however, using the quantitative counting method.

Detection of luciferase activity in transformants
Luciferase activity may be determined as shown in *Figure 17* (9, 18).

Figure 17. Luciferase assay. (Reagent: D-luciferin (from Sigma Chemical Company).)

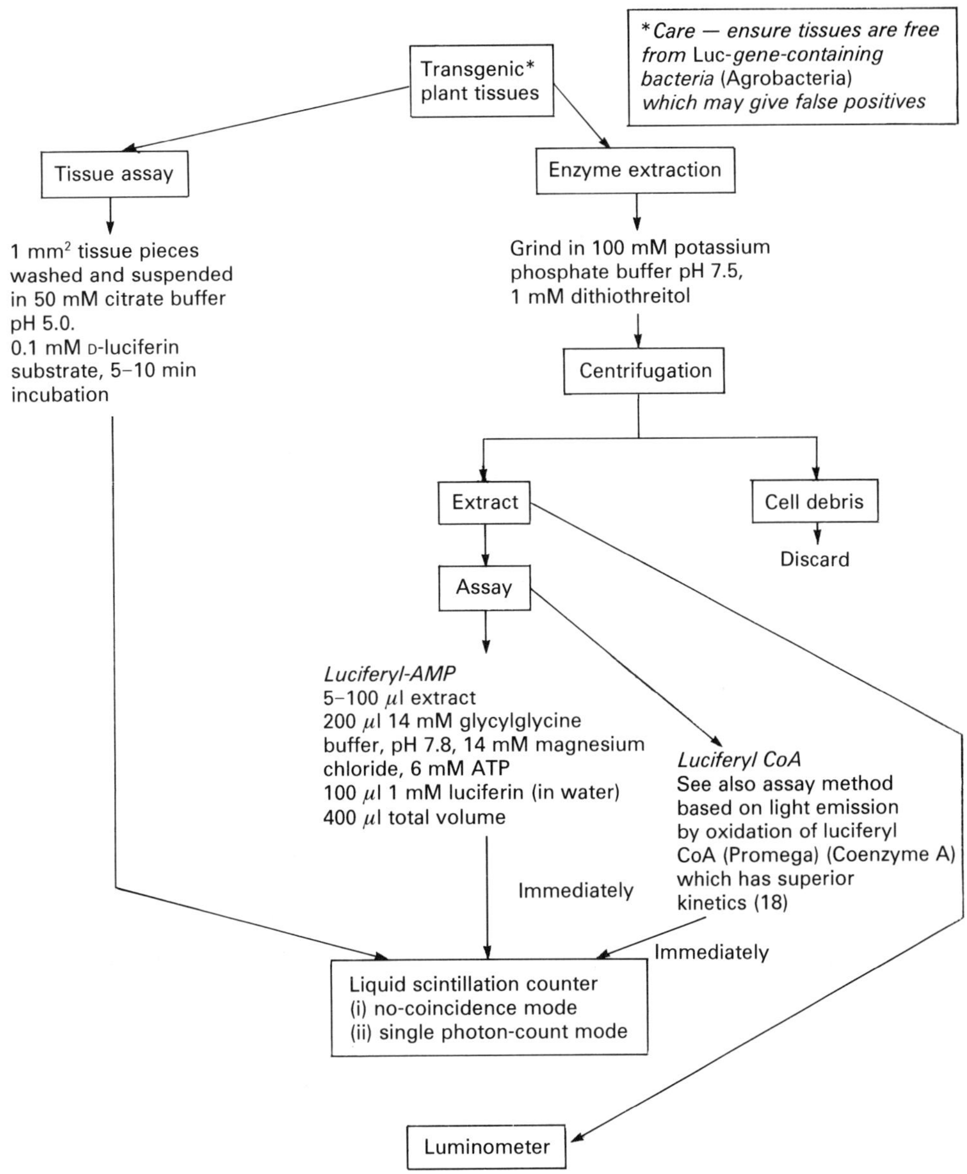

Transformation, Expression Vectors

High-count values will be obtained from tissues expressing the luciferase gene, whereas only background will be generated by untransformed control tissues. It is important to ensure that the tissue is free from bacteria in order to avoid false positive results due to emission of light by *Agrobacterium* containing the luciferase gene. A method for accurate quantitation of luciferase activity in plant extracts using a luminometer is presented in ref. 9.

3.3. Other reporter genes

The expression of the *E. coli lacZ* gene coding for β-galactosidase has been recorded in tobacco crown galls (19). In some of the transformed calli, the β-galactosidase activity was 20-fold higher than the background activity present in untransformed calli. The expression of the kanamycin-resistance gene (*npt*II) and the phosphinothricin-resistance gene (*bar*), described below, can also be quantified using biochemical assays (20, 21). See *Chapter 5* for reporter gene sequences.

4. SELECTABLE MARKERS

A selectable marker gene enables the recovery of transformants in a population of untransformed cells by giving them the ability to emerge in the presence of a selective agent which is toxic for untransformed cells.

4.1. Kanamycin resistance

The *npt*II tene from Tn*5* is the most widely used selectable marker in plant transformation. A kanamycin-resistance gene recovered from pK18 (22) was inserted into the polylinker of pUC19, giving pJIT134 (*Figure 18*). An interesting feature of this gene is that the *Pst*I and *Sph*I sites normally present in the coding region have been removed. The coding sequence has been placed under control of the CaMV 35S promoter in pJIT161 (*Figure 19*).

In some binary Ti vectors (pBIN19, pROK1, pBI101.1,.2,.3) (*Figures 29, 30* and *8*), the *npt*II gene has been shown to be defective and does not encode a fully active aminoglycoside phosphotransferase (46). This has not proved to be a problem for transformation of the Solanaceae (e.g. tobacco, tomato) but may be a problem for transforming other species.

Figure 18. Map for plasmid pJIT134.

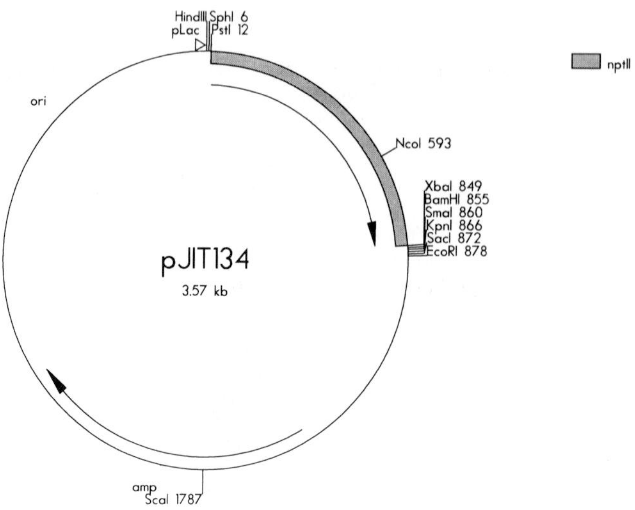

4.2. Hygromycin resistance

The hygromycin-resistance gene (*hph*) coding for hygromycin B phosphotransferase was first used as a selectable marker in tobacco (23, 24). The *hph* gene was recovered from pJR225 (25). A *Stu*I site has been created upstream of the translation start codon in pJIT6 (*Figure 20*). The coding sequence has been placed under control of the 35S promoter in pJIT72 (*Figure 21*).

Figure 19. Map for plasmid pJIT161.

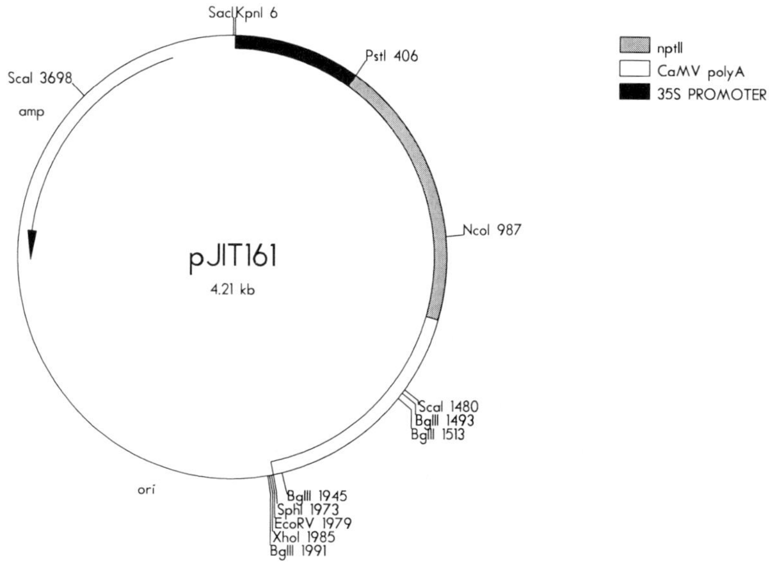

Figure 20. Map for plasmid pJIT6.

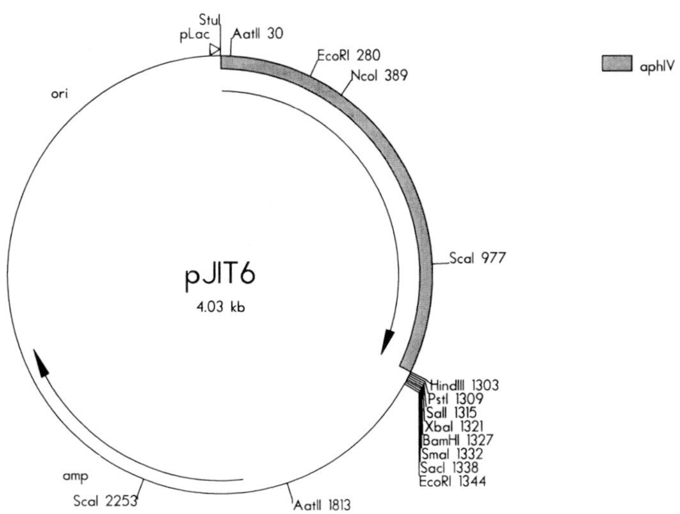

4.3. Bialaphos resistance

The bialaphos-resistance gene (*bar*) from *Streptomyces hygroscopicus* has been used to confer resistance to the herbicide phosphinothricin (21). The *bar* coding sequence can be recovered from pIJ4104 (26) (*Figure 22*). It has been placed under the control of the 35S promoter in pJIT82 (*Figure 23*).

Figure 21. Map for plasmid pJIT72.

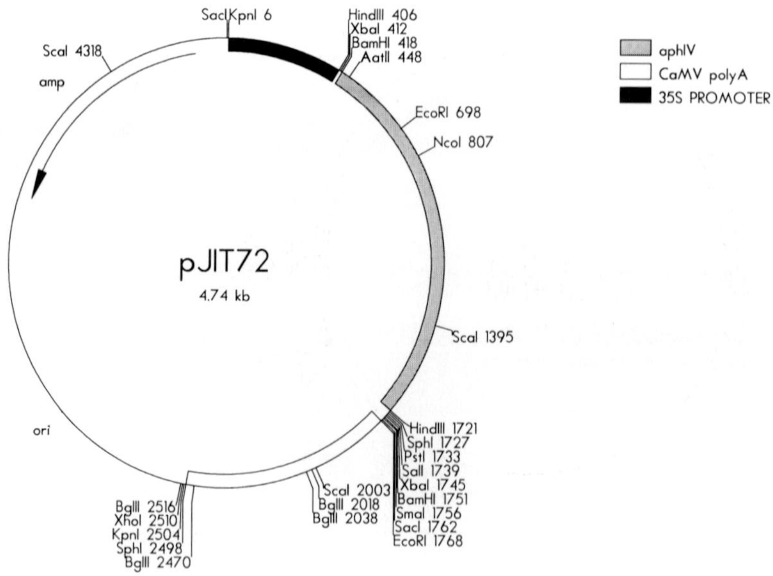

Figure 22. Map for plasmid pIJ4104.

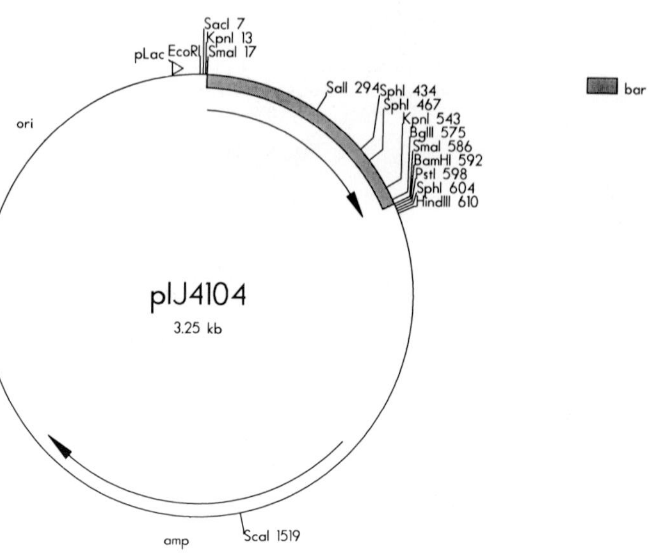

4.4. Sulfonamide resistance

Sulfonamide resistance (*sul*) can be conferred by a mutated dihydropteroate synthase gene present in various R factors. The *sul*I coding sequence has been recovered from a subclone of plasmid R46 and inserted into pUC19, creating pJIT92 (27) (*Figure 24*). As the gene product had to be targeted to chloroplasts to confer resistance to sulfonamides on plant cells, the *sul*I coding sequence was cloned into the expression cassette pJIT117 (6) (*Figure 7*), creating pJIT118 (28) (*Figure 25*).

Figure 23. Map for plasmid pJIT82.

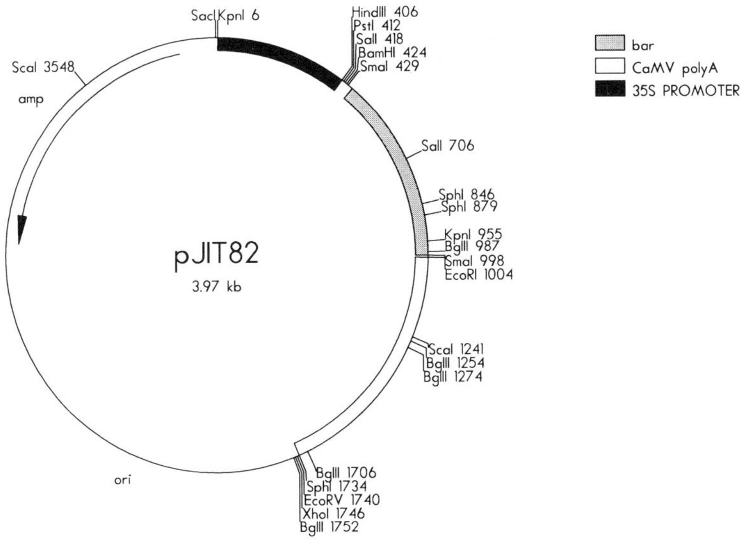

Figure 24. Map for plasmid pJIT92.

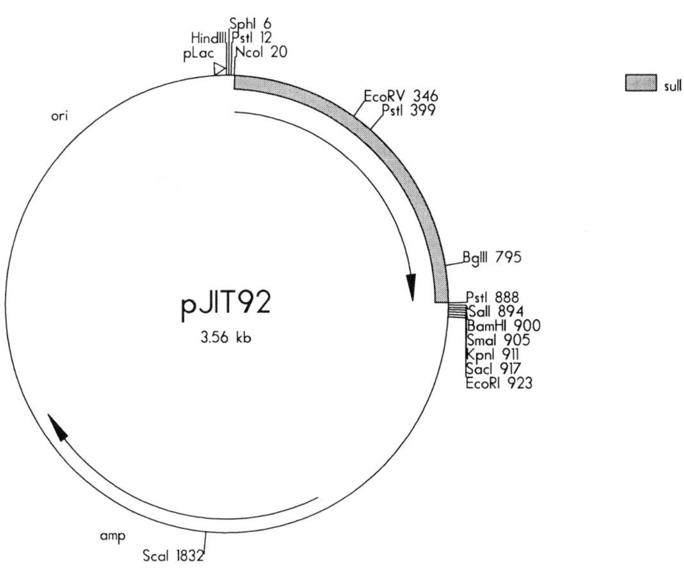

4.5. Streptomycin/spectinomycin resistance

Streptomycin and spectinomycin prevent greening of certain plant tissues at given concentrations. Resistance conferred by an *aad*A gene has been used as a color marker similar to the *spt* streptomycin-resistance gene (29). The *aad*A gene (30) can be recovered as a *Bsp*HI–*Xba*I fragment from pPM19 (31) (*Figure 26*). For expression in plant cells, a chimeric 35S-*aad*A-*ocs* is provided in pPM21 (31) (*Figure 27*).

Figure 25. Map for plasmid pJIT118.

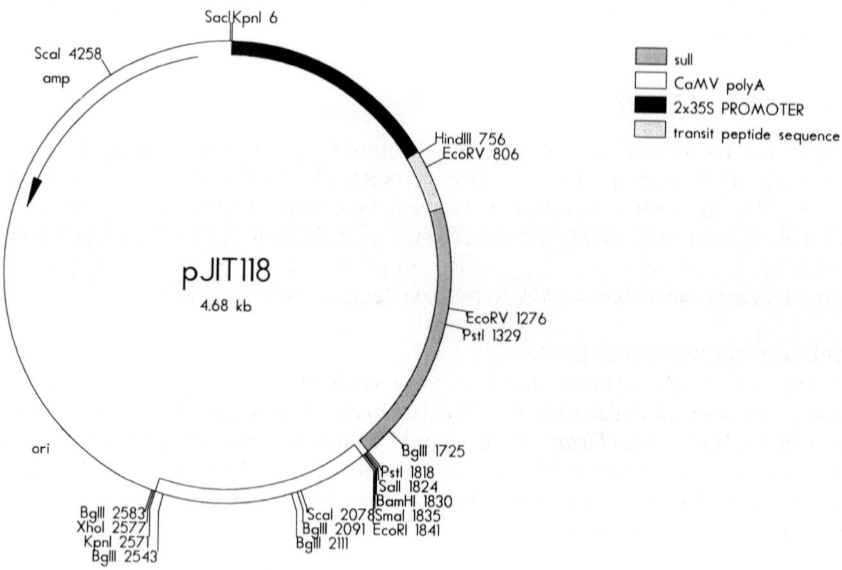

Figure 26. Map for plasmid pPM19.

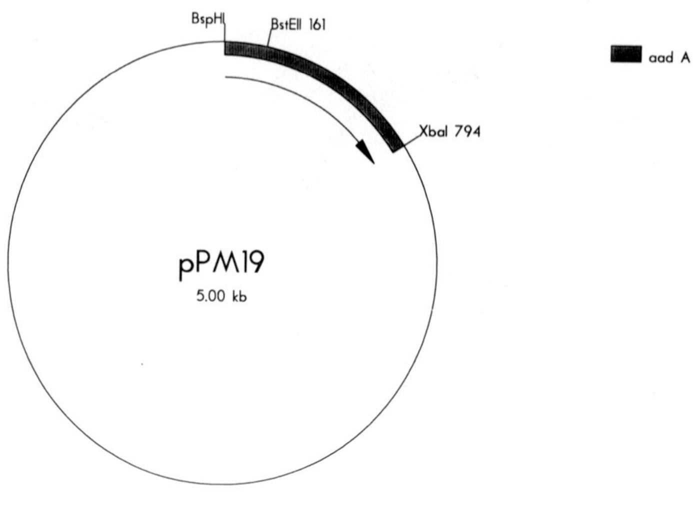

4.6. Chlorsulfuron resistance

Chlorsulfuron-resistance genes code for an acetolactate synthase (ALS) insensitive to inhibition by chlorsulfuron. A mutated *Arabidopsis* ALS gene was used to produce tobacco plants resistant to chlorsulfuron (32). The same mutation was introduced into the maize ALS cDNA. The mutated gene, present in pALS, proved to be an efficient selectable marker for maize transformation (33).

4.7. Methotrexate resistance

Another selectable marker used for cereal transformation is the mouse *dhfr* gene coding for dihydrofolate reductase and conferring resistance to methotrexate. Plasmid pMON806, containing the *dhfr* gene under control of the CaMV 35S promoter (34), has been used to produce various Gramineae cell lines resistant to methotrexate (35). See *Chapter 5* for selectable gene sequences.

5. PHENOTYPIC MARKERS

In contrast with the selectable markers, which require the use of a selective agent to recover transformants, phenotypic markers give transformants distinct features which can be identified easily. The thaumatin II gene has been used as a simple marker to confirm the transformed genotype of potato plants (36). Confirmation of the presence of the gene in the plants was given by the intensely sweet taste developed as a consequence of the presence of thaumatin in transformed tissues, which could be detected in as little as 0.1 g of leaf.

5.1. Anthocyanin regulatory genes

A regulatory gene (*Lc*) (37) involved in the anthocyanin biosynthesis pathway has been used to visualize transformed maize cells after bombardment of various tissues with gold particles coated with pPHI443 (38) (*Figure 28*), a plasmid containing an overexpressed chimeric *Lc* gene. Pigmented cells were seen 48 hours after bombardment. The *Lc* gene and other genes involved in pigment synthesis may prove to be extremely valuable tools for recovering transformants in the absence of any selection.

Figure 27. Map for plasmid pPM21.

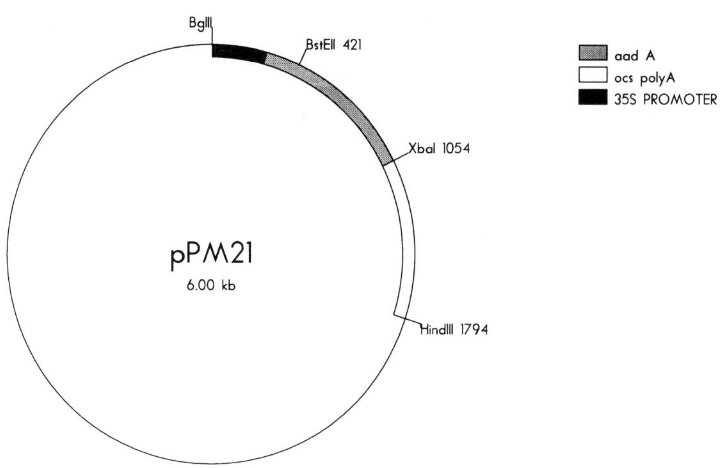

6. BINARY VECTORS

Binary vectors were developed to exploit the ability of *Agrobacterium* strains containing a Ti plasmid depleted of its T-DNA (disarmed) to transfer to plant cells a fragment of DNA comprised between T-DNA borders carried on another plasmid. The plasmid possesses the functions for its mobilization from *E. coli* to *Agrobacterium* and for its replication in these two hosts. Selectable marker gene(s) for the selection of bacterial transformants or exconjugants are present in the vector and a selectable marker gene for selection of transformed plant cells is integrated between the T-DNA borders. The gene to be transferred to plant cells can be cloned into unique restriction site(s) close to the selectable marker, between the T-DNA borders.

6.1. pBIN19

This widely used plasmid (39) (*Figure 29*) contains the broad host-range origin of replication and the mobilization functions from pRK2. One kanamycin-resistance gene is used as a selectable marker in bacteria and a chimeric *nos–npt*II*–nos* gene located between the T-DNA borders provides a selectable marker in plant tissues. Insertion of a sequence into the polylinker can be monitored using the *lac*Z blue/white complementation assay in suitable *E. coli* strains.

6.2. pROK1

An expression cassette comprising the CaMV 35S promoter and the *nos* polyadenylation signal has been introduced into pBIN19 to give pROK1 (40) (*Figure 30*). A unique *Bam*HI site must be used for placing any coding sequence under control of the 35S promoter. The sequence can then be transferred to *Agrobacterium* and to plant cells without any further manipulation.

Figure 28. Map for plasmid pPHI443.

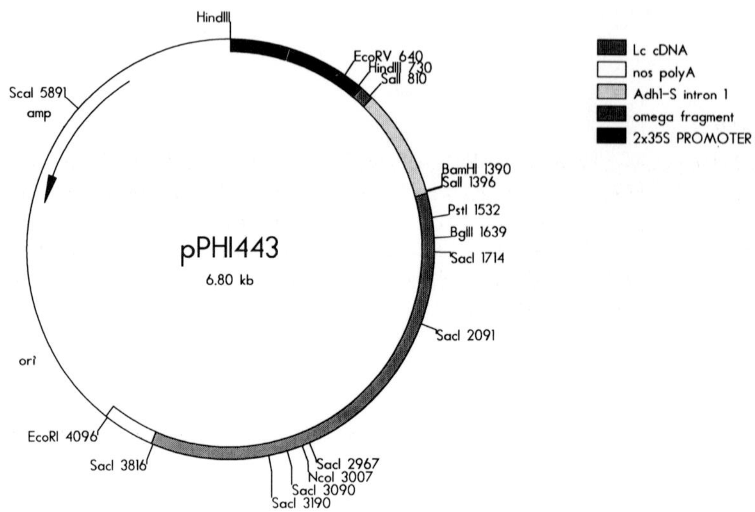

Figure 29. Map for plasmid pBIN19.

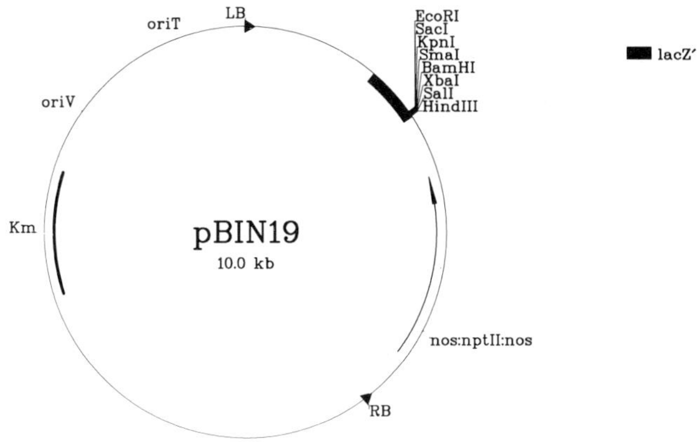

Figure 30. Map for plasmid pROK1.

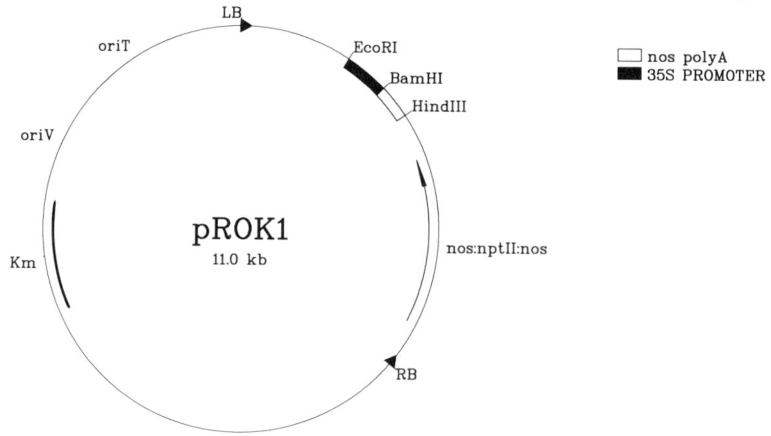

6.3. pGA482

As for pBIN19, pGA482 contains pRK2-derived mobilization and broad host-range replication functions (41) (*Figure 31*). A tetracycline-resistance gene can be used as a selectable marker in bacteria. The addition of a ColEI origin of replication to the broad host-range *ori*V provides a much higher plasmid copy number in *E. coli*. The presence of a lambda-phage *cos* site between the T-DNA borders facilitates the recovery of the T-DNA and surrounding plant DNA sequence from transformants.

6.4. pCGN1547, pCGN1548

The pCGN binary vectors contain a pRiHRI origin of replication for stable maintenance in *Agrobacterium* and a ColEI origin of replication for high copy number in *E. coli* (42) (*Figures 32, 33*). They must be introduced into *Agrobacterium* by transformation because they do not contain mobilization functions. The selectable markers are a gentamycin-resistance gene for bacteria and a chimeric *mas–npt*II–*mas* gene for plants. The *lacZ'* fragment from pUC18 is present between the T-DNA borders for easy identification of recombinants. Other constructs described in the same reference harbour a 35S-*npt*II gene as a selectable marker for plants, in both orientations with respect to the polylinker.

Figure 31. Map for plasmid pGA482.

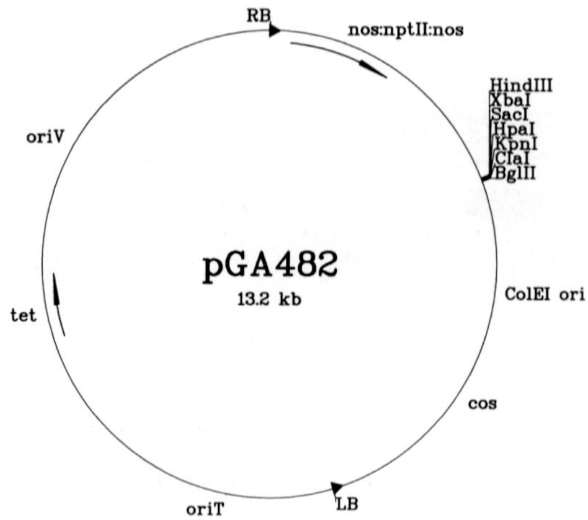

Figure 32. Map for plasmid pCGN1547.

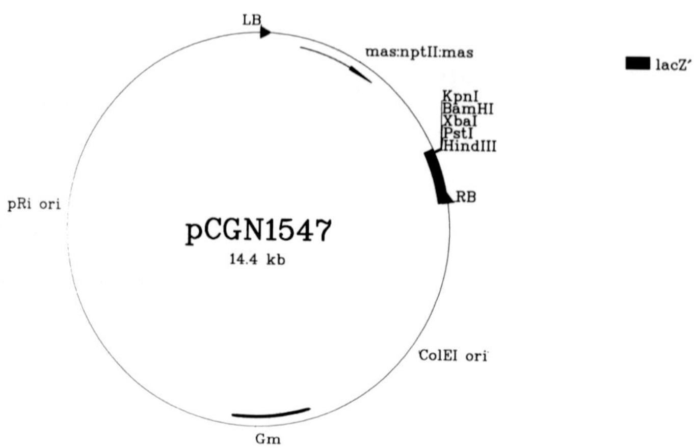

6.5. pUCD2340

The original feature of this binary vector is that the hygromycin-resistance gene is the selectable marker for plants (43) (*Figure 34*). A polylinker is present downstream of the CaMV 35S promoter. The selectable marker for bacteria is the kanamycin/gentamycin-resistance gene. pUCD2340 also contains the pSa broad host-range origin of replication *ori*V, the pBR origin of replication and the sequence for conjugative transfer using a helper plasmid. The lambda *cos* site is present between the T-DNA borders. Related plasmids not having the 35S promoter are described in the same reference (43).

Figure 33. Map for plasmid pCGN1548.

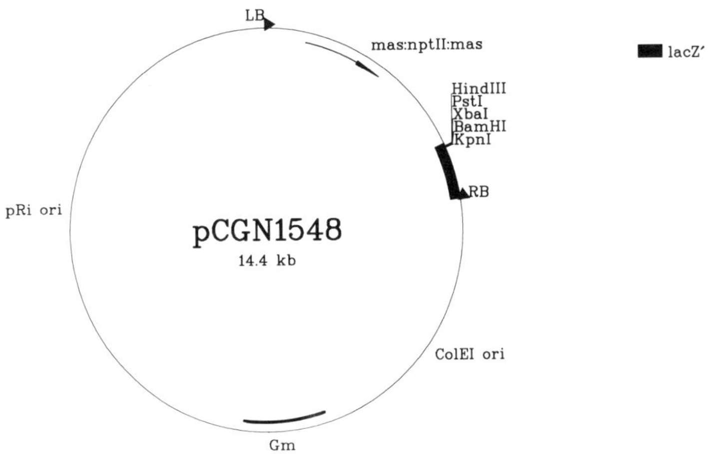

Figure 34. Map for plasmid pUCD2340.

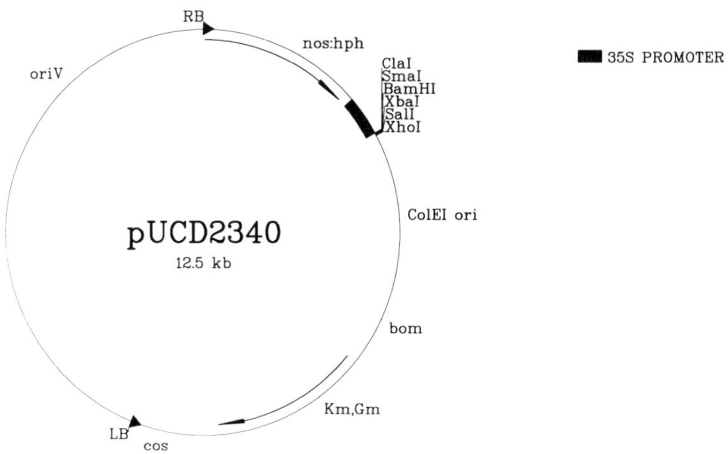

6.6. pC22

This vector was designed to introduce genomic libraries into plants. It contains the ColEI and pRiHRI origins of replication, and the pBR322 *bom* site for mobilization from *E. coli* to *Agrobacterium* (44) (*Figure 35*). Carbenicillin- and spectinomycin/streptomycin-resistance genes can be used as selectable markers in bacteria. A *nos–npt*II gene allows selection of plant transformants. Two unique restriction sites, *Bam*HI and *Xba*I, are located near a *cos* site. Another binary vector for transferring plant genomic libraries from *Agrobacterium* to plants is described in ref. 45.

Figure 35. Map for plasmid pC22.

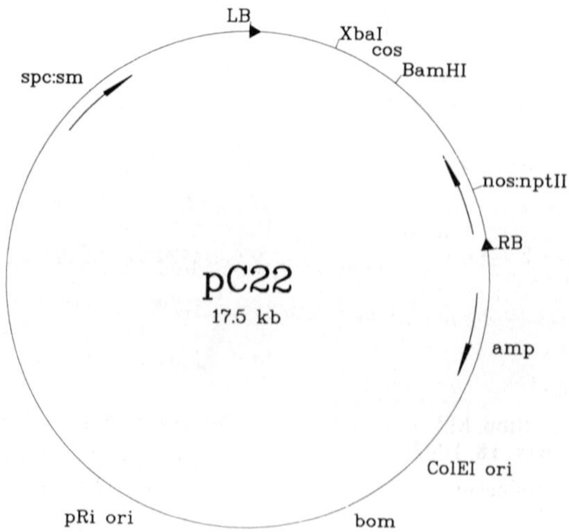

ACKNOWLEDGEMENTS

We thank S. Woolston and J. White for constructing pJIT6 and pJIT82, respectively.

7. REFERENCES

1. Topfer, R., Matzeit, V., Gronenborn, B., Schell, J. and Steinbiss, H.H. (1987) *Nucleic Acids Res.*, **15**, 5890.

2. Pietrzak, M., Shillito, R.D., Hohn, T. and Potrykus, I. (1986) *Nucleic Acids Res.*, **14**, 5857.

3. Guerineau, F., Lucy, A. and Mullineaux, P. (1992) *Plant Mol. Biol.*, **18**, 815.

4. Ainley, W.M. and Key, J.L. (1990) *Plant Mol. Biol.*, **14**, 949.

5. Schardl, C.L., Byrd, A.D., Benzion, G., Altschuler, M.A., Hildebrand, D.F. and Hunt, A.G. (1987) *Gene*, **61**, 1.

6. Guerineau, F., Woolston, S., Brooks, L. and Mullineaux, P. (1988) *Nucleic Acids Res.*, **16**, 11380.

7. Jefferson, R.A., Burgess, S.M. and Hirsh, D. (1986) *Proc. Natl. Acad. Sci. USA*, **83**, 8447.

8. Herrera-Estrella, L., Depicker, A., Van Montagu, M. and Schell, J. (1983) *Nature*, **303**, 209.

9. Ow, D.W., Wood, K.V., DeLuca, M., de Wet, J.R., Helinski, D.R. and Howell, S.H. (1986) *Science*, **234**, 856.

10. Jefferson, R.A., Kavanagh, T.A. and Bevan, M.W. (1987) *EMBO J.*, **6**, 3901.

11. Gallagher, S.R. 'GUS Protocols', Clontech product number V2003-1.

12. Alton, N.K. and Vapnek, D. (1979) *Nature*, **282**, 864.

13. De Wet, J.R., Wood, K.V., Helinski, D.R. and DeLuca, M. (1985) *Proc. Natl. Acad. Sci. USA*, **82**, 7870.

14. Mullineaux, P.M., Guerineau, F. and Accotto, G.P. (1990) *Nucleic Acids Res.*, **18**, 7259.

15. Gorman, C.M., Moffat, L.F. and Howard, B.H. (1982) *Mol. Cell. Biol.*, **2**, 1044.

16. Sleigh, M.J. (1986) *Anal. Biochem.*, **156**, 251.

17. Bradford, M.M. (1976) *Anal. Biochem.*, **72**, 248.

18. Wood, K.V. (1990) In 'Firefly Luciferase: A New Tool for Molecular Biologists', *Promega Notes*, **28**.

19. Matsumoto, S., Takebe, I. and Machida, Y. (1988) *Gene*, **66**, 19.

20. McDonnell, R.E., Clark, R.D., Smith, W.A. and Hinchee, M.A. (1987) *Plant Mol. Biol. Rep.*, **5**, 380.

21. De Block, M., Botterman, J., Vandewiele, M., Dockx, J., Thoen, C., Gossele, V., Movva, N.R., Thompson, C., Van Montagu, M. and Leemans, J. (1987) *EMBO J.*, **6**, 2513.

22. Pridmore, R.D. (1987) *Gene*, **56**, 309.

23. Waldron, C., Murphy, E.B., Roberts, J.L., Gustafson, G.D., Armour, S.L. and Malcolm, S.K. (1985) *Plant Mol. Biol.*, **5**, 103.

24. Van den Elzen, P.J.M., Townsend, J., Lee, K.Y. and Bedbrook, J.R. (1985) *Plant Mol. Biol.*, **5**, 299.

25. Gritz, L. and Davies, J. (1983) *Gene*, **25**, 179.

26. White, J., Chang, S.Y.P., Bibb, M.J. and Bibb, M.J. (1990) *Nucleic Acids Res.*, **18**, 1062.

27. Guerineau, F. and Mullineaux, P. (1989) *Nucleic Acids Res.*, **17**, 4370.

28. Guerineau, F., Brooks, L., Meadows, J., Lucy, A., Robinson, C. and Mullineaux, P. (1990) *Plant Mol. Biol.*, **15**, 127.

29. Maliga, P., Svab, Z., Harper, E.C. and Jones, J.D.G. (1988) *Mol. Gen. Genet.*, **214**, 456.

30. Chinault, A.C., Blakesley, V.A., Roessler, E., Willis, D.G., Smith, C.A., Cook, R.G. and Fenwick, R.G., Jr. (1986) *Plasmid*, **15**, 119.

31. Svab, Z., Harper, E.C., Jones, J.D.G. and Maliga, P. (1990) *Plant Mol. Biol.*, **14**, 197.

32. Haugh, G.W., Smith, J., Mazur, B. and Somerville, C. (1988) *Mol. Gen. Genet.*, **211**, 266.

33. Fromm, M.E., Morrish, F., Armstrong, C., Williams, R., Thomas, J. and Klein, T.M. (1990) *Biotechnology*, **8**, 833.

34. Eichholtz, D.A., Rogers, S.G., Horsch, R.B., Klee, H.J., Hayford, M., Hoffmann, N.L., Braford, S.B., Fink, C., O'Connell, K.M. and Fraley, R.T. (1987) *Somatic Cell Mol. Genet.*, **13**, 67.

35. Hauptmann, R.M., Vasil, V., Ozias-Akins, P., Tabaeizadeh, Z., Rogers, S.G., Fraley, R.T., Horsch, R.B. and Vasil, I.K. (1988) *Plant Physiol.*, **86**, 602.

36. Witty, M. (1989) *Nucleic Acids Res.*, **17**, 3312.

37. Ludwig, S.R., Habera, L.F., Dellaporta, S.L. and Wessler, S.R. (1989) *Proc. Natl. Acad. Sci. USA*, **86**, 7092.

38. Ludwig, S.R., Bowen, B., Beach, L. and Wessler, S.R. (1990) *Science*, **247**, 449.

39. Bevan, M. (1984) *Nucleic Acids Res.*, **12**, 8711.

40. Baulcombe, D.C., Saunders, G.R., Bevan, M.W., Mayo, M.A. and Harrison, B.D. (1986) *Nature*, **321**, 446.

41. An, G. (1986) *Plant Physiol.*, **81**, 86.

42. McBride, K.E. and Summerfelt, K.R. (1990) *Plant Mol. Biol.*, **14**, 269.

43. Zyprian, E. and Kado, C.I. (1990) *Plant Mol. Biol.*, **15**, 245.

44. Simoens, C., Alliotte, T., Mendel, R., Muller, A., Schiemann, J., Van Lijsebettens, M., Schell, J., Van Montagu, M. and Inze, D. (1986) *Nucleic Acids Res.*, **14**, 8073.

45. Olszewski, N.E., Martin, F.B. and Ausubel, F.M. (1989) *Nucleic Acid Res.*, **16**, 10765.

46. Yenofsky, R.L., Fine, M. and Pellow, J.W. (1990) *Proc. Natl. Acad. Sci. USA*, **87**, 3435.

CHAPTER 5

PLANT SELECTABLE GENES, REPORTER GENES AND PROMOTERS

R.R.D. Croy

1. INTRODUCTION

This chapter contains the nucleotide sequences, encoded polypeptides, sequence features and restriction maps for most of the commonly used selectable [**S**] and screenable or reporter genes [**R**] used for plant transformations and gene expression studies (see *Table 1*). Some reporter genes can be assayed histochemically with appropriate substrates [$\mathbf{R^H}$], providing further information on the spatial distribution of gene expression and also for protein targeting studies. Promoter [**P**] and terminator [**T**] sequences, often used in conjunction with selectable marker genes to provide high levels of expression, have also been included. Most of the gene sequences in this section were extracted from the GenBank database (release 72.0) distributed on CD-ROM using the National Biomedical Research Foundation (NBRF) Atlas retrieval program version 1.40 (June 1992). The sequences of some genes described in *Chapter 4* were supplied by F. Guerineau and P. Mullineaux. Raw sequence data were processed using Seqaid II, version 3.5, and formatted using a standard wordprocessing program. Restriction maps were compiled and drawn using the IBM program CLONING, version 2.0, supplied by Marti Alde. Gene and protein properties and sequence features of potential interest are indicated where known or as indicated through sequence processing. The presence of potential N-glycosylation sites within the encoded polypeptides sequence was revealed using a simple scan program for the occurrence of the amino acid sequence N–X–S or N–X–T within the encoded polypeptide sequence. The presence of either sequence does not confirm glycosylation at this site since the localized conformation in the vicinity of the site is an important determinant. Thus Pro in the X position prevents glycosylation and Pro occurring after Ser or Thr also decreases N-glycosylation. The information given for each sequence includes: nucleotide sequence; length; transcription and translation start and stop points; putative glycosylation sites; multiple cloning sites; encoded polypeptide sizes and isoelectric points (calculated). A restriction map for each sequence is also provided, which shows the positions of unique and useful enzyme sites. Two selectable genes which have been used in plants, but not included here, are the mouse dihydrofolate reductase gene conferring methotrexate resistance (1, 2), and the *Arabidopsis* acetolactate synthase gene which confers resistance to chlorsulfuron (3, 4).

2. AMINOGLYCOSIDE ADENYLTRANSFERASE GENE [S]

This is the aminoglycoside adenyltransferase (*aadA*) gene from *Shigella flexneri* 2a plasmid pCN1. It encodes the enzyme aminoglycoside adenyltransferase which provides resistance to trimethoprim, streptomycin, spectinomycin and sulfonamides. The gene encodes a polypeptide of 262 amino acids, molecular weight 29206 and pI 6.81. Overall sequence length = 938 nt (5). Sequence features are shown in bold type and labeled above or below the sequence. **TLS**, translation start; **TLE**, translation end; ■ indicates a TAA (ochre) stop codon. A potential N-linked glycosylation site is present at Asn167.

A restriction map of this gene is shown in *Figure 1*.

Table 1. Gene abbreviations and phenotypes conferred by the gene sequences

Gene	Abbreviations	Phenotype	Refs
Aminoglycoside adenyl-transferase	*aad*	Resistance to trimethoprim, streptomycin, spectinomycin and sulphonamides	5
Bleomycin resistance	*ble*	Resistance to bleomycin	6
Cauliflower mosaic virus promoter	CaMV 5′	Plant promoter used for high-level "constitutive" expression	
Cauliflower mosaic virus promoter and terminator	CaMV 5′ and 3′	A second CaMV 5′ with authentic terminator sequence	7
Chloramphenicol acetyltransferase	*cat*	Synthesis of chloramphenicol acetyltransferase enzyme	8, 9
Dihydrofolate reductase	*dhfr*	Resistance to methotrexate	10, 32
Dihydropteroate synthase	*sul, dhps*	Resistance to the sulphonamide Asulam	13
Enolpyruvylshikimate phosphate synthase	*epsps*	Resistance to glyphosate herbicide	14
β-Galactosidase	*β-gal, LacZ*	Synthesis of galactosidase enzyme	15–18
β-Glucuronidase	*uidA, gus, β-glu*	Synthesis of glucuronidase enzyme	19–23
Hygromycin phospho-transferase	*hph, aphIV, hyg*	Resistance to the aminoglycoside hygromycin B	24
Luciferase (firefly)	*luc*	Synthesis of luciferase enzyme producing bioluminescence (light emission)	25–27
Luciferase (bacterial)	*luxA, luxB*	Synthesis of luciferase enzyme polypeptides producing bioluminescence (light emission)	28, 29
Neomycin phosphatase	*nptII, kan, aphII*	Resistance to neomycin amino-glycosides such as kanamycin	30
Nopaline synthase	*nos, nop,* *nos* 5′ and 3′	Synthesis of nopaline; also promoter and terminator sequences to provide high levels of constitutive expression	11, 31, 32
Octopine synthase	*ocs* 5′ and 3′	Synthesis of octopine; promoter and terminator sequences to provide constitutive expression	11, 33, 34
Phosphinothricin acetyltransferase	*pat, bar*	Resistance to bialaphos herbicide	35

```
  1  CATTCAAGCCGACACCGCTTCGCGGCGCGGCTTAATTCAGGAGTTAAACA
     ..................................................
       ⇦ TLS
 51  TCATGAGGGAAGCGGTGATCGCCGAAGTATCGACTCAACTATCAGAGGTA
     ..M  R  E  A  V  I  A  E  V  S  T  Q  L  S  E  V

101  GTTGGCGTCATCGAGCGCCATCTCGAACCGACGTTGCTGGCCGTACATTT
      V  G  V  I  E  R  H  L  E  P  T  L  L  A  V  H  L

151  GTACGGCTCCGCAGTGGATGGCGGCCTGAAGCCACACAGTGATATTGATT
       Y  G  S  A  V  D  G  G  L  K  P  H  S  D  I  D

201  TGCTGGTTACGGTGACCGTAAGGCTTGATGAAACAACGCGGCGAGCTTTG
     L  L  V  T  V  T  V  R  L  D  E  T  T  R  R  A  L

251  ATCAACGACCTTTTGGAAACTTCGGCTTCCCCTGGAGAGAGCGAGATTCT
      I  N  D  L  L  E  T  S  A  S  P  G  E  S  E  I  L

301  CCGCGCTGTAGAAGTCACCATTGTTGTGCACGACGACATCATTCCGTGGC
       R  A  V  E  V  T  I  V  V  H  D  D  I  I  P  W

351  GTTATCCAGCTAAGCGCGAACTGCAATTTGGAGAATGGCAGCGCAATGAC
     R  Y  P  A  K  R  E  L  Q  F  G  E  W  Q  R  N  D

401  ATTCTTGCAGGTATCTTCGAGCCAGCCACGATCGACATTGATCTGGCTAT
      I  L  A  G  I  F  E  P  A  T  I  D  I  D  L  A  I

451  CTTGCTGACAAAAGCAAGAGAACATAGCGTTGCCTTGGTAGGTCCAGCGG
       L  L  T  K  A  R  E  H  S  V  A  L  V  G  P  A

501  CGGAGGAACTCTTTGATCCGGTTCCTGAACAGGATCTATTTGAGGCGCTA
     A  E  E  L  F  D  P  V  P  E  Q  D  L  F  E  A  L
glycosylation site
551  AATGAAACCTTAACGCTATGGAACTCGCCGCCCGACTGGGCTGGCGATGA
      N  E  T  L  T  L  W  N  S  P  P  D  W  A  G  D  E

601  GCGAAATGTAGTGCTTACGTTGTCCCGCATTTGGTACAGCGCAGTAACCG
       R  N  V  V  L  T  L  S  R  I  W  Y  S  A  V  T

651  GCAAAATCGCGCCGAAGGATGTCGCTGCCGACTGGGCAATGGAGCGCCTG
     G  K  I  A  P  K  D  V  A  A  D  W  A  M  E  R  L

701  CCGGCCCAGTATCAGCCCGTCATACTTGAAGCTAGGCAGGCTTATCTTGG
      P  A  Q  Y  Q  P  V  I  L  E  A  R  Q  A  Y  L  G

751  ACAAGAAGATCGCTTGGCCTCGCGCGCAGATCAGTTGGAAGAATTTGTTC
       Q  E  D  R  L  A  S  R  A  D  Q  L  E  E  F  V
                                              TLE ⇨
801  ACTACGTGAAAGGCGAGATCACCAAGGTAGTCGGCAAATAATGTCTAACA
     H  Y  V  K  G  E  I  T  K  V  V  G  K  ■ ........

851  ATTCGTTCAAGCCGACGCCGCTTCGCGGCGCGGCTTAACTCAAGCGTTAG
     ..................................................

901  AGAGCTGGGGAAGACTATGCGCGATCTGTTGAAGGTGG
     ......................................
```

Figure 1. Restriction map of the *aad* gene.

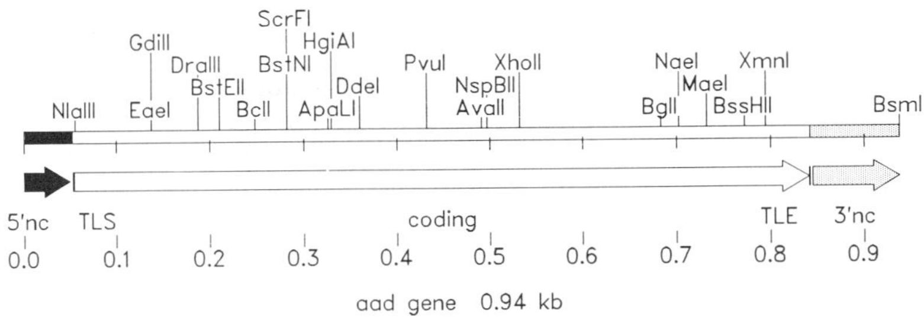

3. BLEOMYCIN-RESISTANCE GENE [S]

The bleomycin-resistance (*ble*) gene was isolated from transposon Tn*5* along with neomycin- and streptomycin-resistance genes. Bleomycin is a glycopeptide antibiotic which causes single- and double-stranded breaks in DNA, and has been proposed as a selection system for transformed plant cells. The regeneration and development of *Nicotiana plumbaginifolia* and *Lycopersicon esculentum* plant tissues are inhibited by low levels of bleomycin (1–10 μg ml^{-1}). The gene encodes a polypeptide of 126 amino acids, molecular weight 14 076 and pI 6.89, which provides resistance against the antibiotic. The nature of the resistance conferred is unknown. Overall sequence length = 415 nt (6). Sequence features are shown in bold type, underlined and labeled above or below the sequence. **TLS**, translation start; **TLE**, translation end; ■ indicates a TGA (umber) stop codon. There are no apparent N-glycosylation sites.

```
     ⇦ TLS
  1  ATGACCGACCAAGCGACGCCCAACCTGCCATCACGAGATTTCGATTCCAC
      M  T  D  Q  A  T  P  N  L  P  S  R  D  F  D  S  T

 51  CGCCGCCTTCTATGAAAGGTTGGGCTTCGGAATCGTTTTCCGGGACGCCG
       A  A  F  Y  E  R  L  G  F  G  I  V  F  R  D  A

101  GCTGGATGATCCTCCAGCGCGGGGATCTCATGCTGGAGTTCTTCGCCCAC
     G  W  M  I  L  Q  R  G  D  L  M  L  E  F  F  A  H

151  CCCGGGCTCGATCCCCTCGCGAGTTGGTTCAGCTGCTGCCTGAGGCTGGA
      P  G  L  D  P  L  A  S  W  F  S  C  C  L  R  L  D

201  CGACCTCGCGGAGTTCTACCGGCAGTGCAAATCCGTCGGCATCCAGGAAA
       D  L  A  E  F  Y  R  Q  C  K  S  V  G  I  Q  E

251  CCAGCAGCGGCTATCCGCGCATCCATGCCCCCGAACTGCAGGAGTGGGGA
     T  S  S  G  Y  P  R  I  H  A  P  E  L  Q  E  W  G

301  GGCACGATGGCCGCTTTGGTCGACCCGGACGGGACGCTCCTGCGCCTGAT
      G  T  M  A  A  L  V  D  P  D  G  T  L  L  R  L  I
                                        TLE ⇨
351  ACAGAACGAATTGCTTGCAGGCATCTCATGAGTGTGTCTTCCCGTTTTCC
       Q  N  E  L  L  A  G  I  S  ■

401  GCCTGAGGTCACTGC
     ...............
```

A restriction map of this gene is shown in *Figure 2*.

Figure 2. Restriction map of the *ble* gene.

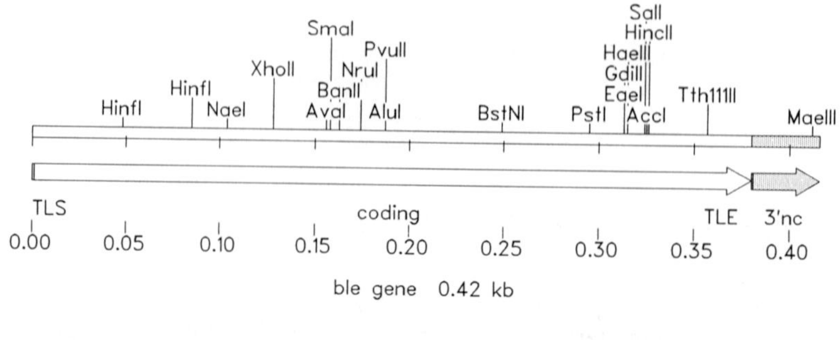

4. CAULIFLOWER MOSAIC VIRUS 35S PROMOTER [P]

This promoter drives the expression of the 35S RNA of cauliflower mosaic virus (CaMV) and provides high levels of expression of heterologous genes in a variety of different cell types of many different dicot and monocot plant species. The promoter is highly active in both transient expression and stably transformed plant cells. The putative 56 bp core promoter sequence lies between positions 338–393; enhancer elements lie upstream from this sequence. Overall sequence length = 461 nt. Sequence features are shown in bold type, underlined and labeled above or below the sequence. **TCS**, transcription start. Sequence data supplied by F. Guerineau and P. Mullineaux, corresponding to CaMV positions 7040–7500.

```
  1  CTACTCCAAAAATGTCAAAGATACAGTCTCAGAAGACCAAAGGGCTATTG
     ..................................................

 51  AGACTTTTCAACAAAGGGTAATTTCGGGAAACCTCCTCGGATTCCATTGC
     ..................................................

101  CCAGCTATCTGTCACTTCATCGAAAGGACAGTAGAAAAGGAAGGTGGCTC
     ..................................................

151  CTACAAATGCCATCATTGCGATAAAGGAAAGGCTATCATTCAAGATGCCT
     ..................................................

201  CTGCCGACAGTGGTCCCAAAGATGGACCCCCACCCACGAGGAGCATCGTG
     ..................................................

251  GAAAAAGAAGACGTTCCAACCACGTCTTCAAAGCAAGTGGATTGATGTGA
     ..................................................
                                         ⇦  56bp core

301  CATCTCCACTGACGTAAGGGATGACGCACAATCCCACTATCCTTCGCAAG
     ..................................................
     promoter                                  ⇨
                   TATA Box                       ⇦ TCS
351  ACCCTTCCTCTATATAAGGAAGTTCATTTCATTTGGAGAGGACACGCTGA
     ..................................................

               35S RNA→
401  AATCACCAGTCTCTCTCTACAAATCTATCTCTCTCTATTTTCTCCATAAA
     ..................................................

451  TAATGTGTGAG
     ...........
```

A restriction map of this promoter is shown in *Figure 3.*

Figure 3. Restriction map of the CaMV 35S promoter.

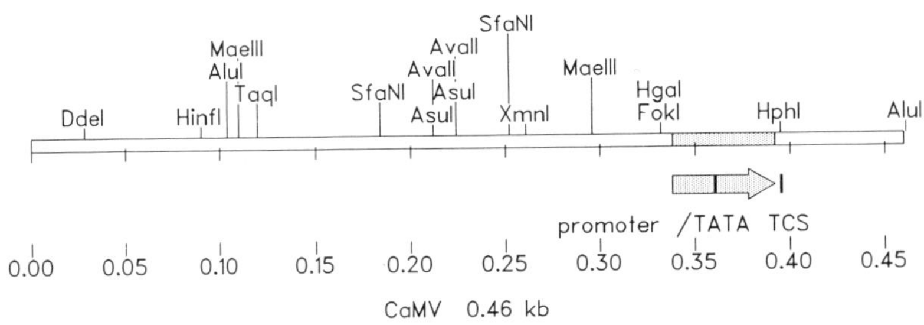

5. CAULIFLOWER MOSAIC VIRUS 35S PROMOTER AND TERMINATOR [P, T]

This is the cauliflower mosaic virus 35S promoter used in plant expression vectors pRT100–pRT104 for transcriptional and translational fusions derived from CaMV strain Cabb B-D. Ths translation start codon in pRT103 and pRT104 is inserted within the consensus sequence for optimal ribosome initiation in eukaryotes. Sequence features are shown in bold type and indicated above or below the sequence. Overall sequence length = 711 nt (7). **mcs**, multiple cloning sites; **TCS**, transcription start; **TCE**, transcription end; **TLS**, translation start; **TLE**, translation end (stop codons present in all three reading frames).

```
     ⇦        mcs        ⇨
  1  AAGCTTGCATGCCTGCAGGTCAACATGGTGGAGCACGACACTCTCGTCTA

 51  CTCCAAGAATATCAAAGATACAGTCTCAGAAGACCAGAGGGCTATTGAGA

101  CTTTTCAACAAAGGGTAATATCGGGAAACCTCCTCGGATTCCATTGCCCA

151  GCTATCTGTCACTTCATCGAAAGGACAGTAGAAAAGGAAGATGGCTTCTA

201  CAAATGCCATCATTGCGATAAAGGAAAGGCTATCGTTCAAGAATGCCTCT

251  ACCGACAGTGGTCCCAAAGATGGACCCCCACCCACGAGGAACATCGTGGA

301  AAAAGAAGACGTTCCAACCACGTCTTCAAAGCAAGTGGATTGATGTGATA

351  TCTCCACTGACGTAAGGATGACGCACAATCCCACTATCCTTCGCAAGAC

                                                ⇦
                                              ⇦ TCS
             TATA box                       ACCTCGAGTGG pRT104
401  CCTTCCTCTATATAAGGAAGTTCATTTCATTTGGAGAGGACCTCGAGGGC pRT100

        mcs                                 ⇨
      ⇦ ⇦ TLS
     CCACCATGGCGAGCTCGGTACCCGGGGATCCTCTAGAGT            pRT104
451  CCATGGGCGAGCTCGGTACCCGGGGATCCTCTAGAGTCCGCAAATCACCA pRT100
     .. M  G  coding ➔.................................
                                           TLE ⇨   TLE
501  GTCTCTCTCTACAAATCTATCTCTCTCTATTTTCTCCAGAATAATGTGTG
     ......................................... ■ .... ■
    ⇨TLE⇨
551  AGTAGTTCCCAGATAAGGGAATTAGGGTTCTTATAGGGTTTCGCTCATGT
      . ■ .............................................

601  GTTGAGCATATAAGAAACCCTTAGTATGTATTTGTATTTGTAAAATACTT

          polyA signal                 TCE ⇨     ⇦
651  CTATCAATAAAATTTCTAATTCCTAAAACCAAAATCCAGTGACCTGCAGG

      mcs        ⇨
701  CATGCAAGCTT
```

A restriction map of the promoter and terminator is shown in *Figure 4*.

Figure 4. Restriction map of CaMV 35S promoter and terminator.

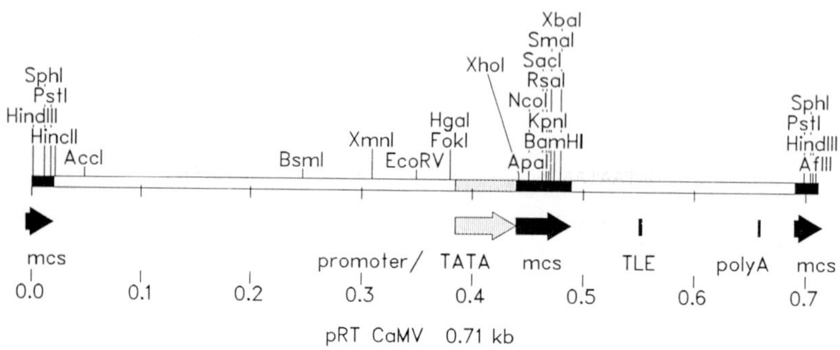

6. CHLORAMPHENICOL ACETYLTRANSFERASE GENE [R]

This is the chloramphenicol resistance (*cat*) gene isolated from transposon Tn*9*. The gene encodes the enzyme chloramphenicol acetyltransferase (CAT) and was the first bacterial gene to be expressed in plants. The enzyme specifically acetylates chloramphenicol antibiotics to the 1-, 3-, and 1,3-acetylated derivatives, which are inactive. Although not used as a selection system in plants, the gene is used frequently as a reporter gene in plant promoter work, since there is a very sensitive CAT assay (see *Chapter 4*). The gene encodes an enzyme of 219 amino acids, molecular weight 25 666 and pI 7.22. Overall sequence length = 839 nt (8, 9). Sequence features are shown in bold type, underlined and labeled above or below the sequence. **TLS**, translation start; **TLE**, translation end; ■ indicates a TAA (ochre) stop codon. A potential N-linked glycosylation site is present at Asn34.

```
                                                 ⇦ TLS
  1  AGGCCTAGCTTGCGAGATTTTCAGGAGCTAAGGAAGCTAAAATGGAGAAA
     ..........................................M  E  K

 51  AAAATCACTGGATATACCACCGTTGATATATCCCAATGGCATCGTAAAGA
      K  I  T  G  Y  T  T  V  D  I  S  Q  W  H  R  K  E
                                          glycosylation site
101  ACATTTTGAGGCATTTCAGTCAGTTGCTCAATGTACCTATAACCAGACCG
       H  F  E  A  F  Q  S  V  A  Q  C  T  Y  N  Q  T

151  TTCAGCTGGATATTACGGCCTTTTTAAAGACCGTAAAGAAAAATAAGCAC
     V  Q  L  D  I  T  A  F  L  K  T  V  K  K  N  K  H

201  AAGTTTTATCCGGCCTTTATTCACATTCTTGCCCGCCTGATGAATGCTCA
      K  F  Y  P  A  F  I  H  I  L  A  R  L  M  N  A  H

251  TCCGGAATTCCGTATGGCAATGAAAGACGGTGAGCTGGTGATATGGGATA
       P  E  F  R  M  A  M  K  D  G  E  L  V  I  W  D

301  GTGTTCACCCTTGTTACACCGTTTTCCATGAGCAAACTGAAACGTTTTCA
     S  V  H  P  C  Y  T  V  F  H  E  Q  T  E  T  F  S

351  TCGCTCTGGAGTGAATACCACGACGATTTCCGGCAGTTTCTACACATATA
      S  L  W  S  E  Y  H  D  D  F  R  Q  F  L  H  I  Y

401  TTCGCAAGATGTGGCGTGTTACGGTGAAAACCTGGCCTATTTCCCTAAAG
       S  Q  E  V  A  C  Y  G  E  N  L  A  Y  F  P  K

451  GGTTTATTGAGAATATGTTTTTCGTCTCAGCCAATCCCTGGGTGAGTTTC
     G  F  I  E  N  M  F  F  V  S  A  N  P  W  V  S  F

501  ACCAGTTTTGATTTAAACGTGGCCAATATGGACAACTTCTTCGCCCCCGT
      T  S  F  D  L  N  V  A  N  M  D  N  F  F  A  P  V

551  TTTCACCATGGGCAAATATTATACGCAAGGCGACAAGGTGCTGATGCCGC
       F  T  M  G  K  Y  Y  T  Q  G  D  K  V  L  M  P

601  TGGCGATTCAGGTTCATCATGCCGTCTGTGATGGCTTCCATGTCGGCAGA
     L  A  I  Q  V  H  H  A  V  C  D  G  F  H  V  G  R
                                                 TLE  ⇨
651  ATGCTTAATGAATTACAACAGTACTGCGATGAGTGGCAGGGCGGGGCGTA
      M  L  N  E  L  Q  Q  Y  C  D  E  W  Q  G  G  A  ■

701  ATTTTTTTAAGGCAGTTATTGGTGCCCTTAAACGCCTGGTGCTACGCCTG
       ................................................
                                                ⇦
751  AATAAGTGATAATAAGCGGATGAATGGCAGAAATTCGCGAGTCTGGTACC
     ..................................................
                 mcs                           ⇨
801  CGGGGATCCTCTAGAGTCGACCTGCAGGCATGCAAGCTT
     .......................................
```

A restriction map of this gene is shown in *Figure 5*.

Figure 5. Restriction map of the *cat* gene.

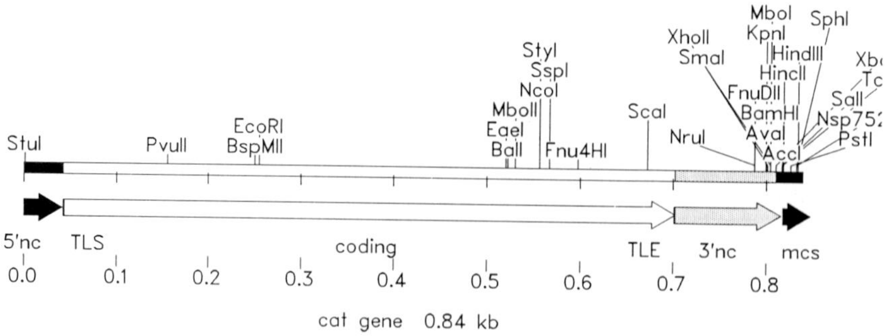

7. DIHYDROFOLATE REDUCTASE [S]

The type II dihydrofolate reductase (*dhfr*) gene from the *Escherichia coli* plasmid R67 encodes an enzyme that is insensitive to methotrexate (amethopterin) and trimethoprim, and therefore confers resistance to these antibiotics. When expressed by a plant promoter, the gene provides resistance to plant tissues growing in the presence of these antibiotics. The antibiotics normally inhibit by binding tightly to dihydrofolate reductase, which functions to reduce dihydrofolate to tetrahydrofolate, an important coenzyme in many metabolic pathways, including amino acid purine and thymidylate synthesis. A second *dhfr* gene, isolated from mouse, has also been used for methotrexate selection in plants (1, 2). Overall sequence length = 293 nt (10–12). The gene encodes an enzyme of 78 amino acids, molecular weight 8437 and pI 7.24. Sequence features are shown in bold type, underlined and labeled above or below the sequence. **TLS**, translation start; **TLE**, translation end; ■ indicates a TGA (umber) stop codon. Two putative N-glycosylation sites are present at Asn21 and Asn49, the latter site is followed by a Pro (see *Introduction*).

```
                                                   ⇦ TLS
  1   CCACATAGAACACCTAGAAGTTCACAAGAAAGGTCGGAAATGGAACGAAG
      ....................................... M  E  R  S

 51   TAGCAATGAAGTCAGTAATCCAGTTGCTGGCAATTTTGTATTCCCATCGA
       S  N  E  V  S  N  P  V  A  G  N  F  V  F  P  S
 glycosylation site
101   ACGCCACGTTTGGTATGGGAGATCGCGTGCGCAAGAAATCCGGCGCCGCC
      N  A  T  F  G  M  G  D  R  V  R  K  K  S  G  A  A
                                      glycosylation site
151   TGGCAAGGTCAGATTGTCGGGTGGTACTGCACAAATTTGACCCCCGAAGG
       W  Q  G  Q  I  V  G  W  Y  C  T  N  L  T  P  E  G

201   CTACGCCGTCGAGTCTGAGGCTCACCCAGGCTCAGTACAGATTTATCCTG
        Y  A  V  E  S  E  A  H  P  G  S  V  Q  I  Y  P
                          TLE ⇨
251   TTGCGGCGCTTGAACGCATCAACTGAGTTCAAGGTTGAAGTGC
      V  A  A  L  E  R  I  N  ■ .................
```

A restriction map of this gene is shown in *Figure 6*.

Figure 6. Restriction map of the bacterial *dhfr* gene.

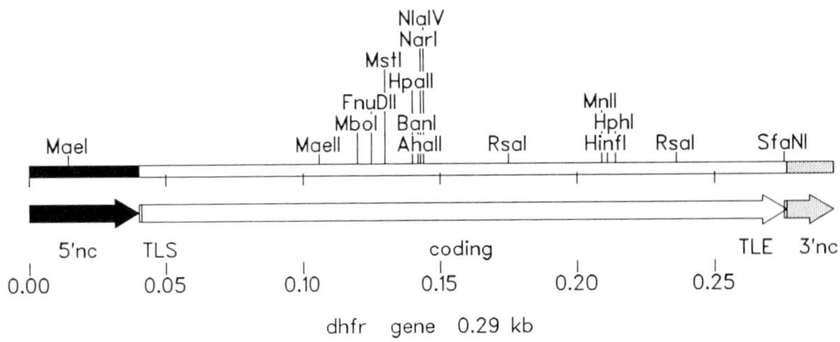

8. DIHYDROPTEROATE SYNTHASE GENE [S]

The dihydropteroate synthase (*sul I*) gene is a sulfonamide-resistance gene from the *E. coli* plasmid R46. The gene confers resistance to the sulfonamide drug Asulam (methyl (4-aminobenzenesulfonyl) carbamate). The sequence shown is that present in pJIT92 (see *Chapter 4*). The gene encodes an enzyme of 279 amino acids, molecular weight 30 097 and pI 7.49. Ile259 is Phe259 in the *sul* gene in plasmid R388. Overall sequence length = 906 nt (13). Sequence features are shown in bold type, underlined and labeled above or below the sequence. **TLS**, translation start; **TLE**, translation end; ■ indicates a TGA (umber) stop codon; **mcs**, multiple cloning sites. A potential N-linked glycosylation site is present at Asn9.

```
     ⇦         mcs          ⇨    ⇦ TLS                    glycosylation site
  1  AAGCTTGCATGCCTGCAGGTCCATGGTGACGGTGTTCGGCATTCTGAATC
     ......................M  V  T  V  F  G  I  L  N

 51  TCACCGAGGACTCCTTCTTCGATGAGAGCCGGCGGCTAGACCCCGCCGGC
     L  T  E  D  S  F  F  D  E  S  R  R  L  D  P  A  G

101  GCTGTCACCGCGGCGATCGAAATGCTGCGAGTCGGATCAGACGTCGTGGA
      A  V  T  A  A  I  E  M  L  R  V  G  S  D  V  V  D

151  TGTCGGACCGGCCGCCAGCCATCCGGACGCGAGGCCTGTATCGCCGGCCG
       V  G  P  A  A  S  H  P  D  A  R  P  V  S  P  A

201  ATGAGATCAGACGTATTGCGCCGCTCTTAGACGCCCTGTCCGATCAGATG
     D  E  I  R  R  I  A  P  L  L  D  A  L  S  D  Q  M

251  CACCGTGTTTCAATCGACAGCTTCCAACCGGAAACCCAGCGCTATGCGCT
      H  R  V  S  I  D  S  F  Q  P  E  T  Q  R  Y  A  L

301  CAAGCGCGGCGTGGGCTACCTGAACGATATCCAAGGATTTCCTGACCCTG
       K  R  G  V  G  Y  L  N  D  I  Q  G  F  P  D  P

351  CGCTCTATCCCGATATTGCTGAGGCGGACTGCAGGCTGGTGGTTATGCAC
     A  L  Y  P  D  I  A  E  A  D  C  R  L  V  V  M  H

401  TCAGCGCAGCGGGATGGCATCGCCACCCGCACCGGTCACCTTCGACCCGA
      S  A  Q  R  D  G  I  A  T  R  T  G  H  L  R  P  E

451  AGACGCGCTCGACGAGATTGTGCGGTTCTTCGAGGCGCGGGTTTCCGCCT
       D  A  L  D  E  I  V  R  F  F  E  A  R  V  S  A

501  TGCGACGGAGCGGGGTCGCTGCCGACCGGCTCATCCTCGATCCGGGGATG
     L  R  R  S  G  V  A  A  D  R  L  I  L  D  P  G  M

551  GGATTTTTCTTGAGCCCCGCACCGGAAACATCGCTGCACGTGCTGTCGAA
      G  F  F  L  S  P  A  P  E  T  S  L  H  V  L  S  N
```

```
601  CCTTCAAAAGCTGAAGTCGGCGTTGGGGCTTCCGCTATTGGTCTCGGTGT
       L  Q  K  L  K  S  A  L  G  L  P  L  L  V  S  V

651  CGCGGAAATCCTTCTTGGGCGCCACCGTTGGCCTTCCTGTAAAGGATCTG
     S  R  K  S  F  L  G  A  T  V  G  L  P  V  K  D  L

701  GGTCCAGCGAGCCTTGCGGCGGAACTTCACGCGATCGGCAATGGCGCTGA
      G  P  A  S  L  A  A  E  L  H  A  I  G  N  G  A  D

751  CTACGTCCGCACCCACGCGCCTGGAGATCTGCGAAGCGCAATCACCATCT
      Y  V  R  T  H  A  P  G  D  L  R  S  A  I  T  I

801  CGGAAACCCTCGCGAAATTTCGCAGTCGCGACGCCAGAGACCGAGGGTTA
     S  E  T  L  A  K  F  R  S  R  D  A  R  D  R  G  L
                TLE ⇨     ⇦                    mcs
851  GATCATGCCTAGCTTGGCTGCAGGTCGACGGATCCCCGGGTACCGAGCTC
      D  H  A  ■ ...................................
                ⇨
901  GAATTC
     ......
```

A restriction map of this gene is shown in *Figure 7*.

Figure 7. Restriction map of the *sul* gene.

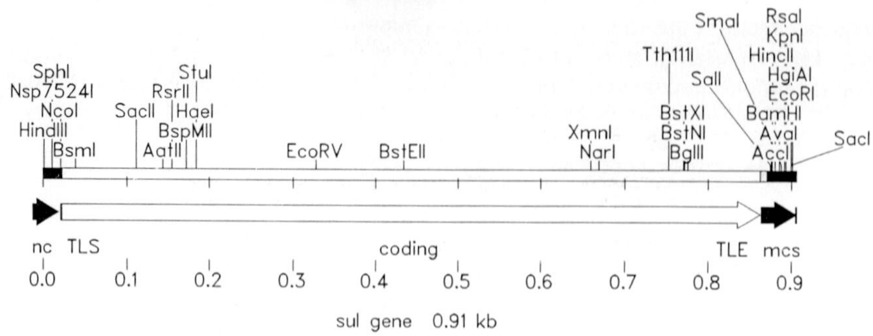

9. ENOLPYRUVYLSHIKIMATE PHOSPHATE SYNTHASE GENE [S]

The 5-enolpyruvylshikimate-3-phosphate synthase (*epsps*) gene (cDNA) sequence has been isolated from a *Petunia hybrida* cell line overproducing the enzyme. This cell line was established by stepwise selection in increasing levels of the herbicide glyphosate. Glyphosate is a potent inhibitor of the enzyme enolpyruvylshikimate phosphate synthase, preventing the synthesis of aromatic amino acids and secondary metabolites derived from the shikimate pathway. The gene, under the control of a high-expression promoter, provides greatly increased tolerance to glyphosate. The gene encodes an enzyme precursor of 516 amino acids, including a putative, 72 amino acid, N-terminal chloroplast transit peptide (ctp) from Met1 to Gln72, which may be removed in the chloroplast. Precursor polypeptide molecular weight is 55 544, and the mature monomeric polypeptide (−ctp) is 47 624 and pI 7.78. Overall sequence length = 1944 nt (14). Sequence features are shown in bold type, underlined and labeled above or below the sequence. **TLS**, translation start; **TLE**, translation end; ■ indicates a TGA (umber) stop codon. A putative N-glycosylation site is present at Asn 465.

```
                                     ⇦ TLS
   1  GAATTCCCTCAATCTTTACTTTCAAGAATGGCACAAATTAACAACATGGC

      ..........................  M  A  Q  I  N  N  M  A
                                  ⇦        ctp start
  51  TCAAGGGATACAAACCCTTAATCCCAATTCCAATTTCCATAAACCCCAAG
        Q  G  I  Q  T  L  N  P  N  S  N  F  H  K  P  Q

 101  TTCCTAAATCTTCAAGTTTTCTTGTTTTTGGATCTAAAAAACTGAAAAAT
      V  P  K  S  S  S  F  L  V  F  G  S  K  K  L  K  N

 151  TCAGCAAATTCTATGTTGGTTTTGAAAAAAGATTCAATTTTTATGCAAAA
       S  A  N  S  M  L  V  L  K  K  D  S  I  F  M  Q  K

 201  GTTTTGTTCCTTTAGGATTTCAGCATCAGTGGCTACAGCACAGAAGCCTT
        F  C  S  F  R  I  S  A  S  V  A  T  A  Q  K  P
                                          ctp end ⇨ ⇦ Mature protein
 251  CTGAGATAGTGTTGCAACCCATTAAAGAGATTTCAGGCACTGTTAAATTG
      S  E  I  V  L  Q  P  I  K  E  I  S  G  T  V  K  L

 301  CCTGGCTCTAAATCATTATCTAATAGAATTCTCCTTCTTGCTGCCTTATC
       P  G  S  K  S  L  S  N  R  I  L  L  L  A  A  L  S

 351  TGAAGGAACAACTGTGGTTGACAATTTACTAAGTAGTGATGATATTCATT
        E  G  T  T  V  V  D  N  L  L  S  S  D  D  I  H

 401  ACATGCTTGGTGCCTTGAAAACACTTGGACTGCATGTAGAAGAAGATAGT
      Y  M  L  G  A  L  K  T  L  G  L  H  V  E  E  D  S

 451  GCAAACCAACGAGCTGTTGTTGAAGGTTGTGGTGGGCTTTTCCCTGTTGG
       A  N  Q  R  A  V  V  E  G  C  G  G  L  F  P  V  G

 501  TAAAGAGTCCAAGGAAGAAATTCAACTGTTCCTTGGAAATGCAGGAACAG
        K  E  S  K  E  E  I  Q  L  F  L  G  N  A  G  T

 551  CAATGCGGCCACTAACAGCAGCAGTTACTGTAGCTGGTGGAAATTCAAGG
      A  M  R  P  L  T  A  A  V  T  V  A  G  G  N  S  R

 601  TATGTACTTGATGGAGTTCCTCGAATGAGAGAGAGACCAATTAGTGATTT
       Y  V  L  D  G  V  P  R  M  R  E  R  P  I  S  D  L

 651  GGTTGATGGTCTTAAACAGCTTGGTGCAGAGGTTGATTGTTTCCTTGGTA
        V  D  G  L  K  Q  L  G  A  E  V  D  C  F  L  G

 701  CGAAATGTCCTCCTGTTCGAATTGTCAGCAAGGGAGGTCTTCCTGGAGGG
      T  K  C  P  P  V  R  I  V  S  K  G  G  L  P  G  G

 751  AAGGTCAAGCTCTCTGGATCCATTAGCAGCCAATACTTGACTGCTCTGCT
       K  V  K  L  S  G  S  I  S  S  Q  Y  L  T  A  L  L

 801  TATGGCTGCTCCACTGGCTTTAGGAGATGTGGAGATTGAAATCATTGACA
        M  A  A  P  L  A  L  G  D  V  E  I  E  I  I  D

 851  AACTAATTAGTGTACCTTATGTCGAGATGACATTGAAGTTGATGGAGCGA
      K  L  I  S  V  P  Y  V  E  M  T  L  K  L  M  E  R

 901  TTTGGTATTTCTGTGGAGCACAGTAGTAGCTGGGACAGGTTCTTTGTCCG
       F  G  I  S  V  E  H  S  S  S  W  D  R  F  F  V  R

 951  AGGAGGTCAGAAATACAAGTCTCCTGGAAAAGCTTTTGTCGAAGGTGATG
        G  G  Q  K  Y  K  S  P  G  K  A  F  V  E  G  D

1001  CTTCAAGTGCTAGCTACTTCTTGGCTGGTGCAGCAGTCACAGGTGGAACT
      A  S  S  A  S  Y  F  L  A  G  A  A  V  T  G  G  T

1051  ATCACTGTTGAAGGTTGTGGGACAAACAGTTTACAGGGGGATGTCAAATT
       I  T  V  E  G  C  G  T  N  S  L  Q  G  D  V  K  F

1101  TGCTGAGGTACTTGAAAAAATGGGAGCTGAAGTTACGTGGACAGAGAACA
        A  E  V  L  E  K  M  G  A  E  V  T  W  T  E  N
```

```
1151  GTGTCACAGTCAAAGGACCTCCAAGGAGTTCTTCTGGGAGGAAGCATTTG
      S  V  T  V  K  G  P  P  R  S  S  S  G  R  K  H  L

1201  CGTGCCATTGATGTGAACATGAATAAAATGCCTGATGTTGCCATGACACT
       R  A  I  D  V  N  M  N  K  M  P  D  V  A  M  T  L

1251  TGCTGTTGTTGCACTTTATGCTGATGGTCCCACAGCTATAAGAGATGTTG
        A  V  V  A  L  Y  A  D  G  P  T  A  I  R  D  V

1301  CTAGCTGGAGAGTCAAGGAAACTGAGCGCATGATCGCCATATGCACAGAA
      A  S  W  R  V  K  E  T  E  R  M  I  A  I  C  T  E

1351  CTTAGGAAGTTAGGAGCAACCGTTGAAGAAGGACCAGACTACTGCATAAT
       L  R  K  L  G  A  T  V  E  E  G  P  D  Y  C  I  I

                        glycosylation site
1401  CACCCCACCGGAGAAACTAAATGTGACCGATATTGATACATACGATGATC
        T  P  P  E  K  L  N  V  T  D  I  D  T  Y  D  D
                          -------

1451  ACAGGATGGCCATGGCTTTTTCTCTTGCTGCTTGTGCAGATGTTCCCGTC
      H  R  M  A  M  A  F  S  L  A  A  C  A  D  V  P  V

1501  ACCATCAATGACCCTGGCTGCACGCGGAAAACCTTCCCTAACTACTTTGA
       T  I  N  D  P  G  C  T  R  K  T  F  P  N  Y  F  D

                                TLE ⇨
1551  TGTACTTCAGCAGTACTCCAAGCATTGAACCGCTTCCCTATATTGCAGAA
        V  L  Q  Q  Y  S  K  H  ■ ......................

1601  TGTAAGTAAGAATATGTGAAGAGTTTAGTTCTTGTACAAGACAGGCTACG
      .................................................

1651  ACTGCCTGGTATCAGAACCACAATGGGTTCCATTTCAGTTCAGAAGGGCA
      .................................................

1701  TTCCAAGGCTTCGAACTCTTTACTTATTTGCGAGTGATGAAATGTATTTG
      .................................................

1751  TTAGAGTTGAGCTTCTTTTTGTCTTTAAGGAATGTACACTAATAGAGTTA
      .................................................

1801  AGAATTACTAGTATGGGCCAGTGTAAGGAGTACTATTACTCTTTGCTTAT
      .................................................

1851  TTTATTGATTGAGTTTTGTCAAGGATCTGGCTTTGTCAAGAATTACTGGT
      .................................................

1901  TAATTTTATTGACAATCTCATGTGTCTAAATGAAATTGTTTGAT
      ............................................
```

A restriction map of this gene is shown in *Figure 8*.

Figure 8. Restriction map of *epsps* gene.

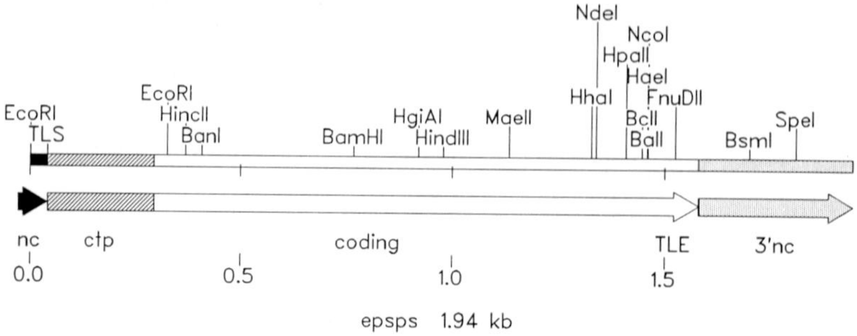

10. β-GALACTOSIDASE GENE [R^H]

This is the β-galactosidase (*β-gal*, *LacZ*) gene isolated from the *E. coli* lactose operon. The gene encodes a large polypeptide subunit of 1024 amino acids with a molecular weight of 116 495 and pI 7.44. Active β-galactosidase consists of a tetramer of identical subunits, molecular weight 465 980, and is responsible for the hydrolysis of β-galactosides to galactose plus another product. The β-galactosidase coding sequence has been used as a reporter gene in plants, although it has limited utility due to endogenous enzyme activity in plant cells and therefore requires a discriminatory assay based on gel electrophoresis to separate plant and bacterial (transgene) enzymes followed by *in situ* fluorescent detection of activity using 4-methylumbelliferyl-β-D-galactoside (4-MUG). The enzyme tolerates N-terminal extensions and is stable to heat, urea, detergents, proteases and fixation for histochemical location in tissues. Overall sequence length = 3118 nt (15–18). Sequence features are shown in bold type, underlined and labeled above or below the sequence. **TLS**, translation start; **TLE**, translation end; ■ indicates a TAA (ochre) stop codon. Five potential N-glycosylation sites are found at Asn102, Asn457, Asn704, Asn725 and Asn958. The potential sequence at Asn118 contains a central Pro which inhibits glycosylation (see *Introduction*).


```
                                   ⇦ TLS
  1  AATTTCACACAGGAAACAGCTATGACCATGATTACGGATTCACTGGCCGT
     .....................M  T  M  I  T  D  S  L  A  V

 51  CGTTTTACAACGTCGTGACTGGGAAAACCCTGGCGTTACCCAACTTAATC
       V  L  Q  R  R  D  W  E  N  P  G  V  T  Q  L  N

101  GCCTTGCAGCACATCCCCCTTTCGCCAGCTGGCGTAATAGCGAAGAGGCC
     R  L  A  A  H  P  P  F  A  S  W  R  N  S  E  E  A

151  CGCACCGATCGCCCTTCCCAACAGTTGCGCAGCCTGAATGGCGAATGGCG
      R  T  D  R  P  S  Q  Q  L  R  S  L  N  G  E  W  R

201  CTTTGCCTGGTTTCCGGCACCAGAAGCGGTGCCGGAAAGCTGGCTGGAGT
       F  A  W  F  P  A  P  E  A  V  P  E  S  W  L  E

251  GCGATCTTCCTGAGGCCGATACTGTCGTCGTCCCCTCAAACTGGCAGATG
     C  D  L  P  E  A  D  T  V  V  V  P  S  N  W  Q  M
                                   glycosylation site
301  CACGGTTACGATGCGCCCATCTACACCAACGTAACCTATCCCATTACGGT
      H  G  Y  D  A  P  I  Y  T  N  V  T  Y  P  I  T  V

351  CAATCCGCCGTTTGTTCCCACGGAGAATCCGACGGGTTGTTACTCGCTCA
       N  P  P  F  V  P  T  E  N  P  T  G  C  Y  S  L

401  CATTTAATGTTGATGAAAGCTGGCTACAGGAAGGCCAGACGCGAATTATT
     T  F  N  V  D  E  S  W  L  Q  E  G  Q  T  R  I  I

451  TTTGATGGCGTTAACTCGGCGTTTCATCTGTGGTGCAACGGGCGCTGGGT
      F  D  G  V  N  S  A  F  H  L  W  C  N  G  R  W  V

501  CGGTTACGGCCAGGACAGTCGTTTGCCGTCTGAATTTGACCTGAGCGCAT
       G  Y  G  Q  D  S  R  L  P  S  E  F  D  L  S  A

551  TTTTACGCGCCGGAGAAAACCGCCTCGCGGTGATGGTGCTGCGTTGGAGT
     F  L  R  A  G  E  N  R  L  A  V  M  V  L  R  W  S

601  GACGGCAGTTATCTGGAAGATCAGGATATGTGGCGGATGAGCGGCATTTT
      D  G  S  Y  L  E  D  Q  D  M  W  R  M  S  G  I  F

651  CCGTGACGTCTCGTTGCTGCATAAACCGACTACACAAATCAGCGATTTCC
       R  D  V  S  L  L  H  K  P  T  T  Q  I  S  D  F

701  ATGTTGCCACTCGCTTTAATGATGATTTCAGCCGCGCTGTACTGGAGGCT
     H  V  A  T  R  F  N  D  D  F  S  R  A  V  L  E  A
```

```
751   GAAGTTCAGATGTGCGGCGAGTTGCGTGACTACCTACGGGTAACAGTTTC
       E  V  Q  M  C  G  E  L  R  D  Y  L  R  V  T  V  S

801   TTTATGGCAGGGTGAAACGCAGGTCGCCAGCGGCACCGCGCCTTTCGGCG
        L  W  Q  G  E  T  Q  V  A  S  G  T  A  P  F  G

851   GTGAAATTATCGATGAGCGTGGTGGTTATGCCGATCGCGTCACACTACGT
      G  E  I  I  D  E  R  G  G  Y  A  D  R  V  T  L  R

901   CTGAACGTCGAAAACCCGAAACTGTGGAGCGCCGAAATCCCGAATCTCTA
       L  N  V  E  N  P  K  L  W  S  A  E  I  P  N  L  Y

951   TCGTGCGGTGGTTGAACTGCACACCGCCGACGGCACGCTGATTGAAGCAG
        R  A  V  V  E  L  H  T  A  D  G  T  L  I  E  A

1001  AAGCCTGCGATGTCGGTTTCCGCGAGGTGCGGATTGAAAATGGTCTGCTG
      E  A  C  D  V  G  F  R  E  V  R  I  E  N  G  L  L

1051  CTGCTGAACGGCAAGCCGTTGCTGATTCGAGGCGTTAACCGTCACGAGCA
       L  L  N  G  K  P  L  L  I  R  G  V  N  R  H  E  H

1101  TCATCCTCTGCATGGTCAGGTCATGGATGAGCAGACGATGGTGCAGGATA
        H  P  L  H  G  Q  V  M  D  E  Q  T  M  V  Q  D

1151  TCCTGCTGATGAAGCAGAACAACTTTAACGCCGTGCGCTGTTCGCATTAT
      I  L  L  M  K  Q  N  N  F  N  A  V  R  C  S  H  Y

1201  CCGAACCATCCGCTGTGGTACACGCTGTGCGACCGCTACGGCCTGTATGT
       P  N  H  P  L  W  Y  T  L  C  D  R  Y  G  L  Y  V

1251  GGTGGATGAAGCCAATATTGAAACCCACGGCATGGTGCCAATGAATCGTC
        V  D  E  A  N  I  E  T  H  G  M  V  P  M  N  R

1301  TGACCGATGATCCGCGCTGGCTACCGGCGATGAGCGAACGCGTAACGCGA
      L  T  D  D  P  R  W  L  P  A  M  S  E  R  V  T  R

1351  ATGGTGCAGCGCGATCGTAATCACCCGAGTGTGATCATCTGGTCGCTGGG
       M  V  Q  R  D  R  N  H  P  S  V  I  I  W  S  L  G
    glycosylation site
1401  GAATGAATCAGGCCACGGCGCTAATCACGACGCGCTGTATCGCTGGATCA
        N  E  S  G  H  G  A  N  H  D  A  L  Y  R  W  I

1451  AATCTGTCGATCCTTCCCGCCCGGTGCAGTATGAAGGCGGCGGAGCCGAC
      K  S  V  D  P  S  R  P  V  Q  Y  E  G  G  G  A  D

1501  ACCACGGCCACCGATATTATTTGCCCGATGTACGCGCGCGTGGATGAAGA
       T  T  A  T  D  I  I  C  P  M  Y  A  R  V  D  E  D

1551  CCAGCCCTTCCCGGCTGTGCCGAAATGGTCCATCAAAAAATGGCTTTCGC
        Q  P  F  P  A  V  P  K  W  S  I  K  K  W  L  S

1601  TACCTGGAGAGACGCGCCCGCTGATCCTTTGCGAATACGCCCACGCGATG
      L  P  G  E  T  R  P  L  I  L  C  E  Y  A  H  A  M

1651  GGTAACAGTCTTGGCGGTTTCGCTAAATACTGGCAGGCGTTTCGTCAGTA
       G  N  S  L  G  G  F  A  K  Y  W  Q  A  F  R  Q  Y

1701  TCCCCGTTTACAGGGCGGCTTCGTCTGGGACTGGGTGGATCAGTCGCTGA
        P  R  L  Q  G  G  F  V  W  D  W  V  D  Q  S  L

1751  TTAAATATGATGAAAACGGCAACCCGTGGTCGGCTTACGGCGGTGATTTT
      I  K  Y  D  E  N  G  N  P  W  S  A  Y  G  G  D  F

1801  GGCGATACGCCGAACGATCGCCAGTTCTGTATGAACGGTCTGGTCTTTGC
       G  D  T  P  N  D  R  Q  F  C  M  N  G  L  V  F  A

1851  CGACCGCACGCCGCATCCAGCGCTGACGGAAGCAAAACACCAGCAGCAGT
        D  R  T  P  H  P  A  L  T  E  A  K  H  Q  Q  Q

1901  TTTTCCAGTTCCGTTTATCCGGGCAAACCATCGAAGTGACCAGCGAATAC
      F  F  Q  F  R  L  S  G  Q  T  I  E  V  T  S  E  Y
```

```
1951  CTGTTCCGTCATAGCGATAACGAGCTCCTGCACTGGATGGTGGCGCTGGA
       L  F  R  H  S  D  N  E  L  L  H  W  M  V  A  L  D

2001  TGGTAAGCCGCTGGCAAGCGGTGAAGTGCCTCTGGATGTCGCTCCACAAG
        G  K  P  L  A  S  G  E  V  P  L  D  V  A  P  Q

2051  GTAAACAGTTGATTGAACTGCCTGAACTACCGCAGCCGGAGAGCGCCGGG
      G  K  Q  L  I  E  L  P  E  L  P  Q  P  E  S  A  G
                                             glycosylation site
2101  CAACTCTGGCTCACAGTACGCGTAGTGCAACCGAACGCGACCGCATGGTC
       Q  L  W  L  T  V  R  V  V  Q  P  N  A  T  A  W  S
                                                       glycosylation site
2151  AGAAGCCGGGCACATCAGCGCCTGGCAGCAGTGGCGTCTGGCGGAAAACC
        E  A  G  H  I  S  A  W  Q  Q  W  R  L  A  E  N

2201  TCAGTGTGACGCTCCCCGCCGCGTCCCACGCCATCCCGCATCTGACCACC
      L  S  V  T  L  P  A  A  S  H  A  I  P  H  L  T  T

2251  AGCGAAATGGATTTTTGCATCGAGCTGGGTAATAAGCGTTGGCAATTTAA
       S  E  M  D  F  C  I  E  L  G  N  K  R  W  Q  F  N

2301  CCGCCAGTCAGGCTTTCTTTCACAGATGTGGATTGGCGATAAAAAACAAC
        R  Q  S  G  F  L  S  Q  M  W  I  G  D  K  K  Q

2351  TGCTGACGCCGCTGCGCGATCAGTTCACCCGTGCACCGCTGGATAACGAC
      L  L  T  P  L  R  D  Q  F  T  R  A  P  L  D  N  D

2401  ATTGGCGTAAGTGAAGCGACCCGCATTGACCCTAACGCCTGGGTCGAACG
       I  G  V  S  E  A  T  R  I  D  P  N  A  W  V  E  R

2451  CTGGAAGGCGGCGGGCCATTACCAGGCCGAAGCAGCGTTGTTGCAGTGCA
        W  K  A  A  G  H  Y  Q  A  E  A  A  L  L  Q  C

2501  CGGCAGATACACTTGCTGATGCGGTGCTGATTACGACCGCTCACGCGTGG
      T  A  D  T  L  A  D  A  V  L  I  T  T  A  H  A  W

2551  CAGCATCAGGGGAAAACCTTATTTATCAGCCGGAAAACCTACCGGATTGA
       Q  H  Q  G  K  T  L  F  I  S  R  K  T  Y  R  I  D

2601  TGGTAGTGGTCAAATGGCGATTACCGTTGATGTTGAAGTGGCGAGCGATA
        G  S  G  Q  M  A  I  T  V  D  V  E  V  A  S  D

2651  CACCGCATCCGGCGCGGATTGGCCTGAACTGCCAGCTGGCGCAGGTAGCA
      T  P  H  P  A  R  I  G  L  N  C  Q  L  A  Q  V  A

2701  GAGCGGGTAAACTGGCTCGGATTAGGGCCGCAAGAAAACTATCCCGACCG
       E  R  V  N  W  L  G  L  G  P  Q  E  N  Y  P  D  R

2751  CCTTACTGCCGCCTGTTTTGACCGCTGGGATCTGCCATTGTCAGACATGT
        L  T  A  A  C  F  D  R  W  D  L  P  L  S  D  M

2801  ATACCCCGTACGTCTTCCCGAGCGAAAACGGTCTGCGCTGCGGGACGCGC
      Y  T  P  Y  V  F  P  S  E  N  G  L  R  C  G  T  R
                                                     glycosylation site
2851  GAATTGAATTATGGCCCACACCAGTGGCGCGGCGACTTCCAGTTCAACAT
       E  L  N  Y  G  P  H  Q  W  R  G  D  F  Q  F  N  I

2901  CAGCCGCTACAGTCAACAGCAACTGATGGAAACCAGCCATCGCCATCTGC
        S  R  Y  S  Q  Q  Q  L  M  E  T  S  H  R  H  L

2951  TGCACGCGGAAGAAGGCACATGGCTGAATATCGACGGTTTCCATATGGGG
      L  H  A  E  E  G  T  W  L  N  I  D  G  F  H  M  G

3001  ATTGGTGGCGACGACTCCTGGAGCCCGTCAGTATCGGCGGAATTCCAGCT
       I  G  G  D  D  S  W  S  P  S  V  S  A  E  F  Q  L
                                                  TLE ⇨
3051  GAGCGCCGGTCGCTACCATTACCAGTTGGTCTGGTGTCAAAAATAATAAT
        S  A  G  R  Y  H  Y  Q  L  V  W  C  Q  K  ■.......

3101  AACCGGGCAGGCCATGTC
      ..................
```

A restriction map of this gene is shown in *Figure 9*.

Figure 9. Restriction map of the *gal* gene.

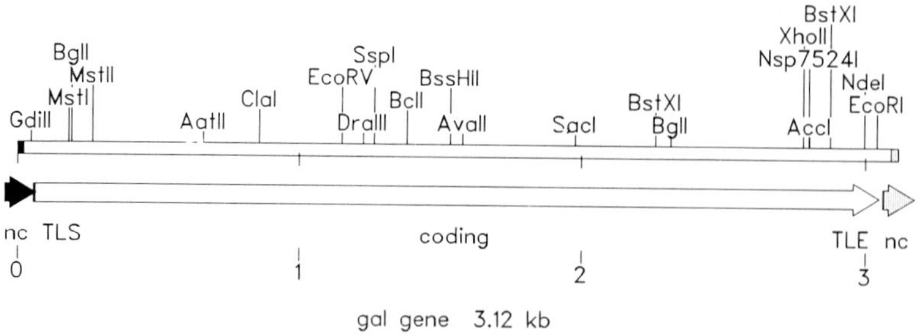

11. β-GLUCURONIDASE GENE [R^H]

This is the β-glucuronidase (*gus, uidA*) gene isolated from *E. coli* strain RAJ201 in plasmid pRAJ220 and inserted in a number of plasmids, e.g. the pBI101 series (see *Chapter 4*). The gene encodes β-glucuronidase (GUS), an enzyme that hydrolyzes β-glucuronides to glucuronic acid and another product. The gene encodes a polypeptide of 602 amino acids, with a subunit molecular weight of 68 266 and pI 7.37. The active enzyme is probably a tetramer of molecular weight 273 064. The enzyme is stable, can be assayed over a broad pH range (pH 5–7.5) and is resistant to temperature ($T_{1/2}$ at 55°C = 2 h). It is easily and sensitively assayed in tissue extracts or histochemically in tissue sections (see *Chapter 4*). It is widely used as a reporter gene in fusions with a variety of plant promoters. The enzyme will tolerate extensions to the N-terminal end (5′ translation fusions). The enzyme is strongly inhibited when glycosylated at Asn358. A mutagenized variant is available from Clontech in which Asn358 has been changed to Ser (pGUSN358→S). A second putative glycosylation site exists at Asn422 but is followed by a Pro (see *Introduction*). Overall sequence length = 1926 nt (19–23). Sequence features are shown in bold type, underlined and labeled above and below the sequence. **TLS**, translation start; **TLE**, translation end; **mcs**, multiple cloning sites; ■ indicates a TGA (umber) stop codon.

```
     ⇦                     mcs                       ⇨
  1  AAGCTTGCATGCCTGCAGGTCGACTCTAGAGGATCCCCGGGTGGTCAGTC
     ..................................................
         ⇦ TLS
 51  CCTTATGTTACGTCCTGTAGAAACCCCAACCCGTGAAATCAAAAAACTCG
     .....M  L  R  P  V  E  T  P  T  R  E  I  K  K  L

101  ACGGCCTGTGGGCATTCAGTCTGGATCGCGAAAACTGTGGAATTGATCAG
     D  G  L  W  A  F  S  L  D  R  E  N  C  G  I  D  Q

151  CGTTGGTGGGAAAGCGCGTTACAAGAAAGCCGGGCAATTGCTGTGCCAGG
      R  W  W  E  S  A  L  Q  E  S  R  A  I  A  V  P  G

201  CAGTTTTAACGATCAGTTCGCCGATGCAGATATTCGTAATTATGCGGGCA
       S  F  N  D  Q  F  A  D  A  D  I  R  N  Y  A  G

251  ACGTCTGGTATCAGCGCGAAGTCTTTATACCGAAAGGTTGGGCAGGCCAG
     N  V  W  Y  Q  R  E  V  F  I  P  K  G  W  A  G  Q

301  CGTATCGTGCTGCGTTTCGATGCGGTCACTCATTACGGCAAAGTGTGGGT
      R  I  V  L  R  F  D  A  V  T  H  Y  G  K  V  W  V
```

```
351   CAATAATCAGGAAGTGATGGAGCATCAGGGCGGCTATACGCCATTTGAAG
        N  N  Q  E  V  M  E  H  Q  G  G  Y  T  P  F  E

401   CCGATGTCACGCCGTATGTTATTGCCGGGAAAAGTGTACGTATCACCGTT
      A  D  V  T  P  Y  V  I  A  G  K  S  V  R  I  T  V

451   TGTGTGAACAACGAACTGAACTGGCAGACTATCCCGCCGGGAATGGTGAT
       C  V  N  N  Q  L  N  W  Q  T  I  P  P  G  M  V  I

501   TACCGACGAAAACGGCAAGAAAAAGCAGTCTTACTTCCATGATTTCTTTA
        T  D  E  N  G  K  K  K  Q  S  Y  F  H  D  F  F

551   ACTATGCCGGGATCCATCGCAGCGTAATGCTCTACACCACGCCGAACACC
      N  Y  A  G  I  H  R  S  V  M  L  Y  T  T  P  N  T

601   TGGGTGGACGATATCACCGTGGTGACGCATGTCGCGCAAGACTGTAACCA
       W  V  D  D  I  T  V  V  T  H  V  A  Q  D  C  N  H

651   CGCGTCTGTTGACTGGCAGGTGGTGGCCAATGGTGATGTCAGCGTTGAAC
        A  S  V  D  W  Q  V  V  A  N  G  D  V  S  V  E

701   TGCGTGATGCGGATCAACAGGTGGTTGCAACTGGACAAGGCACTAGCGGG
      L  R  D  A  D  Q  Q  V  V  A  T  G  Q  G  T  S  G

751   ACTTTGCAAGTGGTGAATCCGCACCTCTGGCAACCGGGTGAAGGTTATCT
       T  L  Q  V  V  N  P  H  L  W  Q  P  G  E  G  Y  L

801   CTATGAACTGTGCGTCACAGCCAAAAGCCAGACAGAGTGTGATATCTACC
        Y  E  L  C  V  T  A  K  S  Q  T  E  C  D  I  Y

851   CGCTTCGCGTCGGCATCCGGTCAGTGGCAGTGAAGGGCCAACAGTTCCTG
      P  L  R  V  G  I  R  S  M  A  V  K  G  Q  Q  F  L

901   ATTAACCACAAACCGTTCTACTTTACTGGCTTTGGTCGTCATGAAGATGC
       I  N  H  K  P  F  Y  F  T  G  F  G  R  H  E  D  A

951   GGACTTACGTGGCAAAGGATTCGATAACGTGCTGATGGTGCACGACCACG
        D  L  R  G  K  G  F  D  N  V  L  M  V  H  D  H

1001  CATTAATGGACTGGATTGGGGCCAACTCCTACCGTACCTCGCATTACCCT
      A  L  M  D  W  I  G  A  N  S  Y  R  T  S  H  Y  P

1051  TACGCTGAAGAGATGCTCGACTGGGCAGATGAACATGGCATCGTGGTGAT
       Y  A  E  E  M  L  D  W  A  D  E  H  G  I  V  V  I
                                    glycosylation site
1101  TGATGAAACTGCTGCTGTCGGCTTTAACCTCTCTTTAGGCATTGGTTTCG
        D  E  T  A  A  A  G  F  N  L  S  L  G  I  G  F

1151  AAGCGGGCAACAAGCCGAAAGAACTGTACAGCGAAGAGGCAGTCAACGGG
      E  A  G  N  K  P  K  E  L  Y  S  E  E  A  V  N  G

1201  GAAACTCAGCAAGCGCACTTACAGGCGATTAAAGAGCTGATAGCGCGTGA
       E  T  Q  Q  P  H  L  Q  A  I  K  E  L  I  A  R  D

1251  CAAAAACCACCCAAGCGTGGTGATGTGGAGTATTGCCAACGAACCGGATA
        K  N  H  P  S  V  V  M  W  S  I  A  N  E  P  D
                           glycosylation site
1301  CCCGTCCGCAAGTGCACGGGAATATTTCGCCACTGGCGGAAGCAACGCGT
      T  R  P  Q  V  H  G  N  I  S  P  L  A  E  A  T  R

1351  AAACTCGACCCGACGCGTCCGATCACCTGCGTCAATGTAATGTTCTGCGA
       K  L  D  P  T  R  P  I  T  C  V  N  V  M  F  C  D

1401  CGCTCACACCGATACCATCAGCGATCTCTTTGATGTGCTGTGCCTGAACC
        A  H  T  D  T  I  S  D  L  F  D  V  L  C  L  N

1451  GTTATTACGGATGGTATGTCCAAAGCGGCGATTTGGAAACGGCAGAGAAG
      R  Y  Y  G  W  Y  V  Q  S  G  D  L  E  T  A  E  K

1501  GTACTGGAAAAAGAACTTCTGGCCTGGCAGGAGAAACTGCATCAGCCGAT
       V  L  E  K  E  L  L  A  W  Q  E  K  L  H  Q  P  I
```

```
1551  TATCATCACCGAATACGGCGTGGATACGTTAGCCGGGCTGCACTCAATGT
        I  I  T  E  Y  G  V  D  T  L  A  G  L  H  S  M

1601  ACACCGACATGTGGAGTGAAGAGTATCAGTGTGCATGGCTGGATATGTAT
      Y  T  D  M  W  S  E  E  Y  Q  C  A  W  L  D  M  Y

1651  CACCGCGTCTTTGATCGCGTCAGCGCCGTCGTCGGTGAACAGGTATGGAA
       H  R  V  F  D  R  V  S  A  V  V  G  E  Q  V  W  N

1701  TTTCGCCGATTTTGCGACCTCGCAAGGCATATTGCGCGTTGGCGGTAACA
        F  A  D  F  A  T  S  Q  G  I  L  R  V  G  G  N

1751  AGAAAGGGATCTTCACTCGCGACCGCAAACCGAAGTCGGCGGCTTTTCTG
      K  K  G  I  F  T  R  D  R  K  P  K  S  A  A  F  L

1801  CTGCAAAAACGCTGGACTGGCATGAACTTCGGTGAAAAACCGCAGCAGGG
       L  Q  K  R  W  T  G  M  N  F  G  E  K  P  Q  Q  G
                 TLE ⇨
1851  AGGCAAACAATGAATCAACAACTCTCCTGGCGCACCATCGTCGGCTACAG
        G  K  Q  ■ .................................

1901  CCTCGGTGGGGAATTGTACCGAGCTC
      ..........................
```

A restriction map of this gene is shown in *Figure 10.*

Figure 10. Restriction map of the *gus* gene.

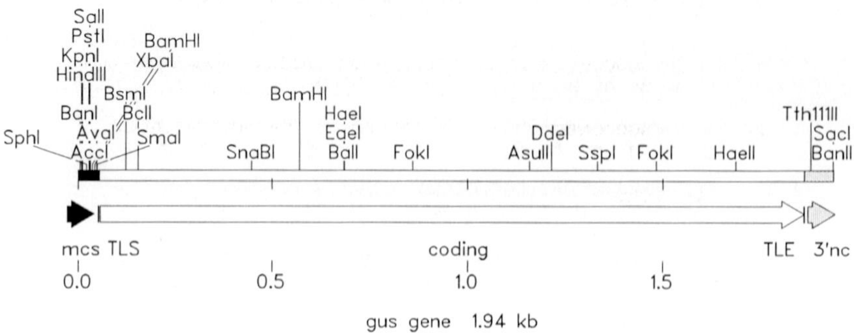

12. HYGROMYCIN PHOSPHOTRANSFERASE GENE [S]

This is the hygromycin B phosphotransferase (*hgh*, *aphIV*) gene isolated from the *E. coli* plasmid pJR225 derived originally from *Klebsiella*. Under a constitutive plant promoter this gene provides resistance in plants to the aminoglycoside antibiotic, hygromycin. The gene encodes an enzyme of 341 amino acids of molecular weight 38 037 and pI 7.47. Overall sequence length = 1523 nt (24). Sequence features are shown in bold type, underlined and labeled above or below the sequence. **TLS**, translation start; **TLE**, translation end; **mcs**, multiple cloning sites; ■ indicates a TAG (amber) stop codon. There are no apparent N-glycosylation sites.

```
   1  TGACGTTGGAGGGGCAAGGTCGCGCTGATTGCTGGGGCAACACGTGGAGC
      ..................................................

  51  GGATCGGGGATTGTCTTTCTTCAGCTCGCTGATGATATGCTGACGCCAAT
      ..................................................

 101  GCCGTTTGGCCTCCGACTAACGAAAATCCCGCATTTGGACGGCTGATCCG
      ..................................................

 151  ATTGGCACGGCGGACGGCGAATGGCGGAGCAGACGCTCGTCCGGGGGCAA
      ..................................................
            ⇦ TLS
 201  TGAGATATGAAAAAGCCTGAACTCACCGCGACGTCTGTCGAGAAGTTTCT
      ...... M  K  K  P  E  L  T  A  T  S  V  E  K  F  L

 251  GATCGAAAAGTTCGACAGCGTCTCCGACCTGATGCAGCTCTCGGAGGGCG
        I  E  K  F  D  S  V  S  D  L  M  Q  L  S  E  G

 301  AAGAATCTCGTGCTTTCAGCTTCGATGTAGGAGGGCGTGGATATGTCCTG
      E  E  S  R  A  F  S  F  D  V  G  G  R  G  Y  V  L

 351  CGGGTAAATAGCTGCGCCGATGGTTTCTACAAAGATCGTTATGTTTATCG
       R  V  N  S  C  A  D  G  F  Y  K  D  R  Y  V  Y  R

 401  GCACTTTGCATCGGCCGCGCTCCCGATTCCGGAAGTGCTTGACATTGGGG
        H  F  A  S  A  A  L  P  I  P  E  V  L  D  I  G

 451  AATTCAGCGAGAGCCTGACCTATTGCATCTCCCGCCGTGCACAGGGTGTC
      E  F  S  E  S  L  T  Y  C  I  S  R  R  A  Q  G  V

 501  ACGTTGCAAGACCTGCCTGAAACCGAACTGCCCGCTGTTCTGCAGCCGGT
       T  L  Q  D  L  P  E  T  E  L  P  A  V  L  Q  P  V

 551  CGCGGAGGCCATGGATGCGATCGCTGCGGCCGATCTTAGCCAGACGAGCG
        A  E  A  M  D  A  I  A  A  A  D  L  S  Q  T  S

 601  GGTTCGGCCCATTCGGACCGCAAGGAATCGGTCAATACACTACATGGCGT
      G  F  G  P  F  G  P  Q  G  I  G  Q  Y  T  T  W  R

 651  GATTTCATATGCGCGATTGCTGATCCCCATGTGTATCACTGGCAAACTGT
       D  F  I  C  A  I  A  D  P  H  V  Y  H  W  Q  T  V

 701  GATGGACGACACCGTCAGTGCGTCCGTCGCGCAGGCTCTCGATGAGCTGA
        M  D  D  T  V  S  A  S  V  A  Q  A  L  D  E  L

 751  TGCTTTGGGCCGAGGACTGCCCCGAAGTCCGGCACCTCGTGCACGCGGAT
      M  L  W  A  E  D  C  P  E  V  R  H  L  V  H  A  D

 801  TTCGGCTCCAACAATGTCCTGACGGACAATGGCCGCATAACAGCGGTCAT
       F  G  S  N  N  V  L  T  D  N  G  R  I  T  A  V  I

 851  TGACTGGAGCGAGGCGATGTTCGGGGATTCCCAATACGAGGTCGCCAACA
        D  W  S  E  A  M  F  G  D  S  Q  Y  E  V  A  N

 901  TCTTCTTCTGGAGGCCGTGGTTGGCTTGTATGGAGCAGCAGACGCGCTAC
      I  F  F  W  R  P  W  L  A  C  M  E  Q  Q  T  R  Y

 951  TTCGAGCGGAGGCATCCGGAGCTTGCAGGATCGCCGCGGCTCCGGGCGTA
       F  E  R  R  H  P  E  L  A  G  S  P  R  L  R  A  Y

1001  TATGCTCCGCATTGGTCTTGACCAACTCTATCAGAGCTTGGTTGACGGCA
        M  L  R  I  G  L  D  Q  L  Y  Q  S  L  V  D  G

1051  ATTTCGATGATGCAGCTTGGGCGCAGGGTCGATGCGACGCAATCGTCCGA
      N  F  D  D  A  A  W  A  Q  G  R  C  D  A  I  V  R

1101  TCCGGAGCCGGGACTGTCGGGCGTACACAAATCGCCCGCAGAAGCGCGGC
       S  G  A  G  T  V  G  R  T  Q  I  A  R  R  S  A  A

1151  CGTCTGGACCGATGGCTGTGTAGAAGTACTCGCCGATAGTGGAAACCGAC
        V  W  T  D  G  C  V  E  V  L  A  D  S  G  N  R
```

```
                                                  TLE ⇨
1201  GCCCCAGCACTCGTCCGAGGGCAAAGGAATAGAGTAGATGCCGACCGAAC
      R  P  S  T  R  P  R  A  K  E  ■ ..................

1251  AAGAGCTGATTTCGAGAACGCCTCAGCCAGCAACTCGCGCGAGCCTAGCA
      ..................................................

1301  AGGCAAATGCGAGAGAACGGCCTTACGCTTGGTGGCACAGTTCTCGTCCA
      ..................................................

1351  CAGTTCGCTAAGCTCGCTCGGCTGGGTCGCGGGAGGGCCGGTCGCAGTGA
      ..................................................

1401  TTCAGGCCCTTCTGGATTGTGTTGGTCCCCAGGGCACGATTGTCATGCCC
      ..................................................
                            ⇦                      mcs
1451  ACGCACTCGGGGGGCCAGTGCCAAGCTTGGGCTGCAGGTCGACTCTAGAG
      ..................................................
                            ⇨
1501  GATCCCCGGGCGAGCTCGAATTC
      .......................
```

A restriction map of this gene is shown in *Figure 11.*

Figure 11. Restriction map of the *hyg* gene.

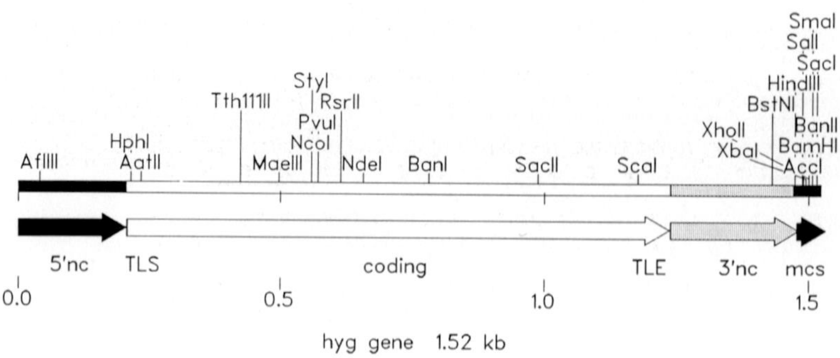

13. LUCIFERASE GENE [R]

The luciferase (*luc*) gene isolated from *Photinus pyralis* (firefly) encodes the enzyme catalyzing the ATP/oxygen-dependent oxidation of the substrate luciferin (e.g. Sigma) which produces emission of light (bioluminescence). As a reporter, the gene is the basis of highly sensitive assays for promoter activity and for protein targeting sequences, involving the measurement of light emission using liquid scintillation counter photomultipliers, luminometers, X-ray film exposure or sensitive camera/film. ATP-based assays give extremely rapid light emission — a flash of light. Promega have developed an assay based on coenzyme A which gives a sustained, more easily measured light emission. The gene encodes a single polypeptide enzyme of 550 amino acids, molecular weight of 60 753 and pI 7.62. Overall sequence length = 1847 nt (25–27). Sequence features are shown in bold type, underlined and labeled above or below the sequence. **TLS**, translation start; **TLE**, translation end; ■ indicates a TAA (ochre) stop codon; **mcs**, multiple cloning sites. Three potential N-glycosylation sites are present at Asn50, Asn119 and Asn197.

⇦ **TLS**

1 **ATG**GAAGACGCCAAAAACATAAAGAAAGGCCCGGCGCCATTCTATCCTCT

M E D A K N I K K G P A P F Y P L

51 AGAGGATGGAACCGCTGGAGAGCAACTGCATAAGGCTATGAAGAGATACG

E D G T A G E Q L H K A M K R Y

glycosylation site

101 CCCTGGTTCCTGGAACAATTGCTTTTACAGATGCACATATCGAGGTGAAC

A L V P G T I A F T D A H I E V **N**

151 ATCACGTACGCGGAATACTTCGAAATGTCCGTTCGGTTGGCAGAAGCTAT

I T Y A E Y F E M S V R L A E A M

201 GAAACGATATGGGCTGAATACAAATCACAGAATCGTCGTATGCAGTGAAA

K R Y G L N T N H R I V V C S E

251 ACTCTCTTCAATTCTTTATGCCGGTGTTGGGCGCGTTATTTATCGGAGTT

N S L Q F F M P V L G A L F I G V

301 GCAGTTGCGCCCGCGAACGACATTTATAATGAACGTGAATTGCTCAACAG

A V A P A N D I Y N E R E L L N S

glycosylation site

351 TATGAACATTTCGCAGCCTACCGTAGTGTTTGTTTCCAAAAAGGGGTTGC

M **N I S** Q P T V V F V S K K G L

401 AAAAAATTTTGAACGTGCAAAAAAAATTACCAATAATCCAGAAAATTATT

Q K I L N V Q K K L P I I Q K I I

451 ATCATGGATTCTAAAACGGATTACCAGGGATTTCAGTCGATGTACACGTT

I M D S K T D Y Q G F Q S M Y T F

501 CGTCACATCTCATCTACCTCCCGGTTTTAATGAATACGATTTTGTACCAG

V T S H L P P G F N E Y D F V P

glycosylation site

551 AGTCCTTTGATCGTGACAAAACAATTGCACTGATAATGAATTCCTCTGGA

E S F D R D K T I A L I M **N S S** G

601 TCTACTGGGTTACCTAAGGGTGTGGCCCTTCCGCATAGAACTGCCTGCGT

S T G L P K G V A L P H R T A C V

651 CAGATTCTCGCATGCCAGAGATCCTATTTTTGGCAATCAAATCATTCCGG

R F S H A R D P I F G N Q I I P

701 ATACTGCGATTTTAAGTGTTGTTCCATTCCATCACGGTTTTGGAATGTTT

D T A I L S V V P F H H G F G M F

751 ACTACACTCGGATATTTGATATGTGGATTTCGAGTCGTCTTAATGTATAG

T T L G Y L I C G F R V V L M Y R

801 ATTTGAAGAAGAGCTGTTTTTACGATCCCTTCAGGATTACAAAATTCAAA

F E E E L F L R S L Q D Y K I Q

851 GTGCGTTGCTAGTACCAACCCTATTTTCATTCTTCGCCAAAAGCACTCTG

S A L L V P T L F S F F A K S T L

901 ATTGACAAATACGATTTATCTAATTTACACGAAATTGCTTCTGGGGGCGC

I D K Y D L S N L H E I A S G G A

951 ACCTCTTTCGAAAGAAGTCGGGGAAGCGGTTGCAAAACGCTTCCATCTTC

P L S K E V G E A V A K R F H L

1001 CAGGGATACGACAAGGATATGGGCTCACTGAGACTACATCAGCTATTCTG

P G I R Q G Y G L T E T T S A I L

1051 ATTACACCCGAGGGGGATGATAAACCGGGCGCGGTCGGTAAAGTTGTTCC

I T P E G D D K P G A V G K V V P

1101 ATTTTTTGAAGCGAAGGTTGTGGATCTGGATACCGGGAAAACGCTGGGCG

F F E A K V V D L D T G K T L G

```
1151  TTAATCAGAGAGGCGAATTATGTGTCAGAGGACCTATGATTATGTCCGGT
      V  N  Q  R  G  E  L  C  V  R  G  P  M  I  M  S  G

1201  TATGTAAACAATCCGGAAGCGACCAACGCCTTGATTGACAAGGATGGATG
       Y  V  N  N  P  E  A  T  N  A  L  I  D  K  D  G  W

1251  GCTACATTCTGGAGACATAGCTTACTGGGACGAAGACGAACACTTCTTCA
       L  H  S  G  D  I  A  Y  W  D  E  D  E  H  F  F

1301  TAGTTGACCGCTTGAAGTCTTTAATTAAATACAAAGGATATCAGGTGGCC
      I  V  D  R  L  K  S  L  I  K  Y  K  G  Y  Q  V  A

1351  CCCGCTGAATTGGAATCGATATTGTTACAACACCCCAACATCTTCGACGC
       P  A  E  L  E  S  I  L  L  Q  H  P  N  I  F  D  A

1401  GGGCGTGGCAGGTCTTCCCGACGATGACGCCGGTGAACTTCCCGCCGCCG
       G  V  A  G  L  P  D  D  D  A  G  E  L  P  A  A

1451  TTGTTGTTTTGGAGCACGGAAAGACGATGACGGAAAAAGAGATCGTGGAT
      V  V  V  L  E  H  G  K  T  M  T  E  K  E  I  V  D

1501  TACGTCGCCAGTCAAGTAACAACCGCGAAAAAGTTGCGCGGAGGAGTTGT
       Y  V  A  S  Q  V  T  T  A  K  K  L  R  G  G  V  V

1551  GTTTGTGGACGAAGTACCGAAAGGTCTTACCGGAAAACTCGACGCAAGAA
       F  V  D  E  V  P  K  G  L  T  G  K  L  D  A  R

1601  AAATCAGAGAGATCCTCATAAAGGCCAAGAAGGGCGGAAAGTCCAAATTG
      K  I  R  E  I  L  I  K  A  K  K  G  G  K  S  K  L
      TLE ⇨
1651  TAAAATGTAACTGTATTCAGCGATGACGAAATTCTTAGCTATTGTAATAT
      ■.................................................

1701  TATATGCAAATTGATGAATGGTAATTTTGTAATTGTGGGTCACTGTACTA
      ..................................................

                 polyA signal                     ⇦
1751  TTTTAACGAATAATAAAATCAGGTATAGGTAACTAAAAAGGAATTCGAGC
      ..................................................

        mcs                                           ⇨
1801  TCGGTACCCGGGGATCCTCTAGAGTCGACCTGCAGGCATGCAAGCTT
      ...............................................
```

A restriction map of this gene is shown in *Figure 12.*

Figure 12. Restriction map of the *luc* gene.

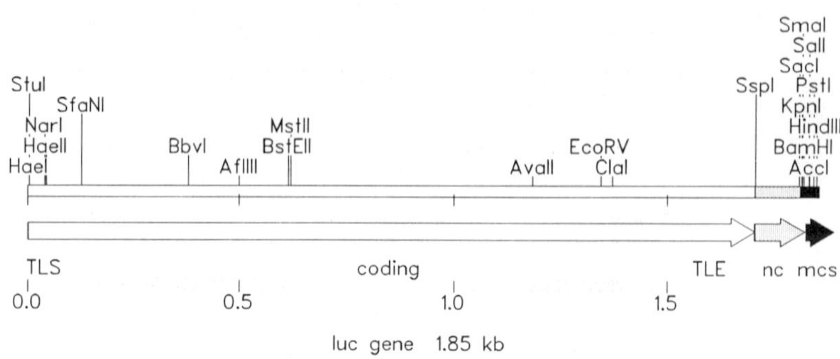

14. LUCIFERASE A AND B GENES [R]

The luciferase (*luxA* , *luxB*) genes isolated from *Vibrio harveyi* in plasmid pTB7, are part of a single transcriptional unit in the bacterial genome. The structural genes encode the LUX A or α-subunit and LUX B or β-subunit of the enzyme. The enzyme is a heterodimer of molecular weight 76 578, composed of one of each of the polypeptides, and catalyzes the oxidation of long-chain fatty aldehydes such as decyl aldehyde (Sigma D7384 = *n*-decyl aldehyde; capric aldehyde or decanal). Sequences of *luxA* and *luxB* coding regions are shown (28, 29), as used for the construction of cassettes for expression in plants. Binary vector pCV701 *lux*A&B (11.8 kb) contains the two genes transcribed in opposite directions from the promoters of T_R-DNA genes 1 and 2. The two-gene component system has been proposed for the study of the coordination of gene expression.

The *luxA* gene encodes a polypeptide of 355 amino acids, molecular weight 40 152 and pI 7.23. Overall sequence length = 1068 nt (28, 29). Sequence features are shown in bold type and labeled above or below the sequence. **TLS**, translation start; **TLE**, translation end. ■ indicates a TAA (ochre) stop codon. Two potential N-linked glycosylation sites are present at Asn256 and Asn330. A third site at Asn159 has a central Pro which inhibits glycosylation (see *Introduction*).

The *luxB* gene encodes a polypeptide of 324 amino acids, molecular weight 36 426 and pI 7.08. Overall sequence length = 975 nt. Sequence features are shown in bold type and labeled above or below the sequence. **TLS**, translation start; **TLE**, translation end. ■ indicates a TAA (ochre) stop codon. A potential N-linked glycosylation site is present at Asn173.

Luciferase A gene and α-subunit

```
      ⇦ TLS
  1   ATGAAATTTGGAAACTTCCTTCTCACTTATCAGCCACCTGAGTTATCTCA
       M  K  F  G  N  F  L  L  T  Y  Q  P  P  E  L  S  Q

 51   GACCGAAGTGATGAAGCGATTGGTTAATCTGGGCAAAGCGTCTGAAGGTT
        T  E  V  M  K  R  L  V  N  L  G  K  A  S  E  G

101   GTGGTTTCGACACAGTTTGGTTACTAGAGCACCACTTCACTGAGTTTGGG
      C  G  F  D  T  V  W  L  L  E  H  H  F  T  E  F  G

151   TTGTTAGGGAATCCTTATGTTGCTGCCGCACATCTATTAGGTGCGACAGA
       L  L  G  N  P  Y  V  A  A  A  H  L  L  G  A  T  E

201   AAAGCTCAATGTTGGCACCGCAGCTATTGTATTGCCGACTGCCCATCCGG
        K  L  N  V  G  T  A  A  I  V  L  P  T  A  H  P

251   TTCGACAAGCAGAAGACGTAAACCTACTGGATCAAATGTCAAAAGGACGA
      V  R  Q  A  E  D  V  N  L  L  D  Q  M  S  K  G  R

301   TTCCGTTTTGGTATTTGTCGCGGTTTGTACGACAAAGATTTCCGTGTCTT
      ·F  R  F  G  I  C  R  G  L  Y  D  K  D  F  R  V  F

351   TGGTACAGACATGGATAACAGCCGAGCCTTAATGGACTGTTGGTATGACT
        G  T  D  M  D  N  S  R  A  L  M  D  C  W  Y  D

401   TGATGAAAGAAGGCTTCAATGAAGGCTATATCGCGGCGGATAACGAACAT
      L  M  K  E  G  F  N  E  G  Y  I  A  A  D  N  E  H

451   ATTAAGTTCCCGAAAATCCAACTGAATCCATCGGCTTACACACAAGGTGG
       I  K  F  P  K  I  Q  L  N  P  S  A  Y  T  Q  G  G
                                -  -  -
501   CGCTCCTGTTTATGTCGTCGCGGAGTCAGCGTCAACGACAGAATGGGCTG
        A  P  V  Y  V  V  A  E  S  A  S  T  T  E  W  A

551   CAGAGCGTGGCCTACCAATGATTCTAAGCTGGATCATCAACACTCACGAG
      A  E  R  G  L  P  M  I  L  S  W  I  I  N  T  H  E
```


Selectable Genes

```
 601  AAGAAAGCGCACGTTGATCTTTACAATGAAGTTGCGACTGAACATGGCTA
       K  K  A  H  V  D  L  Y  N  E  V  A  T  E  H  G  Y

 651  CGATGTGACTAAGATTGACCACTGTTTGTCTTACATCACCTCCGTCGATC
        D  V  T  K  I  D  H  C  L  S  Y  I  T  S  V  D

 701  ATGACTCAAATAAAGCCAAAGATATTTGCCGCAACTTCTTGGGTCATTGG
      H  D  S  N  K  A  K  D  I  C  R  N  F  L  G  H  W
                     glycosylation site
 751  TACGACTCATACGTGAATGCCACAAAGATTTTTGACGACTCTGACCAAAC
       Y  D  S  Y  V  N  A  T  K  I  F  D  D  S  D  Q  T
                      -------

 801  AAAAGGCTACGACTTCAATAAAGGTCAATGGCGTGATTTTGTCTTGAAAG
        K  G  Y  D  F  N  K  G  Q  W  R  D  F  V  L  K

 851  GCCACAAAGACACCAATCGCCGAATTGATTACAGCTACGAAATCAACCCA
      G  H  K  D  T  N  R  R  I  D  Y  S  Y  E  I  N  P

 901  GTAGGGACGCCTGAAGAGTGTATCGCGATTATCCAGCAAGATATCGATGC
       V  G  T  P  E  E  C  I  A  I  I  Q  Q  D  I  D  A
                                           glycosylation site
 951  GACGGGTATTAACAATATTTGTTGTGGTTTTGAAGCAAACGGTTCTGAAG
      T  G  I  N  N  I  C  C  G  F  E  A  N  G  S  E  E
                                           -------

1001  AAGAAATTATCGCATCTATGAAGCTATTCCAGTCTGATGTGATGCCATAT
       E  I  I  A  S  M  K  L  F  Q  S  D  V  M  P  Y
                TLE ⇨
1051  CTCAAAGAAAAACAGTAA      1068
       L  K  E  K  Q  ■
```

Luciferase B gene and *β*-subunit

```
   ⇦ TLS
   1  ATGAAATTTGGATTATTCTTCCTCAATTTTATGAATTCAAAGCGTTCTTC
       M  K  F  G  L  F  F  L  N  F  M  N  S  K  R  S  S

  51  TGATCAAGTCATCGAAGAAATGTTAGATACCGCACATTACGTAGATCAGT
        D  Q  V  I  E  E  M  L  D  T  A  H  Y  V  D  Q

 101  TGAAGTTTGACACGTTGGCTGTTTACGAAAACCATTTCTCGAACAATGGT
      L  K  F  D  T  L  A  V  Y  E  N  H  F  S  N  N  G

 151  GTGGTTGGTGCCCCATTAACAGTGGCTGGTTTTTTACTTGGTATGACAAA
       V  V  G  A  P  L  T  V  A  G  F  L  L  G  M  T  K

 201  GAACGCCAAAGTGGCTTCGTTGAATCACGTTATTACTACACATCATCCAG
        N  A  K  V  A  S  L  N  H  V  I  T  T  H  H  P

 251  TACGTGTGGCGGAAGAAGCGTGTCTACTCGACCAAATGAGTGAAGGCCGT
      V  R  V  A  E  E  A  C  L  L  D  Q  M  S  E  G  R

 301  TTTGTCTTTGGCTTTAGTGATTGTGAAAAGAGTGCAGATATGCGCTTCTT
       F  V  F  G  F  S  D  C  E  K  S  A  D  M  R  F  F

 351  TAATCGACCAACGGATTCTCAGTTTCAGTTGTTCAGTGAGTGTCACAAGA
        N  R  P  T  D  S  Q  F  Q  L  F  S  E  C  H  K

 401  TCATCAATGATGCATTCACTACTGGGTACTGCCATCCAAACAATGATTTT
      I  I  N  D  A  F  T  T  G  Y  C  H  P  N  N  D  F

 451  TATAGTTTTCCTAAAATCTCTGTTAACCCACACGCGTACACTGAAGGCGG
       Y  S  F  P  K  I  S  V  N  P  H  A  Y  T  E  G  G
                     glycosylation site
 501  TCCTGCGCAATTTGTGAATGCGACGAGCAAAGAAGTGGTTGAATGGGCGG
        P  A  Q  F  V  N  A  T  S  K  E  V  V  E  W  A
                         -------

 551  CTAAGTTAGGGCTTCCACTCGTGTTTAAATGGGACGACTCAAACGCGCAA
      A  K  L  G  L  P  L  V  F  K  W  D  D  S  N  A  Q
```

601 AGAAAAGAATACGCCGGTTTGTACCACGAAGTTGCTCAGGCACATGGTGT
R K E Y A G L Y H E V A Q A H G V

651 CGATGTTAGTCAGGTTCGACACAAGCTGACGCTGCTGGTCAACCAAAATG
D V S Q V R H K L T L L V N Q N

701 TAGATGGTGAAGCAGCAAGGGCAGAAGCACGCGTGTATTTGGAAGAGTTT
V D G E A A R A E A R V Y L E E F

751 GTCCGTGAATCTTACCCAAATACCGACTTTGAGCAAAAAATGGTAGAGCT
V R E S Y P N T D F E Q K M V E L

801 GTTGTCAGAAAATGCCATCGGTACTTATGAAGAAAGTACTCAGGCAGCGC
L S E N A I G T Y E E S T Q A A

851 GAGTTGCGATTGAGTGTTGTGGTGCCGCGGACCTATTGATGTCTTTTGAG
R V A I E C C G A A D L L M S F E

901 TCGATGGAAGATAAAGCGCACGAAAGAGCGGTTATCGATGTGGTAAACGC
S M E D K A H E R A V I D V V N A

TLE ⇨

951 CAACATCGTCAAATACCACTCG**TAA**
N I V K Y H S ■ 324

Restriction maps of these genes are shown in *Figure 13*.

Figure 13. Restriction maps of the *luxA* and *luxB* genes.

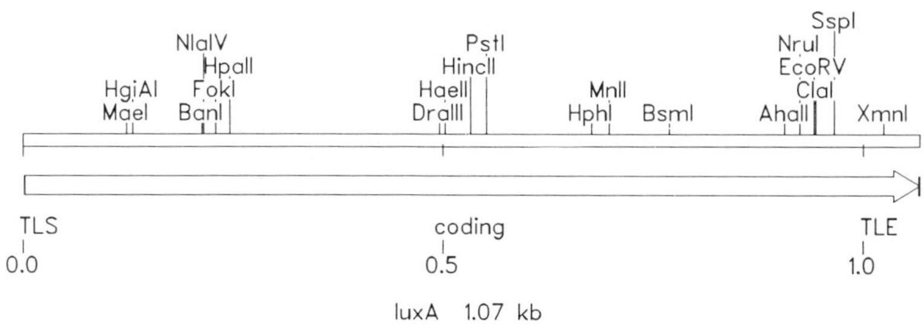

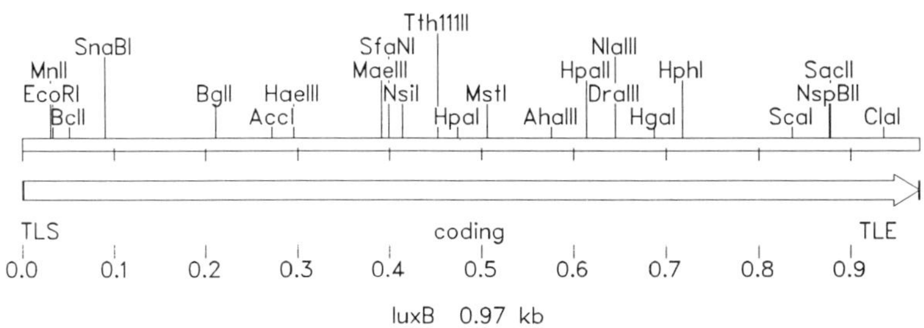

15. NEOMYCIN PHOSPHOTRANSFERASE GENE [S, R]

This is the neomycin phosphotransferase (*npt*II, *aph3′*II) gene isolated from transposon Tn*5*. Under a constitutive plant promoter this gene confers resistance in plants to the aminoglycoside antibiotics neomycin, kanamycin and G418. The enzyme phosphorylates these antibiotics, rendering them inactive. The gene encodes an enzyme of 280 amino acids, molecular weight 29 052 and pI 6.93. Overall sequence length = 1250 nt (30). Sequence features are shown in bold type, underlined and labeled above or below the sequence. **TLS**, translation start; **TLE**, translation end; ■ indicates a TGA (umber) stop codon. There are no apparent N-glycosylation sites.

```
  1  TAGACTGGGCGGTTTTATGGACAGCAAGCGAACCGGAATTGCCAGCTGGG
     ..................................................

 51  GCGCCCTCTGGTAAGGTTGGGAAGCCCTGCAAAGTAAACTGGATGGCTTT
     ..................................................

101  CTTGCCGCCAAGGATCTGATGGCGCAGGGGATCAAGATCTGATCAAGAGA
     ..................................................

                              ⇦ TLS
151  CAGGATGAGGATCGTTTCGCATGATTGAACAAGATGGATTGCACGCAGGT
     .....................M  I  E  Q  D  G  L  H  A  G

201  TCTCCGGCCGCTTGGGTGGAGAGGCTATTCGGCTATGACTGGGCACAACA
      S  P  A  A  W  V  E  R  L  F  G  Y  D  W  A  Q  Q

251  GACAATCGGCTGCTCTGATGCCGCCGTGTTCCGGCTGTCAGCGCAGGGGC
       T  I  G  C  S  D  A  A  V  F  R  L  S  A  Q  G

301  GCCCGGTTCTTTTTGTCAAGACCGACCTGTCCGGTGCCCTGAATGAACTC
     R  P  V  L  F  V  K  T  D  L  S  G  A  L  N  E  L

351  CAAGACGAGGCAGCGCGGCTATCGTGGCTGGCCACGACGGGCGTTCCTTG
      Q  D  E  A  A  R  L  S  W  L  A  T  T  G  V  P  C

401  CGCAGCTGTGCTCGACGTTGTCACTGAAGCGGGAAGGGACTGGCTGCTAT
       A  A  V  L  D  V  V  T  E  A  G  R  D  W  L  L

451  TGGGCGAAGTGCCGGGGCAGGATCTCCTGTCATCTCACCTTGCTCCTGCC
     L  G  E  V  P  G  Q  D  L  L  S  S  H  L  A  P  A

501  GAGAAAGTATCCATCATGGCTGATGCAATGCGGCGGCTGCATACGCTTGA
      E  K  V  S  I  M  A  D  A  M  R  R  L  H  T  L  D

551  TCCGGCTACCTGCCCATTCGACCACCAAGCGAAACATCGCATCGAGCGAG
       P  A  T  C  P  F  D  H  Q  A  K  H  R  I  E  R

601  CACGTACTCGGATGGAAGCCGGTCTTGTCGATCAGGATGATCTGGACGAA
     A  R  T  R  M  E  A  G  L  V  D  Q  D  D  L  D  E

651  GAGCATCAGGGGCTCGCGCCAGCCGAACTGTTCGCCAGGCTCAAGGCGCG
      E  H  Q  G  L  A  P  A  E  L  F  A  R  L  K  A  R

701  GATGCCCGACGGCGAGGATCTCGTCGTGACCCATGGCGATGCCTGCTTGC
       M  P  D  G  E  D  L  V  V  T  H  G  D  A  C  L

751  CGAATATCATGGTGGAAAATGGCCGCTTTTCTGGATTCATCGACTGTGGC
     P  N  I  M  V  E  N  G  R  F  S  G  F  I  D  C  G

801  CGGCTGGGTGTGGCGGACCGCTATCAGGACATAGCGTTGGCTACCCGTGA
      R  L  G  V  A  D  R  Y  Q  D  I  A  L  A  T  R  D

851  TATTGCTGAAGAGCTTGGCGGCGAATGGGCTGACCGCTTCCTCGTGCTTT
       I  A  E  E  L  G  G  E  W  A  D  R  F  L  V  L

901  ACGGTATCGCCGCTCCCGATTCGCAGCGCATCGCCTTCTATCGCCTTCTT
     Y  G  I  A  A  P  D  S  Q  R  I  A  F  Y  R  L  L
```

```
                  TLE ⇨
 951  GACGAGTTCTTCTGAGCGGGACTCTGGGGTTCGAAATGACCGACCAAGCG
       D  E  F  F  ■ ..................................

1001  ACGCCCAACCTGCCATCACGAGATTTCGATTCCACCGCCGCCTTCTATGA
      ..................................................

1051  AAGGTTGGGCTTCGGAATCGTTTTCCGGGACGCCGGCTGGATGATCCTCC
      ..................................................

1101  AGCGCGGGGATCTCATGCTGGAGTTCTTCGCCCACCCCAAAAGGATCTAG
      ..................................................

1151  GTGAAGATCCTTTTTGATAATCTCATGACCAAAATCCCTTAACGTGAGTT
      ..................................................
```

A restriction map of this gene is shown in *Figure 14.*

Figure 14. Restriction map of the *npt* gene.

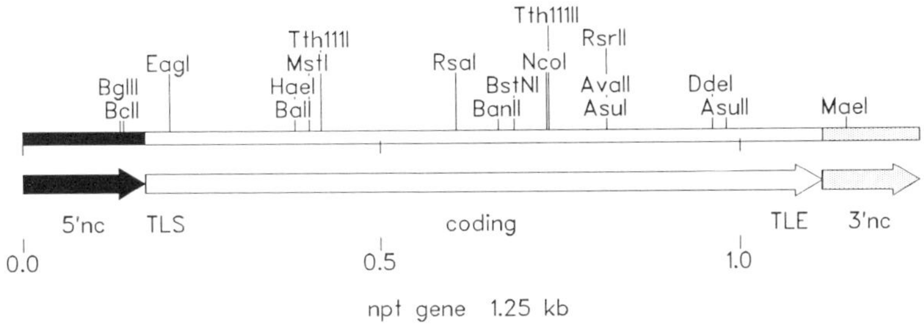

16. NOPALINE SYNTHASE GENE [R, P, T]

This is the nopaline synthase (*nos*, *nop*) gene from *Hin*dIII fragment 23 of the T-DNA in the *Agrobacterium tumefaciens* Ti plasmid T37. The enzyme, nopaline synthase, is responsible for the biosynthesis of the opine, nopaline, from α-ketoglutarate and amino acids arginine or ornithine, in plant cells transformed with T-DNA from Ti plasmids such as pTiT37 and pTiC58. The sequence shown includes the 5′ upstream sequence of the gene up to the right-hand border (RHB) of the T-DNA, ending at adenine base number 61. This region contains the *nos* promoter, which is like a eukaryotic promoter and is used in many gene fusions for high-level 'constitutive' expression in transgenic plant tissues (see *Chapter 4*). The 3′ 'terminator' region is also commonly used in many gene fusions to provide an efficient plant gene expression termination sequence and polyadenylation signals. The nopaline synthase coding sequence has also been used as a reporter gene for plant promoter studies. The gene encodes a polypeptide of 413 amino acids, molecular weight of 45 468 and pI 7.41. Overall sequence length = 2337 nt (11, 31, 32). Sequence features are shown in bold type, underlined and indicated above or below the sequence. **TCS**, transcription start; **TLS**, translation start; **TLE**, translation end; ■ indicates a TAA (ochre) stop codon; **poly A signal**, putative polyadenylation signal. A potential N-glycosylation site is present at Asn391. Another potential site at Asn358 has a central Pro which inhibits glycosylation (see *Introduction*).

```
   1  TATCCGATTATTCTAATAAACGCTCTTTTCTCTTAGGTTTACCCGCCAAT
      ..................................................
                 ⇦ right hand border of T-DNA
  51  ATATCCTGTCAAACACTGATAGTTTAAACTGAAGGCGGGAAACGACAATC
      ..................................................

 101  TGATCATGAGCGGAGAATTAAGGGAGTCACGTTATGACCCCCGCCGATGA
      ..................................................

 151  CGCGGGACAAGCCGTTTTACGTTTGGAACTGACAGAACCGCAACGTTGAA
      ..................................................

 201  GGAGCCACTCAGCCGCGGGTTTCTGGAGTTTAATGAGCTAAGCACATACG
      ..................................................
                                           CAAT-Box
 251  TCAGAAACCATTATTGCGCGTTCAAAAGTCGCCTAAGGTCACTATCAGCT
      ..................................................
                                            TATA-Box
 301  AGCAAATATTTCTTGTCAAAAATGCTCCACTGACGTTCCATAAATTCCCC
      ..................................................
                    ⇦ TCS
 351  TCGGTATCCAATTAGAGTCTCATATTCACTCTCAATCCAAATAATCTGCA
      ..................................................
      ⇦ TLS
 401  ATGGCAATTACCTTATCCGCAACTTCTTTACCTATTTCCGCCGCAGATCA
       M  A  I  T  L  S  A  T  S  L  P  I  S  A  A  D  H

 451  CCATCCGCTTCCCTTGACCGTAGGTGTCCTCGGTTCTGGTCACGCGGGGA
        H  P  L  P  L  T  V  G  V  L  G  S  G  H  A  G

 501  CTGCATTAGCGGCTTGGTTCGCCTCCCGGCATGTTCCCACGGCGCTGTGG
      T  A  L  A  A  W  F  A  S  R  H  V  P  T  A  L  W

 551  GCACCAGCAGATCATCCAGGATCGATCTCAGCAATCAAGGCCAATGAAGG
       A  P  A  D  H  P  G  S  I  S  A  I  K  A  N  E  G

 601  AGTTATCACCACCGAGGGAATGATTAACGGTCCATTTAGGGTCTCAGCCT
        V  I  T  T  E  G  M  I  N  G  P  F  R  V  S  A

 651  GTGATGACCTTGCCGCAGTTATTCGCTCCAGCCGTGTACTGATTATTGTA
      C  D  D  L  A  A  V  I  R  S  S  R  V  L  I  I  V

 701  ACCCGTGCGGACGTTCACGACAGCTTCGTCAACGAACTCGCCAACTTCAA
       T  R  A  D  V  H  D  S  F  V  N  E  L  A  N  F  N

 751  CGGCGAACTCGCAACAAAGGATATTGTCGTCGTGTGCGGCCATGGCTTCT
        G  E  L  A  T  K  D  I  V  V  V  C  G  H  G  F

 801  CCATCAAGTACGAGAGACAGCTGCGATTCAAGCGAATATTCGAGACGGAT
      S  I  K  Y  E  R  Q  L  R  F  K  R  I  F  E  T  D

 851  AATTCGCCCATAACGTCTAAGCTATCGGATCAAAAAAAATGTAACGTCAA
       N  S  P  I  T  S  K  L  S  D  Q  K  K  C  N  V  N

 901  CATCAAGGAAATGAAAGCGTCTTTCGGACTGTCATGTTTCCCAATTCATC
        I  K  E  M  K  A  S  F  G  L  S  C  F  P  I  H

 951  GCGATGATGCTGGCGTGATTGATCTACCCGAAGATACCAAGAACATCTTT
      R  D  D  A  G  V  I  D  L  P  E  D  T  K  N  I  F

1001  GCCCAGCTATTTTCCGCTAGAATCATCTGCATCCCGCCGTTGCAAGTGCT
       A  Q  L  F  S  A  R  I  I  C  I  P  P  L  Q  V  L
```

1051 ATTCTTTTCCAACTATATCACTCATGCGGTTCCGGCAGTCATGAACATCG
F F S N Y I T H A V P A V M N I

1101 GAAGACTCCGCGACCCAGCCAATTCTCTTACTAAAAGAGCTGAGAAGTGG
G R L R D P A N S L T K R A E K W

1151 CTTCTTGAACTAGACGAGCGAACCCCACGAGCCGAGAAGGGCTTTTTCTT
L L E L D E R T P R A E K G F F F

1201 TTATGGTGAAGGATCCAACACTTACGTTTGCAACGTCCAAGAGCAAATAG
Y G E G S N T Y V C N V Q E Q I

1251 ACCACGAACGCCGGAAGGTTGCCGCAGCGTGTGGATTGCGTCTCAATTCT
D H E R R K V A A A C G L R L N S

1301 CTCTTGCAGGAATGCAATGATGAATATGATACTGACTATGAAACTTTGAG
L L Q E C N D E Y D T D Y E T L R

1351 GGAATACTGCCTAGCACCGTCACCTCATAACGTGCATCATGCATGCCCTG
E Y C L A P S P H N V H H A C P

1401 ACAACATGGAACATCGCTATTTTTCTGAAGAATTATGCTCGTTGGAGGAT
D N M E H R Y F S E E L C S L E D

1451 GTCGCGGCAATTGCAGCTATTGCCAAAATCGAAATACCCCTCACGCATGC
V A A I A A I A K I E I P L T H A

1501 ATTCATCAATATTATTCATGCGGGGAAAGGCAAGATTAATCCAACTGGCA
F I N I I H A G K G K I **N** **P** **T** G

glycosylation site

1551 AATCATCCAGCGTGATTGGTAACTTCAGTTCCAGCGACTTGATTCGTTTT
K S S S V I G **N F S** S S D L I R F

TLE ⇨

1601 GGTGCTACCCACGTTTTCAATAAGGACGAGATGGTGGAG**TAA**AGAAGGAG
G A T H V F N K D E M V E ■ ....

poly A signal

1651 TGCGTCGAAGCAGATCGTTCAAACATTTGGC**AATAAA**GTTTCTTAAGATT
....

1701 GAATCCTGTTGCCGGTCTTGCGATGATTATCATATAATTTCTGTTGAATT
....

poly A signal

1751 ACGTTAAGCATGT**AATAAT**TAACATGTAATGCATGACGTTATTTATGAGA
....

end of mRNA ⇨

1801 TGGGTTTTTATG**A**T**T**AGA**GTCCCGC**AATTATACATTTAATACGCGATAGA
....

1851 AAACAAAATATAGCGCGCAAACTAGGATAAATTATCGCGCGCGGTGTCAT
....

1901 CTATGTTACTAGATCGATCAAACTTCGGTACTGTGTAATGACGATGAGCA
....

1951 ATCGAGAGGCTGACTAACAAAAGGTATGCCCAAAAACAACCTCTCCAAAC
....

2001 TGTTTCGAATTGGAAGTTTCTGCTCATGCCGACAGGCATAACTTAGATAT
....

2051 TCGCGGGCTATTCCCACTAATTCGTCCTGCTGGTTTGCGCCAAGATAAAT
....

```
2101  CAGTGCATCTCCTTACAAGTTCCTCTGTCTTGTGAAATGAACTGCTGACT
      ..................................................

2151  GCCCCCCAAGAAAGCCTCCTCATCTCCCAGTTGGCGGCGGCTGATACACC
      ..................................................

2201  ATCGAAAACCCACGTCCGAACACTTGATACATGTGCCTGAGAAATAGGCC
      ..................................................

2251  TACGTCCAAGAGCAAGTCCTTTCTGTGCTCGTCGGAAATTCCTCTCCTGT
      ..................................................

2301  CAGACGGTCGTGCGCATGTCTTGCGTTGATGAAGCTT
      .....................................
```

A restriction map of this gene is shown in *Figure 15*.

Figure 15. Restriction map of the *nos* gene, promoter and terminator.

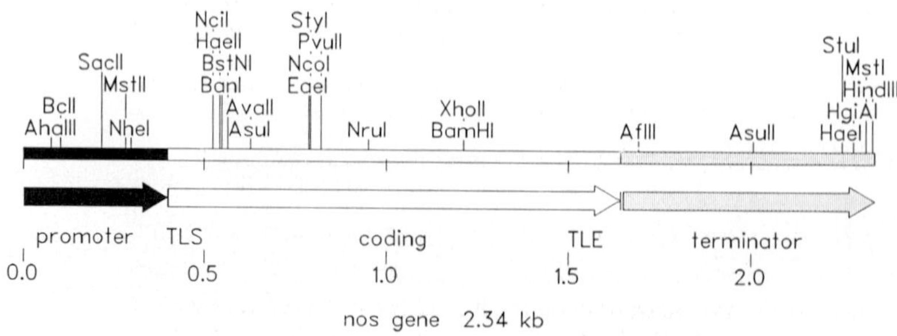

17. OCTOPINE SYNTHASE GENE [R, P, T]

The octopine synthase (*ocs*, *oct*) gene was the first reporter gene system used in plants. The enzyme, octopine synthase, is responsible for the biosynthesis of the opine, octopine, from pyruvate and amino acids such as arginine, lysine or ornithine, in plant cells transformed with T-DNA from Ti plasmids such as pTiAch5. As with the *nos* gene, the *ocs* promoter and terminator sequences have been used in chimeric gene constructs to provide high levels of foreign gene expression in transgenic plants. The *ocs* gene encodes a polypeptide of 365 amino acids, molecular weight 38 802 and pI 7.43. Overall sequence length = 1582 nt (33, 34). Sequence features are shown in bold type, underlined and labeled above or below the sequence. **G–H Box**, 'Goldberg–Hogness' box; **TCS**, transcription start; **TLS**, translation start; **TLE**, translation end; ■ indicates a TGA (umber) stop codon; **poly A signal**, putative polyadenylation signal. A potential N-glycosylation site is present at Asn74.

```
   1  TAGGTGAAGGTGGAAATTAATGATCCATCTTGAGACCACAGGCCCACAAC
      ..................................................

  51  AGCTACCAGTTTCCTCAAGGGTCCACCAAAAACGTAAGCGCTTACGTACA
      ..................................................

 101  TGGTCGATAAGAAAAGGCAATTTGTAGATGTTAACATCCAACGTCGCTTT
      .................................................

 151  CAGGGATCCTTTTTACCGACAACTCATCCACATTGATGGTAGGCAGAAAG
      ..................................................
```

```
                                                      G-H Box
 201  TTAAAGGATTATCGCAAGTCAATACTTGCCCATTCATTGATCTATTTAAA
      ..................................................
                                  ⇦ TCS
 251  GGTGTGGCCTCAAGGATAATCGCCAAACCATTATATTTGCAATCTACCAA
      ..................................................
      ⇦ TLS
 301  ATGGCTAAAGTGGCAATTTTGGGTGCGGGAAACGTGGCTCTTACTCTTGC
       M  A  K  V  A  I  L  G  A  G  N  V  A  L  T  L  A

 351  AGGTGATCTCGCCCGGAGGCTCGGCCAGGTGTCCTCAATCTGGGCGCCAA
        G  D  L  A  R  R  L  G  Q  V  S  S  I  W  A  P

 401  TCTCCAACAGGAACAGCTTCAACTCTGTGAGGTCCCTTGGCTCCTTGGAG
      I  S  N  R  N  S  F  N  S  V  R  S  L  G  S  L  E

 451  CTAGTAGGGCCGGACTATGGAGGCGACTTTCAGCCGCAATTGGAGGATGA
       L  V  G  P  D  Y  G  G  D  F  Q  P  Q  L  E  D  D

 501  TCTTGAAACAGCGATTTCAGGCGCGGCGTTCATTTTTCTTACGGTCCCGA
        L  E  T  A  I  S  G  A  A  F  I  F  L  T  V  P
                                                 glycosylation site
 551  CCATGGGCCAGCAAGGAATTCTTTGCGAGTTGGCGAACTTCAATCTGAGC
      T  M  G  Q  Q  G  I  L  C  E  L  A  N  F  N  L  S

 601  AGCTCGGTCCTCGTAGCATTGCCCGGTAGCGCAACGTCTCTGGCATGCAA
        S  S  V  L  V  A  L  P  G  S  A  T  S  L  A  C  K

 651  GCAGACTTTAACTCCAGCTTTTGCACCTATCGCCGTCATCGAAGCAACGA
        Q  T  L  T  P  A  F  A  P  I  A  V  I  E  A  T

 701  CATCTCCCTATGCATGCCGCCGTGTCAATGCACAGGTGCTAATGCTAAGT
      T  S  P  Y  A  C  R  R  V  N  A  Q  V  L  M  L  S

 751  GTGAAGAGAACGTTCGAAGTTGCGTCAACTCAGGCTTTGAGCGAAGAGGT
       V  K  R  T  F  E  V  A  S  T  Q  A  L  S  E  E  V

 801  TAGGGGGGGCTTCGAGATTCTCTTTCCAAATCGGCTTCAGTGGTATCAAA
        R  G  G  F  E  I  L  F  P  N  R  L  Q  W  Y  Q

 851  ATCCCGCGTCGATATTTTTTTCAAATACCAACCCTGTTGCTCACCCTGCT
      N  P  A  S  I  F  F  S  N  T  N  P  V  A  H  P  A

 901  GGAATTCTCGCTGCAAAAGATACCATCGAGCAAGGGATATCGCCGATCCC
       G  I  L  A  A  K  D  T  I  E  Q  G  I  S  P  I  P

 951  AAAATTCTACCGGAAGTTTGTCCCACAAGCCATCACGCGTGTTACCGCAA
        K  F  Y  R  K  F  V  P  Q  A  I  T  R  V  T  A

1001  TAGACGAGGAACGCCTTACGATTGTTAATGCGCTTGGACTTGAGTCCGAG
      I  D  E  E  R  L  T  I  V  N  A  L  G  L  E  S  E

1051  ACGGATTTCGCATACTGCAAAAAATGGTATGGTGGGCACGCCAGCAACGC
       T  D  F  A  Y  C  K  K  W  Y  G  G  H  A  S  N  A

1101  AAGAGAATTCTATGAGACCTTTGAGGGCTACGCTGATATCGAAACTCCAA
        R  E  F  Y  E  T  F  E  G  Y  A  D  I  E  T  P

1151  GAAACATGAACCATCGCTACCTTAGTGAGGACGTGAAGCACATTTTGGTG
      R  N  M  N  H  R  Y  L  S  E  D  V  K  H  I  L  V

1201  CTCTGGGTTGAGATAGCTGAAGTGATCGGCGTGCAAGTTCCAGAAATGAA
       L  W  V  E  I  A  E  V  I  G  V  Q  V  P  E  M  K

1251  ATCTGTAGTGCAGGAGGCAAGCGACGTGCTGAACGAGGACCTCTCTCATA
        S  V  V  Q  E  A  S  D  V  L  N  E  D  L  S  H

1301  CTGGTAGGGGGCTATCGTCTCTCAACCTGGAAGGTTCAAACGCGAATGCC
      T  G  R  G  L  S  S  L  N  L  E  G  S  N  A  N  A
```

```
                                        TLE ⇨
1351  ATAGTCAGGGCTCTCAATGGAGTTTGAATCAAATCTTCCAGCTGCTTTAA
       I  V  R  A  L  N  G  V  ■ .......................

1401  TGAGATATGCGAGACGCCTATGATCGCATGATATTTGCTTTCAATTCTGT
      ..................................................

1451  TGTGCACGTTGTAAAAAACCTGAGCATGTGTAGCTCAGATCCTTACCGCC
      ..................................................
                        poly A signal
1501  GGTTTCGGTTCATTCTAATGAATATATCACCCGTTACTATCGTATTTTTA
      ..................................................
       poly A signal  end of mRNA ⇨
1551  TGAATAATATTCTCCGTTCAATTTACTGATTG
      ................................
```

A restriction map of this gene is shown in *Figure 16.*

Figure 16. Restriction map of the *ocs* gene, promoter and terminator.

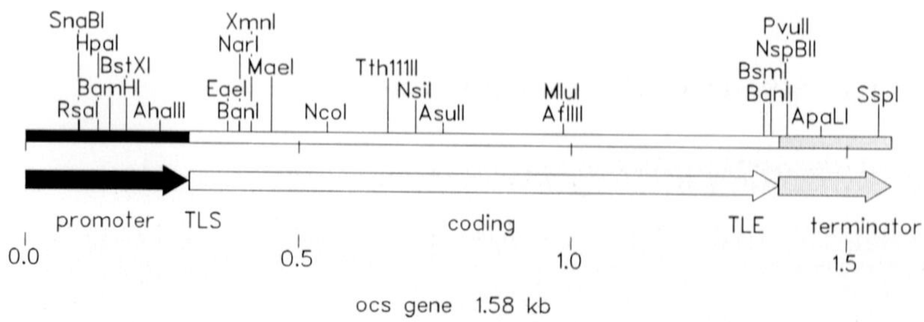

18. PHOSPHINOTHRICIN ACETYLTRANSFERASE GENE [S]

Phosphinothricin acetyltransferase (*pat, bar*) gene, isolated from *Streptomyces hygroscopicus*, provides resistance to the herbicide bialaphos (Basta). The sequence shown is modified upstream from the translation start, to that required for optimal translational initiation in eukaryotes. It also contains additional multiple restriction sites to facilitate promoter and terminator fusions. The gene encodes an enzyme of 183 amino acids, molecular weight 20 639 and pI 7.62. Overall sequence length = 615 nt (35). Sequence features are shown in bold type, underlined and labeled above or below the sequence. **TLS**, translation start; **TLE**, translation end; **mcs**, multiple cloning sites; ■ indicates a TGA (umber) stop codon. There are no apparent N-glycosylation sites.

```
      ⇦          mcs          ⇨          ⇦ TLS
  1   GAATTCGAGCTCGGTACCCGGGGATCTACCATGAGCCCAGAACGACGCCC
      ............................. M  S  P  E  R  R  P

 51   GGCCGACATCCGCCGTGCCACCGAGGCGGACATGCCGGCGGTCTGCACCA
        A  D  I  R  R  A  T  E  A  D  M  P  A  V  C  T

101   TCGTCAACCACTACATCGAGACAAGCACGGTCAACTTCCGTACCGAGCCG
      I  V  N  H  Y  I  E  T  S  T  V  N  F  R  T  E  P

151   CAGGAACCGCAGGAGTGGACGGACGACCTCGTCCGTCTGCGGGAGCGCTA
       Q  E  P  Q  E  W  T  D  D  L  V  R  L  R  E  R  Y

201   TCCCTGGCTCGTCGCCGAGGTGGACGGCGAGGTCGCCGGCATCGCCTACG
        P  W  L  V  A  E  V  D  G  E  V  A  G  I  A  Y
```

```
251  CGGGCCCCTGGAAGGCACGCAACGCCTACGACTGGACGGCCGAGTCGACC
     A  G  P  W  K  A  R  N  A  Y  D  W  T  A  E  S  T

301  GTGTACGTCTCCCCCCGCCACCAGCGGACGGGACTGGGCTCCACGCTCTA
      V  Y  V  S  P  R  H  Q  R  T  G  L  G  S  T  L  Y

351  CACCCACCTGCTGAAGTCCCTGGAGGCACAGGGCTTCAAGAGCGTGGTCG
       T  H  L  L  K  S  L  E  A  Q  G  F  K  S  V  V

401  CTGTCATCGGGCTGCCCAACGACCCGAGCGTGCGCATGCACGAGGCGCTC
     A  V  I  G  L  P  N  D  P  S  V  R  M  H  E  A  L

451  GGATATGCCCCCCGCGGCATGCTGCGGGCGGCCGGCTTCAAGCACGGGAA
      G  Y  A  P  R  G  M  L  R  A  A  G  F  K  H  G  N

501  CTGGCATGACGTGGGTTTCTGGCAGCTGGACTTCAGCCTGCCGGTACCGC
       W  H  D  V  G  F  W  Q  L  D  F  S  L  P  V  P
                                    TLE ⇨    ⇦           mcs
551  CCCGTCCGGTCCTGCCCGTCACCGAGATCTGATGACCCGGGGGATCCCTG
     P  R  P  V  L  P  V  T  E  I  ■ ................
                           ⇨
601  CAGGCATGCAAGCTT
     ...............
```

A restriction map for this gene is shown in *Figure 17.*

Figure 17. Restriction map of the *pat* gene.

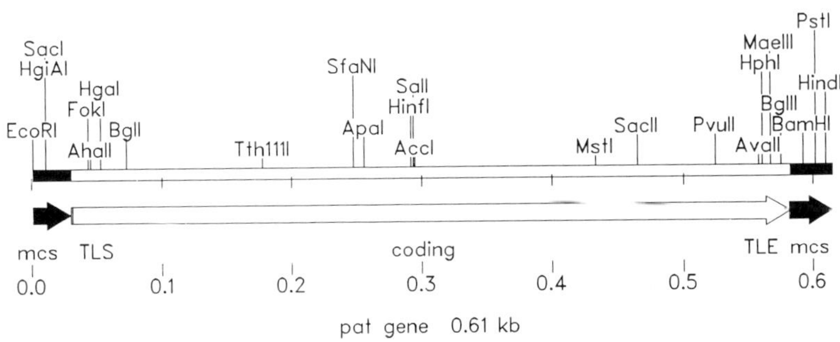

19. REFERENCES

1. Hauptmann, R.M., Vasil, V., Ozias-akins, P., Tabaeizadeh, Z., Rogers, S.G., Fraley, R.T., Horsch, R.B. and Vasil, I.K. (1988) *Plant Physiol.*, **86,** 602.

2. Eichholtz, D.A. *et al.* (1987) *Som. Cell Mol. Genet.*, **13,** 67.

3. Mazur, B.J., Chui, C.-F. and Smith, J.K. (1987) *Plant Physiol.*, **85,** 1110.

4. Haughn, G.W., Smith, J., Mazur, B.J. and Somerville, C. (1988) *Mol. Gen. Genet.*, **211,** 266.

5. Chinault, A.C., Blakesley, V.A., Roessler, E., Willis, D.G., Smith, C.A., Coog, R.G. and Fenwick, R.G. (1986) *Plasmid*, **15,** 119.

6. Hille, J., Verheggen, F., Rocluink, P., Franssen, H., Van Kammen, A. and Zabel, P. (1986) *Plant Mol. Biol.*, **7,** 171.

7. Topfer, R., Matzeit, V., Gronenborn, B., Schell, J. and Steinbiss, H. (1987) *Nucleic Acids Res.*, **15,** 5890.

8. Alton, N.K. and Vapnek, D. (1979) *Nature*, **282,** 864.

9. Shaw, W.V. (1983) *Crit. Rev. Biochem.*, **14,** 1.

10. Brisson, N. and Hohn, T. (1984) *Gene*, **28,** 271.

11. Herrera-Estrella, L., Depicker, A., Van Montagu, M. and Schell, J. (1983) *Nature*, **303,** 209.

12. Fling, M.E. and Elwell, L.P. (1985) *J. Bacteriol.*, **141**, 779.

13. Guerineau, F. and Mullineaux, P. (1989) *Nucleic Acids Res.*, **17**, 4370.

14. Shah, D.M., Horsch, R.B., Klee, H.J., Kishore, G.M., Winter, J.A., Tumer, N.E., Hironaka, C.M., Sanders, P.R., Gasser, C.S., Aykent, S., Siegel, N.R., Rogers, S.G. and Fraley, R.T. (1986) *Science*, **233**, 478.

15. Calos, M.P. (1978) *Nature*, **274**, 762.

16. Maizels, N.M. (1973) *Proc. Natl. Acad. Sci. USA*, **70**, 3585.

17. Herrera-Estrella, L. and Simpson, J. (1988) in *Plant Molecular Biology — A Practical Approach* (C. Shaw, ed.). IRL Press, Oxford, Chapter 6, p. 131.

18. Teeri, T.H. and Herrera-Estrella, L. (1991) *Plant Mol. Biol. Manual*, **B15**, 1.

19. Jefferson, R.A., Burgess, S.M. and Hirsch, D. (1986) *Proc. Natl. Acad. Sci. USA*, **83**, 8447.

20. Jefferson, R.A., Kavanagh, T.A. and Bevan, M.W. (1987) *EMBO J.*, **6**, 3901.

21. Jefferson, R.A. (1987) *Plant Mol. Biol. Rep.*, **5**, 387.

22. Iturriaga, G., Jefferson, R.A. and Bevan, M.W. (1989) *Plant Cell*, **1**, 381.

23. Jefferson, R.A. and Wilson, K.J. (1991) *Plant Mol. Biol. Manual*, **B14**, 1.

24. Gritz, L. and Davies, J. (1983) *Gene*, **25**, 178.

25. Gould, S.J. and Subramani, S. (1988) *Anal. Biochem.*, **175**, 5.

26. Ow, D.W., Wood, K.V., DeLuca, M., De Wet, J.R., Helinski, D.R. and Howell, S.H. (1986) *Science*, **234**, 856.

27. Howell, S.H., Ow, D.W. and Schneider, M. (1989) *Plant Mol. Biol. Manual*, **B8**, 1.

28. Baldwin, T.O., Berends, T., Bunch, T.A., Holzman, T.F., Rausch, S.K., Shamansky, L., Treat, M.L. and Zeigler, M.M. (1984) *Biochemistry*, **23**, 3663.

29. Koncz, C., Olsson, O., Langridge, W.H.R., Schell, J. and Szalay, A.A. (1987) *Proc. Natl. Acad. Sci. USA*, **84**, 131.

30. Pridmore, R.D. (1987) *Gene*, **56**, 309.

31. Bevan, M., Barnes, M. and Chilton, M. (1983) *Nucleic Acids Res.*, **11**, 369.

32. Herrera-Estrella, L., De Block, M., Messens, E., Hernalsteens, J.-P., Van Montagu, M. and Schell, J. (1983) *EMBO J.*, **2**, 987.

33. De Greve, H., Dhaese, P., Seurinck, J., Lemmers, M., Van Montagu, H. and Schell, J. (1982) *J. Mol. Appl. Genet.*, **1**, 499.

34. Herrera-Estrella, L., Teeri, T.H. and Simpson, J. (1988) *Plant Mol. Biol. Manual*, **B1**, 1.

35. White, J., Chang, S.-Y.P., Bibb, M.J. and Bibb, M.J. (1990) *Nucleic Acids Res.*, **18**, 1062.

CHAPTER 6

CHARACTERISTICS OF PLANT PRE-mRNA INTRONS AND TRANSPOSABLE ELEMENTS

C.G. Simpson, D.J. Leader and J.W.S. Brown (Plant Introns)

T. Franklin (Plant Transposable Elements)

Plant Introns

1. INTRODUCTION

The majority of eukaryotic mRNAs are produced from precursor mRNAs (pre-mRNAs) by the excision of intervening sequences (introns) in a process called splicing. The basic mechanism of splicing has been established in animal and yeast systems and proceeds through two transesterification reactions which require a number of intron sequence signals (1–4). In the first step of the reaction, cleavage occurs at the 5′ splice site with concomitant covalent attachment of the 5′ side of the intron to an internal adenosine branchpoint located upstream of the 3′ splice site. The second step of the reaction involves cleavage at the 3′ splice site and ligation of the two exon (coding) sequences. Splicing depends on mechanisms which recognize these intron sequence signals, allowing the formation of a large ribonucleoprotein complex called the spliceosome in which splicing occurs. The major components of the spliceosome are small nuclear ribonucleoprotein particles (snRNPs) which are complexes of uracil-rich small nuclear RNAs (UsnRNAs) and proteins which interact with the intron splicing signals and mediate the excision and ligation events (4–8).

Compared with yeast and animals, relatively little is known about splicing in plants. Sequence similarity of isolated UsnRNAs (9) and intron sequence signals suggest that the basic mechanism of splicing is similar to the animal system but variations do exist. First, although plant introns are essentially similar to yeast and animal introns in terms of 5′ and 3′ splice sites, they generally have a less pronounced polypyrimidine stretch (9–12). Secondly, although plant introns contain putative branchpoint sequences (10), their role in plant splicing remains to be demonstrated. Finally, plant introns have a high AU content. The importance of AU-rich sequences has been demonstrated particularly in dicotyledonous (dicot) plant tissues where only AU-rich introns (>59%) are accurately and efficiently spliced (13, 15). Monocotyledonous (monocot) cells appear to be more flexible in their ability to splice efficiently both monocot and dicot introns; this highlights a major difference in the processing of pre-mRNA in monocot and dicot plant tissues (13–16). Different splicing efficiencies between monocot and dicot tissues are likely to be reflected in differences in the intron sequence and structure. This chapter aims to present the important features of plant pre-mRNA introns.

An extensive collection of plant genomic sequences was collated from the EMBL version 30, and EMBL daily update (to June 19, 1992) databases via the SEQNET facility at Daresbury. Files containing plant introns were extracted by searching the database for 'joining' data and '*Liliopsida*' for monocot genes and '*Magnoliopsida*' for dicot genes. Mitochondrial and chloroplast intron data were discarded and the nuclear intron data preserved. From the data of each entry, the first 6 nt and last 20 nt of intron sequence were tabulated along with 3 nt of flanking exon sequences (*Tables 1* and *2*). The size and AU content (AU%) of each intron

▶ p. 235

Table 1. Compilation of monocot intron data

Plant species	Gene	Intron	5′ splice site	3′ splice site	Size (bp)	AU (%)	Ref.
Avena sativa	Type 3 phytochrome gene (*phy3*)	1	CAG : GTACCT	GGTTCAATTACTTGAGGCAG : GCG	1269	58	18
		2	AGG : GTGAGT	TAATTTCATTGATGAAGCAG : GTA	203	69	18
		3	AAA : GTACCC	TTTGTCATTTATGGCCACAG : ATC	341	69	18
		4	TAG : GTATAT	CTTGTCTTTATTGTCATCAG : GAT	138	64	18
		5	ATG : GTATAA	TTGTTCTGTTCTTCATGCAG : TTC	264	64	18
	12S globulin seed storage protein gene	1	AAG : GTTAGT	ACAAATGTGTTTTCTTTTAG : GTA	117	73	19
		2	AAG : GTAACT	TCATCGATATAAAATTTTAG : GAG	126	74	19
		3	CAG : GTATTT	ACAAATAATTGCTATTGCAG : AAT	100	73	19
Hordeum vulgare	Acyl carrier protein I (*Acl1*)	1	CAG : GTACCG	AATTTATTGGTTCATTGCAG : GTG	1083	63	20
		2	GCG : GTAGGC	TTCATTCGTTTCTTGAGCAG : GCC	568	64	20
		3	ACG : GTAAAT	ACACCGATTATGTTCTGTAG : GTT	86	49	20
	Acyl carrier protein III (*Acl3*)	1	AAG : GTATGC	GTTTGTTCCTTGAATTCCAG : GCA	1184	58	20
		2	TCA : GTGAGC	TTCTGAATATGTGCCAACAG : GCA	640	63	20
		3	ACG : GTAACT	TGTCATGCTCCTGTATGCAG : GTT	174	60	20
	Alcohol dehydrogenase gene *Adh2*	1	AAG : GTTGGA	AATTTGGTTGGTATTGGCAG : CGG	104	59	21
		2	AAC : GTACCA	TGCTGCTTGGATTTGGGCAG : GGC	84	57	21
		3	AGG : GTATCC	TATGATTTGTATGCTCACAG : CAT	178	65	21
		4	CCG : GTAAGA	TTGTCACGCCTCCAATTCAG : GGC	105	61	21
		5	GCT : GTAAGT	TCTGACACATGCATGATCAG : GCC	96	63	21
		6	AAG : GTACGC	GCATGTCCATTGTTTTTCAG : CTA	134	55	21
		7	GAG : GTATAC	CAATGGTGGTGCCGCCACAG : GTG	113	55	21
		8	GAC : GTACAT	TGGCCTGATGATGGGAGCAG : GGG	86	56	21

Alcohol dehydrogenase gene *Adh3*	1	AAG : GTTTGC	AATTGTACGGCTGAGTGCAG : CGG	97	55	21
	2	GCC : GTTTTT	GTTCTGCTGGATTTGTGCAG : GGG	94	60	21
	3	AGG : GTATGT	AAATAGCTTCTTTATTGCAG : CAT	377	62	21
	4	CTG : GTAAGA	GAGTTGCCATTTTTCTTCAG : GAC	96	62	21
	5	GCT : GTGAGT	TATGCTTCTGACATGGGCAG : GCC	92	59	21
	6	AAG : GTACAG	TGTACTACATTATTTTTCAG : CTA	88	68	21
	7	GAG : GTCTGT	TGTTCAATGATGCTCTTCAG : GTG	85	62	21
	8	GAT : GTACGT	TTGCAATCTGGTGGATGCAG : GGG	95	62	21
Alpha-amylase 2 gene	1	CAG : GTACGC	AGCTGGAATGGATGTTGCAG : GGG	80	46	22
	2	AAG : GTCCTC	AACGATCGACCTGGGTGCAG : GTT	233	59	22
	3	ATC : GTAAGC	TTTGTCTTGGTGATTTGCAG : TTC	92	58	22
Alpha-amylase type B (*Amy46*) gene	1	CAG : GTAAGA	GGTTCTGCTGCGTTCGACAG : GGT	103	47	23
	2	ATC : GTGAGC	AAATAATTGTGATTCTTCAG : TTC	106	66	23
Alpha-amylase type B isozyme (*Amy6-4*) gene	1	CAG : GTAAGA	TTGGGTTCTGCATGCGACAG : GGC	80	56	24
	2	ATC : GTGAGT	AAATAATTTTGGTTCTTCAG : TTC	106	68	24
(Type A) high pI isozyme of alpha-amylase gene	1	CAG : GTACAC	AGCTGGAATTGATGTTGCAG : GGG	100	54	25
	2	AAG : GTCCCT	AACGATGGACTTGGGTGCAG : GTT	253	59	25
	3	ATC : GTAAGC	CTTGTCTTGTCGGTTTGCAG : TTC	96	54	25
(1,3;1,4)-beta-glucanase	1[a]	AAA : GTGATT		2514	56	26
	1[b]	CTG : GTGCAA	GAGGTTATATTACATTGCAG : GCG	55[d]	67	26 67
CHS gene for chalcone synthase	1	TGT : GTAAGT	TCTGCCGGGTTGATTTGCAG : GTG	1289	63	27
Nitrate reductase	1	CAG : GTACGT	TTCGTTGTTGTGGTTTGTAG : GCG	1625	56	28
Thiol protease aleurain	1	CAG : GTCCGT	CGTTCGATCCGCTTTCGCAG : GTA	122	31	29
	2	ACC : GTACGA	CTTTTGCTGTCGTCCCGCAG : GCT	95	32	29
	3	ACC : GTAAGT	TGATGTTGTCATTGCTGCAG : AAA	586	57	29
	4	CAG : GTAACT	TCATATACTTATTCTTGCAG : CAC	1334	64	29
	5	CTG : GTAAGG	TGCTTTTTTTTCTTGGTAAG : AAT	94	66	29
	6	GAT : GTGAGT	AGTTTTGCTGAAACTACTAG : GAC	163	64	29
	7	GTG : GTGAGT	GACTGTTGTAAATGATGCAG : CTA	98	51	29
Ltp1 gene for lipid transfer protein	1	CAG : GTGATT	CACGCATGATTGACCTGCAG : GAT	133	68	30

Table 1. Continued

Plant species	Gene	Intron	5′ splice site	3′ splice site	Size (bp)	AU (%)	Ref.
Hordeum vulgare (continued)	Gene for LTP 1	1	CAG : GTGATT	CACGCATGATTGACCTGCAG : GAT	133	68	31
	Gene for type A alpha-amylase (*Amy32b*)	1	CAG : GTACAC	AGCTGGAATTGATGTTGCAG : GGG	100	54	29
		2	AAG : GTCCCT	AACGATGGACTTGGGTGCAG : GTT	253	59	29
		3	ATC : GTAAGC	CTTGTCTTGTCGGTTTGCAG : TTC	96	54	29
	nar7 gene for NAD(P)H-bispecific nitrate reductase	1	AAG : GTACTT	GTTTGTCTGGATGTTTTCAG : CTT	118	45	32
		2	CCG : GTGAGT	GGTGCTGTTTCTTTTATCAG : GCG	176	48	32
	Paz1 gene for protein z	1	AAG : GTGACC	TGTATTGTTCCTTGGTCTAG : ACT	334	64	33
	Peroxidase BP 2A (*Prx6*) gene	1	CAG : GTACCT	TGATTCATGAATGCATGCAG : GGG	77	52	34
	Rubisco activase genes	1	CCG : GTACGT	ACTATGTCATACTTGTGCAG : GCG	2222	60	35
		2	CCG : GTGCGT	ATGTAATGTAATGTATACAG : GCT	121	60	35
		3	AAG : GTACCT	GACCTTGAATCGAAATTCAG : ATC	151	71	35
		4	CAA : GTAAGA	ACCTGTGCTTTTTGCCACAG : CCC	250	69	35
		5	TTG : GTATAT	CATGCATCGATCCGATACAG : ACT	141	79	35
		6[a]	ACG : GTTAGA		231	58	35
		6[b]	GAG : GTGAAC	TTAAATTCGAGGTAAAGGAG : CAC	181	60	35
	Seed imbibition protein (*Sip1*) gene	1	TAG : GTACGC	AGGCCGTTCTCGTGTTGCAG : AAG	76	46	36
		2	CCG : GTACGT	CGGAATTGCTGTTGATGCAG : AGA	363	63	36
		3	GCG : GTATAA	GTGCGCGCGTGTGAGAGCAG : GGG	95	56	36
		4	CAA : GTAAAT	GTTGCTTGTGTGGAATGCAG : GGC	94	55	36
		5	TCG : GTAGTA	TGGCGTATGTATTTATGCAG : GTG	106	59	36
		6	CAT : GTAATT	CTGACTCTGCATTGACGCAG : AGC	85	61	36
		7	CAG : GTGCAG	TCTGCTCTTCTTTCTCTCAG : CTT	98	53	36

	Waxy locus encoding starch synthase	1	CAG : GTACGC	ACACTGACGATCACGTCCAG : TCG	559	56	37
		2	GCC : GTAAGC	CTGACGGCGTGTCTGTTCAG : GCC	89	52	37
		3	GAG : GTAGGT	CGACTGGCTGGATTTCGCAG : ATC	84	51	37
		4	AAG : GTGCGT	CTGCTGCTTACAATTTCTAG : GTC	126	57	37
		5	ACG : GTAAGA	CCCCGTGTCGTGCCGTGCAG : GGG	136	51	37
		6	AAG : GTTTTG	GAACCATCTGCTTTTGCTAG : GTG	106	68	37
		7	ACC : GTGAGC	CTAACGAGGCAAAGTGACAG : GCG	92	46	37
		8	CTT : GTACGT	AGTGATCGGATGAATTGCAG : GGC	94	48	37
		9	ACG : GTAAAC	AGTTTCTTCCAAATTTTAAG : CCG	82	55	37
		10	GAC : GTATGC	CCTTTCTTCTTGGTGGCCAG : TGC	97	55	37
		11	AAG : GTACAT	GGTCTGGGCGTATGCTGCAG : GGA	76	55	37
	BKIN12 gene for protein kinase	1	AAG : GTGTGC	TTCTCATATCTTGTTTCCAG : CAA	296	61	38
		2	CAG : GTGCCT	ATTAATATTTTCTTGAACAG : ATT	1339	64	38
		3	GAG : GTATTC	TAGTGAGTGTTTCTGCACAG : ATT	88	58	38
		4	AAG : GTGAGA	CTCCTTGTTTGGTGTTCCAG : GGA	122	62	38
		5	ATG : GTAACC	ATATATGTTCATTTCTGAAG : ATT	848	66	38
		6	GAG : GTATAT	AGACAATTAATGTTTTCCAG : GAA	75	69	38
		7	ATG : GTTAGT	CTGAATGGATGGTTCTCTAG : GAT	86	65	38
		8	CAG : GTCTTG	CTAATGCATCACTTATCCAG : CAG	77	57	38
		9	CAG : GTGTGC	TCTCAATTTTTTTCTACAG : CTT	398	64	38
	BKIN9 gene for protein kinase, exons 1 & 2	1	AAG : GTATGC	TTCTGATCCCTTGTTTCCAG : CAA	291	59 ($N=1$)	38
	Gene for CP29 precursor for core chlorophyll *a*/*b* binding (*CAB*) protein of photosystem II (PSII)	1	ACG : GTGAGT	ATGTCTTCTATTTGTTTCAG : GCC	104	63	39
		2	AGA : GTAAGT	CTTTCCCTTTGGACTTGCAG : CTA	209	57	39
		3	CAA : GTAAGT	CCTACACCTCGATTCTGCAG : GTA	117	59	39
		4	AAG : GTAATT	TCATAAACAACATCCTGCAG : ACC	95	58	39
		5	CTG : GTCAGT	TTCGTCAACTTCTCCGGCAG : GAA	85	45	39
Lemna gibba	Chlorophyll *a*/*b* apoprotein gene	1	CTG : GTTAGA	GCGGCGGGCTTCCTGATCAG : GTA	84	52	40
	Gene *NPR1* (negatively phytochrome regulated 1)	1	CGA : GTACGG	ACGGGAATCCCCTCCGCCAG : ACC	22[d]	32	41
		2	TCG : GTACGT	CTCTAGAAAATGTGATCGAG : TCG	76	63	41

Table 1. Continued

Plant species	Gene	Intron	5′ splice site	3′ splice site	Size (bp)	AU (%)	Ref.
Oryza sativa	RSs2 gene for sucrose-UDP glucoslytransferase (isozyme 2)	1	AAG : GTAATA	GTGGTTGGTTGTGCACGCAG : GCT	1445	56	42
		2	CAG : GTTCGT	TGACACTGTTATTTGCGCAG : GCT	726	67	42
		3	CAG : GTTTGA	AATTGTTTTTATCATGCAG : GAG	95	65	42
		4	AGG : GTATGA	ATCTCTGATGTATATTTCAG : CAC	120	80	42
		5	ATG : GTAAGG	AATATATATTTTTCTGCAG : ACC	125	70	42
		6	CAG : GTACTG	CATTTGATTGGAAAATACAG : GTT	119	72	42
		7	CAG : GTAGAA	TGTGCTATGATTACCGAAAG : GTT	84	74	42
		8	ATT : GTAAGT	CCTTCTTTGGTCTTGAACAG : GTC	81	63	42
		9	GAT : GTAGAT	CTCTTTCCCATGGATGTCAG : GAT	87	63	42
		10	CAT : GTACGC	CCTTGTTCCATGATATGCAG : TGT	83	63	42
		11	AAA : GTAATA	GGCATTTTCTGGTTTTTCAG : CAA	104	66	42
		12	CAA : GTAAGA	TAGAATACTCTCTGCAACAG : GTT	86	58	42
		13	CAG : GTAACT	ATCTTGTCTTGCGTTTCCAG : CCC	98	69	42
		14	GAA : GTATGC	TTCTCCAACCTTTGTTTCAG : ATA	94	62	42
	Alcohol dehydrogenase (*adh2*) gene	1	AAG : GTCCGT	GGGTTTTGGGGTTTTTGCAG : CGG	98	54	43
		2	AAG : GTAATT	TTCTGCTTTGATCTTCGCAG : GGG	112	56	43
		3	AGG : GTATGT	AAACAATTTTTTGCTCACAG : CAT	608	66	43
		4	CAG : GTAAGA	GGGTTTTGCTGTTCTTGCAG : GTT	115	64	43
		5	GCT : GTAATG	GTAAAATTACATGATTATAG : GCT	105	69	43
		6	AAG : GTACAA	ACTTGTTTTCATCTCTTCAG : CTA	94	67	43
		7	CAG : GTACAT	CTCTGCAATGGTGATCTCAG : GTG	98	62	43
		8	GAT : GTCAGT	AATCTGAAACCTGAATCCAG : GGA	130	62	43
		9	AAG : GTAAAA	CTGGTGTTTGATGTCTGAAG : GAG	136	65	43

ADP-glucose pyrophosphorylase gene	1	GAG : GTAACC	CTCTTCTTTTGTTGGTGCAG : AGT	735	57	44
	2[a]	GTT : CAGGTA	TTTTGTTTTATATTCTGACC : TCA	642	68	44
	3	ATG : GTATGC	TGCCTATTTTCGTCTTGTAG : GTT	114	64	44
	4	AGG : GTATGG	TCTATTTTTGTTGTCTGCAG : GTA	187	67	44
	5	CAG : GTATGC	TACCATCAAAATATCTGCAG : TTT	111	69	44
	6	AAA : GTAAGT	TTCACTTTTCGTCCTTGTAG : ACC	122	69	44
	7	GAG : GTAAAA	TTCCTTTTTCTGGTTACCAG : ACT	84	68	44
	8	AAG : GTATTC	CAATTGACTCATGGATGCAG : ATA	658	67	44
	9	GAA : GTAAGT	GCCGTTGCCTTGCCCTGCAG : GTA	257	61	44
					($N=5$)	
Alpha-amylase (*RAmy1B*) gene, exons 1 and 2 (5′ end)	1	CAG : GTAAGA	TGTGTTTGATGATGATGCAG : GGT	101	55	45
Alpha-amylase (*RAmy3D*) gene	1	CAG : GTACGT	TTATAACATGCATCATGCAG : GGT	124	65	45
	2	ATC : GTAAGC	ATGAACGCTTCTTGTGGCAG : TTC	88	57	45
Alpha-amylase (*RAmy3E*) gene	1	CAG : GTAGTT	TGTTTTTGCCGTACGTGCAG : GGG	143	58	45
	2	ATC : GTAAGT	ATGTGTACGTGTGTATGCAG : TTC	85	58	45
Cytochrome *c* (*OsCc-1*) gene	1	AAG : GTGAGC	CCATGTTTGTACTCTTTCAG : GTC	817	58	46
	2	AAG : GTATGA	CTGTTTTTCTCATTATGCAG : TAC	748	64	46
Emp1 gene	1	AGG : GTATGC	GCCTCAGCACCGGCGACGAG : TCC	107	68	47
GluA-3 gene for glutelin	1	AAG : GTTTGT	ATGGCAATCCACACTTGCAG : GCA	80	72	48
	2	AGG : GTAATT	GTTTGTTTATGGATTGTTAG : GAT	111	72	48
	3	CAG : GTAAAA	CAATGTTCTCTCTATTGCAG : AAC	77	70	48
GluB-1 gene for glutelin	1	AAG : GTTTGT	GTGAACATTATAATTTTTAG : GGA	111	71	49
	2	AAG : GTAATT	TTCATTGATAAAAATTTCAG : GAG	110	78	49
	3	CAG : GTATTT	TTCTCTACTGATTTCTACAG : AAT	94	72	49
GluB-2 gene for glutelin	1	AAG : GTTCGT	ATGAACAATATAATTTTCAG : GGA	121	72	49
	2	AAG : GTAACT	TTCATTGATAAAAATTTCAG : GAG	114	80	49
	3	CAG : GTATTT	ATGTCTATTGATTTCTACAG : AAT	91	79	49
GluB-3 gene for glutelin	1	AAG : GTTTGT	GTGAACATTTTAATTTGTAG : GGA	129	71	49
	2	AAG : GTATCT	TTCATTGATAAAAGTTTCAG : GAA	133	70	49

Table 1. Continued

Plant species	Gene	Intron	5′ splice site	3′ splice site	Size (bp)	AU (%)	Ref.
Oryza sativa (continued)	Glutelin gene	1	AAG : GTTTGT	CATGCATGCCATACTTTCAG : GGA	89	69	50
		2	AGG : GTAATT	ACCTCGTAATCAATTGTTAG : GAT	103	74	50
		3	CAG : GTAAAT	GTGTATCTGTTCTATTGCAG : AAT	83	67	50
	Constitutive GOS2 gene	1	AGG : GTATGT	TGAAGCTGTAATCGGGATAG : TTA	960	63	51
		2	TCG : GTATGC	GTTTTCCTGAACATGTACAG : ATC	85	58	51
		3	CAG : GTGTGT	ATCAATGCTTCACTCTGTAG : GTC	91	63	51
		4	CAG : GTAGGG	TAACATGCTATGTCCCGCAG : GCC	106	59	51
	Shoot-specific GOS5 gene for a putative chloroplast transit peptide	1	CAG : GTACAT	CTCATAATTTCTTGCTACAG : GCC	125	58	52
		2	CAG : GTTAGT	CGTGTGGGTGATCGATGCAG : AGC	78	67	52
	Root-specific GOS9 gene	1	ATG : GTTAGC	TTGATGTGTTTTATTTGCAG : AGC	707	73	53
		2	ACG : GTACAT	CCGCACGTCCTTGCATACAG : ATT	158	72	53
	Oryzacystatin gene	1	GCC : GTAAGC	CTTGATACTACCATGCGCAG : AAT	336	58	54
		2	AAG : GTCTAG	GTCAAATCTACTAATTTCAG : GCC	371	75	54
	Gene for proliferating cell nuclear antigen (PCNA)	1	CCA : GTGAGT	ATTTGTGCTTGGATGAATAG : ACC	85	58	55
		2	CTG : GTGTGT	TAAACTGCTTGGCATTCCAG : TTA	96	58	55
		3	AAG : GTAATT	GTTTTTTTGGGTCATTGCAG : CAA	717	64	55
	phy 18 gene for phytochrome	1	CAG : GTACCC	ACTTATATTATCTTGTTCAG : CAG	2585	62	56
		2	AAG : GTGGGT	ACATTTCTTTGATTGATCAG : GCA	222	66	56
		3	CAA : GTACTT	TCCTGCTGCTGTGTTCACAG : GTC	281	66	56
		4	TAG : GTATGT	TGTGGATCAATGTTTTTCAG : GAT	102	66	56
		5	ATG : GTATGT	AAGAGTTGTGTTTTCTGCAG : TTC	747	66	56
	RAmy3A gene for alpha-amylase	1	CAG : GTGCTT	TCTGTAACATTTTTGTGCAG : GGT	126	59	57
		2[b]	AAG : GCCAGC	TTTCCCGTGCATGCACGCAG : GCT	409	68	57
		3	ATC : GTACGC	TTTGGTTTTGTGTTGTTTAG : TTC	91	65	57

	RAmy3B gene for alpha-amylase	1	CAG : GTAAGC	CATGACTGTTCTTGGTGTAG : GGG	76	64	57
		2	ATT : GTGAGT	TTTATTGACTATGTGTGCAG : TTC	99	63	57
	RAmy3C gene for alpha-amylase	1	CAG : GTGAGT	TTGACATGATGTTGGTGCAG : GGA	128	64	57
		2	ATT : GTGAGT	CATTCCGTCTGTCTGTACAG : TTC	90	60	57
	rbcS gene for ribulose 1,5-bisphosphate carboxylase/ oxygenase small subunit	1	CAG : GTAATT	ATTATCGTTTATTCATTGAG : GTG	92	67	58
	UDP-glucose starch glycosyl transferase gene	1	AAG : GTGGAG	GCTTCCACTGGTGATTTCAG : GGT	96	64	59
		2	CAG : GTCAGT	TGTATTGATGCTACCTGCAG : GCA	99	61	59
		3	ATG : GTGAGT	CGGTTTGTTACTGACGACAG : GTG	92	62	59
		4	AAG : GTCTAT	CTGAATCTTTTTCACTGCAG : GTT	90	58	59
		5	CGG : GTATGA	CAAATCCGGTTAAATTTCAG : GTA	121	60	59
	Osamy-c gene for alpha-amylase	1	CAG : GTAAGA	GTTGTTTGATGATGATGCAG : GGT	102	56	60
		2	AAG : GTCGGT	GCTTGAATGCTGATCTGCAG : GCT	89	48	60
		3	ATC : GTGAGT	ATTCTGCAAAACTCTTTCAG : TTC	88	61	60
	Serine carboxypeptidase III gene	1	GAG : GTAGAT	GATTGGGGCGTTGCTTGCAG : GAT	130	63	61
		2	ACG : GTAACG	ATAAAACTTTCTTCATGCAG : ATT	759	64	61
		3	CAG : GTACTA	TTGCTTGATTCCCATGTCAG : GTG	607	66	61
		4	AAG : GTATGA	TTTACCTATTTTTCTTGTAG : GGT	85	69	61
		5	GTG : GTCAGT	CACACCTACTGATTTTTCAG : GTA	220	65	61
		6	AAT : GTAAGC	ATATTAATTCTTCTTTACAG : TAC	85	64	61
		7	TTG : GTAAGG	AACCTTTTCTATGTATGTAG : GAA	812	64	61
		8	AAG : GTTTTT	TTGCTTCTATGCTAATGCAG : GTC	397	68	61
	Osamy-B gene for alpha-amylase	1[b]	CAG : GCTCAG	TTGACATGATGTTGGTGCAG : GGA	139	58	62
	RAc2 gene for actin	1	AAG : GTAATA	ACTGTTTTCATTTTTCATAG : GCT	113	70	63
		2	CTG : GTAAGT	TGATATTTTTTGTCAAGTAG : GTA	266	70	63
		3	CAG : GTATGC	CTTTCCTTTTTTTTGTGCAG : ATG	92	67	63
Oryza sativa cv. *japonica*	Alpha-amylase (*amy2A*) gene	1	CAG : GTACGC	AGCTGTTGCGTTCTCTTCAG : GGG	81	49	64
		2	AAG : GTCGGT	ATGTGGCGATGTGCGTGAAG : GGT	1503	66	64
		3	ATC : GTAAGT	CTGTATATATGTACATGTAG : TTC	544	74	64

Table 1. Continued

Plant species	Gene	Intron	5′ splice site	3′ splice site	Size (bp)	AU (%)	Ref.
Pennisetum glaucum	Alcohol dehydrogenase gene	1	AAG : GTATCT	TGGCGGCTTATCTGTATTAG : GGA	102	51	65
		2	AGG : GTACGT	AAATTTGTTGATTTCTGCAG : TAT	497	66	65
		3	CTG : GTAAGT	TGTGTCATTTCTCTGTTTAG : GTC	83	64	65
		4	GCT : GTAAGT	TGCATATATCCATGATTTAG : GCT	100	58	65
		5	AAG : GTGAGA	TATTTCTATCTATTTTTCAG : CTA	74	73	65
		6	GAG : GTAATA	TCTAGTTCATTATGGTTTAG : GTA	81	63	65
		7	GAT : GTATGT	CAGCAATGCACTTTATCCAG : GGC	93	62	65
		8	AAG : GTAAAT	TCTTGGCTTGCATGCAGGAG : CTG	139	55	65
	Alcohol dehydrogenase (*Adh1*) gene	1	AAG : GTTCGT	TTGCGTTCAATTTATTGCAG : CCG	442	60	66
		2	AAG : GTATCT	TGGCGGCTTATCTGTATTAG : GGA	102	51	66
		3	AGG : GTACGT	AAATTTGTTGATTTCTGCAG : TAT	497	66	66
		4	CTG : GTAAGT	TGTGTCATTTCTCTGTTTAG : GTC	82	63	66
		5	GCT : GTAAGT	TGCATATATCCATGATTTAG : GCT	100	58	66
		6	AAG : GTGAGA	TATTTCTATCTATTTTTCAG : CTA	74	73	66
		7	GAG : GTAATA	TCTAGTTCATTATGGTTTAG : GTA	81	63	66
		8	GAT : GTATGT	CAGCAATGCACTTTATCCAG : GGC	93	62	66
		9	AAG : GTAAAT	TCTTCTTGGCTTGCATGCAG : GAG	136	56	66
Saccarum hybrid	Phosphoenolpyruvate carboxylase (*SCPEPCD1*) gene	1	ACG : GTATGT	TTTCCCTTTCTCATGAATAG : GTG	796	60	67
		2	CAG : GTTAGT	CCTCACCTTTCAATGTGCAG : GAT	340	66	67
		3	GAG : GTAATT	GTTCCATTTCTTATCTACAG : ATT	136	74	67
		4	ATG : GTGCAT	TAAACTGATATCCACAACAG : GGA	118	64	67
		5	GAG : GTAAAA	ATTCTTTTTATCGCTTGCAG : TTA	76	71	67
		6	TAG : GTGCAG	TGACATGCTTTTCCATGCAG : AGT	1164	66 ($M = 1$)	67
		7	CAG : GTATGT	GAATCTTTTTTCTCTTGCAG : TTC	223	69	67
		8	CTT : GTAAGT	TGTGTCTTTCTTTCAAATAG : GCA	310	69	67
		9	CAG : GTGATT	TGTGGCCTGCTGAATTTCAG : GTT	284	58	67

Saccarum officinarum	Protein gene	1	CAG : GTAACA	AAACACGTACATCGATGCAG : GTG	95	49	68
Sorghum vulgare	NADP malate dehydrogenase gene	1	CAA : GTGAGT	GTTGATGCAGGCAGGTGCAG : GAT	89	52	69
		2	GCG : GTGAGT	TTGGGAGTTGGGACACACAG : GAA	497	53	69
		3	AAA : GTAACT	TTAACACAAAAAAAATTCAG : CTT	95	69	69
		4	AAG : GTGAAT	CTCAAAAACTGCAACTGTAG : GTG	256	62	69
		5	CAG : GTTTCT	TTTTTTTTAAATTGTGTCAG : GGG	254	64	69
		6	TAA : GTACGG	GTGGCTTGCCCTTGATGCAG : TGC	79	58	69
		7	CAG : GTGACT	CACACAATCTCCTTTTGTAG : CTA	88	61	69
		8	CAG : GTGAAG	GCACTAACTTCTCTGTGTAG : GTT	88	64	69
		9	AAG : GTATGA	TCATTCTTATATCTAAACAG : CGT	86	66	69
		10	GGG : GTAAGT	AAGTAATAAAAAAAACACAG : GTT	80	69	69
		11	AAG : GTATGA	TCGTAAAATTCTTCCAACAG : GGT	84	68	69
		12	AAG : GTAAGG	GGCCCTGTACATATTTTCAG : AGT	94	59	69
		13	GAG : GTATGT	TTCTGTTTGTGCAAATGCAG : GGG	92	61	69
	PEPC gene for phospho-enolpyruvate carboxylase	1	ATG : GTAACT	TGAGTTCTCTTGCTGATTAG : GTT	1171	64	70
		2	GAG : GTTGGT	GATCCCAACATTCTTACCAG : GAT	82	59	70
		3	GAG : GTATTT	CTCCTCTCTCCTTTTTCAAG : ATC	104	68	70
		4	ATG : GTGCAC	CAATGTTTCTACAAATGCAG : GAA	180	62	70
		5	GAG : GTTGCT	ATAACTTGAAATGCTTGCAG : TTG	85	68	70
		6	TAG : GTAGTT	CTTCTTCATTTCTGCTGTAG : AGT	103	66	70
		7	CAG : GTTGAG	GTAGATGCCAACTTTTGCAG : CTG	91	66	70
		8	CTC : GTGCAG	TGTGATATTTGTGTTGACAG : GCA	240	65	70
		9	CAG : GTCGCT	CTGTTTTGGACTTCTATCAG : GTT	103	60	70
	NADH II gene for NADP-malate dehydrogenase	1	CAA : GTGAGT	TCTGATGAGAACCCCAGCAG : GCA	136	48	71
		2	GCG : GTGAGT	GAATGGATTGGGACACACAG : GAA	425	50	71
		3	AAA : GTAAAC	TTAACCCAAAAAAAATTCAG : CTC	97	67	71
		4	AAG : GTGAAA	CTCAAAAACTGCAACTGCAG : GCG	166	64	71
		5	CAG : GTTTCT	AAATTTTGAATTGTGTTCAG : GGG	290	63	71
		6	TAA : GTACGG	TGGGTTGACACTTGATGCAG : TGC	77	62	71
		7	CAG : GTGACT	CACACAATCTCCTTTTGTAG : ATA	89	60	71
		8	CAG : GTGAAG	GCACTAACTTCCCTGTGTAG : GTT	89	64	71
		9	AAG : GTGTGT	TCATTCCTATATCTAAACAG : CGT	102	60	71
		10	GGG : GTAAGT	GAAGCAATAAAAAAACACAG : GTT	84	67	71
		11	AAG : GTACGA	TCATAAAATTCTTCCAACAG : GGT	73	66	71
		12	AAG : GTAGGG	TGACCCTGCAATATTTTCAG : AGT	92	59	71
		13	GAG : GTATGC	TTTTTGTTTGTGCAATGCAG : GGG	103	60	71

Table 1. Continued

Plant species	Gene	Intron	5′ splice site	3′ splice site	Size (bp)	AU (%)	Ref.
Sorghum vulgare (continued)	*pepC* gene for PEP carboxylase	1	TTC : GTAAGC	CTCTCTCTTCGCGTGCGCAG : GTC	104	40	72
		2	CAG : GTGGGT	GATTTGGTTTTGTTGTTCAG : GAT	134	53	72
		3	GAG : GTAAAA	TTGTTTGACAATCTTCGCAG : ATT	1098	63	72
		4	ATG : GTAACT	ATGCTGTATTGTTTTTATAG : GAA	280	69	72
		5	GAG : GTACTG	AACACAATATTCTATTCTAG : CTC	92	60	72
		6	TAG : GTAACA	AATTGAGCGTTTGCTTGCAG : AAT	1086	66	72
		7	CAG : GTTTGC	CACCACCTTCCACGTGGCAG : TTT	82	61	72
		8	TCG : GTAGGT	TCGATCTTATATTCTTTCAG : GCT	461	64	72
		9	CAG : GTAAAG	CTGTTGCCATTTCGCTGCAG : GTT	191	67	72
Triticum aestivum	*Alpha-Amy2/34* gene 5′-region	1	CAG : GTACGC	GGCGCGAAATCATGTTGCAG : GGG	122	54	73
	Alpha-Amy2/46 gene 5′-region	1	CAG : GTACGC	GGCGCGAAACCATGCTGCAG : GGG	122	49	73
	Alpha-Amy2/53 gene 5′-region	1[b]	CAG : GCACGC	TGCCTTAATGGTTGTTGCAG : GGG	101	57	73
	Alpha-Amy2/54 gene 5′-region	1	CAG : GTACGC	GAGCTTAATGGATGTTGCAG : GGG	85	58	73
	Alpha-Amy2/8 gene 5′-region	1	CAG : GTACGC	ATGGCATAATCTTGTTGCAG : GGG	100	61	73
	Alpha-amylase gene	1	CAG : GTGAGA	TTTGGTGTTTCGTTGGCCAG : GGG	85	60	74
		2	ATC : GTAAGT	ACTTGGTTTTGCGCGCGCAG : TTC	104	55	74
	FBP gene for chloroplast fructose-1,6-bisphosphatase	1	GAG : GTAAGT	GTACTGTACCGCGCGCGCAG : GTG	87	55	75
		2	ACC : GTATGT	TGCTCGCCTCGAGCGTGCAG : CTT	89	45	75
		3	GCG : GTACGT	TGTCGATGCGTTTTGTGCAG : GTC	86	58	75
	Storage protein (*Gbl1*) gene	1	AAG : GTACGT	GCATGCCTGCATGCATGCAG : GCG	78	45	76
		2[b]	CAG : GCAAGT	CACGTGGGTATCCCATGCAG : TTT	87	53	76
		3	AAG : GTACGT	TTCTGAACGTGATCATGCAG : ACC	66[d]	58	76
		4	CCG : GTGCGT	GATGCATGTGTCAACGGCAG : GGT	96	51	76

Gibberellin responsive protein gene	1	CAG : GTCAGT	TGAATGGATTCCTTGCGCAG : GAT	181	43	77
	2	AAG : GTAGGC	ATGCTACTTGCTGCCTGCAG : ATT	246	60	77
	3	CAG : GTACTG	GTTCTTAACTCCCATTTCAG : GTC	791	62	77
	4	AAG : GTAAGA	CTCTCTTATTTGTATTATAG : GGA	94	68	77
	5	GTG : GTCAGT	ACGACATGCTTATGTTTCAG : GTA	344	69	77
	6	AAT : GTAAGT	TGCTAATGGCCTCTTTACAG : TAT	107	68	77
	7	TCG : GTAAGG	ATTGTTTATGTTTTGAGCAG : GAA	315	66	77
	8	AAG : GTTTGT	TTCTGTTCGCGCGAATGCAG : GTT	255	61	77
gstA1 gene for glutathione-S-transferase	1	AAC : GTAAGC	CGTGTTTTTCTTCCTTGCAG : CCG	90	53	78
	2	TCG : GTGAGC	CGATGAGTGAGTGAGTGCAG : AGT	121	60	78
gstA2 gene for glutathione-S-transferase	1	AAC : GTAAGC	GACAGAGATGCTACTTGCAG : CCG	77	51	79
	2	TCA : GTACAC	AAATCGATGAGTGAGTGCAG : AGT	101	61	79
Alpha-amylase gene (*alpha-Amy3*)	1	CAG : GTGAGA	TTTGGTGTTTCGTTGGCCAG : GGG	85	60	74
	2	ATC : GTAAGT	ACTTGGTTTTGCGCGCGCAG : TTC	104	55	74
Peroxidase	1	TTT : GTCCAA	TACGTTTATGTTGGCTACAG : GGC	114	64	80
	2	GCC : GTAAGA	CTGCTATCTTTCACATACAG : CTT	143	64	80
PRK gene for ribulose-5-phosphate	1	AAT : GTGAGT	AACAGAAATTTTTAAACCAG : GTA	257	64	81
	2	CAG : GTGAGC	AATCACTGGCACTTTTGCAG : AGA	87	66	81
	3	TTG : GTACTG	CTGACCACTTGGTATGACAG : ATC	91	65	81
	4	GAG : GTACTC	TGTTTGCTTGTTTATTTCAG : GTA	741	58	81
TthV gene for wheat type V thionin	1	CCG : GTAAAG	GATCTACTTCGTTTTGGCAG : ATG	320	71	82
	2	CAG : GTAAAT	CTATCCTATTTTGGCACCAG : CTG	98	71	82
Peroxidase	1	TTT : GTCCAA	TACGTTTATGTTGGCTACAG : GGC	114	64	83
	2	GCC : GTAAGA	CTGCTATCTTTCACATACAG : CTT	143	64	83
Gene for sedoheptulose-1,7-bisphophatase	1	AAG : GTGTGT	ATGGGCAATGCGTGCGGCAG : GCT	85	49	84
	2	CTG : GTTCGT	TTTTTTTTGCGTGGTGGCAG : GAG	122	44	84
	3	GAG : GTGAGA	TCGCTTGTATATATATGCAG : GCG	88	41	84
	4	AAG : GTACTT	TTTTCGGTCGCGTCTCGCAG : GCG	456	52	84
	5	AAG : GTTTTT	GTCTACTACAAATGCTGCAG : GTA	94	56	84
	6	AAG : GTGGTA	TGGAAAATGTATCCTGTCAG : CTT	103	57	84
	7	CAG : GTACAT	TGGTGCGTGCATGCATGCAG : ATC	626	52	84

Table 1. Continued

Plant species	Gene	Intron	5′ splice site	3′ splice site	Size (bp)	AU (%)	Ref.
Zea mays	Transposable element activator (*Ac*) major transcript	1	AAG : GTATAT	GAAAAATCTGTTTATAACAG : GGT	107	71	85
		2	CAG : GTATGT	TCAAATGTCAATTATTGTAG : GTA	71	75	85
		3	AAG : GTGAAT	GCATATTTGTGTCATTCAAG : GTG	89	76	85
		4	GAG : GTATTT	TTATTTTTGTGTATTAACAG : GAT	387	59	85
	Actin 1 gene (*MAc1*)	1	AAG : GTTGTT	CAACTCCTCAATATTTACAG : GCC	84	62	86
		2	CTG : GTAAGA	TCAAATATCTCTGTGTGCAG : GTA	107	59	86
		3	CAG : GTCTTC	ATGGGCAACTGTGTTGGCAG : ATG	885	63	86
	Alcohol dehydrogenase (*ADH-1 C-m* allele) gene	1	AAG : GTCCGC	GCTTTCCTGGACCCGTGCAG : CTG	534	57	87
		2	AAG : GTATCT	TGGCGGCTTATCTGTCTTAG : GGG	97	56	87
		3	AGG : GTATGT	AAAATTCTTGATTTTGCCAG : TAT	416	65	87
		4	CCG : GTAAGT	CAAATTCTTTATCTGTTTAG : GTC	86	66	87
		5	GCC : GTAAGT	ATTACCCTGCGCATGGTTAG : GCT	93	58	87
		6	AAG : GTACAG	CCATCTGTCCCATTTTTCAG : CTA	341	62	87
		7	GAG : GTCTGT	ATAATTCCTTTATGGTCTAG : GTA	87	62	87
		8	GAT : GTAAGT	CCGCCGCCCTCGTGATCCAG : GGC	91	53	87
		9	AAG : GTAAAT	TTCTGCTTGGCTCTGCACAG : GAG	100	55	87
	Alcohol dehydrogenase 1 gene (*Adh1-1F*)	1	AAG : GTCCGC	GCTTTCCTGGACCCGTGCAG : CTG	535	57	88
		2	AAG : GTATCT	TGGCGGCTTATCTGTCTTAG : GGG	97	56	88
		3	AGG : GTATGT	AAATTTCTTGATTTTGCCAG : TAT	415	65	88
		4	CCG : GTAAGT	CAAATTCTTTCTCTGTTTAG : GTC	86	64	88
		5	GCC : GTAAGT	TTACCCTGCGCATGGTTAAG : GCT	94	62	88
		6	AAG : GTACAG	CCATCTGTCCCATTTTTCAG : CTA	344	62	88
		7	GAG : GTCTGT	ATAATTCCTTTATGGTCTAG : GTA	87	62	88
		8	GAT : GTAAGT	CTGCCGCGCTCGTGATCCAG : GGC	91	53	88
		9	AAG : GTAAAT	AATTCTGCTTGGCTGCACAG : GAG	98	55	88

Alcohol dehydrogenase 1 gene (*Adh1-1S*)	1	AAG : GTCCGC	GCTTTCCTGGACCCGTGCAG : CTG	534	57	89
	2	AAG : GTATCT	TGGCGGCTTATCTGTCTCAG : GGG	97	55	89
	3	AGG : GTATGT	AAATTTCTTGATTTTGCCAG : TAT	419	65	89
	4	CTG : GTAAGT	CAAATTCTTTCTCTGTTTAG : GTC	86	65	89
	5	GCC : GTAAGT	TTACCCTGCGCATGGTTAAG : GCT	94	61	89
	6	AAG : GTACAG	CCATCTGTCCCATTTTTCAG : CTA	342	62	89
	7	GAG : GTCTGT	ATAATTCCTTTATGGTCTAG : GTA	87	62	89
	8	GAT : GTAAGT	CCGCCGCCCTCGTGATCCAG : GGC	91	53	89
	9	AAG : GTAAAT	GTTCCTGCAATTCTGCACAG : GAG	90	58	89
Gene for mitochondrial adenine nucleotide translocator (*ANT-G1*)	1	CAG : GTAGCC	TCTTCAAATCTTTGTCCCAG : GCT	174	64	90
	2	CAG : GTTTGT	TATGTTTCTCTCTCTTGCAG : GAC	80	60	90
ANT-G2 gene for adenine nucleotide translocator	1	CAG : GTAGCC	TAAAATTAATACTTGCCTAG : GCT	94	68	90
	2	CAG : GTTTGC	GATGTTTCTCTCTTTTGCAG : GAC	81	63	90
Aux311 gene for auxin-binding protein	1	GAG : GTGCAC	TGATATTATTATGTATACAG : ATA	1612	54	91
	2	GAG : GTAATG	CATGAAATGTTTTCCTGCAG : GTG	340	65	91
	3	CAG : GTAAAG	ACATCATCTTTCTCTTGCAG : GTT	172	70	91
	4	GAT : GTGAGC	CAATGACATCCTCGTTGCAG : ATT	1525	59	91
Bz-McC gene for UDP-glucose flavonoid glycosyl-transferase	1	AGG : GTGCGT	TGCATTTCCATCGTTCGCAG : CAG	100	50	92
Bz-W22 gene for UDP-glucose flavonoid glycosyl-transferase	1	AGG : GTGCGT	TGCATTTCCATCGTTCGCAG : CCG	105	50	92
c2 gene for chalcone synthase	1	TGT : GTGAGT	TGATTTTGACGTGCATGCAG : GCG	1524	50	93
Chloroplast glyceraldehyde-3-phosphate dehydrogenase gene	1	CAG : GTATGC	TTCTTTGATGTGTACTCCAG : GGG	134	46	94
	2	GCG : GTACGT	ATCACATCAACAACAACCAG : GTC	182	42	94
	3	TCG : GTGCGT	ATATATATGCATGGATCAAG : GCA	119	56	94
Glb1-0 gene for vicilin-like storage protein (truncated)	1	AAG : GTACGT	TGTTGTTTTGTCGTGTCTAG : GCG	99	41	95
	2	CAG : GTCATT	CCCCATCGTTCCTTTGGCAG : TTC	100	52	95
	3	AAG : GTACAA	CGATCGCGTTGCGCGTGCAG : ACC	91	42	95
	4[b]	GAG : CACGAC	TGTATGTACACACGTGGCAG : GGG	140	47	95

Table 1. Continued

Plant species	Gene	Intron	5′ splice site	3′ splice site	Size (bp)	AU (%)	Ref.
Zea mays (continued)	*Glb1-L* gene for vicilin-like embryo storage protein	1	AAG : GTACGT	TGTTGTTGCCTCGTGACTAG : GCG	99	39	95
		2	CAG : GTCATT	GTCCTTCGTTCCTTTGGCAG : TTC	114	54	95
		3	AAG : GTACAA	CGATCGCGTTGCGCGTGCAG : ACC	82	45	95
		4	GCG : GTGCGT	TGTATATACACACGTGGCAG : GGG	108	51	95
	Glb1-S gene for vicilin-like embryo storage protein	1	AAG : GTACGT	TGTTGTTGCCTCGTGACTAG : GCG	87	40	95
		2[b]	GGC : GCCGGC	GTCCTTCGTTCCTTTGGCAG : TTC	214	45	95
		3	AAG : GTACAA	CGATCGCGTTGCACGTGCAG : ACC	90	43	95
		4	GCG : GTGCGT	TGTATATACACACGTGGCAG : GGG	100	51	95
	ABA-inducible gene for glycine-rich protein (ABA = abscisic acid)	1	AAG : GTTTGC	GTTTTCATGTGATTATGCAG : GTC	147	50	96
	Glutathione S-transferase gene (*GST-I*), exons 2 and 3	1[c]		GGTTATCTATCTCTCTGCAG : CCG	—	—	97
		2	TCG : GTATGA	TCCATCTGTGTGCTATATAG : AAT	669	62	97
	O-methyltransferase (*OMT*)	1	CTG : GTGAGT	GATCACTGCCACTCTGCCAG : GTA	916	55	98
	Pdc mRNA for pyruvate decarboxylase	1	CAG : GTGCTG	CTTCTCTCTACTCTCGGCAG : GCC	90	48	99
		2	CAG : GTACAA	ACGTGTGGCTGGCTGCCCAG : GTG	106	42	99
		3	AAG : GTCTGT	CTCCACCGTGTGCCCTGCAG : GTT	94	39	99
	rab28 gene	1	GCG : GTACGA	GTTCTTTGTCGTACGTGCAG : AAC	129	51	100
	rbcS gene for ribulose-1,5-bisphosphate carboxylase/oxygenase small subunit	1	CAG : GTAACA	AAACACACGTACTACGTCAG : GTG	163	53	101
	Sucrose synthase	1	GGG : GTATGC	CCGTATAGCTCGAATTGCAG : TAT	1014	57	102
		2	CAG : GTGGGC	GTGAAATACCACTTCTACAG : GTA	511	66	102
		3	CAG : GTAACA	TTTCCCTTGTCTGCATATAG : GAA	87	57	102
		4	ACA : GTAAGT	ATCCAGTCCTTTTTTACCAG : ATC	79	70	102
		5	ACG : GTGAGC	ACTGCTGTTTTCTGTTACAG : ACG	150	65	102
		6	TAG : GTGAAT	GCCACTATGATCTGTGTTAG : GTT	98	64	102

	7	CAG : GTACAA	CACTTGCAGTCGCTTTGCAG : GTT	75	57	102
	8	ATT : GTATGT	CATGATCTTATTTGTTGCAG : GTT	95	61	102
	9	GAG : GTATAC	CCATACATTCTGTGCTGCAG : GAT	83	60	102
	10	CAG : GTCTGT	GGTTTTGTACATACTTGCAG : TGT	109	75	102
	11	AAG : GTAGAA	TTTCTGTGTTGTTTCTGCAG : CAA	80	69	102
	12	CAA : GTGAGT	TGGTGTTTACTTGCTTCCAG : GTT	134	57	102
	13	CAG : GTATAT	GTTTTTTTTGTGTGGGTAG : CCT	128	61	102
	14	GAA : GTATGC	ACAAGCTTTGGATTGCTCAG : GTA	162	58	102
	15	CTG : GTAAGC	CATCTCTTTCTGGAATCCAG : GCA	119	49	102
Autonomous transposable element *En-1* mosaic protein gene	1	AGC : GTAAGC	TATTTTGTGAATTATTGCAG : ATG	4434	59	103
	2	GAG : GTACTT	TAAAATTGTTTACATTTCAG : GGC	66 [d]	67	103
	3	CAG : GTATTC	TAACAAACTTTCATTTGTAG : GTA	90	77	103
	4	TGG : GTAAGT	GGTGATCTTCAAATTTGCAG : GGC	74	66	103
	5	ATG : GTATGT	TTTTTAATCATGTGTTGCAG : GCG	71	65	103
	6	CTG : GTAAGT	CTTGGATGATTTTTTCATAG : GCT	70	67	103
	7	GAG : GTATAT	TATATATGATACTCTTACAG : GTT	86	72	103
	8	CAG : GTATCA	GATGAATAATCTATTTGCAG : ATG	73	71	103
	9	CAG : GTACAA	GAGTGTGTCTTCTGTTTCAG : GGT	182	67	103
	10	TCA : GTACGT	AATGTTTGTTTCTTTTAAAG : GGT	142	70	103
Alpha-3-tubulin gene	1	CAG : GTGCGA	CACCTTTCTATTTCACGCAG : CCT	847	59	104
	2	CAA : GTAATT	TCATGCTCTTGTTCCTGTAG : TTG	88	57	104
	3	CAG : GTAGGA	TGCTTTAAACTGTTGTGCAG : GTG	99	58	104
Waxy (*wx* +) locus for UDP-glucose starch glycosyl transferase	1	CAG : GTTCTG	TTTCTCTCTCTCCTACGCAG : TGG	139	48	105
	2	GCC : GTAAGC	TTGTTCGGGCATGCATGCAG : GCG	131	44	105
	3	GAG : GTACGG	GCAACTGCAAATGCATGCAG : ATC	94	49	105
	4	AGG : GTGAGA	GGCCTGGTCGCTGGTTTCAG : GTT	96	49	105
	5	CAG : GTCAGG	AAATGCATGCTGTTCTGCAG : GCA	81	57	105
	6	ACG : GTAAGA	TCTTCGTCCATCCATACAAG : GGG	123	54	105
	7	AAG : GTTGCC	CCATCTTCACGTACTACCAG : ACC	137	53	105
	8	CGG : GTCTGT	ACCTGATTGCATTATTGCAG : CTA	97	55	105
	9	ACG : GTGAGC	GACGGTGGTGTCCGGTTCAG : GCC	110	40	105
	10	CTG : GTACGT	CGCGATGGATAATGCTGCAG : GGC	95	42	105
	11	ACG : GTACGA	TCCCGCTGCGACTCTTGCAG : CCC	111	57	105
	12	GAC : GTAAGC	TCGTCGTCCTCTCTTCCCAG : TGT	97	54	105
	13	AAG : GTACGT	TTAATTTGCGAAATGCGCAG : GGC	121	34	105

Table 1. Continued

Plant species	Gene	Intron	5′ splice site	3′ splice site	Size (bp)	AU (%)	Ref.
Zea mays (continued)	*whp* (white pollen) gene for chalcone synthase	1	TGT : GTGAGT	GCCATGGTCATGCCATGCAG : GCG	2157	56	106
	Gene for lactate dehydrogenase	1	CAG : GTTCGT	TCAATCAATTTGTTTTGCAG : GCG	295	61	107
	Triosephosphate isomerase gene	1	TGC : GTAATT	NNNNNTCCTGATGCGTGCAG : AAT	730	— ($N=193$)	108
		2	TTG : GTACGG	TGGTATTGATTGCATTGCAG : AGG	290	55	108
		3	CAG : GTTAGT	CAAAGTCATTATTAATGCAG : TGC	325	64	108
		4	GAA : GTATGA	CTCTTCTGCTTGGATGGCAG : TTT	208	60	108
		5	CTG : GTACCT	AATGGCTGTTTGTTTTACAG : AGA	457	66	108
		6	GAA : GTAAGT	AATTCGTATTATGTTCCCAG : GTG	149	66	108
		7	GAG : GTACAT	CACGTGCCTCCCTGCTACAG : GCT	93	55	108
		8	AAG : GTAATG	CAACCCTATCTCGTCTGCAG : CCT	124	60	108

[a] 5′ and 3′ splice site conserved, GT and AG absent.
[b] 5′ splice site conserved, GT absent.
[c] No 5′ splice site given.
[d] Short intron, below predicted minimum functional length (14).

Table 2. Compilation of dicot intron data

Plant species	Gene	Intron	5′ splice site	3′ splice site	Size (bp)	AU (%)	Ref.
Actinidia deliciosa	Actinidin gene	1	GTG : GTATGT	TATTAATCATTGGAATGAAG : GGG	412	77	109
		2	TTA : GTAAGT	GGTGATTTTTGTTTATGCAG : CAA	91	74	109
		3	TCG : GTAAGA	TTTTCGAATTCTCTCTTTAG : GGC	913	66	109
		4	ATG : GTAAAA	TACATTCTTTGATGCGACAG : AGC	218	73	109
Antirrhinum majus	Chalcone synthase gene	1	TGT : GTAAGA	CATTTAATTTGAATTATCAG : GCG	721	69	110
		2	CAG : GTACGT	TATTTATTATCCAACACTAG : GGT	211	72	110
	Deficiens gene (*def A*)	1	TAC : GTACAC	GTTTGTTTGTTTTTAAACAG : GAC	101	74	111
		2	GAG : GTTTCC	TGCGGTGAAAATATGTGCAG : AAA	179	73	111
		3	CAG : GTACAT	TTTAACTTTATAATGTGCAG : GCA	134	79	111
		4	AAG : GTTTTA	CAACGCTTCGGTCGTAACAG : TAC	915	72	111
		5	AAG : GTATGT	TCCTTTGTTTATTTATGCAG : GTT	102	71	111
		6	TTT : GTAAGC	CAATGATGTTTAATTCGTAG : GAT	261	70	111
	TAM1 gene	1	GAG : GTATTG	TTTGTTTGCTGCAAATTTAG : TTT	138	72	112
		2	CAG : GTATTG	GCTTTGACTTCATTCTGCAG : TTC	1827	67	112
		3	TCG : GTAAGT	TTACTTGCTTTGATTTGCAG : CTT	2472	67	112
		4	AAG : GTACTT	CTATAAAATTGCACATGCAG : AAA	4505	64	112
		5	GAG : GTACAA	TAACTAGTGTATCTTCGTAG : ATT	489	69	112
		6	ATG : GTGAGC	TACTTTTACATCATTAACAG : TAA	423	69	112
		7	AAG : GTTATC	TAACTTTTATATTTTTTCAG : GCA	71	82	112
		8	ATG : GTATGA	ACTTCTTCAAACTACTGCAG : AAA	73	74	112
		9	GTG : GTAGTG	ATGATTTTATATTTGCGCAG : AAG	81	70	112
		10	ATT : GTAAGT	TATAATTTCATTATGTATAG : GTA	290	70	112
		11	CGG : GTAAAT	TAATATATTTTGACATGTAG : GTT	87	74	112
		12	TTG : GTAATT	ATTCCTAATATGATACGCAG : GTT	1893	66	112

Table 2. Continued

Plant species	Gene	Intron	5′ splice site	3′ splice site	Size (bp)	AU (%)	Ref.
Antirrhinum majus (continued)	Squamosa gene	1	TTG : GTTAGT	CCATAATGGTTCATAAGCAG : CAT	3910	70	113
		2	CCT : GTAAGT	AGTTTTATGATTTTGTGTAG : GCG	140	68	113
		3	TAG : GTAAAT	TTCTTAATTTCTTCTGTCAG : GCA	222	62	113
		4	AAA : GTATTA	GTTCGTATGATCCAATGCAG : AAC	121	74	113
		5	AAG : GTTAGT	TTAATCTGTTTTTTTGACAG : GAG	132	67	113
		6	AAG : GTAACT	ATTATTGTTCTGTGATATAG : ATC	151	68	113
		7	CGG : GTAATT	GAAGAGGTCTATTGATGCAG : AAA	184	66	113
	Tap 1 gene	1	ATG : GTATAC	TCTAACATTTGTTGTTGCAG : ATG	365	69	114
Arabidopsis thaliana	Actin-1 gene (*AAc1*)	1	AAG : GTGTGT	GTTCTTTTGTATTGGAATAG : GCT	137	74	115
		2	CCG : GTGAGT	CTTGAATTGTGGATGTGCAG : GTA	100	59	115
		3	CAG : GTACAC	CTATGTGTTACCTTGTGCAG : ATG	111	62	115
	Proline rich protein gene (*APG*)	1	TAG : GTACGT	TGGTATCCTAATTCCAATAG : CCA	165	68	116
		2	GCG : GTAAGT	ATTTGATTTATATAACTCAG : AAT	89	75	116
		3	TTG : GTACGT	TAAATGTATAAATTGGGCAG : CAA	120	75	116
		4	ATG : GTAAAT	TTTAAAATATTTGCATGCAG : GAT	88	78	116
	Tonoplast intrinsic protein gene (*α-TIP*)	1	TCG : GTTCGT	TTTTTTTTTCTTCTCCTTAG : ATA	91	63	117
		2	ATG : GTAACA	AAGAGTTTGTGTTCATGCAG : AGA	119	68	117
	Plasma membrane H^+-ATPase gene	1	TTG : GTATGG	GTGTTCATATTATATTGCAG : GAG	1011	72	118
		2	AAG : GTATAG	ATGGTGGTTTCAATAAATAG : GAA	82	70	118
		3	AAG : GTCCCT	TTATGACCAGGTTCTAAGAG : ATG	221	69	118
		4	GAG : GTAATA	TCATTGTTTGTTTCTTTCAG : GCA	110	73	118
		5	CAG : GTAACT	TGATTCTTTGATGATTTCAG : ATT	80	71	118
		6	CAG : GTGAGA	TTTGTGTGTCTATGTTTCAG : ACT	92	70	118
		7	CTG : GTTTAT	TGTTTCCAATGGTTTTATAG : GTG	75	69	118
		8	CAG : GTTTGT	TCTTTACCAAATTCATTCAG : AGC	95	67	118
		9	ACA : GTGAGT	TTAAATGTTCCTTTAAACAG : ATA	95	69	118
		10	GTG : GTAAGT	CACTTTGTGCCTTTTCACAG : TTT	86	72	118

	11	ATG : GTGAGT	ATCATTAACTTACGCAACAG : GAA	80	66	118
	12	CCG : GTATGT	ACTTTGTTTGTTTCCTATAG : AGA	72	68	118
	13	TTG : GTAAGC	CATAAACTTTGACCTTGTAG : ATT	71	70	118
	14	AGG : GTACGT	CTGTGTCTTTGAAAAAACAG : ACT	85	69	118
	15	AAG : GTAAAT	CGTTGGGTTTTTGTTTACAG : GCT	68[d]	69	118
Plama membrane H^+-ATPase gene	1	CTG : GTATAC	TTTGCTTTTGATTTTTTCAG : GAA	1014	66	119
	2	AAG : GTCTGA	TGTGCAATGAAAATTGACAG : GAA	99	67	119
	3	AAG : GTATAT	GTGATGTTTCTCTGTTTCAG : GTT	108	69	119
	4	GAG : GTATAA	TTTGATTGGGATTTTCACAG : GCT	87	64	119
	5	CAG : GTGATG	TGATTTGATTCTCCCAACAG : ATC	79	71	119
	6	CAG : GTACTT	CTTCTTATTCCCTTTTGTAG : GTG	103	65	119
	7	CTG : GTAAAT	TTGAACACATATGTGGCCAG : GTG	89	58[e]	119
	8	CAG : GTTATA	CTTTGTTTTTTTATTGCAG : CAT	98	70	119
	9	ACT : GTGCGT	ATTTTTCTCTCTTGTTACAG : ATC	99	66	119
	10	GTG : GTATGC	CGTTGCAAACGTTTTTGCAG : TTT	94	63	119
	11	ACG : GTGAGC	TTGCTATCTTGATCTTATAG : GTA	103	70	119
	12	TCG : GTACAT	ATTTCACATTGGATTCACAG : GAC	96	71	119
	13	CTG : GTAAGT	ATTCATTCTTCCAAATGCAG : ATT	96	66	119
	14	TAG : GTAATA	AATTGTTATATGTTAAACAG : GCT	91	75	119
Gene for P-glycoprotein (*atpqp1*), homologous to mammalian *mdr* genes	1	AAG : GTACTA	CTTGTTTTTGGTTATTGCAG : TAT	69	71	120
	2	CAG : GTTAGA	ATATGGTTTCGTTGTTGCAG : AGA	85	69	120
	3	AAA : GTAAGA	TTTTTTTTTGGGAATCTCAG : CTT	73	70	120
	4[a]	CAG : GTAGAG		486	62	120
	4[b]	ATT : GTAAGC	TTTTTTATTTTTGTTTAAAG : GGC	99	69	120
	5	CAG : GTTAAC	CGGTGTTATCGTCTCTGCAG : GAC	87	67	120
	6	CAG : GTCTGA	GTAAATGTGAATAATGTCAG : GTT	328	67	120
	7	AAG : GTATAC	TGTTAATCTATTTGTTGCAG : ACC	83	72	120
	8	CAG : GTTTGT	TTGTCTTGTCTTTTGAACAG : AAA	77	64	120
Auxin-inducible gene (*Aux2-11*)	1	AAA : GTAAGT	TGCTATTTCTTGTTTTATAG : GGC	87	70	121
	2	GGA : GTAAGT	GTGTTTTGTTGTTATTGCAG : GAT	89	74	121
Auxin-inducible gene (*Aux2-27*)	1	TTG : GTTAGT	CTTTATATATGAAAATGAAG : ATG	117	78	121
	2	GGA : GTAAGT	ATGAAACTATTTGTGTGCAG : AAT	90	76	121
	3	CAT : GTAACA	AACCTCTAGGTACCGATTAG : ATG	379	68	121

Table 2. Continued

Plant species	Gene	Intron	5′ splice site	3′ splice site	Size (bp)	AU (%)	Ref.
Arabidopsis thaliana (continued)	Beta-tubulin-4 gene	1	AAG : GTTAGG	TTTTGTTAGTTTGGATTTAG : GTT	131	69	122
		2	CCT : GTGAGT	GTATTTTGTGTTTTCATTAG : TTG	88	65	122
	Calmodulin-3 gene (*cam-3*)	1	ATG : GTACGT	AATTCTTGATTTGCTTGTAG : GTT	492	67	123
	Chalcone synthase gene (*CHS*)	1	TGT : GTACGT	CAAGTATTTACTATTCACAG : GCG	86	70	124
	Receptor kinase gene (*Ark1*)	1	TCG : GTTAGT	ATGCCTCGTTTTTATTGCAG : AGG	411	73	125
		2	ATG : GTAAAA	TTAACGCCTGAATGAAACAG : TTG	103	78	125
		3	AAG : GTAAAA	AATCACATGTGTGGTAATAG : GGA	76	82	125
		4	TTG : GTTAGT	TGGTTGTGTTGGATGTGTAG : ACC	85	66	125
		5	TTA : GTAAGC	GTTTATATTTTGTTATGCAG : CGG	103	74	125
		6	TTT : GTAAGC	ATTACATCCTTTATGCACAG : GTG	88	72	125
	Enolase gene	1	GAG : GTTAGT	CTCTGATGATGTTGGTTTAG : GTT	355	63	126
		2	AAG : GTGAGA	TTATTTGTTTATATCACTAG : GCT	97	67	126
		3	AAG : GTGAGT	GCTTTTATTATTTCTTGCAG : GAC	104	65	126
		4	AAG : GTACTT	CGCCTTCTATCTGTGTTTAG : CTT	88	65	126
		5	AAG : GTAGAA	TGTGTGGTATTTGGGTGCAG : CAC	131	59	126
		6	AAG : GTAAAT	GTTCTCATATGTTTTTCCAG : TCT	146	68	126
		7	AGG : GTCTTG	TCCCATTCAATTTTTTCCAG : GTT	87	64	126
		8	GAG : GTGAAA	CGCTCTTTGTTTCTCTGCAG : AAC	95	60	126
		9	AAG : GTTTGT	ATACTCTTTTCTTATTCCAG : AGA	97	66	126
		10	ACT : GTAAGT	GGATCTTTCTGTCTATACAG : GGA	88	64	126
		11	CAG : GTTTAA	TGTGTCTTGTTTCTCAACAG : CTT	83	65	126
	Glyceraldehyde-3-phosphate dehydrogenase gene (*gapA*)	1	GAA : GTAAGT	GGATATGGTTGGTATCTTAG : GGA	85	67	127
		2	TCG : GTATGC	ATGTGTTTGTAACATTGCAG : GTA	82	56 [e]	127

Glyceraldehyde-3-phosphate dehydrogenase gene (*gapB*)	1	AAG : GTAACT	TGAGTTCTTCGATGCTGTAG : CTG	79	75	127
	2	AAT : GTAAGT	ACATTTGATGAAATTTGCAG : GCA	70	69	127
	3	GAG : GTGAAA	TATATGGTTATGATCATCAG : GGA	91	64	127
	4	TAG : GTCCGT	GTTTTGGTATTTCTTTGCAG : CAA	84	64	127
	5	TTG : GTACAT	ACATGAACTGAACTCCACAG : GAA	78	69	127
	6	CAA : GTAAGG	TTTGAAATGAATTGTTTCAG : AGG	104	67	127
	7	CAG : GTGAGA	TTTGCGTGTTCGTGTTTTAG : CCA	98	69	127
Glyceraldehyde-3-phosphate dehydrogenase gene (*gapC*)	1	TGG : GTTAGT	TCGATTCTACAATGGGTTAG : TCT	105	63	127
	2	ACG : GTGAGT	TTGATTTCTGTTATCGTTAG : GAT	89	65	127
	3	ATG : GTTAGT	TCTGTGTTTGATTGCTGCAG : ACC	363	63	127
	4	CAG : GTACAT	AGTTTTTGGTATTTGTGAAG : GAA	94	68	127
	5	AAG : GTTTGT	GTGACTTGTTGTATTTTCAG : GGT	110	75	127
	6	AAG : GTAAAA	TCTTTGCTATCATTGTGTAG : GTA	95	74	127
	7	CTG : GTAAAA	GGTTTGCTTTATGATTTCAG : CTA	94	70	127
	8	CAA : GTAAGC	TTTTTACCACCAATACATAG : GGA	89	67	127
Glycine-rich cell wall protein gene	1	GAG : GTATGC	GTAAATGTGGTTATGTCTAG : GTG	686	80	128
Tonoplast intrinsic protein gene (γ-*TIP*)	1	TTG : GTATGT	TAACCTTTTGAAAATTTTAG : GCT	91	69	129
Histone *H1-1flk* gene	1[c]		AGCATTCTGGGCAGGAGCAG : ACT	—	—	130
	2	TTG : GTAAGC	TGTTTTGTGAGAATATATAG : GTT	89	64	130
	3	TAG : GTGAGT	GTTTTTGGCTGAGATTGTAG : GAG	89	67	130
	4	AAG : GTTAGT	GAAGTGTGATTGTGAAACAG : CTG	256	64	130
Histone *H1-1* gene	1	GAG : GTAAAA	CTTTGTTGATGCGATTTCAG : ATG	104	69	131
Histone *H1-2flk* gene	1[c]		GTGAAATGGAAATAAAACAG : GTT	—	—	132
	2	TAG : GTGAGA	TTGGTTTACTATATATGCAG : GAG	79	70	132
	3	ATG : GTGAAC	AACAAATGTGGTTTATGCAG : CTG	74	65	132
81 kDa heat-shock protein gene (*HSP81-1*)	1	GAT : GTAAGT	TTATATGAATGGTTTTGTAG : GCT	315	62	133
	2	CAG : GTAACG	GACTCTCTATTATTTTTCAG : ATT	83	64	133
	3	CAG : GTAAGG	AAGTGTGTTTATTCATGCAG : CTT	88	64	133
Isocitrate lyase gene, 5′ end	1	CAG : GTAAAA	CATGAATATATGTATCACAG : AAA	625	72	134
	2	AAG : GTAAAA	GTTACAATTGGACTAATCAG : GTA	276	78	134

Table 2. Continued

Plant species	Gene	Intron	5′ splice site	3′ splice site	Size (bp)	AU (%)	Ref.
Arabidopsis thaliana (continued)	Meri-5 gene	1	TAC : GTGAGT	ACATTTTTGTGTATTTGCAG : CTA	92	73	135
		2	CAT : GTAAAG	TCTGTTCTGTTTTCTATCAG : ATT	100	64	135
	Myb gene	1	CTG : GTACTC	TAAATTTGATTCGTTGATAG : GGC	105	72	136
		2	TAG : GTACAT	TATTTTTTTCCCAAATTCAG : ATG	765	75	136
	Nitrate reductase gene fragment (*NIA1*)	1	AAG : GTAAAC	TTACTTGGTTTTGTCTACAG : CTT	79	68	137
	Oleosin gene	1	CAA : GTAAGC	TACCTATTGATTGTGAATAG : GTA	240	68	138
	posF21 DNA-binding protein gene	1	GAG : GTAATT	GTTTTTGTTTCATACAATAG : GAT	202	64	139
		2	CAG : GTTAGT	TTTTCTTCCTGTTTGTATAG : AGA	134	66	139
		3	ATG : GTGAGT	TGAATCTTCTCTATTAACAG : AAC	92	63	139
		4	CAC : GTAAGT	ACGGGTGCCTTGGCCTACAG : GGT	127	58[e]	139
	Peroxidase gene (basic) (*prxEa*)	1	CGT : GTAAGG	AAAATATATGGTTTTTGTAG : GGT	141	77	140
		2	TTG : GTATAT	AAAGTCTATGTTTTATATAG : TCG	81	80	140
		3	CCG : GTAAAA	GTTTTGTCAAAACGTTTTAG : GTG	600	76	140
	Peroxidase gene (neutral) (*prxCa*)	1	AAT : GTAAGA	ATATTTTGGACTTTTGAAAG : GGT	560	74	140
		2	TTG : GTATGC	GGTTAATTATTATACTATAG : GCA	464	74	140
		3	CCG : GTAATT	TATGGTCTTCCTTTTGTTAG : GGG	90	74	140
	Pyrroline carboxylate reductase gene (*proc1*)	1	GAA : GTAAGT	GATCTTCAAAAAAATGACAG : GTT	81	68	141
		2	TTG : GTATAT	GTACCTTTCATATGATGTAG : TTA	93	74	141
		3	CAG : GTTTTG	TCTCTCCTTTTTGTTTTTAG : GAA	125	67	141
		4	CAG : GTATAG	ATTATGTTGTGTGTATATAG : TTA	115	68	141
		5	CAG : GTACAG	AAATTTCAAACCTTTTACAG : TGG	78	64	141
		6	ACC : GTCAGT	TTTTTTGTTTCCTATTACAG : GTT	106	65	141
	Photosystem II 10 kDa polypeptide gene	1	TCG : GTTCGT	GTTAATCGATGTGTGTGTAG : GAA	91	68	142
		2	AAG : GTTTGT	AACAATGACGATAATCGCAG : GGA	85	68	142
		3	CAG : GTGAAT	TTTTAATGTTTGGGTTGCAG : TCC	97	69	142

RNA polymerase II largest subunit gene (*RPB1*)	1	ATC : GTAATG	GCTAATGGCTTTTTTTCAG : AGG	189	60	143
	2	GAG : GTATGT	AGGAGCATAAGTTTAAGCAG : GCT	112	64	143
	3	AGG : GTTCGT	GGCATGTGATATTCTTGCAG : GTT	77	62	143
	4	GAG : GTAAGT	TATGCTTATTGTTGATGCAG : GAT	85	61	143
	5	AGG : GTAATG	TTGGCTTTTATTTTCTTCAG : GCT	82	66	143
	6	AAG : GTTAGT	CTAACAATTTAGTTTGGCAG : ATT	74	68	143
	7	AAG : GTAGCA	GATATCTTATCCATTTTCAG : GAT	94	67	143
	8	TTG : GTAAGT	CATTTTTACTTGGCTTCCAG : GGA	176	66	143
	9	CAG : GTATAT	CTCATAAACTTTCCTTGTAG : GTT	85	68	143
	10	CAG : GTAAGT	ATTAAAAATTTACTCTGCAG : CCC	113	71	143
	11	CAG : GTTTGA	TGTTATTAAATCTGTCTCAG : CCC	106	64	143
	12	TAG : GTAACG	GGATATCGTGTAACCCGCAG : GTC	88	55[e]	143
Plastid ribosomal protein CL15 gene (*rpL15*)	1	AAG : GTAATC	GGAATTTGAAATGTGTACAG : GGA	71	70	144
	1	GAG : GTAAAG	TTACTGTGTTCTTTAAACAG : GTA	292	66	144
	3	AAG : GTTAGT	TGATTGGTCTTGGATTTCAG : AAT	94	67	144
Chloroplast ribosomal protein CL9 gene (*rpl9*)	1	AAG : GTAAAA	GCTTGTTTTGTGTCTTTAAG : GTA	334	68	145
	2	CAA : GTAATT	TGTGTGTGTTTTTCCCCCAG : GGA	156	63	145
	3	AGG : GTAATC	GCGTGGAACCTTTTTTGCAG : GTG	141	70	145
	4	ATC : GTAAGT	AACTTTATATTTTCCTGCAG : TGT	83	67	145
	5	AAA : GTAACA	TCTGTGGCTGTGTTTTGCAG : GGA	151	64	145
	6	AAG : GTATTT	TTGGTGTTTTGAACTTGCAG : GAA	248	62	145
Sucrose synthase gene, partial sequence	1[c]		CACTTCTCATCAAGCTACAG : TGA	—	—	146
	2	AAG : GTTTGT	TTGGTTGTTATCTTTATTAG : ATA	227	57[e]	146
	3	CAG : GTCTGC	ACTCAAGTTTTCGCATTCAG : GAA	98	64	146
	4	CCA : GTAAGT	AGTCTCAACCTTTTCATTAG : TGC	85	65	146
	5	CGT : GTTAGT	TCTCTCTGTATAATGTGCAG : CCT	95	66	146
	6	TGA : GTAAAG	GGCAATGTTCTTTGGTGAAG : ACT	104	68	146
	7	ATG : GTTACT	TCTACTGTTCTGACTTGCAG : GTT	90	63	146
	8	ATT : GTATAT	TTCTATGCTGGCTTATATAG : GTA	103	69	146
	9	GAG : GTTAGC	ACTTCTATCAATGTGACTAG : GAT	196	64	146
					($N = 1$)	
	10	CAG : GTTGAC	CTTTACCATTTCGGATGCAG : TGT	85	65	146
	11	AAG : GTAAAA	CAGAAACTATTTTGTCACAG : CAA	82	62	146
	12	TGT : GTAAGT	ATCTTCTTGTGTTGATGCAG : TGG	108	68	146
	13	CAG : GTAACA	TCTCTCTTTTTCCACCACAG : CCT	83	65	146
	14	AAG : GTTGGC	AAGGATTTTTGCATGTATAG : GTA	88	61	146
	15	TTG : GTTAGT	TTCTTTTCTTCGGTTTGCAG : GCC	75	67	146

Table 2. Continued

Plant species	Gene	Intron	5′ splice site	3′ splice site	Size (bp)	AU (%)	Ref.
Arabidopsis thaliana (continued)	Topoisomerase II gene (*top2*), partial sequence	1	TCG : GTAAGT	CATAGTGTACACAATTGTAG : GGT	72	67	147
		2	TTG : GTAATT	AACATGTTAATGTTGGTTAG : GTA	74	69	147
	Tryptophan synthase beta-subunit gene (*trpB*)	1	CAG : GTAGAG	TGTGAATGAATCTTTGTTAG : AGA	250	64	148
		2	GAG : GTGATG	TTGGGTGTCTGGTGCTATAG : GTG	83	65	148
		3	AGG : GTAAGT	GGTTTGATTTTCTCTTACAG : ATT	92	61	148
		4	AAG : GTAATG	CTTTGGTTTGTGGTCTGTAG : CGT	96	65	148
	Tryptophan synthase beta-subunit gene (*TSB2*)	1	CAG : GTATGC	AATGTTTGATGAATCTGTAG : AGA	375	66	149
		2	GAG : GTGATG	TCTAAGTCTGATGCTTGTAG : GTA	75	63	149
		3	CGG : GTAAGA	GTAAAAGTTGATGTTATCAG : ATT	87	64	149
		4	AAG : GTACTT	ATTCTAATCTATGCGTGTAG : CTT	91	66	149
	Alpha-1-tubulin gene	1	CAG : GTCTTT	TAATTGGATTTGTTTTGCAG : TGA	160	70	150
		2	CTG : GTTATT	ATATTTTTCCATTGATGCAG : TTG	512	67	150
		3	CAG : GTAACC	AACTCTTTTCTTGGTCTCAG : GTT	72	65	150
		4	CTG : GTAAAG	CGAGCGGTTTGCATGATCAG : CAA	85	67	150
	Alpha-tubulin gene	1	GAG : GTATAA	AGTCTTTTGCGTTGTTGCAG : TGA	82	61	151
		2	CTG : GTGAGA	AATGCTTCTTGTAATTGCAG : TTG	77	62	151
		3	CAG : GTACTC	ATATTCCTCTGTGATGTCAG : GTT	111	65	151
		4	CTG : GTATGA	GCATGGGTTTTGTTTTGCAG : GTG	91	66	151
	Beta-1-tubulin gene	1	AAG : GTATAT	TTAACATTTCTGTATTACAG : GGT	793	67	152
		2	GCT : GTAAGT	TATATATATACATTTTGCAG : TTG	88	73	152
	Ubiquitin extension protein gene (*UBQ1*)	1	AAG : GTATCA	TTACTTGTTTCTCCTTTAAG : GAA	87	68	153
		2	AAG : GTACGA	TTGTTATCATTTGTAAACAG : AAT	85	67	153
		3	CAA : GTAAGT	AATTTCGCATCTTTTGGTAG : GTG	379	68	153
		4	CAG : GTTTGT	CAAACTTTTTTTATTTGCAG : TTG	76	72	153
	Ubiquitin extension protein gene (*UBQ2*)	1	AAG : GTATAT	TCGTTTTTGCTCTGTTTCAG : GAA	78	73	153
		2	AAG : GTCTCT	TTTTGTTTACTATAAAACAG : AGT	103	72	153
		3	CAA : GTAAGA	TGTTAACCTTGTTGGTTTAG : GTG	235	67	153
		4	CAG : GTTTGG	TTTTCTGTTTATTTTTGCAG : TTG	91	74	153

Dehydrin; salt-stress induced gene (*XERO*)	1	TCG : GTATGC	GACGATTTTCATGCATGAAG : GAG	340	76	154
SNF1-related protein kinase (*AKin10*) gene	1	TTG : GTAAGT	CTCTCTCTCTTTTTTTGTAG : AGA	417	67	155
	2	AAG : GTACAA	TTATTTGGTCTTTGATGCAG : TGA	122	65	155
	3	CAG : GTAGTT	GCTACTTGGCTTTGATGTAG : ATA	96	73	155
	4	GAG : GTAAGT	AGATCTATAAATTGTTTCAG : GTA	91	65	155
	5	AAG : GTACAT	TTTCTCTTTTCGTCCTATAG : GGA	115	69	155
	6	AAG : GTAAGC	GGCGGGGTCTTCATTTGTAG : ATT	94	61	155
	7	GAT : GTATGT	TCTTACCGACTGTCTTCCAG : GGC	117	68	155
	8	ATG : GTTAGT	TCATTGTTTGATTTAATCAG : GAA	73	68	155
	9	CAG : GTTTTG	GTAACAATTTGATCGCACAG : TCT	85	74	155
	10	CAG : GTAAAT	TTATGTTGATGGTGAAACAG : TTG	212	65	155
Cytochrome *cAt-Cc-1*	1	AAG : GTACGA	CTTTTCTGACCACTTTGTAG : ACC	78	53[e]	156
	2	CTC : GTACGT	CCTCCGCTAACACGACACAG : TTC	59[d]	42[e]	156
ATS1 gene for glycerol-3-phosphate acyltransferase	1	AAG : GTTAGT	ATCTTCATTTCCATCGTCAG : ATC	408	65	157
	2	GCA : GTAAAT	GTCTTACTGATGGTTTTTAG : GTT	122	67	157
	3	GAG : GTACTT	TTGCGCTGTGAATATTTCAG : GAT	97	65	157
	4	CAA : GTAAGC	CCCTTGAATATGTTATGCAG : AAA	152	66	157
	5	CAG : GTCATC	ACTCAACTGTTTAATTACAG : GGA	103	66	157
	6	ATT : GTGAGC	ATTTGTTTCTTTGGTTGCAG : AAA	72	58[e]	157
	7	AAG : GTATCA	GGGATTTCTACTGCTTTCAG : GAA	96	56[e]	157
	8	AAG : GTTAAT	AATGTTGATGAACCCTGTAG : GTC	102	67	157
	9	CCT : GTAAGT	ACTAATCTAGTAATGTTCAG : GCA	75	67	157
	10	CAG : GTATTG	CTTCATTGTGCATCTGCCAG : GTT	95	57[e]	157
	11	GAG : GTCTGT	ACATTCTTGTGGTATGGCAG : GCG	90	62	157
	12	AGG : GTAACA	CATTGTTCTTTCATTTGCAG : GAT	94	67	157
ats1A gene for ribulose-1,5-biphosphate carboxylase small subunit (*rbcS*)	1	CAG : GTCATT	TGGAAAAATCGAAAAAATAG : GTG	106	76	158
	2	GAG : GTAATT	GTAACATGTGGTTTGTATAG : CAC	136	76	158
ats1B gene for ribulose-1,5-biphosphate carboxylase small subunit (*rbcS*)	1	AAG : GTAAAT	TTGGGAATTGGGAATTATAG : GTG	90	70	159
	2	GAG : GTAATA	GTTTTCTTATGTGTTTATAG : CAC	171	72	159
ats2B gene for ribulose-1,5-biphosphate carboxylase small subunit (*rbcS*)	1	AAG : GTAAAT	TTGGGAATTCGAATTTATAG : GTG	90	71	159
	2	GAG : GTAATA	ATGGTTTTCTTTATATATAG : CAC	143	71	159

Table 2. Continued

Plant species	Gene	Intron	5′ splice site	3′ splice site	Size (bp)	AU (%)	Ref.
Arabidopsis thaliana (continued)	*ats3B* gene for ribulose-1,5-biphosphate carboxylase small subunit (*rbcS*)	1	AAG : GTAATA	ATTGGATTTGGGGATTATAG : GTG	90	73	159
		2	GAG : GTAATA	GTTTTCTTATGTGTTTATAG : CAC	170	72	159
	Chlorophyll *a*/*b*-binding protein	1	TGG : GTACGC	TTGGGTACAACATTGTGCAG : TGA	313	67	160
		2	GTT : GTAAGT	CTTGGTGGGTGGTGTTGTAG : CCT	87	66	160
		3	ACG : GTAAGC	TTGCTTGGTGGGTGTAACAG : GGC	91	65	160
	Carboxypeptidase Y-like protein gene	1	AAG : GTTTAT	GTTTGCTTGGTGTGTGTTAG : CAT	195	66	161
		2	CAG : GTATTA	ATAGAGAGTTTGTCTTGCAG : GTT	87	66	161
		3	CAG : GTTAGG	GTTTGCTGATGTTACCTCAG : GCT	190	63	161
		4[b]	AAG : GCATGT	TCACAAGTATAACCTTGCAG : GGA	74	62	161
		5	GTG : GTACTT	CCGTGTTTTTGTTTTACCAG : GAA	86	69	161
		6	AAC : GTAAGC	GTTCACATATCTTTGTGTAG : TAT	168	65	161
		7	TCG : GTAAGT	GGGAAACAAAATTGGAGCAG : GTA	90	69	161
		8	AAG : GTCGGC	ATTTAAACATAAAAACACAG : GTG	106	66	161
	Dihydroflavonol 4-reductase gene	1	CCG : GTACGT	GGTTTTTTTGGATTGTCAAG : GTA	91	73	162
		2	GAG : GTGAGT	CTTTTGTATTTGCGAATCAG : AAC	79	68	162
		3	TGG : GTAAGA	CATATATATTTGTGTGGCAG : ATG	86	77	162
		4	CTC : GTACGT	AATGTATTATAATGTTTTAG : GGA	79	70	162
		5	AAC : GTAAGA	AACAAAACTCTATATAACAG : GTT	90	66	162
	Embryogenesis abundant protein	1	AAG : GTCATT	TATTTATTATATGAAGGGAG : GAG	101	67	163
	Gene encoding glycine-rich protein *Atgrp6*	1	CAA : GTAAAT	AAGTATGTGATTGTATGTAG : GAA	260	70	164
	Gene encoding glycine-rich protein *Atgrp7*	1	TAA : GTAAAA	CACTTGTGTGTGTGATGTAG : ACA	186	75	164
	Gene encoding glycine-rich protein *Atgrp8*	1	TAA : GTAAGA	TACTTTTTTGTGTGATGCAG : ACG	336	73	164

	Gene encoding glycine-rich protein *putg1*	1	CAG : GTGGTT	ACTTGTGAGTATATATGTAG : ACG	290	73	164
	Putative serine/threonine protein kinase	1	AAA : GTGAGT	TTGTTTTCAATGTGTGGCAG : ATT	157	64	165
	Ribulose bisphosphate carboxylase/oxygenase activase (*rca*) gene	1	CCG : GTACAC	TTTTTGTGTGTGTGTTGCAG : TTG	484	65	166
		2	GCA : GTAAGT	CTTTACTTCTTGAAATATAG : GTA	91	70	166
		3	AAG : GTATAA	CTGTTTTTTTTGTCTCATAG : GTT	99	75	166
		4	CAA : GTGAGT	AAGTTTGAATCTTTTTGCAG : CCC	91	60	166
		5	TCG : GTAAGA	TGTGTTTTTGGTGTGGACAG : ATT	81	62	166
		6[a]	ACG : GTAAAA		124	66	166
		6[b]	CTC : GTTTGT	TAAACTTACTTGACTTGCAG : GAA	113	66	166
	RNA polymerase II fifth largest subunit	1	CAG : GTTTTC	TCGTTTTGTGGTATGGACAG : TGG	575	66	167
		2	AAG : GTAATG	TTTTCCTAATATTATCACAG : ATG	85	74	167
		3	CAT : GTAAGC	CCTCTCTTTTATCACCACAG : TTG	99	64	167
		4	TCA : GTGTGT	TTCCAATGCTTCTCTTGCAG : GTT	93	66	167
		5	ATT : GTATGC	TTTGTTTTTTGATTTTGCAG : TTT	95	69	167
Arachis hypogaea	Resveratrol synthase gene	1	TCT : GTATGT	ATTTCATATTATTATTTCAG : GTG	369	81	168
Armoracia rusticana	Peroxidase isoenzyme C gene (*HRPC1*)	1	AAT : GTAAAT	GATATATGGATTTTTGAAAG : GGT	350	81	169
		2	TTG : GTATGC	CAAAATGGTTATTATTATAG : GCG	519	72	169
		3	CCG : GTAATT	ATCTCTCTTTCTTTTGTTAG : GGG	84	77	169
	Peroxidase isoenzyme C gene (*HPRC2*)	1	AAT : GTAAGA	ATATTTTGGATGTTTGAAAG : GGT	835	73	169
		2	TTG : GTATGC	TGGTTACTATTATATTATAG : GCA	428	73	169
		3	GTA : GTTAAC	TATGGTCTTTCTTTTGTTAG : GTG	91	78	169
	Peroxidase isoenzyme C gene (*prxC2*)	1	AAT : GTAAGA	TATATTTAATTTTGAAACAG : GGA	1230	73	170
		2	TTG : GTATTC	TGAAAATGAAAATAAAATAG : GCG	161	78	170
		3	CTG : GTACAT	TGATATTTTTCTTTGGTTAG : GTG	120	80	170
	Peroxidase isoenzyme C gene (*prxC3*)	1	AGT : GTAAGG	ATATATATATAATTTTGTAG : GGT	166	73	170
		2	TTG : GTATGT	TCTTTATATATATGTTATAG : TCG	131	76	170
		3	CTG : GTAAAA	AGTTATTGTCAAAATGGTAG : GTG	494	78	170
Bertholletia excelsa	2S albumin gene1	1	CAG : GTTAGT	CGAAATGAACCCTGGTGCAG : ATG	89	71	171
	2S albumin gene2	1	CAG : GTTAAT	TCAAACGAACCCGGGTGCAG : ATG	87	67	171

Table 2. Continued

Plant species	Gene	Intron	5′ splice site	3′ splice site	Size (bp)	AU (%)	Ref.
Brassica napus	*Bp19*	1	TTC : GTAAGT	ATATACATGCATGCATGCAG : AGG	137	70	172
	Bp4A and *Bp4C* genes	1	ATG : GTTTGT	GAAATTTTATATCTTACAG : GAA	765	81	173
		1	AAG : GTTTGT	GAAATTTTAATATCTTACAG : GAA	766	80	173
	cruA gene for cruciferin	1	AAG : GTACGT	TTTGGTTTGTTACATTATAG : GAG	228	72	174
		2	AGG : GTATAT	GTTTTCTGTTTTATCTGTAG : CCA	467	71	174
	Napin seed storage protein	1[f]	GCG : GTGGTC		—	—	175
	Oleosin *BN-III*	1	TAA : GTAAGT	GTGTGAATTGTGATGAATAG : GTA	306	70	176
	rbcS gene for the small subunit of ribulose-1,5-bisphosphate carboxylase	1	AAG : GTAAAT	ATTGGAAATACGTTTCGTAG : GTG	162	67	177
		2	GAG : GTAATA	GAATTCTCACGTGTGTGTAG : CAC	93	71	177
	rbcSF1 gene for ribulose-1,5-biphosphate carboxylase/oxygenase small subunit	1	AAG : GTAAAT	AAATGTGATTTAATTTGTAG : GTG	173	66	178
		2	GAG : GTAATA	GAATTCTCACGTGTGTGTAG : CAC	93	70	178
Casuarina glauca	*HbII* gene for hemoglobin	1	AAA : GTAAGC	AAGTATTGGTCTGATTTGAG : GAT	108	68	179
		2	ATG : GTAAGA	CTTGAATGTATGTATTGCAG : ACC	739	75	179
		3	GAG : GTAGTA	CCTTTGTAAATGGCTCACAG : GTA	115	62	179
Cucumis sativus	*hpr-A* gene for NADH-dependent hydroxypyruvate reductase	1	GAA : GTATTT	CATTCTTTTTCATGTTCTAG : ATT	97	70	180
		2	CAG : GTATCT	TGTCTGGTTGGTTGTTTCAG : TTG	214	71	180
		3	CCT : GTAAGT	TTCAATTTTTACTTGTTTAG : GGA	98	73	180
		4	TTT : GTATGA	AAGTAATTTCTTATCAATAG : GTT	95	72	180
		5	ATG : GTACTT	TTGTTGTTATATTCTTGCAG : GTA	173	72	180
		6[b]	CAG : GCATGT	AAGTGCTTGTTTAATTGCAG : CCT	73	66	180
		7	GTG : GTATGA	GTTTTCCTTTGAATTTATAG : ATA	90	67	180
		8	AAG : GTTAAG	AAACTATGTTATAATGGCAG : GAT	122	67	180
		9	GAG : GTATTT	GCTTGTTTTGTCACTCTTAG : GAT	326	66	180
		10	AAG : GTAAAT	GTTTCATAACCATCATGTAG : TGG	79	65	180
		11	TTG : GTATGC	CTTGCTTCTTTTAACCACAG : GGA	78	64	180

Cucurbita pepo	1-aminocyclopropane-1-carboxylate synthase	1	CAG : GTTTGG	TTTTTTGTGAATTTCGATAG : CTT	87	70	181
		2	AAT : GTACGA	TAAAATTTCTTTCTTGTCAG : GCA	83	70	181
		3	AGG : GTGAGT	GTTAGTATGTTGTTTTTCAG : ATT	465	74	181
		4	AAG : GTTTGT	AATGTATTAATCAAATTCAG : GTT	86	76	181
Daucus carota	ATP synthase b subunit	1	ACT : GTAATG	TTGCAATCGTGTGTGTCCAG : ATA	303	67	182
		2	TAA : GTAAGA	CATATTGTTCCGAAATGCAG : AAA	125	70	182
		3	AAG : GTAATT	ATATCCATGTCGCATCGCAG : GTT	175	70	182
		4	ATG : GTTTGT	GGTTCTGTTTCTATACTCAG : GTG	94	65	182
		5	CAG : GTTTGT	CATTTTCCATCAACTTTTAG : GCT	98	66	182
		6	CAG : GTAACA	CCATCTTGCCCTTTTTTTAG : GCC	96	68	182
		7	CAG : GTCTGT	TTTATTTTTATAAATTGCAG : ATA	84	74	182
		8	CAG : GTAATT	AATTGCTAAAATTTTGACAG : GGA	101	70	182
	DC8 gene for an embryonic-specific 66 kDa protein	1	AAG : GTCAGT	TGAGGGGCTGTATGATGTAG : GAA	120	73	183
		2	TAG : GTGAGC	GTGTGTAAATGACCTGACAG : GGG	116	67	183
	Gene for elongation factor 1A	1	AAG : GTGAGG	TCTTCTGGCTTGTGTTATAG : ATG	93	71	184
Dianthus caryophyllus	*CARSR12* gene; ethylene-regulated; phytohormone; senescence-related protein	1	GAG : GTAATT	TGGGTTGTTAATTGTTACAG : ATG	1125	72	185
		2	AAG : GTAAAT	ATGTGTAATTTTTGATACAG : TAT	466	77	185
		3	TGG : GTAATT	ACTCTGTTTTAAATATGTAG : GGG	874	72	185
		4	AAA : GTAAGC	TTTTTATTTTTGTTATGAAG : GAA	490	74	185
		5	CAG : GTAACC	CTCATATAATAATGTAACAG : ATT	810	75	185
		6	GTT : GTACGT	TTGGTATCTGATTGTGACAG : ATT	618	70	185
		7	ATG : GTATGT	ATTTTTGAATGTTTTTCAAG : GTA	88	66	185
		8	ATG : GTAAAC	CCTTTTATATTGTTCCTCAG : TTT	187	67	185
		9	ATG : GTAAAT	TTAATGTTAATTATGTTTAG : GTT	113	75	185
		10	GAG : GTAAAA	GCTATAACTTCCTTTTCTAG : GCA	87	70	185
		11[a]			—	—	185
		12[b]		TTTCTGTTTGTACTTAACAG : TGT	—	—	185
		13	AAG : GTTACT	AAAAATCGACTTGTGTGCAG : GCC	91	75	185
		14	GCG : GTTAGT	AACTGTCGAGTCGAATGCAG : AAC	83	63	185
		15	AAG : GTTAGT	TTTTGTGTGTTGGTGTGCAG : ATC	125	65	185
		16	AAG : GTATGA	TTATCTGAACTTTTCGACAG : ACG	89	66	185
Glycine max	Actin	1	AG : GTTAGT	TTTATTTTAATATGGAACAG : GCA	88	69	186
		2	CTG : GTTTGT	ATCAATTCCTTTTAAAACAG : GTA	81	68 ($N = 1$)	186
		3	CAG : GTGATT	AAAATGTTGTGGTTTTGCAG : ATG	79	68	186

Table 2. Continued

Plant species	Gene	Intron	5′ splice site	3′ splice site	Size (bp)	AU (%)	Ref.
Glycine max (continued)	Auxin-regulated protein gene, last 2 exons	1	AAG : GTAACT	ATTGTGATTTGCGAAATCAG : ATT	388	72	187
	Alpha′-type beta conglycinin storage protein gene	1	ACG : GTTAGT	ATTTGTTTTGTTAATTACAG : GGA	203	80	188
		2	GAG : GTACTA	ATCTTTTTCTTTTCCTGCAG : AGT	87	78	188
		3	GAC : GTAAGC	TTCCGCGCTTGATTTTATAG : ACC	85	67	188
		4	GAG : GTAAGC	TTCTTTGTTCACAAATATAG : GGA	115	71	188
		5	CAG : GTACAT	TGGTGATTGAAAATTTGAAG : GTT	132	71	188
	Gmachs1 gene for chalcone synthase	1	TGT : GTAAGA	TAATTTATTATTTCGATCAG : GTG	122	79	189
	Gene 3 for chalcone synthetase protein	1	TGT : GTAAGA	TAATTTATTATTTCGATCAG : GTG	122	78	190
	GH3 gene for auxin-regulated protein	1	CAG : GTACAT	TACCGTTAATTAAATTGCAG : TTC	521	71	191
		2	TCA : GTATGT	CATGTTGTTACATTACACAG : ATA	105	66	191
	Leghemoglobin gene (gene *Lbc*)	1	TTC : GTAAGT	GTTTGATTAAAAATAAATAG : GAT	169	82	192
		2	TTG : GTAAGT	ATTATTTTTCGAATTTGTAG : GTG	234	82	192
		3	GTG : GTATGA	ATGTTTTTTATATTTTGTAG : GTG	285	74	192
	Glycinin subunit *A-2B-1a* gene	1	AAG : GTACGT	GATATTGATGTATGGTGCAG : AAT	625	75	193
	vspA gene	1	TGG : GTATTA	TGCAATATTTTATGTGATAG : GGT	331	70	194
		2	CAA : GTCAGT	TCTATTTTTGGTTGGTGCAG : GGA	180	69	194
	Glycinin; *Gy1* gene	1	AAG : GTCCAT	TTGTTTTGTAATATGAATAG : GTA	228	78	195
		2	AGG : GTGAGC	GTTTTGACAATTCTTGGCAG : AGA	291	71	195
		3	AAG : GTACGT	TGCATTGATGTACGATGCAG : AAT	381	77	195
	Glycinin; *Gy2* gene	1	AAG : GTCCAT	TGTGTAATATATATGAATAG : GTA	238	75	196
		2	AGG : GTGAGA	GTTTTGACAATTCTTGGCAG : AGA	292	70	196
		3	AAG : GTACGT	GATATTGATGTATGGTGCAG : AAT	624	75	196

Glycinin: *Gy3* gene	1	AAG : GTTTTT	TGATTGTTTAATGTGAATAG : GTA	617	76	197
	2	AGG : GTACGT	ATAAAATATACACTTGGCAG : AGA	312	73	197
	3	AAG : GTACTT	ACACTTGATGTATGGTGCAG : AAT	439	77	197
Heat shock protein (*Gmhsp26-A*) gene	1	AAG : GTAACT	ATTGTGATTTGCGAAATCAG : ATT	388	72	198
Lipoxygenase; *lox-3* gene	1	ATG : GTTCCT	TCTTTCTTTTCTTTGGGTAG : CCA	448	65	199
	2	CAG : GTCAAG	GTTAACAAAACATATTGCAG : ACA	191	76	199
	3	AAG : GTTAGG	CTGCTATTTTCATTAAACAG : ATC	323	73	199
	4	AAG : GTATCT	ATTTGTTATTATTTGAGCAG : TGA	122	70	199
	5	AAG : GTTGAG	CTTATAATTTATTATTGCAG : GAG	104	65	199
	6	GAG : GTATAT	AATTGTATTTGGTTGGAAAG : GCA	131	79	199
	7	TTG : GTATAC	TATCTTTTGATGGATAATAG : GTT	93	84	199
	8	AAG : GTAAAT	ATTAATGTCATGTGATATAG : AGG	107	79	199
Ngm-16 gene coding for nodulin 16	1	AGG : GTGAGC	TGGTTCGTGTGTTGATTAAG : AGC	495	71	200
PAL1 gene for phenylalanine ammonia lyase	1	CAG : GTAACT	TTTTTATTCTTCCTTTTCAG : ATT	1519	73	201
Peribacteroid membrane protein (*nod24*)	1[b]	AGG : GCAAGT	TAAGGGTTAATGTGTTCCAG : CAG	639	76	202
	2	CTG : GTGGTG	ATTTAATTAATGTGTTCCAG : CAG	127	69	202
	3	GTG : GTGGTG	ATTTAATTAATGTGTTCCAG : CAG	126	69	202
	4	GTG : GTGGTG	CTTAATTAATGTGTTTGCAG : CAG	127	69	202
RNA polymerase II	1	TAG : GTGAGT	CTAATTTATTTATTTTCCAG : TCC	325	71	143
	2	CAG : GTAAGT	CTCATGTCTGACTATAACAG : TAG	67[d]	66	143
	3	TAG : GTATGC	TTCTGTTTGATCATTTGCAG : TCC	112	70	143
	4[c]		AGCCGGGTGAAAATGTTCAG : GAC	—	—	143
RNA polymerase II	1	TAG : GTGAGT	TCTAATTTGTTATTTTCCAG : CCC	263	67	143
	2	CAG : GTAAAT	TTTTTTTCCCTTTTATCTAG : CCC	168	74	143
	3	TAG : GTATAT	GTCTGTTTGATCATTTGCAG : TCC	101	66	143
	4	GAG : GTAAAC	AGTCCGGTGAAAATGTTCAG : GAC	421	65	143
RNA polymerase II *RPB1-C* gene largest subunit	1	TAG : GTGATT	GTATAATTTATTATTTTCAG : CCC	271	70	143
	2	CAG : GTAAGT	GTTCCTTTTTTTCCTTCTAG : TCC	131	67	143
	3	TAG : GTATGC	TTCTGTTTGATCATTTGCAG : TCC	111	66	143

Table 2. Continued

Plant species	Gene	Intron	5′ splice site	3′ splice site	Size (bp)	AU (%)	Ref.
Glycine max (continued)	Gene encoding ribulose-1,5-bisphosphate carboxylase small subunit	1	CAG : GTAAGA	CTTTTGATGCAAATTACTAG : GTG	193	70	203
		2	GAG : GTCAAT	TTTGTTGTGTTTGTATATAG : CAC	290	72	203
						(*N* = 5)	
	Ribulose-1,5-bisphosphate carboxylase small subunit (*SRS4*)	1	CAG : GTATGA	TCATGCAAATTAATTACTAG : GTG	203	70	204
		2	GAG : GTCAAT	GTTGATCTGCTTGTTTGCAG : CAC	284	70	204
	Beta-tubulin (*S-beta-1*) gene	1	AGG : GTACTA	GTTTGTTTGATGATCAACAG : GGT	508	72	205
		2	GTT : GTAAGT	GTGTGCTTTAACAATTGCAG : TTG	72	60	205
	Beta-tubulin (*S-beta-2*) gene	1	AAG : GTATGG	GATGTTTCTTCTTGTTGCAG : GTT	723	61	205
		2	GCT : GTAAGC	TGTTTCATTTTTGTGTGCAG : TTG	13	68	205
	sc514 gene for lipoxygenase	1	ATG : GTTTGT	GTTTTGTATGGTTTTTGCAG : GCA	624	73	206
		2	GAT : GTAAGC	TTAAATGAAATATATTGTAG : ACA	74	76	206
		3	CAG : GTTAGA	TATTGTTTGAGCGGTAACAG : ATC	306	77	206
		4	AAG : GTACCT	ATTAATTATGATTTGAGCAG : TTA	100	75	206
		5	CAA : GTAAGT	TCATTCTTCAATAATGTCAG : GAA	130	72	206
		6	GAG : GTAATG	AATTAAATTTTGCGGAACAG : GCA	204	80	206
		7	TTG : GTAATT	TTTTTTTGTTGGGGTTACAG : GTT	118	69	206
		8	GAG : GTATTT	TGTTTTGGGGCTTGATTCAG : AGG	202	73	206
	tefS1 gene for elongation factor EF-1a	1	AAA : GTAAAT	CTTGTCCCATTTTGCAACAG : ATG	86	72	207
	Urease gene partial	1[c]		TTGAATAATTTCTTCATTAG : GTG	—	—	208
		2	GAG : GTATAA	TCATTTAGACCAGTTTGCAG : GCG	361	65	208
		3	CCG : GTAATG	TAGGTTTCTCAATCTTGTAG : GTG	128	70	208
		4[g]	CAG : GTGGAC		—	—	208
	Vegetative storage protein (*vspB*)	1	TGG : GTACGT	TACGAATAATTTATTGATAG : GGT	786	72	209
		2	CAA : GTATGT	TCTATTTTTGGTTGGTGTAG : GGA	163	74	209

	Gmachs2 gene for chalcone synthase	1	TGT : GTAAGA	TAATTTATTGTTTGGATCAG : GTG	137	76	210
	Catalase	1	AAG : GTTTGC	GTGTTTTGTTTTCCATTTAG : AAT	91	70	211
		2	GAG : GTATGG	AATTGAATTGGGGCATGCAG : GTC	78	72	211
		3	GAG : GTAAAT	TGACTGCTACTGGTTTATAG : GTC	99	68	211
		4	AAG : GTTAGT	AGAAATGTGACTTATTGCAG : TGC	403	71	211
		5	CAG : GTTCCT	CTGTTTTCTGATTTCTCCAG : GCA	98	68	211
		6	CAG : GTGAGT	TGGATGGTATTGCTACGCAG : GCT	515	63	211
Gossypium hirsutum	*cab* gene for chlorophyll *a*/*b* binding protein	1	CTG : GTACGT	TTCTGATTACTTATAAACAG : GTA	90	70	212
	Legumin A D-genome alloallele gene	1	AAG : GTATAA	GTTTAAATTTGTTTTTTCAG : GAA	76	74	213
		2	AGA : GTATAT	TCTCTCTTGTAATGTTATAG : AGA	89	78	213
		3	AAC : GTAAGC	TGTAACATTTGTTTTTGCAG : AGA	85	75	213
	Putative desiccation protectant (*Lea4-D9*)	1	AAG : GTATAA	AAATGTATGATGATGAAAAG : GGA	102	75	214
	Late embryogenesis abundant protein (*Lea*)	1	AAG : GTAACC	TTAAAGGAGAGGTTACTCAG : GTG	119	69	215
	Set *5B Lea* gene for seed protein D-11	1	TCA : GTAAGT	TGCATGCATGCACCATGCAG : TCT	102	68	216
		2	n/d				216
	Set *5A Lea* gene for seed protein D-113	1	AAG : GTTAAT	TGAAATGGATGATGTGTTAG : GTG	96	73	216
	Set *5B Lea* gene for seed protein D-19	1	AAG : GTATAA	AAATGTATGATGATGAAAAG : GGA	102	75	216
	Embryogenesis abundant protein gene, exon 1	1	AAG : GTGACC	TTGATCGAATGGTGTTTTAG : GGA	111	60	217
	Set *5B Lea* gene for seed protein D-34	1	CAG : GTGTTT	TTCGTTCGTTTCATTGTTAG : GTT	83	72	216
		2	ACG : GTACAA	ATTATATTTATACTTTGTAG : GGT	94	70	216
	Lea4-A gene	1	AAG : GTATAA	AAATGTATGATGATGAAAAG : GGA	101	75	218
	Set *5A Lea* gene for seed protein D-7	1	CAT : GTAATA	ATTATGTTCTTTTATGGTAG : GAA	87	75	216

Table 2. Continued

Plant species	Gene	Intron	5′ splice site	3′ splice site	Size (bp)	AU (%)	Ref.
Gossypium hirsutum (continued)	Late embryogenesis-abundant protein 2-D (*LeaA2-D*)	1	AAG : GTGACC	TTGATCGAATGGTGTTTTAG : GGA	111	60	219
	rbcS gene for ribulose-1,5-bisphosphate carboxylase, small subunit	1	CAG : GTACTT	ATCTTGGCTTTTGAGTATAG : GTG	101	72	220
		2	GAG : GTGTTT	TGTTTTATATGTATGTGCAG : GAG	92	72	220
	Storage protein (alpha-globulin A) gene	1	ACG : GTGAAT	TAAATTATTTACTTTCTCAG : GAA	89	74	221
		2	CAG : GTATTT	TTATCTGTATTTTGCTGCAG : AAA	80	75	221
		3	AAT : GTATTA	TTCATGTTTTGTCGTATTAG : ACA	87	83	221
		4	AAA : GTACTT	ATCATCCCTTTTTTTACAG : GGA	93	66	221
	Storage protein (late embryogenesis abundant)	1	CAG : GTGTTT	TTCGTTCGTTTCATTGTTAG : GTT	83	72	222
		2	ACG : GTACAA	ATTATATTTATACTTTGTAG : GGT	94	70	222
	Group 4 late embryogenesis-abundant protein (*Lea14-A*) gene	1	AAG : GTAATA	TATGTGTAATTTGGGGACAG : AGG	81	69	223
	5A *Lea* mRNA for seed protein D-29	1	CAT : GTAATA	ATTATGTTCTTTTATGGTAG : GAA	87	75	224
	Lea2 gene for late embryogenesis-abundant protein	1	AAG : GTGACC	TTGATCGAATGGTGTTTTAG : GGA	111	60	225
	Lea5 gene for late embryogenesis-abundant protein	1	AAG : GTAACC	TTAAAGGAGAGGTTACTCAG : GTG	119	69	226
Helianthus annuus	*HaG5* gene for 2 S albumin storage protein	1	AAG : GTATAA	CTAATGACGATTTCCAACAG : GTT	191	78	227
	GUbB1 gene for hexaubiquitin protein	1	AAG : GTATCT	GTTTGTATTTTTTAATCCAG : ATG	963	70	228

	GUbB2 gene for polyubiquitin protein	1	AAG : GTATCT	TTTGTATTTTTATTATCCAG : ATG	931	69	228
	sf2 gene for an anther-specific protein	1	CAG : GTGCAT	TTTTCGCATGACGTGAACAG : GAA	2611	67	229
Hevea brasiliensis	*HMGR1* gene for 3-hydroxy-3-methylglutaryl coenzyme A reductase 1	1	CAA : GTATGC	TCTCTTCTTAATGATTGCAG : GTC	132	72	230
		2	CAG : GTAAAT	TATGTGTCTCAATTGGCTAG : GAA	379	68	230
		3	GAG : GTTATC	TTAATTTGTTTTTCTGTCAG : GTG	94	69	230
Lycopersicon esculentum	*LE-ACC2* gene for 1-amino-cyclopropane-1-carboxylic acid synthase	1	CAG : GTAATT	GTGTTAAATTTCTTTTGTAG : CTT	98	76	231
		2	AAA : GTACAT	TAACGAGTTTATTTAATCAG : GCG	85	79	231
		3	AGC : GTAAGT	TTGGCAATTGACTAATGCAG : ATT	881	75	231
	Acid phosphatase-1 gene	1	TGG : GTATGA	TATTGATCTCAATATTGTAG : ATT	823	65	232
		2	GAG : GTGAGT	ATGATTCTGATTCTATGCAG : AGG	1133	65	232
	Type I chlorophyll *a*/*b*-binding protein gene	1	CGG : GTAAGT	GGATAACTTTATAAATGCAG : AGA	107	72	233
		2	GTT : GTAAGC	TTTCTTCGCAATTTTTGTAG : CCC	87	74	233
		3	ATG : GTTAGA	AATCTCGATTATTTAGTTAG : GTC	89	74	233
	CAB-8 gene for type III chlorophyll *a*/*b* binding polypeptide of photosystem I	1	AAG : GTAATT	CACTTGATTCTATTTTACAG : CAA	330	68	234
		2	AAG : GTAAGA	GATTGTTTGCTGATCCACAG : TCT	362	68	234
	ATP-dependent protease (*CD4A*) gene	1	TAG : GTACCC	TGATTGGTTTTGTGATGCAG : GGC	103	61	235
		2[b]	CAG : GCAAGT	CTTGTAAAATAAATAAATAG : GTG	251	61	235
		3	GAG : GTACAT	TGGTCATATTATATTTTCAG : GGG	83	67	235
		4	AAG : GTCACA	TTGCTTTGGGTGATGGTCAG : GTC	633	65	235
		5	CAG : GTTCGT	TGCTTGGTTTCACTTGTCAG : TGT	179	66	235
		6	CAG : GTATTA	GAGTTGAACCTTTATTCCAG : TGA	302	68	235
		7[b]	CAG : GCATAG	GTTATATTGATACTATCTAG : CTC	82	60	235
		8	AAG : GTACTG	CCTCTTTATGCATTGGACAG : GCT	109	64	235
	ATP-dependent protease (*CD4B*) gene	1	TAG : GTATCC	TTATAAACTGGTTGGTTCAG : GGC	220	65	236
		2[b]	CAG : GCAAGT	TATTCTTCTTCTAAAAATAG : GTG	121	64	236
		3	GAG : GTAATC	TGTCCATTTTTGGTTTGTAG : GGA	70	74	236
		4	AAG : GTACGT	CTTTTTTTGGTGATAATTAG : GTG	508	65	236
		5	CAG : GTAAGT	ATTTACTTGGCTTCTTGCAG : TGT	209	68	236
		6	CAG : GTATCT	CTTGCGTTTTCACATTTCAG : TGA	315	68	236
		7[b]	CAG : GCATGA	TGCTCTGTTGTTGCCTCCAG : CTC	84	58[e]	236
		8	AAG : GTACTG	GGTTTTGGATTAATGGACAG : GCT	126	70	236

Table 2. Continued

Plant species	Gene	Intron	5′ splice site	3′ splice site	Size (bp)	AU (%)	Ref.
Lycopersicon esculentum (continued)	Ethylene-responsive fruit ripening gene *E8*	1	CGG : GTAAGT	TATGTCATGTTTCAATGCAG : GGA	421	65	237
		2	CAG : GTTAGT	TTTTTAATTATTTTTTGCAG : CTC	76	75	237
	Enolase gene	1[c]		GGGTTGGCCTTTTTTGACAG : CTT	—	—	238
		2	AAG : GTACAT	GAGCTGCCTTTTCTTTGCAG : CAT	314	69	238
		3[b]	CAG : GCATGT	TAATTAGTTGTTATTTGCAG : GAA	73	73	238
		4	AAG : GTTTCA	GCAGTTTACTCTAATTCTAG : GCT	533	65	238
		5	CAG : GTAAAG	TATGGCTTTTGAAAATTCAG : GAG	86	69	238
		6	AAA : GTAAGC	TAAGACTATTTTCCTTTTAG : GTT	81	68	238
		7	GAG : GTATAT	CAATTGTTTCATCTTTGCAG : AAC	310	66	238
		8	AAG : GTAAAA	GTATCTGTGCACTGGTGCAG : AGA	172	65	238
		9	AAG : GTACTT	GCTTCTAATTTTGTGTGCAG : GTT	88	67	238
		10	CAG : GTAAGG	TCTTTGGGTATTTGGTGCAG : TGG	104	61	238
		11	ACG : GTAATT	TATCTCGTGGATTCTTTTAG : GGA	94	70	238
		12	CAG : GTACTT	TTTCTCGTTCATTCCATCAG : CTC	95	59	238
	ETH.1 gene, involved in ethylene biosynthesis	1	GAG : GTAATC	TGCTTATGTGATCATTGTAG : TTG	278	74	239
		2	CAG : GTACAT	TGTTGTACCAAATGATGCAG : AGA	106	73	239
		3	GAG : GTACAA	ATGAAATGTTATAAAAACAG : GTG	89	74	239
	Fruit-specific protein gene	1	TAG : GTTCAC	CTTAATCATAATATATACAG : TTG	714	73	240
	GAST1	1	CAG : GTGAGC	GTGTGTGGTTGTGTGTTCAG : GTT	118	70	241
		2	CTG : GTATGA	TTTATATGTGGTGTTTGCAG : TAT	90	70	241
		3	AAG : GTATAA	GTGTTGTTGTAATTTTTCAG : ATT	126	68	241
	GTom5 gene for phytoene synthase	1	TAG : GTTAGC	TGATATTTGACTTTGATAAG : GAA	127	75	242
		2	ATG : GTGAGG	ATTATGATCTGTTTAAACAG : TTA	425	66	242
		3	CAG : GTTAGT	CCTTCCAATATATGTTGCAG : CCA	309	67	242
		4	AGA : GTAAGT	ATATATTGCTTTCCCTGCAG : TGC	520	66	242
		5	CCT : GTAAGC	TCGTTACTCGGTGTTTCCAG : GTA	689	63	242
	Ethylene-related gene (*GTOMA*)	1	GAG : GTACAA	ATTTTTTTTTAAATGTATAG : TTG	334	77	243
		2	TAG : GTAAAC	TATTGTTTTATATGATGTAG : GGA	825	72	243
		3	GAG : GTAAAA	AAGCTCAATCCTTTAAACAG : GTA	398	70	243

3-Hydroxy-3-methyl glutaryl coenzyme A reductase (*hmg2*) gene	1	CCA : GTAAGT	CGGTTGTGTTTGAATGGCAG : ATC	82	66	244
	2	CTG : GTATTA	GTTATCTATAACTGTGACAG : GGA	201	68	244
	3	GAG : GTATGT	TGGTATTTGCTTCGATACAG : GTT	415	68	244
Threonine deaminase (*Td*) gene	1	AGG : GTAAGT	ATAATTTGGTTATTGGTTAG : GTG	574	74	245
	2	AAG : GTAATA	TGTTTTTTTTTTTATACAG : ATT	173	79	245
	3	AAG : GTACTA	TAAAATTCTGCGTACGTCAG : GTA	285	64	245
	4	AAG : GTTAGC	TTCTTTTGTTGTCAAAATAG : GAT	136	71	245
	5	CTT : GTAAGT	ATTTTAACTTTGAATTACAG : GTG	560	75	245
	6	CAG : GTAATT	TTATATTATTGATTTTGCAG : GGT	102	86	245
	7	TTG : GTAATG	TTATTTTATTTTCTTAATAG : GTT	87	89	245
	8	CAG : GTAAAT	TTATTTTATTGCCATGACAG : GGA	85	80	245
ABA-regulated LE25 (*le25*) gene	1	AAG : GTTTCC	AACCCATCGTATTTGTGTAG : GCG	507	72	246
lhbC1 gene for LHCII type III	1	ATG : GTATAT	TTTTCCTCTCATCTTATTAG : AGT	201	66	247
	2	GAG : GTCAGT	AGTGACATTGAACTATATAG : GTT	107	62	247
rbcS2 gene for ribulose-1,5-bisphosphate carboxylase/oxygenase small subunit	1	CAG : GTAACT	TTTTTTGGATATGTGTATAG : GTG	999	77	248
	2	GAG : GTCAAT	AATGTTTTTTTTATGTGTAG : CAC	98	82	248
	3	AAG : GTCAGT	TTAAAATTTTGAATCCGCAG : GTA	435	71	248
rbcS3A gene for ribulose-1,5-bisphosphate carboxylase/oxygenase small subunit	1	CAG : GTTTGT	AACTATTTATAATTGTATAG : GTG	92	77	248
	2	GAG : GTCAAC	TTGTGGGATATCATGTGCAG : CAC	87	64	248
rbcS3B gene for ribulose-1,5-bisphosphate carboxylase/small subunit	1	CAG : GTTTGT	AACTATTTATAATTGTATAG : GTG	92	77	248
	2	GAG : GTTAAC	TTGTGGGATATCATGTGCAG : CAC	82	70	248
rbcS3C gene for ribulose-1,5-bisphosphate carboxylase/oxygenase small subunit	1	CAG : GTTTGT	AACTATTTATAATTGTATAG : GTG	92	77	248
	2	GAG : GTCAAC	TTGTGGGATATCATGTGCAG : CAC	87	64	248
Rbcs-1 gene for ribulose-1,5-bisphosphate carboxylase/oxygenase small subunit	1	CAG : GTTATT	GGAATGTTTGACTAATGTAG : GTG	95	77	249
	2	GAG : GTCAAT	GTAATATATGTATGTAACAG : CGC	96	79	249
Rbcs-3A gene for ribulose-1,5-bisphosphate carboxylase/oxygenase small subunit	1	CAG : GTTTGT	AACTATTTATAATTGTATAG : GTG	92	77	249
	2	GAG : GTCAAC	TTGTGGGATATCATGTGCAG : CAC	87	64	249

Table 2. Continued

Plant species	Gene	Intron	5′ splice site	3′ splice site	Size (bp)	AU (%)	Ref.
Lycopersicon esculentum (continued)	Fruit-specific protein gene	1	TAG : GTTCAC	CTTAATCATAATATATACAG : TTG	714	73	250
	Phenylalanine ammonia-lyase (*PAL5*) gene	1	CAG : GTAAAT	TTATGATTTATTTTCTTCAG : GTT	710	77	251
	Prosystemin gene, exons 1–11	1	AAG : GTATCA	ATCATGGTATTATCCTCCAG : GAG	147	82	252
		2	AAG : GTAACT	TTTTCCATCAAAATTTACAG : GGA	952	78	252
		3	TAA : GTATTT	ATTATAATATTATCCAGCAG : GAG	109	84	252
		4	GAG : GTAACT	ATTATTAAAGGATATTACAG : GGA	528	79	252
		5	AAG : GTATAA	CTTGATATATTATTCTCCAG : GAG	157	83	252
		6	GAG : GTAACT	TATTTTCTTCAAAATTACAG : GAA	123	83	252
		7	AAG : GTATAT	TGACCTTATATATTCTCCAG : GAG	166	81	252
		8	GAG : GTAACT	TTTTTTCTTCAAAATTACAG : GAA	642	72	252
		9	AAG : GTATCA	ATCATGATATTATCATCCAG : GGG	144	83	252
		10	GAG : GTAACT	CTTTTTTCTTTCAAATACAG : AAA	607	76	252
Lycopersicon peruvianum	Proteinase inhibitor I gene	1	TTT : GTTAGT	TGAAATGTTTTTTCTGACAG : CTT	579	73	253
		2	AAG : GTAAAA	ATTATTATGTGCATGTGTAG : GAA	513	75	253
Medicago sativa	Glutamine synthetase gene	1	ATG : GTTAGA	TTTTTCTCTCATTATGACAG : GAT	714	72	254
		2	AGG : GTAATT	ATGCTTTTTTTGGTGCGAG : ACT	113	77	254
		3	CTA : GTATGA	TGGGGTTGGATTCCTTACAG : CCC	124	78	254
		4	TTG : GTAAGT	TTGTGTTTAATTAAATTCAG : GTT	445	70	254
		5	ATG : GTATCT	GTTGTATGATTTGTGATTAG : GTA	93	69	254
		6	CAG : GTGAAA	TTTTATAATTTGCTCAATAG : GGA	286	75	254
		7	CAA : GTAAGT	CTATTGTTTTTTTAATGTAG : TGG	114	80	254
		8	GAG : GTAGGT	CTTTGTTTATGTTCCAATAG : AGG	128	67	254
		9	AAG : GTTTGC	TTTTTTTAATGCAAAACTAG : GGT	135	73	254
		10	CAG : GTAATG	GCATCCTTATAATGCTGTAG : CAC	90	72	254
		11	TGG : GTAAGC	GTGGCTTGTGTTATTTGAAG : GGT	177	64	254
	Leghemoglobin gene	1	TAT : GTAAGT	GGCTATTTTGATTAAAATAG : AAT	110	78	255
		2	ATG : GTGAGT	TTCTACATTTTGAATTATAG : GTG	491	79	255
		3	GCG : GTAAAT	TAATGTTTTTGTGCCTGTAG : GTG	430	78	255

	Nodulin-25 gene	1	ATG : GTGAGT	ATTGTTTAATGTTATTGTAG : ATT	885	77	256
		2	TTG : GTAAGG	TTTATACATACTATATACAG : CTG	469	78	256
		3	GTG : GTGAGT	ATTATTTACTGTTATTTTAG : ATT	1990	77	256
		4	CGG : GTAAGA	TTGTATATACAATATTACAG : TTG	568	79	256
		5	GTG : GTGAGT	TTTTTCAATTGATATTGCAG : CTT	922	75	256
		6	CTG : GTAAGG	TCTATTTGGTTTTAATGCAG : ATG	685	76	256
		7	ACG : GTAAGG	CTTCGTGTTTGGTTTGTTAG : ATG	1203	72	256
		8	CAG : GTTATT	CTTCATGTTTGGCTTGCTAG : ATG	89	74	256
		9	CAG : GTTATT	CTTCGTGTTTGGCTTGCTAG : ATG	722	62	256
		10	CAG : GTTATT	CTTTGTGTTTGACTTGCTAG : ATG	89	71	256
		11	CAG : GTTATT	ATTTCTCTTTATTACTGTAG : TGT	276	76	256
		12	TTG : GTGAGA	TTTCCAATCAATTATTGCAG : CTA	212	69	256
Medicago truncatula	*Mtlb1* gene for leghemoglobin	1	TGT : GTAAGT	ATTTTATTTTATTAAAATAG : CAT	129	81	257
		2	CTG : GTAAGT	TTTTAACTTTTAAATTATAG : GTG	1091	73	257
		3	GTG : GTATGA	ATGTTTTTGGTGTCGTACAG : GTG	719	76	257
	Mtlb2 gene for leghemoglobin	1	TAT : GTAAGT	GGTTATTTTGATTGAAATAG : TAT	108	75	257
		2	ATG : GTGAGT	TTCTACTTTTTGAATTATAG : GTG	451	78	257
		3	GTG : GTAAAT	ATGTATTTGGTGTCTTGCAG : GTG	301	78	257
Mesembryanthemum crystallinum	*ppc2* gene for phosphoenol-pyruvate carboxylase	1	ACA : GTATGT	TTGTGGATCAATTGATGTAG : GTC	569	69	258
		2	CAG : GTATGT	CTTTTTACTTTAAATATCAG : GAT	104	72	258
		3	GAG : GTAAAA	TTCTTCTGATGGCTTTGCAG : ATC	1454	67	258
		4	ATG : GTATGT	CTTCTTTGGCTTTGGTACAG : GTA	531	65	258
		5	GAG : GTACAA	AACTGCTGTCTACATGGCAG : TTA	107	63	258
		6	TAG : GTACCT	ATCTGGCTTGTTACATGCAG : AGT	325	63	258
		7	CAG : GTATAT	GATCTTTTATATTGATTCAG : TTC	261	61	258
		8	CTG : GTAAGC	CTCTTGCTTTTTCTCCACAG : GCC	379	59	258
		9	AAG : GTAAAA	TTGTCTCAAATTATGGCCAG : ATT	806	67	258
Nicotiana plumbaginifolia	*atp2-1* gene for mitochondrial ATP synthase beta subunit	1	ACC : GTAAGT	TTCCCTTCTTGTGGCAACAG : GTT	997	67	259
		2	TTA : GTAAGT	ATCTGTTAAAATGGCTACAG : CTA	93	67	259
		3	AAG : GTACTT	ATTGTTGTGCTTTTGGTCAG : GTT	145	61	259
		4	ATG : GTTAGG	GGCTTGACTATGTTATTCAG : GTG	157	61	259
		5	CAA : GTTAGT	TGTGACCTCAACCATTTCAG : AGT	90	66	259
		6	CAG : GTTGGC	TATGCATTTTATATTGATAG : GCT	96	68	259
		7	CAG : GTATAA	TCATTTCTTTTGGATGCCAG : ATC	88	66	259
		8	CAG : GTAATA	GTGTTAATTTCTTTTGACAG : GGA	1279	65	259

Table 2. Continued

Plant species	Gene	Intron	5′ splice site	3′ splice site	Size (bp)	AU (%)	Ref.
Nicotiana plumbaginifolia (continued)	Beta-(1,3)-glucanase gene for a vacuolar isoform	1	CAG : GTTTCT	TTCTAATTCATATTGAGCAG : GGG	976	72	260
	Beta-(1,3)-glucanase regulator genes	1	TAG : GTATGC	TTTTTATTCTCATGGAGAAG : GAG	669	69	261
		2	TAG : GTACTA	TTTGATTATATCTTCAACAG : GTA	198	76	261
	Plasma-membrane H$^+$-ATPase (*pma1*) gene	1	TTG : GTAAAA	GCTTTTTTTAATGATTACAG : GAA	105	68	262
		2	AAG : GTACTT	TAATATATGGTGTGTTATAG : GAT	479	69	262
		3	GGA : GTAAGA	ACTGGGTGACATGTCTGTAG : GGA	89	58 [e]	262
		4	AAG : GTACCA	TTTGCCATCGTGTTATGGAG : GTT	159	60	262
		5	CAG : GTGATG	ACTCATTTTTATACTTGCAG : TCT	89	64	262
		6[b]	AAG : GCGAGT	TTATTTGATGTTTTTATCAG : GTT	102	72	262
		7	CAG : GTTTGT	TTTCATTATCTACTGTGCAG : GGT	114	63	262
		8	GAG : GTGAGC	TGTTCCTGTTAATTGTTCAG : GTT	97	60	262
		9	GAG : GTAACT	AATAGTTGCCATTAATATAG : GCA	75	67	262
		10	CAG : GTTTGT	AATATGTGATATGGATATAG : ATC	87	70	262
		11	CAG : GTATAT	TATTCTAATTCTACATGCAG : GAA	208	62	262
		12	CAG : GTATAT	AATTTTTATCTTTTTGGCAG : GGG	223	66	262
		13	CCG : GTACAT	TCTCTTGGTCATTGCTTTAG : AGC	90	66	262
		14	ACG : GTACTT	TGTCTTCTTTGCATTGACAG : ATA	149	66	262
		15	GTG : GTAAGT	TTTCTTCAATTTTATGTCAG : CTT	105	70	262
		16	ACG : GTGGGT	CTAGACATTTGCAATGTCAG : GCA	739	62	262
		17	CCT : GTAAGC	TCATTTTGGTGAATTGCCAG : CAT	197	65	262
		18	CTG : GTGAGA	GTGAACTAAATTATGAACAG : GTT	75	61	262
		19	AGG : GTATGA	CTTTTTCTACTTCTCTACAG : ATT	97	66	262
		20	TAG : GTATAT	GCTACCATCTTTCATTACAG : ACT	430	64	262
	Plasma-membrane H$^+$-ATPase (*pma2*), exons 13–21	13	CTG : GTATTT	TGTCTTGCTCATTTCTTTAG : AGC	105	70	263
		14	ACG : GTATAT	TCTTTTTTTGGCATTGGCAG : ATA	199	65	263
		15	GTG : GTAAGT	TCTTCTATTTAACGTCGCAG : CTC	91	63	263
		16	ACG : GTTGGT	CTAAGTATTTATAATCTCAG : GTA	718	65 ($N=1$)	263

		17	CCG : GTGAGC	TCAGTTTGGTCAATTTCCAG : CGA	634	62	263
		18	TTG : GTGAGC	GGTTACTAAACTGTGAACAG : GTT	83	63	263
		19	AGG : GTAAGC	GGTTTTTCATGGTACTGCAG : ATT	88	65	263
		20	TAG : GTAAGA	AGCTACATTGTTCATTGCAG : GCA	262	61	263
	Plasma-membrane H^+-ATPase (*pma3*) gene, exons 1–9	1	TTG : GTTATA	TTTTATTTTTATGTTTGCAG : GAA	92	71	264
		2	AAG : GTGTCT	TTATTGGTGTACTATTACAG : GAG	528	70	264
		3	GGA : GTAAGA	TATGTGGACTCTTCTTCTAG : GGA	91	67	264
		4	AAG : GTACTT	TTACACTTGTCTTTGTGGAG : GTC	138	64	264
		5	CAG : GTGATT	GCCTGTCTCTCAACTTGCAG : TCT	88	64	264
		6[b]	AAG : GCAAGT	ATGTTTATGGTGTTCGTCAG : GTT	100	61	264
		7	CAG : GTTTGT	TGATCAATTTGTCTATACAG : GGC	94	65	264
		8	GAG : GTGAGT	TGTTCTTGTTAAATTTTCAG : GTA	96	66	264
Nicotiana tabacum	Auxin-related protein (*pGNT1*)	1	AAG : GTAAAT	AAGCCAAGAGCGTTTTGCAG : GTG	508	71	265
	Auxin-induced protein (*pGNT35*)	1	AAG : GTAAAT	AAGTTATGAGCGTTTTGCAG : GTG	513	68	265
	CAB36 gene for chlorophyll *a/b* binding protein	1	TTG : GTAATT	AACAATCCTTGATAAAATAG : ATA	568	73	266
	CHN50 gene for endochitinase	1	CAG : GTATGC	GAACATGTAACTGTTGACAG : GAG	272	68	267
		2	ACA : GTAAGC	TACTTTTTACATAAATACAG : CAA	173	65	267
	Gene for endochitinase	1	CTG : GTAAGT	GACCATGTAACCGTTGACAG : GAG	274	67	268
		2	ACA : GTAAGT	GTATTTTGACATAAATACAG : CAA	269	70	268
	Glucan beta-1,3-glucanase gene	1	CAG : GTTTCT	TTGTAATTCTTATTGAGCAG : GGG	789	71	269
	Glucan beta-1,3-glucosidase gene	1	CAG : GTTTCT	TTCTAATTCTTATTGAGCAG : GGG	747	71	269
	Acidic beta-1,3-glucanase gene	1	TAG : GTCTCT	CTCTTTTTCTTACAATCCAG : GGG	341	68	270
	Glycine rich protein DNA	1	ACC : GTAAGC	AATTTTCATAATTGGTGCAG : CAA	555	74	271
	ltp1 gene for lipid transferase	1	CAA : GTACCT	TTGTTGGACTTGTGATGCAG : GGT	980	70	272
	NeIF-5A3 gene for initiation factor 5A(3)	1	AAG : GTCTCT	GTCGGGTTTTTTGTTGAAG : GTA	1141	66	273
		2	GAT : GTAAAT	CGACAATATTCTCTTTGCAG : GTT	93	68	273
		3	TTT : GTATGT	CACTTGTGATATTGTTTCAG : GTG	131	67	273
		4	CTG : GTAAGC	TTTCTAAATGATGAGCACAG : ATC	345	66	273

Table 2. Continued

Plant species	Gene	Intron	5′ splice site	3′ splice site	Size (bp)	AU (%)	Ref.
Nicotiana tabacum (continued)	Gene for photosystem II oxygen evolving complex 23 kDa polypeptide	1	CTG : GTATGT	TGTGATACATATATATACAG : CAA	522	63	274
		2	CAA : GTAATT	TTCTTTTTTTGTCTGAGCAG : GTG	105	65	274
		3	GAG : GTAAGT	GTGGGAATTTGTTGCTGCAG : GGT	487	61	274
	PG13 (*ASF–1*/*G13*) gene leucine-zipper DNA-binding protein	1	GAG : GTGCAA	TACATAATTAATCTGTCTAG : ACA	76	76	275
		2	AAG : GTTACA	ACCCCTAGCTTCATATACAG : GTG	425	64	275
		3	AAG : GTATCG	ATTCACATGTCGTTACTCAG : GCC	130	69	275
		4	CAG : GTAAAG	GCAAACATTTTTTGTGACAG : GGT	129	71	275
		5	CAG : GTGTGT	CTAACTTGTCGAGTCCGCAG : GAA	1324	63	275
		6	AAG : GTGATG	TGACCCTATGGTACCTATAG : GTT	1093	63	275
		7	CAG : GTAATA	ATACTTGTCTCATTCTACAG : GCA	358	64	275
	Pectate lyase	1	AAG : GTATTG	ATTATATTACGTCTCAACAG : GTG	98	74	276
		2	GAG : GTATAC	TACTATGTCTGTATTGATAG : GTA	97	70	276
	Gene for ribulose-1,5-bisphosphate carboxylase small subunit	1	CAG : GTAATT	TCTATTTTGGTGGAATATAG : GTG	93	75	277
		2	GAG : GTCAAT	ATGTTATTTTGCATGTGCAG : CAC	156	76	277
		3	CAG : GTCAGT	TGAATCTGGTACTGATGCAG : ATA	145	70	277
	Anther-specific gene *TA-29* and stem-specific gene *TSJT1*	1	AAG : GTATTG	TTTTGTTTTGTTTACATTAG : GTC	185	71	278
		2	AGG : GTGAGT	AAGTGCTGCCAATTGGGCAG : GTG	973	62	278
	Phenylalanine ammonia lyase (*tpa1*) gene	1	TAG : GTAAAC	TGTTGACATTGAATTGGCAG : GTT	1932	74	279
	Chitinase gene 50 for class I chitinase C	1	CTG : GTAAGC	ACAATGGACCCTTTTGGCAG : GAG	114	67	280
		2	ATA : GTGAGT	TTTTTTGATGTTTAATTCAG : CAA	170	66	280
	Chitinase gene 50 for class I chitinase B	1	CAG : GTATGC	GAACATGTAACTGTTGACAG : GAG	272	68	281
		2	ACA : GTAAGC	TACTTTTTACATAAATACAG : CAA	173	65	281
Papaver somniferum	*mlp* gene for major latex protein	1	TTG : GTATGC	ATGTGTTTGTCTATGCAGAG : GGG	115	71	282

Parasponia andersonii	Hemoglobin gene	1	CAA : GTAAGT	GGTTGATTGAAACTTTGCAG : GAT	122	75	283
		2	ATG : GTAAAG	CTGGTACTTGTCCTGGAAAG : ACA	326	66	283
		3	GAG : GTACTA	ATGGGAAAAACGATTTGTAG : GTC	159	70	283
Persia americana	*cel2* gene for cellulase, 3′ region	1	CAG : GTACCT	CCAAATCGACCTCTTACCAG : GTT	86	59 ($N=1$)	284
Petroselinum crispum	*PR2* gene for pathogenesis-related protein 2	1	AAG : GTATTT	AAATGTCCGTTTTTTTGTAG : TGA	195	71	285
Petunia sp.	rbcS gene (*ssu8*) for ribulose-1,5-bisphosphate carboxylase small subunit	1	CAG : GTACTT	AAATTCTCTGTTGAGTATAG : GTG	594	70	286
		2	GAG : GTCAAG	GTCTTGTTTTATATGTGCAG : CAC	120	71	286
		3	AAG : GTTAGT	TTGAATATGCTCTGTGATAG : GTA	156	72	286
	rbcS gene (*ssu11A*) for ribulose-1,5-bisphosphate carboxylase small subunit	1	CAG : GTACGT	TTTTGTTTTGTGGGATGTAG : GTG	217	71	286
		2	GAG : GTTAAG	GTTTGGTTTTATATGTGTAG : CAC	150	73	286
Petunia hybrida	ADH1 gene for alcohol dehydrogenase 1	1	AAG : GTGAAT	TATGTGAATGTATACTGTAG : CTG	654	68	287
		2	AAG : GTAATT	GATCCTACCTTTTGTTGCAG : GGG	322	66	287
		3	AGG : GTATGA	TTCCATTCATGTGGCTGTAG : AAT	102	67	287
		4	CAG : GTAGGA	TGGAGATTATTTTTCTGAAG : GAC	99	74	287
		5	GCT : GTGAGT	TTTCTTGACCAAATTATTAG : GCT	92	65	287
		6	ATG : GTGAGT	CATCTTTTAAATTCTTACAG : CCA	234	62	287
		7	CAG : GTTTGT	GTTCCAGCAACAATTTTCAG : GTA	338	62	287
		8[a]	GAT : GTAAAT		375	64	287
		8[b]	AAG : GTGATG	TGATTTGAAATCATTTTCAG : GAA	117	68	287
	Cab gene for chlorophyll *a*/*b* binding protein	1	TTG : GTAAGT	ATTGTTGTTTGATTAAACAG : GTA	106	75	288
	chsA gene for chalcone synthase	1	TGT : GTAAGA	CTTTTTTGGTCAATTTGCAG : GTG	1347	69 ($N=3$)	289
	chsB gene for chalcone synthase	1	TTT : GTGAGT	TTTTCAACAAATAATTTCAG : GCA	3777	71	289
	chsD gene for chalcone synthase	1	TCT : GTAAGT	TTTATTTTATTATATTGCAG : GTG	685	74	289
	chsF gene for chalcone synthase	1	TGT : GTAAGA	ACATTTTTCATATGTTGTAG : TTG	564	73 ($N=9$)	289

Table 2. Continued

Plant species	Gene	Intron	5′ splice site	3′ splice site	Size (bp)	AU (%)	Ref.
Petunia hybrida (continued)	*chsG* gene for chalcone synthase	1	TGT : GTAAGA	CATATATATGGTTCACGCAG : GTG	2439	71	289
	chsJ gene for chalcone synthase	1	TGT : GTAAGA	AAGTTTATCATGTAATGCAG : GTG	729	71	289
	hsp70. gene	1	TCG : GTATGT	GTAATTTTCTGTGGTCTCAG : ATG	622	70	290
Petunia inflata	*S1* protein gene	1	AAG : GTAAAT	TCGTTTACTAACTTTTACAG : GAT	114	73	291
	S3 protein gene	1	AAA : GTACTT	TTTGCTAACATATACAGAAG : GAT	92	74	292
Phaseolus vulgaris	Endopeptidase	1	GTG : GTAAGC	TTGATTAATGTCTCGTGTAG : GTA	325	68	293
		2	AAG : GTGCAT	AAAAATCTTGCATGTGGTAG : GTG	90	70	293
		3	GAG : GTAAAT	TTGTATAAAAAAATTTGCAG : GGA	149	73	293
	Leghemoglobin	1	CTC : GTAAGT	AATATATATAAAAATAATAG : GAT	78	77	294
		2	TTG : GTAAGT	TGATATTTTTGAAATTGTAG : GTG	95	80	294
		3	CTG : GTAGCG	TGGGTATGTATGAATTGCAG : GTG	88	73	294
	loxA gene for lipoxygenase	1	ATG : GTTTGT	AATCTGTATACTTTTTTCAG : GGC	1285	78	295
		2	AAG : GTAAGT	CATGGTTAAATGTGTTGCAG : ACA	130	68	295
		3	AAG : GTTAGA	TTTTGTTTAACTTCTAATAG : ACC	196	73	295
		4	GAG : GTATTG	TTCCAAATGATTTGGAGCAG : TTA	130	72	295
		5	CAA : GTAAGC	TGCCGTTTAACGAATGGTAG : GAG	100	63	295
		6	GAG : GTGAGT	ATTATATTTTTGTGCAACAG : GCA	95	69	295
		7	TTG : GTATTA	TTATTTGTTCGTGATCTCAG : GTT	86	71	295
		8	GAG : GTAATG	ATTATTCACCCTTGATTCAG : AGG	94	72	295
	Beta-type phaseolin storage protein gene	1	GTG : GTAAGT	TGAGTTTTTTATAATTTCAG : GGA	72	71	296
		2	CAT : GTACTG	GATAAATGTTTGTCCTGTAG : GAA	88	69	296
		3	AAT : GTAAGA	TGTTTGCATGATTTTTATAG : AGC	124	71	296
		4	GAG : GTAAAT	TTTTTTGTTAACAAATTTAG : GGA	128	74	296
		5	CAG : GTATAT	TTGTGATTGTAAATATGAAG : GTA	103	77	296

	rbcS cDNA (containing intronic sequences) for small subunit of ribulose-1,5-biphosphate carboxylase/oxygenase	1	CAG : GTAAGA	GATGGAAATGAAATGAGTAG : GTG	98	69	297
		2	GAG : GTGAAT	ATGTTGATGTGTGTTGATAG : CAT	109	72	297
	Glutamine synthetase gamma subunit gene, 5′ end	1	ATG : GTTAGT	TGTATTTTCTCTATCTTTAG : GGT	205	60	298
	Nitrate reductase gene	1	AAG : GTATTT	CTCTGTTTCAAAATTTGTAG : ATT	122	69	299
		2	CCG : GTCAGT	TGATGTTGATGTTGATGCAG : GGG	280	65	299
		3	ATG : GTAACT	ATAAGGTTTCCAATGTGCAG : GGA	659	70	299
Pisum sativum	Alcohol dehydrogenase	1	GAG : GTAAAC	AACTTGTATTATACTTGTAG : CTG	88	69	300
		2	AAG : GTAAAT	CTTTTTTCCCCTTGTTTCAG : GGT	92	70	300
		3	AGG : GTATGC	GTTGGGATAATGTTGATCAG : GAT	151	74	300
		4	CTG : GTATAT	ATTTGCATTGATGCTCATAG : GTC	92	71	300
		5	GCT : GTATGT	CTCTTTAAACCATAATATAG : GCT	111	70	300
		6	TAG : GTACAA	TATTCTTTTCATGTTGGCAG : CTA	105	65	300
		7	CAG : GTTAGT	ATGTGAAAAGTGGTTTTCAG : GTA	120	77	300
		8	GAT : GTATGC	GACTTGTTATTATTTCTAAG : GGT	93	72	300
		9	GGG : GTAAGA	TACTGTGATACTTACTGCAG : GAG	112	67	300
	Albumin 1 gene	1	TAG : GTACGT	GTTGACTTTGCCAATTGCAG : GTA	82	68	301
	Cab II gene for chlorophyll *a*/*b*-binding protein	1	TTG : GTAAGT	TGATCTGTTGTGAAAACCAG : GTA	355	68	302
	Ribosomal protein (*CL22*) gene	1	AAG : GTACTC	AATTTAATGCTGCACTGCAG : TTC	304	67	303
	Convicilin (*cvcB*) gene	1	ATG : GTAATG	TTGGTTAATATGTATTACAG : GGA	151	84	304
	ELIP gene for plastic early-light-induced protein	1	GAG : GTAATT	TTACTTATTAAATTATTCAG : GGA	110	77	305
		2	AAG : GTGACC	AAATTATGATGAAATGGCAG : ATG	117	74	305
	legA gene for legumin	1	AAG : GTTACT	TTCTACTATACCAATTACAG : GTA	88	72	306
		2	AGG : GTGAGC	TATCTATCTATGTTTGACAG : AGA	88	68	306
	legJ gene for minor legumin	1	AGA : GTAAGT	ACGTAAATATGTGTATGCAG : GTA	138	74	307
		2	AGG : GTAACT	GTGTATGTATGTATATGCAG : AAT	98	78	307
	legK gene for minor legumin	1	AGA : GTAAGT	ATGTAAATATGTGTATGCAG : GTA	81	70	307
		2	AGG : GTGTGT	ATGTATATATGTATATGCAG : AAT	105	78	307

Table 2. Continued

Plant species	Gene	Intron	5′ splice site	3′ splice site	Size (bp)	AU (%)	Ref.
Pisum sativum (continued)	*rbcS-3A* gene for ribulose-1,5-bisphosphate carboxylase (RBC) small subunit	1	CAG : GTCAGA	CCCTTTTGTTGAATATTTAG : GTG	470	73	308
		2	GAG : GTTTCA	TTTTGACAACTTGGTTGCAG : AAA	87	71	308
	rbcS-3C gene for ribulose-1,5-bisphosphate carboxylase (RBC) small subunit	1	CAG : GTGACA	CGTTTATGTTGAATATTTAG : GTG	83	76	308
		2	GAG : GTTTCA	TTTTGACTGTTTGGTTGCAG : AAA	86	72	308
	rib-1,5-bisphosphate carboxylase (*rbcS*) small subunit	1	CAG : GTGACA	CGTTATTGTTGAATATTTAG : GTG	79	75	309
		2	GAG : GTTTCA	TTTTGACTGTTTGGTTGCAG : AAA	86	72	309
	mRNA for an 'unknown seed protein'	1	CTA : GTTAGT	TTTGGTATTTAATGTGGCAG : ATG	147	76	310
	gTUB1 gene for beta-tubulin 1	1	AGG : GTATGG	TTTTTTCTTCCTGTGTGCAG : GTT	620	69	311
		2	GCT : GTAAGC	GATTTTCTGTGTTGATGCAG : TTG	107	65	311
	PsNOD6 gene for nodulin	1	GGG : GTCAGC	TTTCTACTTTTTGATTACAG : CAA	103	80	312
Populus trichocarpachitinase	Gene *gwin6.2b*	1	ACG : GTAGGG	ATGCATGCTCCTACATAAAG : GAT	172	68	313
		2[b]	ATA : TTGCCA	TTTATTAACATGTGGTGCAG : GAA	278	65	313
Sesbania rostrata	*Srglb3* gene for leghemoglobin	1	CTT : GTAAGT	TTTGGTTTAATTAAAAATAG : TAT	105	76	314
		2	CTG : GTAAGT	TTATATTTATTGAATTATAG : GTG	334	81	314
		3	GTG : GTATGA	CATGGATCAATCTTTTGTAG : GTG	511	71	314
	Srg1b2 gene for leghemoglobin	1	CTT : GTAAGT	TTTGGTTTAATTAAAAATAG : TAT	105	76	315
		2	CTG : GTAAGT	TTATATTTATTGAATTATAG : GTG	307	82	315
		3	GTG : GTATGA	CATGGATCATTCTTTTGTAG : GTG	240	76	315
Solanum tuberosum	Granule-bound starch synthase	1	GTG : GTAACT	TTGTTTTTGTTCATCTGTAG : CTT	231	67	316
		2	GCA : GTAAGT	TCAGTCTTTTTATCATTTAG : GCC	84	71	316
		3	GAG : GTACAT	ATCCGCTACTTTATCTGCAG : GTC	84	63	316
		4	AAA : GTAAGT	GGTGCTTTTACTCTTTTAAG : GTT	101	65	316

Gene	Intron	5′ splice site	3′ splice site	Length	AU%	Ref.
	5	CAA : GTAAGT	TTTTAACCTTGTTTTCTCAG : GCA	83	71	316
	6	ATG : GTAATT	TATTTTGCCTATTCCTGCAG : GAG	80	68	316
	7	AAG : GTAAAA	AAATTTCACATTGCCTCCAG : GTC	90	68	316
	8	TGG : GTATGT	GAGTTTTAAAAATTTTGCAG : ATA	83	70	316
	9	ACT : GTAAGA	TAATCAAAATCTCTATACAG : GTC	93	72	316
	10	CTT : GTAAGT	CTACTCATCCTATGCATCAG : GGA	89	58[c]	316
	11	ACA : GTAAGA	CATAATCACATGTGAGTCAG : GTG	163	69	316
	12	GAA : GTATGT	TGCTTCTTGAATTTGGTTAG : TGC	104	66	316
	13	AAG : GTAGGT	GCCAACTGGTGTTCATGCAG : GAA	85	56	316
Inhibitor II-chloramphenicol acetyltransferase gene (IIK)	1	TTG : GTAAGA	TATATTATATTTGTTTGTAG : GAT	117	84	317
Wound-inducible proteinase inhibitor I	1	CTT : GTTAGT	AATTGTTTTTTTTTTTACAG : CCT	479	63	318
	2	ATG : GTAAAG	TTATGTGTATATATATGTAG : GAA	415	75	318
Light-inducible tissue-specific *ST-LS1*	1	ATG : GTATGT	GAAGTGGGCTTTGTTACTAG : GAA	838	68	319
	2	AAG : GTTTGT	CAAAACATGGTGATGTTTAG : GGA	189	80	319
	3	CAG : GTTTGT	GTGTGATTTTGAAAATGCAG : TCC	96	77	319
	4	GAG : GTATGA	CACATGTAATATAATTGCAG : GTA	132	69	319
PAL-1 gene for phenylalanine ammonia lyase	1[b]	AAA : CTACTT	TGAATTATGATATTTTGCAG : GTT	111	72	320
Patatin class I gene	1	CAG : GTATCG	TACTTTTCTTTTCGAGTCAG : GAA	376	73	321
	2	TAG : GTACAT	GGTTACATTATATTATGCAG : TGG	727	73	321
	3	AAT : GTAAGT	TTTAAAATGCATGCATGCAG : TTA	138	80	321
	4	CCG : GTACTA	GTTGTACGTACGACGTGCAG : GCG	82	67	321
	5	CAA : GTAAGT	TTATGTATATTTAATTCCAG : GAA	346	75	321
	6	GAG : GTAAAA	AATAGTTTATTTCGTTGTAG : GTT	119	76	321
Patatin	1	CAG : GTATCG	TATTTTTCTTTTCGAGTCAG : GAA	389	75	322
	2	TAG : GTACAT	GGTTACATTATATTATGCAG : TGG	337	75	322
	3	AAT : GTAAGT	TATTCTTTTTTAAAATGCAG : TTA	454	70	322
	4	CCG : GTACGT	GTGTTGTACGTGCAATGCAG : GCA	288	63	322
	5	CAA : GTAAGT	ACGAGTATATTTAATTCCAG : GAA	794	67	322
	6	GAG : GTAAAA	AATAGTTTATTTCATTGTAG : GTT	816	73	322

Plant Introns

Table 2. Continued

Plant species	Gene	Intron	5′ splice site	3′ splice site	Size (bp)	AU (%)	Ref.
Solanum tuberosum (continued)	(Tetraploid variety Maris piper) patatin	1	CAG : GTATCG	TACTTTTCTTTTCGAGTCAG : GAA	376	75	323
		2	TAG : GTACAT	GGTTACATTATCTTATGCAG : TGG	725	74	323
		3	AAT : GTAAGT	TATTTTTAAAATGCATGCAG : TTA	147	80	323
		4	CCG : GTACTA	GTTGTACGTACGACGTGCAG : GCG	85	71	323
		5	CAA : GTAAGT	TTATGTATATTTAATTCCAG : GAA	328	74	323
		6	GAG : GTAAAA	AAAAGTTTATTTCGTTGTAG : GTT	118	78	323
	Proteinase inhibitor II	1	TTG : GTAAGA	TATATTATATTTGTTTGTAG : GAT	117	83	324
	PoAc100 gene for actin (exons 2, 3 & 4)	1	CAG : GTGCGA	GCTTTTGGCCATCTCAACAG : GTA	291	65	325
		2	CAG : GTAATA	TTTCTTTGAACGACTGGCAG : ATG	133	62	325
	PoAc101 gene for actin	1	AAG : GTTAGT	TTTGACCTATTTTCATGCAG : GCT	255	67	325
		2	CAG : GTGCAA	GCTTTTGGCCATTTCAACAG : GTA	296	65	325
		3	CAG : GTAATA	TTTCTTTGAACGACTGGCAG : ATG	133	62	325
	PoAc58 gene for actin	1	AAG : GTAAGC	GATGATGTTTCAATATGTAG : GCT	207	62	325
		2	CCG : GTAAGA	ATTCATTAAGGTTAATACAG : GTA	127	55[c]	325
		3	CAG : GTTTGC	GTTTTCTATTTTGTCTGCAG : ATG	307	64	325
	PoAc71 gene for actin	1	AAG : GTAACA	ATATTTTTTCTTTCTTGCAG : GCA	87	72	325
		2	CTG : GTAAGA	TGATTCATGTTTTCTTGCAG : GTA	172	69	325
		3	CAG : GTGAGC	ATCATTGTTTGAATTTGTAG : ATG	456	63	325
	PoAc75 gene for actin	1	AAG : GTAATG	AGTATATGTTTCATACATAG : GCT	89	72	325
		2[c]		GATTCTTGTCTGCCTTGTAG : GTA	—	—	325
		3	CAG : GTAAGC	TTGTGTTCTTTGAAATGTAG : ATG	190	63	325
	PoAc85c gene for actin (exons 2 & 3)	1	CCG : GTAAGC	ATTCATTAAGGTTTATACAG : GTA	137	58[c]	325
	PoAc97 gene for actin	1	AAG : GTAAGC	TATATTGGTTTATTATCTAG : GCT	120	66	325
		2	CCG : GTAAGA	ATTCTTTTTCTTCTCCACAG : GTA	201	71	325
		3	CAG : GTCTCG	GCTTCTCTTATTTTCTGTAG : ATG	126	56[c]	325

	Pathogenesis-related protein (*prp1*) gene	1	AAG : GTATAT	AAGTTTTGTTTGTTGGGCAG : GGG	407	74	326
	4-Coumarate:CoA ligase (*St4Cl-1*) gene	1	CAG : GTAAAT	AAATTATTAAATTAAAATAG : GGT	501	73	327
		2	AAG : GTATGC	CAATGATTATTTCTTTGTAG : GTT	649	73	327
		3	CCC : GTAAGT	TATTAAATTGTGTTGGACAG : AAT	110	72	327
		4	CAG : GTTTGT	TGTTTTTCAAAATTTTGCAG : GTG	93	74	327
	Pathogenesis-related protein (*STH-2*) gene	1	AAG : GTAATA	ATTTTTGGGAATAATTGCAG : GTT	359	71	328
	Wound-induced genes *WIN1* and *WIN2*	1	CAA : GTGCTT	ACTTTTATATATATGTGCAG : GTG	82	74	329
		1	AGG : GTAAGA	ATCGCAAATTATGTATACAG : GTG	688	73	329
Spinacia oleracea	*NIR* gene for nitrite reductase	1	ACT : GTAAGC	AAAATGGTGGATGTTATTAG : ATG	529	70	330
		2	TCT : GTAAGT	TGTTTTGAACATTGGTTCAG : GCC	93	63	330
		3	CTT : GTGAGT	CATTCTCCAAACAATGGCAG : GGT	93	67	330
	AtpC gene for chloroplast ATP synthase gamma-subunit	1	CAG : GTACGG	GATTATATGTCTTCTTACAG : GTA	1066	62	331
		2	AAG : GTAATA	CTTTCTCATAATTCTATCAG : GTA	665	69	331
	atpD gene for ATP synthase delta-subunit	1	CCG : GTATAA	CGTTTTTGTTGATTATGCAG : GTG	554	70	332
Theobroma cacoa	*csv* gene for seed vicilin	1	ACG : GTAAAT	GTAAATTATTTCTGTGTCAG : GAA	90	68	333
		2	GAG : GTTTTC	TGTACTTATAATTGCTGCAG : TTA	90	69	333
		3	AAT : GTATAT	ATTTTTATCTTGTTTGTTAG : ACA	102	76	333
		4	CAG : GTACTT	GCTCATGAATCCTTTTGCAG : GGA	112	73	333
Trema tomentosa	Hemoglobin gene	1	CAA : GTAAGT	GTTGATTTGAAACTTTGCAG : GAT	127	75	334
		2	ATG : GTAAAG	CTCTGGTACTGCCTGGAAAG : ACG	340	65	334
		3	GAG : GTACTA	ATGGGAAAAACGGTTTGTAG : GTC	159	70	334
Vicia faba	Legumin B gene *LeB4*	1	AGA : GTAAGT	TGTAAATATGTGTTTTTCAG : GTA	95	74	335
		2	AGG : GTACGT	ATGCATGTATGTATATGCAG : AAT	100	78	335
	USP gene for an unknown seed protein	1	TGC : GTAAGC	GTCATGAATTTATTTTGCAG : CTG	81	73	336
		2	CTA : GTTAGT	TTGTAAAATTAACATGGCAG : GTG	110	76	336

Table 2. Continued

Plant species	Gene	Intron	5′ splice site	3′ splice site	Size (bp)	AU (%)	Ref.
Vicia faba (continued)	Unspliced mRNA of USP Vf30.1 gene for unknown seed protein	1	TGC : GTAAGC	GTCATGAATTTATTTTGCAG : CTG	81	73	337
	Vicilin gene	1	GTG : GTAATT	ATGTCATTTGTAATTTGCAG : GAA	145	81	338
		2	CAG : GTAATA	ATTTTTTTTCCTTTCAACAG : TCT	171	76	338
		3	AAT : GTAAGC	ATATATTGTAAATATTGTAG : ACC	100	77	338
		4	GAG : GTATAT	TCTCTAAATTTTAATCTTAG : GGA	255	81	338
		5	CAG : GTATAA	ATGTGATTGAAAATTTAAAG : GTG	109	81	338
Vigna mungo	Gene for sulfhydryl-endopeptidase	1	GTG : GTAAGC	GTTTTCTTTGTCTTGTGTAG : GTA	291	67	339
		2	AAG : GTGCAT	AAAAATCTTTCATGTGGTAG : GTG	93	68	339
		3	GAG : GTAAAT	TAATATAAAAAATTTTGCAG : GGA	150	75	339
Vigna radiata	ACC synthase fragment gene	1	CAG : GTATAA	CTTATCTTAAATTTTTGCAG : CTT	92	73	340
	ACC synthase fragment gene	1	CCG : GTACAC	ATTTCTGATAACTTTTGCAG : CTG	105	71	341
	ACC synthase fragment gene	1	CAA : GTATGT	GCTCGTTTTCTCTTTTCCAG : CTA	110	66	342

[a] 5′ and 3′splice site conserv ed, GT and AG absent.
[b] 5′ splice site conserved, GT absent.
[c] No 5′ splice given.
[d] Short intron, below predicted minimum functional length (14).
[e] Below a minimum expected AU content (13, 15).
[f] 3′ splice site conserved, AG absent.
[g] No 3′ splice site given.

were determined by running the COMPOSITION program in the UWGCG software (17) and were tabulated (*Tables 1* and *2*).

A comparison of 442 monocot and 892 dicot introns from 110 and 277 genes, respectively, was used to compile consensus sequences for the 5′ splice site and 3′ splice sites in both monocot and dicot plants (*Table 3*). This analysis produced very similar 5′ splice site (AAG:GTAAGT) and 3′ splice site (T_{16}GCAG:GT) consensus sequences for both plant classes and to the vertebrate splice site consensus (2). The GT dinucleotide at the 5′ end of the intron and the AG at the 3′ end of the intron were conserved in virtually all cases. There were, however, a few exceptions to 5′ GT usage, with approximately 1% of plant introns containing a GC dinucleotide (*Tables 4* and *5*). A difference between the plant and animal 3′ consensus exists in the low abundance of C nucleotides upstream of the 3′ splice site in plants. The less predominant polypyrimidine tract in plant introns compared to animal introns has been noted previously (10, 13). In plant introns, while Ts predominate in the nucleotide positions upstream of the 3′ splice site (*Tables 3, 4* and *5*, positions 10–25), they only occur at the 40–50% level. In monocot introns the remaining three nucleotides in these positions were present in approximately equal amounts (*Table 4*), while in dicots the A content was greater than the remaining G and C nucleotides (*Table 5*). This preference for A or T in dicots is reflected in the AU content of dicot introns as a whole.

Dicot introns show an average AU content of 70%, with a distribution between 42 and 89%, compared to monocots which show an average AU content of 60% over a distribution of 31% to 80%. The significance of the higher AU content in dicot introns is enhanced when comparing the number of monocot and dicot introns below 59%, which has been recognized as a minimal AU content for efficient splicing in dicot tissue (9, 15). Dicot introns show 14 out of 882 (1.6%) introns present below 59% while 168 out of 440 (38.2%) monocot introns were below this AU level. In many dicot introns the AU content shows some bias towards U richness. The numbers of T and A bases were calculated for all the introns isolated from *Lycopersicon esculentum* in the database and were found to be 39% T and 31% A. This shows a limited bias towards U nucleotides in the introns analyzed, but the significance of such a preference remains to be investigated.

Apart from AU and splice site sequence content, plant introns generally vary in length between 70 nt and over a 1000 nt long, with a minimal intron size between 70 and 73 nt (14). The largest plant introns in this study in both monocot and dicot introns were from autonomous element genes (4434 nt for intron 1 of *Zea mays* autonomous transposable element *En-1* mosaic protein gene; and 4505 nt for intron 4 of *Antirrhinum majus* TAM1 gene), which highlights the ability of plants to splice reasonably large introns. Only four monocot and three dicot introns of less than 70 nt in length were found; the ability of these introns to splice accurately and efficiently would need to be tested directly. However, the majority of plant introns (56%) were found within a size range of 80–139 nt (*Figure 1*), above the minimal intron length, suggesting a preference for introns of this size.

This survey provides comprehensive information on plant introns and the major splicing signals. The analysis confirms some of the requirements for splicing found under experimental conditions, such as the AU content, and identifies natural plant introns which do not fall within the limits of these requirements, such as a GC 5′ splice site. Introns and their regulation are of vital importance to accurate and efficient expression of genes; this chapter provides easy access to this important information.

Table 3. Plant splice site consensus

Monocot consensus

Position																															
1	2	3	4	5	6	7	8	9	10	11	12	13	14	15	16	17	18	19	20	21	22	23	24	25	26	27	28	29	30	31	32
A_{40}	A_{65}	G_{78}:	G	T_{98}	A_{64}	A_{44}	G_{59}	T_{53}	T_{37}	T_{36}	T_{34}	T_{40}	T_{37}	T_{44}	T_{42}	T_{42}	T_{48}	T_{43}	T_{37}	T_{45}	T_{46}	T_{41}	T_{41}	T_{68}	G_{51}	C_{80}	A	G :	G_{60}	T_{40}	N

Dicot consensus

Position																															
1	2	3	4	5	6	7	8	9	10	11	12	13	14	15	16	17	18	19	20	21	22	23	24	25	26	27	28	29	30	31	32
A_{36}	A_{62}	G_{79}:	G	T_{99}	A_{69}	A_{58}	G_{50}	T_{53}	T_{43}	T_{50}	T_{48}	T_{51}	T_{49}	T_{52}	T_{51}	T_{54}	T_{50}	T_{51}	T_{44}	T_{48}	T_{48}	T_{49}	T_{48}	T_{64}	G_{42}	C_{63}	A	G :	G_{57}	T_{44}	A_{32}
C_{31}									A_{25}	A_{24}	A_{22}	A_{22}	A_{22}	A_{21}	A_{21}	A_{20}	A_{24}	A_{23}	A_{23}	A_{24}	A_{26}	A_{21}	A_{23}			T_{32}			A_{21}		

Table 4. Percentage occurrence of single nucleotides and pairs of nucleotides at each monocot splice site position

Position	A	C	G	T	N	W	S	R	Y	M	K	No. of sequences
1	40	34	19	7	0	47	53	59	41	74	26	441
2	65	14	7	14	0	79	21	72	28	79	21	441
3	7	8	78	7	0	14	86	85	15	15	85	441
4	0	0	100	0	0	0	100	100	0	0	100	441
5	0	1	0	98	0	99	1	0	100	2	98	441
6	64	8	19	9	0	73	27	83	17	73	27	441
7	44	26	7	22	0	66	34	51	49	70	30	441
8	17	10	59	13	0	31	69	77	23	27	73	441
9	17	20	10	53	0	70	30	27	73	37	63	441
10	22	23	18	37	0	59	41	40	60	45	55	440
11	24	19	20	36	0	60	39	45	55	43	57	440
12	24	22	20	34	0	58	42	44	56	46	54	440
13	21	20	18	40	0	61	38	39	60	42	58	440
14	20	23	20	37	0	57	42	40	60	43	57	440
15	20	20	16	44	0	63	37	36	64	40	60	440
16	20	19	20	42	0	62	38	40	60	39	61	440
17	18	20	19	42	0	60	40	38	62	39	61	440
18	16	19	16	48	0	65	35	32	68	35	65	440
19	19	20	18	43	0	62	38	36	64	39	61	440
20	22	21	20	37	0	59	41	42	58	43	57	440
21	17	22	17	45	0	62	38	34	66	39	61	440
22	17	16	20	46	0	64	36	37	63	34	66	440
23	15	22	21	41	0	56	44	37	63	38	62	440
24	19	18	22	41	0	60	40	41	59	37	63	440
25	10	11	11	68	0	78	22	21	79	21	79	440
26	15	13	51	21	0	36	64	67	33	28	72	440
27	3	80	1	15	0	18	82	4	96	84	16	440
28	100	0	0	0	0	100	0	100	0	100	0	440
29	0	0	100	0	0	0	100	100	0	0	100	440
30	15	13	60	11	0	27	73	75	25	29	71	440
31	17	20	23	40	0	57	43	40	60	38	62	440
32	20	24	25	31	0	51	49	45	55	44	56	440

Number of intron sequences analyzed was 442.
Ambiguous nucleotide code: A, adenine; C, cytosine; G, guanosine; T, thymine; N, any nucleotide; W, adenine or thymine; S, cytosine or guanosine; R, adenine or guanosine (purine); Y, cytosine or thymine (pyrimidine); M, adenine or cytosine; K, guanosine or thymine.

Table 5. Percentage occurrence of single nucleotides and pairs of nucleotides at each dicot splice site position

Position	A	C	G	T	N	W	S	R	Y	M	K	No. of sequences
1	36	31	17	15	0	51	49	54	46	68	32	884
2	62	10	8	20	0	82	18	70	30	72	28	884
3	9	2	79	9	0	18	82	88	12	12	88	884
4	0	0	100	0	0	0	100	100	0	0	100	884
5	0	1	0	99	0	99	1	0	100	1	99	884
6	69	4	11	15	0	84	16	81	19	74	26	884
7	58	14	3	25	0	84	16	61	39	72	28	884
8	23	9	50	18	0	41	59	73	27	32	68	884
9	22	16	10	53	0	75	25	31	69	38	62	884
10	25	14	18	43	0	68	32	43	57	39	61	887
11	24	11	15	50	0	73	27	39	61	35	65	887
12	22	14	16	48	0	70	30	38	62	36	64	887
13	22	13	14	51	0	72	28	36	64	35	65	887
14	22	11	18	49	0	71	29	40	60	33	67	887
15	21	13	14	52	0	73	27	35	65	34	66	887
16	21	11	17	51	0	71	29	38	62	32	68	887
17	20	11	15	54	0	74	26	35	65	31	69	887
18	24	11	15	50	0	74	26	39	61	36	64	887
19	23	13	14	51	0	73	27	37	63	35	65	887
20	23	14	19	44	0	67	33	43	57	37	63	887
21	24	11	17	48	0	72	28	41	59	35	65	887
22	26	11	14	48	0	75	25	40	60	38	62	887
23	21	12	18	49	0	70	30	39	61	33	67	887
24	23	12	17	48	0	71	29	40	60	35	65	887
25	16	7	13	64	0	80	20	29	71	23	77	887
26	30	7	42	21	0	51	49	72	28	37	63	887
27	4	63	1	32	0	36	64	5	95	67	33	887
28	100	0	0	0	0	100	0	100	0	100	0	887
29	0	0	100	0	0	0	100	100	0	0	100	887
30	21	11	57	11	0	32	68	78	22	32	68	887
31	22	14	20	44	0	66	34	42	58	36	64	887
32	32	14	26	28	0	60	40	58	42	46	54	887

Number of intron sequences analyzed was 892.
Ambiguous nucleotide code: A, adenine; C, cytosine, G, guanosine; T, thymine; N, any nucleotide; W, adenine or thymine; S, cytosine or guanosine; R, adenine or guanosine (purine); Y, cytosine or thymine (pyrimidine); M, adenine or cytosine; K, guanosine or thymine.

Figure 1. Size distribution of plant introns. The number of monocot intron sequences analyzed was 441 and 882 dicot intron sequences were analyzed.

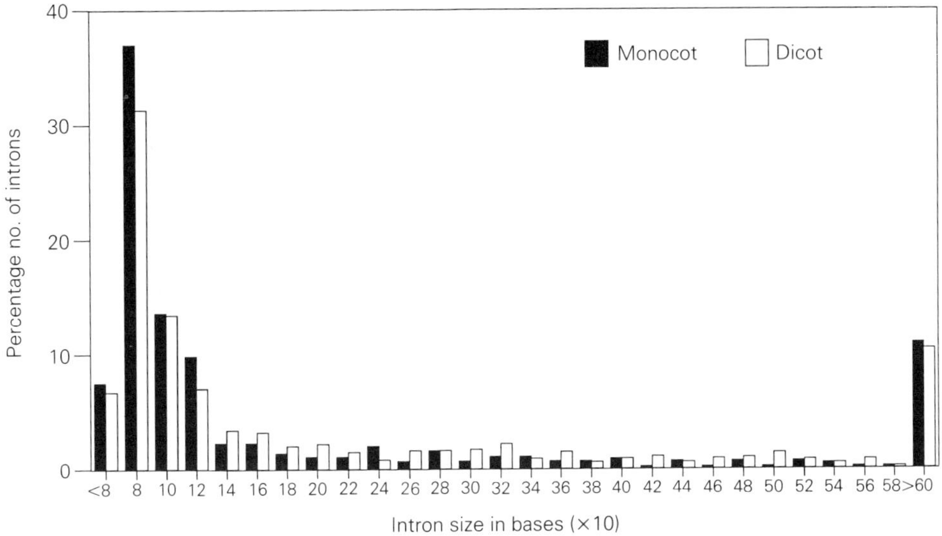

ACKNOWLEDGEMENTS

We thank Susan Simpson for writing the computer programs used to analyze the tabulated data and Frank Wright for guidance in setting up the plant intron search. This work was supported by the AFRC and SOAFD.

2. MOBILE GENETIC ELEMENTS IN HIGHER PLANTS

T. Franklin

Mobile genetic elements are segments of DNA which appear to be able to move to new positions within the genome. They have been detected genetically through the abnormalities they produce in the activities and structures of the genes near or at the sites into which they insert. Transposable elements have been well characterized in bacteria, fungi, viruses, insects nematodes and higher plants.

In *Tables 6–10* the transposable elements identified in higher plants have been grouped on the basis of similarity in structure. The first group, in *Table 6*, share homology within their terminal inverted repeats. The second group, in *Table 7*, also share homology within their terminal repeats. In *Table 8* all the elements have a retrotransposon-like structure, while the group in *Table 9* does not fall into any of the above groups. Finally, *Table 10* lists those transposable elements which have been characterized but not cloned and studied at the molecular level.

The first column of each table gives the abbreviated name of each transposable element, followed by the plant species, and the Latin genera is given for each. The transposition activity is given as either (i) 'autonomous' — fully functional and able to mobilize non-functional related elements, (ii) 'non-autonomous' — only fully functional when a related

▶ p. 242

Table 6. Transposable elements showing homology within their terminal inverted repeats

Name	Plant species	Transposition activity	Size (bp)	Terminal repeat size (bp)	Target site size (bp)	Refs
Ac	*Zea mays*	Autonomous	4563	11 IR	8	360, 365
Ds	*Zea mays*	Non-autonomous	400–4563	11 IR, 2 DR, 3 IR	8	347, 358, 362
rUq	*Zea mays*	Non-autonomous	~400	11 IR, 2 DR, 3 IR	8	348
Tam3	*Antirrhinum majus*	Autonomous	3000	12 IR	8	376
Tat1	*Arabidopsis thaliana*	Non-autonomous	431	13 IR	5	363
Tpc1	*Petroselinum crispum*	Non-autonomous	927	11/12 IR	8	354
Ips-r	*Pisum sativum*	Non-autonomous	800	12 IR	8	344

Table 7. Transposable elements showing homology within their terminal repeats

Name	Plant species	Transposition activity	Size (bp)	Terminal repeat size (bp)	Target site size (bp)	Refs
Spm/En	*Zea mays*	Autonomous	8400	13 IR	3	357, 364
dSpm/I	*Zea mays*	Non-autonomous	Variable	13 IR	3	372
Tam1	*Antirrhinum majus*	Autonomous	14900	13 IR	3	345
Tam2	*Antirrhinum majus*	Non-autonomous	5187	13 IR	3	377
Tgm 1–7	*Glycine max*	Autonomous and non-autonomous	1600–12000	13 IR	3	369
Pis1	*Pisum sativum*	Non-autonomous	2500	12 IR	8	375

Table 8. Transposable elements with retrotransposon-like structure

Name	Plant species	Transposition activity	Size (bp)	Terminal repeat size (bp)	Target site size (bp)	Refs
Tnt1	*Nicotiana tabacum*	Retrotransposon	5334	610 DR	—	252
Wis-2	*Triticum aestivum*	Retrotransposon	8000	> 500 DR	5	253
Bs1	*Zea mays*	Retrotransposon	3203	304 DR	5	355, 356
Cin 4–7	*Zea mays*	Retrotransposon	1100	No structure	7	373, 374

Table 9. Identified transposable elements not given in *Tables 6–8*

Name	Plant species	Transposition activity	Size (bp)	Terminal repeat size (bp)	Target site size (bp)	Refs
rCy (Mu7)	*Zea mays*	Non-autonomous	2200	200 IR	9	371
rDt	*Zea mays*	Non-autonomous	704	14 IR	8	346
Mu1	*Zea mays*	Non-autonomous	1367	215/213 IR	9	343
Mu3	*Zea mays*	Non-autonomous	1824	200 IR	9	349, 361

Table 10. Transposable elements not yet studied

Name	Plant species	Transposition activity	Refs
Bg	*Zea mays*	Autonomous	359
Cy/Mu	*Zea mays*	Autonomous	370
Dt	*Zea mays*	Autonomous	350, 366
Fcu	*Zea mays*	Autonomous	351
Mrh	*Zea mays*	Autonomous	367
Mut	*Zea mays*	Autonomous	368

autonomous element is present and, finally, (iii) 'retrotransposon' — where the mode of transposition is thought to occur through a reverse transcribed RNA intermediate. The size of each element is given in base pairs, although where a range of sizes occurs either this range is stated or it is described as variable. The sizes of terminal repeats are given in base pairs and described as either inverted (IR) or direct (DR). The size of the characteristic target site duplication produced by the insertion or transportion in the 'host' DNA is also given in base pairs. Finally, the references are numbered and given at the ends of the tables. This list is not exhaustive and new transposable elements are being discovered all the time (see also *Chapter 3, Table 21*).

3. REFERENCES

U, Unpublished data is given along with the [database:identification code].

1. Green, M.R. (1986) *Ann. Rev. Genet.*, **20**, 671.

2. Padgett, R.A., Grabowski, P.J., Konarska, M.M., Seiler, S. and Sharp, P.A. (1986) *Ann. Rev. Biochem.*, **55**, 1119.

3. Sharp, P.A. (1987) *Science*, **235**, 766.

4. Krainer, A.R. and Maniatis, T. (1988) in *Transcription and Splicing* (B.D. Hames and D.M. Glover, eds). IRL Press, Oxford, p. 131.

5. Steitz, J.A., Black, D.L., Gerke, V., Parker, K.A., Krämer, A., Frendeway, D. and Keller, W. (1988) in *Structure and Function of Major and Minor Small Nuclear Ribonucleoprotein Particles* (M.L. Birnstiel, ed.). Springer Verlag, Berlin, p. 115.

6. Guthrie, C. and Patterson, B. (1988) *Ann. Rev. Genet.*, **22**, 387.

7. Lührmann, R., Kastner, B. and Bach, M. (1990) *Biochim. Biophys. Acta*, **1087**, 265.

8. Ruby, S.W. and Abelson, J. (1991) *Trends in Genetics*, **7**, 79.

9. Goodall, G.J., Kiss, T. and Filipowicz, W. (1991) *Oxford Surv. Plant Mol. Cell Biol.*, **7**, 255.

10. Brown, J.W.S. (1986) *Nucleic Acids Res.*, **14**, 9549.

11. Hanley, B.A. and Schuler, M.A. (1988) *Nucleic Acids Res.*, **16**, 7159.

12. Sinibaldi, R.M. and Mettler, I.J. (1992) *Prog. Nucleic Acid Res. Mol. Biol.*, **42**, 229.

13. Goodall, G.J. and Filipowicz, W. (1989) *Cell*, **58**, 473.

14. Goodall, G.J. and Filipowicz, W. (1990) *Plant Mol. Biol.*, **14**, 727.

15. Goodall, G.J. and Filipowicz, W. (1991) *EMBO J.*, **10**, 2635.

16. Keith, B. and Chua, N.-H. (1986) *EMBO J.*, **5**, 2419.

17. Devereux, J., Haeberli, P. and Smithies, O. (1984) *Nucleic Acids Res.*, **12**, 387.

18. Hershey, H.P., Barker, R.F., Idler, K.B., Murray, M.G. and Quail, P.H. (1987) *Gene*, **61**, 339.

19. Shotwell, M.A., Boyer, S.K., Chesnut, R.S. and Larkins, B.A. (1990) *J. Biol. Chem.*, **265**, 9652.

20. Hansen, L. U [pirembl:hvacl1].

21. Trick, M., Dennis, E.S., Edwards, K.J.R. and Peacock, W.J. (1988) *Plant Mol. Biol.*, **11**, 147.

22. Knox, C.A.P., Sonthayanon, B., Chandra, G.R. and Muthukrishnan, S. (1987) *Plant Mol. Biol.*, **9**, 3.

23. Khurseed, B. and Rogers, J. *J. Biol. Chem.* [pirembl:hvamy46].

24. Rogers, J.C. (1985) *J. Biol. Chem.*, **260**, 3731.

25. Whittier, R.F., Dean, D.A. and Rogers, J.C. U [pirembl:hvamygo1].

26. Slakeski, N., Baulcombe, D.C., Devos, K.M., Ahluwalia, B., Doan, D.N.P. and Fincher, G.B. (1990) *Mol. Gen. Genet.*, **224**, 437.

27. Rohde, W.E., Dorr, S., Salamini, F. and Becker, D. (1991) *Plant Mol. Biol.*, **16**, 1103.

28. Schnorr, K.M., Juricek, M., Culley, D. and Kleinhofs, A. (1991) *Mol. Gen. Genet.*, **227**, 411.

29. Whittier, R.F., Dean, D.A. and Rogers, J.C. *Nucl. Acids Res.* [pirembl:hvleu].

30. Linnestad, C., Loenneborg, A., Kalla, R. and Olsen, O.A. (1991) *Plant Physiol.*, **97**, 841.

31. Skriver, K., Leah, R., Mueller-Uri, F. and Mundy, J. U [pirembl:hvltp1g].

32. Miyazaki, J., Juricek, L., Angelis, Schnorr, K.M., Kleinhofs, A. and Warner, R.L. (1991) *Mol. Gen. Genet.*, **228**, 329.

33. Brandt, A., Hejgaard, J. and Svendsen, I. (1990) *Eur. J. Biochem.*, **194**, 499.

34. Theilade, B. and Rasmussen, S.K. U [pirembl:hvprx6a].

35. Rundle, S.J. and Zielinski, R.E. (1991) *J. Biol. Chem.*, **266**, 4677.

36. Heck, G.R., Dorsett, C. and Ho, T.H. U [pirembl:hvsip1a].

37. Rhode, W., Becker, D. and Salamini, F. (1988) *Nucleic Acids Res.*, **16**, 7185.

38. Halford, N.G., Vincente-Carbajosa, J., Sabelli, P.A., Shewry, P.R., Hannappel, U. and Kreis, M. *Plant Journal* [daily:hvbkin12g].

39. Sorensen, A.B., Jensen, B.F. and Gausing, K. (1992) *Plant Physiol.*, **98**, 1538.

40. Karlin-Neumann, G.A., Kohorn, B.D., Thorner, J.P. and Tobin, E.M. (1985) *J. Mol. Appl. Genet.*, **3**, 45.

41. Okubara, P.A., Williams, S.A., Doxsee, R.A. and Tobin, E.M. U [daily:lgnpr1].

42. Yu, W.P., Wang, A.Y., Juang, R.H., Sung, H.Y. and Su, J.C. (1992) *Plant Mol. Biol.*, **18**, 139.

43. Xie, Y. and Wu, R. (1990) *Gene*, **87**, 185.

44. Anderson, J.M., Larsen, R., Laudencia, D., Kim, W.T., Morrow, D., Okita, T.W. and Preiss, J. (1991) *Gene*, **97**, 199.

45. Huang, N., Koizumi, N., Reinl, S. and Rodriguez, R.L. (1990) *Nucleic Acids Res.*, **18**, 7007.

46. Kemmerer, E.C., Lei, M. and Wu, R. (1991) *Mol. Biol. Evol.*, **8**, 212.

47. Litts, J.C., Erdman, M.B., Chaudrey, S.R., Huang, N., Karrer, E.E., Noueiry, A., Quatrano, R.S. and Rodrigues, R.L. U [pirembl:osemp1g].

48. Takaiwa, F. and Oono, K. (1991) *Jpn. J. Genet.*, **66**, 161.

49. Takaiwa, F., Oono, K., Wing, D. and Kato, A. (1991) *Plant Mol. Biol.*, **17**, 875.

50. Takaiwa, F., Ebinuma, H., Kikuchi, S. and Oono, K. (1987) *FEBS Lett.*, **221**, 43.

51. De Pater, B.S. and Schilperoort, R.A. U [pirembl:osgos2g].

52. De Pater, B.S., Hensgens, L.A.M. and Schilperoort, R.A. (1990) *Plant Mol. Biol.*, **15**, 399.

53. De Pater, B.S. and Schilperoort, R.A. (1992) *Plant Mol. Biol.*, **18**, 161.

54. Kondo, H., Emori, Y., Abe, K., Suzuki, K. and Arai, S. (1989) *Gene*, **81**, 259.

55. Suzuka, I., Hata, S., Matsuoka, M., Kosugi, S. and Hashimoto, J. (1991) *Eur. J. Biochem.*, **195**, 571.

56. Kay, K.S.A., Keith, B., Shinozaki, K., Chye, M.L. and Chua, N.H. (1989) *Nucleic Acids Res.*, **17**, 2865.

57. Sutliff, T.D., Huang, N., Litts, J.C. and Rodriguez, R. (1991) *Plant Mol. Biol.*, **16**, 579.

58. Xie, Y. and Wu, R. (1988) *Nucleic Acids Res.*, **16**, 7749.

59. Shimada, H. and Tada, Y. (1991) *Gene*, **98**, 243.

60. Kim, J.K. and Wu, R. (1992) *Plant Mol. Biol.*, **18**, 399.

61. Washio, K. and Ishikawa, K. U [daily:oscbp34].

62. Goldman, S., Mawal, Y. and Wu, R. U [daily:ososamybg].

63. Reece, K.S., McElroy, D. and Wu, R. (1990) *Plant Mol. Biol.*, **14**, 621.

64. Huang, N., Reinl, S.J. and Rodriguez, R.L. U [daily:osamy2a].

65. Gaut, B.S. and Clegg, M.T. (1991) *Proc. Natl. Acad. Sci. USA*, **88**, 2060.

66. Gaut, B.S. and Clegg, M.T. (1991) *Proc. Natl. Acad. Sci. USA*, **88**, 2060.

67. Henrik, A.H., Martin, T. and Sun, S.S. U [daily:shscpepcd].

68. Tang, W. and Sun, S. U [daily: sobcs1a].

69. Luchetta, P., Cretin, C. and Gadal, P. *Gene* [pirembl:svmdh].

70. Lepiniec, L., Santi, S., Keryer, E., Amiet, V., Vidal, J., Gadal, P. and Cretin, C. (1991) *Plant Mol. Biol.*, **17**, 1077.

71. Luchetta, P., Cretin, C. and Gadal, P. U [daily:svnadpmg].

72. Lepiniec, L. U [daily:svpepcgx].

73. Huttly, A.K., Martienssen, R.A. and Baulcombe, D.C. (1988) *Mol. Gen. Genet.*, **214**, 232.

74. Baulcombe, D.C., Huttly, A.K., Martienssen, R.A., Barker, R.F. and Jarvis, M.G. (1987) *Mol. Gen. Genet.*, **209**, 33.

75. Lloyd, J.C., Raines, C.A., John, U.P. and Dyer, T.A. (1991) *Mol. Gen. Genet.*, **225**, 209.

76. Litts, J.C., Colwell, G.W., Chakerian, R.L. and Quatrano, R.S. U [pirembl:tagbl1a].

77. Baulcombe, D.C., Barker, R.F. and Jarvis, M.G. (1987) *J. Biol. Chem.*, **262**, 13726.

78. Dudler, R., Hertig, C., Rebmann, G., Bull, J. and Mauch, F. (1991) *Mol. Plant Microb. Interact.*, **4**, 14.

79. Mauch, F., Hertig, C., Rebmann, G., Bull, J. and Dudler, R. (1991) *Plant Mol. Biol.*, **16**, 1089.

80. Hertig, C., Rebmann, G., Bull, J., Mauch, F. and Dudler, R. (1991) *Plant Mol. Biol.*, **16**, 171.

81. Lloyd, J.C., Horsnell, P.H., Dyer, T.A. and Raines, C.A. (1991) *Plant Mol. Biol.*, **17**, 167.

82. Garcia-Olmedo, F. U [pirembl:tatthvg].

83. Hertig, C., Rebmann, G., Bull, J., Mauch, F. and Dudler, R. (1991) *Plant Mol. Biol.*, **16**, 171.

84. Raines, C.A., Lloyd, J.C., Willingham, N., Potts, S. and Dyer, T.A. (1992) *Eur. J. Biochem.*, **205**, 1053.

85. Kunze, R., Stochaj, U., Laufs, J. and Starlinger, P. (1987) *EMBO J.*, **6**, 1555.

86. Shah, D.M., Hightower, R.C. and Meagher, R.B. (1983) *J. Mol. Appl. Genet.*, **2**, 111.

87. Osterman, J.C. and Dennis, E.S. (1989) *Plant Mol. Biol.*, **13**, 203.

88. Sachs, M.M., Dennis, E.S., Gerlach, W.L. and Peaccock, W.J. (1986) *Genetics*, **113**, 449.

89. Dennis, E.S., Sachs, M.M., Gerlach, W.L., Finnegan, E.J. and Peacock, W.J. (1985) *Nucleic Acids Res.*, **13**, 727.

90. Bathgate, B., Baker, A. and Leaver, C.J. (1989) *Eur. J. Biochem.*, **183**, 303.

91. Yu, L.X. and Lazarus, C.M. (1991) *Plant Mol. Biol.*, **16**, 925.

92. Ralston, E.J., English, J.J. and Dooner, H.K. (1988) *Genetics*, **119**, 185.

93. Franken, P., Niesbach-Kloesgen, U., Weydemann, U., Marechal-Drouard, L., Saedler, H. and Wienand, U. (1991) *EMBO J.*, **10**, 2605.

94. Quigley, F., Martin, W.F. and Cerff, R. U [pirembl:zmg3pd].

95. Belanger, F.C. and Kriz, A.L. (1991) *Genetics*, **129**, 863.

96. Gomez, J., Sanchez-Martinez, D., Stiefel, V., Rigau, J., Puigdomenech, P. and Pages, M. (1988) *Nature*, **334**, 262.

97. Shah, D.M., Hironaka, C.M., Wiegand, R.C., Harding, E.I., Krivi, G.G. and Tiemeier, D.C. (1986) *Plant Mol. Biol.*, **6**, 203.

98. Collazo, P., Montoliu, L., Puigdomenech, P. and Rigau, J. U [pirembl:zmomth].

99. Kelley, P.M., Godfrey, K., Lal, S.K. and Alleman, M. (1991) *Plant Mol. Biol.*, **17**, 1259.

100. Pla, M., Gomez, J., Goday, A. and Pages, M. (1991) *Mol. Gen. Genet.*, **230**, 394.

101. Lebrun, M., Waksman, G. and Freyssinet, G. (1987) *Nucleic Acids Res.*, **15**, 4360.

102. Werr, W., Frommer, W.B., Maas, C. and Starlinger, P. (1985) *EMBO J.*, **4**, 1373.

103. Pereira, A., Cuypers, H., Gierl, A., Schwarz-Sommer, Z. and Saedler, H. (1986) *EMBO J.*, **5**, 835.

104. Montoliu, L., Puigdomenech, P. and Rigau, J. (1990) *Gene*, **94**, 201.

105. Kloesgen, R.B.,, Gierl, A., Schwarz-Sommer, Z. and Saedler, H. (1986) *Mol. Gen. Genet.*, **203**, 237.

106. Franken, P., Niesbach-Kloesgen, U., Weydemann, U., Marechal-Drouard, L., Saedler, H. and Wienand, U. (1991) *EMBO J.*, **10**, 2605.

107. Good, A.G. and Paetkau, D.H. *Plant Mol. Biol.* [daily:zmlacdehg].

108. Marchionni, M. and Gilbert, W. (1986) *Cell*, **46**, 133.

109. Snowden, K.C. and Gardner, R.C. (1990) *Nucleic Acids Res.*, **18**, 6684.

110. Sommer, H. and Saedler, H. (1986) *Mol. Gen. Genet.*, **202**, 429.

111. Schwarz-Sommer, Z., Hue, I., Huijser, P., Flor, P., Hansen, R., Tetens, F., Loennig, W.E., Saedler, H. and Sommer, H. (1992) *EMBO J.*, **11**, 251.

112. Nacken, W.K.F., Plotrowiak, R., Saedler, H. and Sommer, H. U [pirembl:amtam1tr].

113. Huijser, P., Klein, J., Loennig, W.E., Meijer, H., Saedler, H. and Sommer, H. (1992) *EMBO J.*, **11**, 1239.

114. Nacken, W.K.F., Huijser, P., Beltran, J.P., Saedler, H. and Sommer, H. U [pirembl:amtap1g].

115. Nairn, C.J., Winsett, L. and Ferl, R.J. (1988) *Gene*, **65**, 247.

116. Roberts, M.R., Foster, G.D., Blundell, R.P. and Robinson, S.W. U [pirembl:atapg].

117. Chrispeels, M.J., Hofte, H.R., Ludevid, D. and Hubbard, L. U [pirembl:atatip].

118. Pardo, J.M. and Serrano, R. (1989) *J. Biol. Chem.*, **264**, 8557.

119. Harper, J.F., Manney, L., DeWitt, N.D., Yoo, M.H. and Sussman, M.R. (1990) *J. Biol. Chem.*, **265**, 13601.

120. Dudler, R. and Hertig, C. U [pirembl:atatpgp1].

121. Conner, T.W., Goekjian, V.H., LaFayette, P.R. and Kay, J.L. (1990) *Plant Mol. Biol.*, 15, 623.

122. Marks, M.D., West, J. and Weeks, D.P. (1987) *Plant Mol. Biol.*, **10**, 91.

123. Perera, I. and Zielinski, R.E. U [pirembl:atcam3].

124. Feinbaum, R.L. and Ausubel, F.M. (1988) *Mol. Cell. Biol.*, **8**, 1985.

125. Tobias, C. M., Howlett, B. and Nasrallah, J.B. U [pirembl:atdna].

126. Van der Straeten, D., Rodrigues-Pousada, R.A., Goodman, H.M. and Van Montagu, M. (1991) *Plant Cell*, **3**, 719.

127. Shih, M.C., Heinrich, P. and Goodman, H.M. (1991) *Gene*, **104**, 133.

128. Quigley, F., Villiot, M.L. and Mache, R. (1991) *Plant Mol. Biol.*, **17**, 949.

129. Chrispeels, M.J., Hofte, H.R., Ludevid, D. and Hubbard, L. U [pirembl:atgtip].

130. Gantt, J.S. and Lenvik, T.R. U [pirembl:ath11flk].

131. Gantt, J.S. and Lenvik, T.R. U [pirembl:ath11gene].

132. Gantt, J.S. and Lenvik, T.R. U [pirembl:ath12flk].

133. Takahashi, T., Naito, S. and Komeda, Y. U [pirembl:athsp81].

134. Bernhard, W.R. and Matile, P. U [pirembl:aticla].

135. Medford, J., Elmer, J. and Klee, H.J. *Plant Cell* [pirembl:atmeri5g].

136. Oppenheimer, D.G., Herman, P.L., Sivakumaran, S., Esch, J. and Marks, M.D. (1991) *Cell*, **67**, 483.

137. Cheng, C., Dewdney, J., Nam, H., Den, Boer, B.G.W. and Goodman, H.M. (1988) *EMBO J.*, **7**, 3309.

138. Van Rooijen, G.J., Terning, L.I. and Moloney, M.M. U [pirembl:atol].

139. Aeschbacher, R.A., Shrott, M., Potrykus, I. and Saul, M.W. U [pirembl:atposf21].

140. Intrapruk, C., Higashimura, N., Yamamoto, K., Okada, N., Shinmyo, A. and Takano, M. (1991) *Gene*, **98**, 237.

141. Verbruggen, N., Villarroel, R. and Van Mongatu, M. U [pirembl:atproc1a].

142. Gil-Gomez, G., Marrero, P.F., Haro, D., Ayte, J. and Hegardt, F.G. (1991) *Plant Mol. Biol.*, **17**, 517.

143. Dietrich, M.A., Prenger, J. and Guilfoyle, T.J. (1990) *Plant Mol. Biol.*, **15**, 207.

144. Thompson, M.D., Jacks, C.M., Lenvik, T.R. and Gantt, J.S. U [pirembl:atrpcl15g].

145. Thompson, M.D., Jacks, C.M., Lenvik, T.R. and Gantt, J.S. U [pirembl:atrpcl9g].

146. Chopra, S., Dal-Favero, J., Dolferus, R. and Jacobs, M. (1992) *Plant Mol. Biol.*, **18**, 131.

147. Gerhold, D., Parsons, A. and Hadwiger, L.A. U [pirembl:attop2a].

148. Berlyn, M.B., Last, R.L. and Fink, G.R. (1989) *Proc. Natl. Acad. Sci. USA*, **86**, 4604.

149. Fink, G.R., Radwanski, E.R., Mahoney, D.J., Bissinger, P.H. and Last, R.L. (1991) *Plant Cell*, **3**, 345.

150. Ludwig, S.R., Oppenheimer, D.G., Silflow, C.D. and Snustad, D.P. (1988) *Plant Mol. Biol.*, **10**, 311.

151. Ludwig, S.R., Oppenheimer, D.G., Silflow, C.D. and Snustad, D.P. (1987) *Proc. Natl. Acad. Sci. USA*, **84**, 5833.

152. Oppenheimer, D.G., Haas, N., Silflow, C.D. and Snustad, D.P. (1988) *Gene*, **63**, 87.

153. Callis, J., Raasch, J.A. and Vierstra, R.D. (1990) *J. Biol. Chem.*, **265**, 12486.

154. Rouse, D.T. U [pirembl:atxero].

155. Le Guen, L., Thomas, M., Bianchi, M., Halford, N.G. and Kreis, M. U [daily:atakin10a].

156. Kemmerer, E.C. and Wu, R. U [daily: atatcc1a].

157. Nishida, I., Tasaka, Y., Shiraishi, H. and Murata, N. U [daily:atatsa1].

158. Krebbers, E., Seurinck, J., Herdies, L., Cashmore, A.R. and Timko, M.P. *Plant Mol. Biol.* [daily:atatsg].

159. Krebbers, E., Seurinck, J., Herdies, L., Cashmore, A.R. and Timko, M.P. *Plant Mol. Biol.* [daily:atatsgs].

160. Jensen, P.E., Kristensen, M., Hoff, T., Lehmbeck, J., Stummann, B.M. and Henmngsen, K.W. U [daily:atcab6p].

161. Bradley, D. U [daily:atcpylp].

162. Shirley, B.W., Hanley, S. and Goodman, H.M. (1992) *Plant Cell*, **4**, 333.

163. Finkelstein, R. U [daily:atgpro].

164. de Oliveira, D.E., Franco, L.O., Simoens, C.H., Seurinck, J., Coppieters, J., Botterman, J. and Van Montagu, M. U [daily:atgrpg].

165. Hayashida, N., Mizoguchi, T., Yamaguchi-Shinozaki, K. and Shinozaki, K. U [daily:atpk5].

166. McClung, C.R., Marcotte, B.L., Wernecke, J.M. and Ogren, W.L. U [daily:atrboalp].

167. Ulmasov, T.N. and Guilfoyle, T.J. U [daily:atrnapola].

168. Schroeder, G., Brown, J.W.S. and Schroeder, J. (1988) *Eur. J. Biochem.*, **172**, 161.

169. Fujiyama, K., Takemura, H., Shibayama, S., Kobayashi, K., Choi, J.K., Shinmyo, A., Takano, M., Yamada, Y. and Okada, H. (1988) *Eur. J. Biochem.*, **173**, 681.

170. Fujiyama, K., Takemura, H., Shinmyo, A., Okada, H. and Takano, M. (1990) *Gene*, **89**, 163.

171. Gander, E.S., Holmstroem, K.O., De Paiva, G.R., De Castro, L.A.B., Carneiro, M. and Grossi De Sa, M.F. (1991) *Plant Mol. Biol.*, **16**, 437.

172. Albani, D., Altosaar, I., Arnison, P.G. and Fabijanski, S.F. (1991) *Plant Mol. Biol.*, **16**, 501.

173. Albani, D., Robert, L.S., Donaldson, P.A., Altosaar, I., Arnison, P.G. and Fabijanski, S.F. (1990) *Plant Mol. Biol.*, **15**, 605.

174. Shaw, C.H. and Ryan, A.J. (1989) *Nucleic Acids Res.*, **17**, 3584.

175. Ericson, M.L., Roedin, J., Lenman, M., Glimelius, K., Josefsson, L.G. and Rask, L. (1986) *J. Biol. Chem.*, **261**, 14576.

176. Keddie, J.S., Hubner, G., Slocombe, S.P., Jarvis, R.P., Cummins, I., Edwards, E.W., Shaw, C.H. and Murphy, D.J. U [pirembl:bnoleo].

177. Fiebig, C. and Link, G. U [pirembl:bnrbcs].

178. Nantel, A.M., Lafleur, F., Boiven, R., Baszczynski, C.L. and Bellemare, G. (1991) *Plant Mol. Biol.*, **16**, 955.

179. Christensen, T., Dennis, E.S., Peacock, J.W., Landsmann, J. and Marcker, K.A. (1991) *Plant Mol. Biol.*, **16**, 339.

180. Schwart, B.W., Sloan, J.S. and Becker, W.M. (1991) *Plant Mol. Biol.*, **17**, 941.

181. Huang, P., Parks, J.E., Rottman, W.H. and Theologis, A. (1991) *Proc. Natl. Acad. Sci. USA*, **88**, 7021.

182. Apuya, N.R. and Zimmerman, J. U [pirembl:dcatpsbs].

183. Franz, G., Hatzopoulus, P., Jones, T.J., Krauss, M. and Sung, Z.R. (1989) *Mol. Gen. Genet.*, **218**, 143.

184. Apuya, N.R. and Zimmerman, J. U [pirembl:dcef1a].

185. Raghothma, K.G., Lawton, K.A., Goldsbrough, P.B. and Woodson, W.R. (1991) *Plant Mol. Biol.*, **17**, 61.

186. Shah, D.M., Hightower, R.C. and Meagher, R.B. (1982) *Proc. Natl. Acad. Sci. USA*, **79**, 1022.

187. Hagen, G., Uhrhammer, N. and Guilfoyle, T.J. (1988) *J. Biol. Chem.*, **263**, 6442.

188. Doyle, J.J., Schuler, M.A., Godette, W.D., Zenger, V., Beachy, R.N. and Slightom, J.L. (1986) *J. Biol. Chem.*, **261**, 9228.

189. Akada, S., Kung, S.D. and Dube, S.K. (1991) *Plant Mol. Biol.*, **16**, 751.

190. Akada, S., Kung, S.D. and Dube, S.K. (1990) *Nucleic Acids Res.*, **18**, 5899.

191. Hagen, G., Martin, G., Li, Y. and Guilfoyle, T. (1991) *Plant Mol. Biol.*, **17**, 567.

192. Hyldig-Nielsen, J.J., Jensen, E., Paludan, K., Wiborg, O., Garrett, R., Joergensen, P. and Mercker, K.A. (1982) *Nucleic Acids Res.*, **10**, 689.

193. Marco, Y.A., Thanh, V.H., Tumer, N.E., Scallon, B.J. and Nielsen, N.C. (1984) *J. Biol. Chem.*, **259**, 13436.

194. Rhee, Y. and Staswick, P.E. U [pirembl:gmglymax].

195. Sims, T.L. and Goldberg, R.B. (1989) *Nucleic Acids Res.*, **17**, 4386.

196. Thanh, V.H., Tumer, N.E. and Nielsen, N.C. (1989) *Nucleic Acids Res.*, **17**, 4387.

197. Cho, T.J. and Nielsen, N.C. (1989) *Nucleic Acids Res.*, **17**, 4388.

198. Czarnecka, E., Nagao, R.T., Key, J.L. and Gurley, W.B. (1988) *Mol. Cell. Biol.*, **8**, 1113.

199. Yenofsky, R.L., Fine, M. and Liu, C. (1988) *Mol. Gen. Genet.*, **211**, 215.

200. Nirunsuksini, W. and Sengupta-Gopalan, C. (1990) *Plant Mol. Biol.*, **15**, 835.

201. Frank, R.L. and Vodkin, L.O. (1991) *DNA Seq.*, **1**, 335.

202. Katinakis, P. and Verma, D.P.S. (1985) *Proc. Natl. Acad. Sci. USA* , **82**, 4157.

203. Berry-Lowe, S.L., McKnight, T.D., Shah, D.M. and Meagher, R.B. (1982) *J. Mol. Appl. Genet.*, **1**, 483.

204. Grandbastien, M.A., Berry-Lowe, S., Shirley, B.W. and Meagher, R.B. (1986) *Plant Mol. Biol.*, **7**, 451.

205. Guiltinan, M.J., Ma, D.P., Barker, R.F., Bustos, M.M., Cyr, R.J., Yadegari, R. and Fosket, D.E. (1987) *Plant Mol. Biol.*, **10**, 171.

206. Shibata, D., Kato, T. and Tanaka, K. (1991) *Plant Mol. Biol.*, **16**, 353.

207. Aguilar, F., Montandon, P.E. and Stutz, E. (1991) *Plant Mol. Biol.*, **17**, 351.

208. Krueger, R.W., Holland, M.A., Chisolm, D. and Polacco, J.C. (1987) *Gene*, **54**, 41.

209. Rhee, Y. and Staswick, P.E. U [pirembl: gmvspba].

210. Akada, S., Kung, S.D. and Dube, S.K. U [daily:gmachs2].

211. Isin, S.H. and Allen, R.D. *Plant Mol. Biol.* U [daily:gmcatalg].

212. Sagliocco, F., Kapazoglou, A. and Dure III, L. *Plant Mol. Biol.* [pirembl:ghcab].

213. Galau, G.A., Wang, H.Y. and Hughes, D.W. U [pirembl:ghdgala].

214. Galau, G.A., Wang, H.Y. and Hughes, D.W. U [pirembl:ghdpaa].

215. Hughes, D.W. and Galau, G.A. U [pirembl:ghlea].

216. Baker, J., Steele, C. and Dure, L. (1988) *Plant Mol. Biol.*, **11**, 277.

217. Hughes, D.W. and Galau, G.A. U [pirembl:ghlea2].

218. Galau, G.A., Wang, H.Y. and Hughes, D.W. U [pirembl:ghlea4a].

219. Galau, G.a., Wang, H.Y. and Hughes, D.W. U [pirembl:ghleaa2d].

220. Sagliocco, F., Kapazoglou, L. and Dure III, L. (1991) *Plant Mol. Biol.*, **17**, 1275.

221. Dure III, L. U [pirembl:ghspa].

222. Dure III, L. U [pirembl:ghsph].

223. Galau, G.A., Wang, H.Y.C. and Hughes, D.W. U [daily:ghlea14a].

224. Baker, J., Steele, C. and Dure, L. (1988) *Plant Mol. Biol.*, **11**, 277.

225. Hughes, D.W. and Galau, G.A. U [daily: ghlea2dna].

226. Hughes, D.W. and Galau, G.A. U [daily: ghlea5g].

227. Allen, R.D., Cohen, E.A., Vonder Haar, R.A., Adams, C.A., Ma, D.P., Nessler, C.L. and Thomas, T.L. (1987) *Mol. Gen. Genet.*, **210**, 211.

228. Binet, M.N., Weil, J.H. and Tessier, L.H. (1991) *Plant Mol. Biol.*, **17**, 395.

229. Domon, C., Evrard, J.L., Pillay, T.N. and Steinmetz, A. (1991) *Mol. Gen. Genet.*, **229**, 238.

230. Chye, M.L., Kush, A., Tan, C.T. and Chua, N.H. (1991) *Plant Mol. Biol.*, **16**, 567.

231. Rottmann, W.H., Peter, G.F., Oeller, P.W., Keller, J.A., Shen, N.F., Nagy, B.P., Taylor, L.P., Campbell, A.D. and Theologis, A. (1991) *J. Mol. Biol.*, **222**, 937.

232. Erion, J.L., Ballo, B., Fox, T.W. and May, L. U [pirembl:leapp1a].

233. Pichersky, E., Hoffman, N.E., Bernatzky, R., Piechulla, B., Tanksley, S.D. and Cashmore, A.R. (1987) *Plant Mol. Biol.*, **9**, 205.

234. Pichersky, E., Brock, T.G., Nguyen, D., Hoffman, N.E., Piechulla, B., Tanksley, S.D. and Green, B.R. (1989) *Plant Mol. Biol.*, **12**, 257.

235. Gottesman, S. and Pichersky, E. *Proc. Natl. Acad. Sci. USA* [pirembl:lecd4a].

236. Gottesman, S. and Pichersky, E. *Proc. Natl. Acad. Sci. USA* [pirembl:lecd4b].

237. Deikman, J. and Fischer, R.L. (1988) *EMBO J.*, **7**, 3315.

238. Van Der Straeten, D., Rodrigues-Pousada, R.A., Goodman, H.M. and Van Montagu, M. (1991) *Plant Cell*, **3**, 719.

239. Koeck, M., Hamilton, A.J. and Grierson, D. (1991) *Plant Mol. Biol.*, **17**, 141.

240. Pear, J.R., Ridge, N., Rasmussen, R., Rose, R.E. and Houck, C.M. (1989) *Plant Mol. Biol.*, **13**, 639.

241. Shi, L., Gast, R.T., Gopalraj, M. and Olszewski, N.E. *Plant Journal* [pirembl:legast1].

242. Ray, J.A., Moureau, P., Bird, C., Bird, A., Grierson, D., Maunders, M., Fray, R.G., Truesdale, M., Bramley, P. and Schuch, W. U [pirembl:legtom5].

243. Holdsworth, M.J., Schuch, W. and Grierson, D. (1987) *Nucleic Acids Res.*, **15**, 10600.

244. Yang, Z., Park, H.S., Lacy, G.H. and Cramer, C.L. (1991) *Plant Cell*, **3**, 397.

245. Samach, A., Hareeven, D., Gutfinger, T., Ken-Dror, S. and Lifschitz, E. (1991) *Proc. Natl. Acad. Sci. USA*, **88**, 2678.

246. Cohen, A., Plant, A.L., Moses, M.S. and Bray, E.A. *Plant Physiol.* [pirembl:lele25].

247. Schwart, E., Stasys, R., Aebersold, R., McGrath, J.M., Green, B.R. and Pichersky, E. (1991) *Plant Mol. Biol.*, **17**, 923.

248. Sugita, M., Manzara, T., Pichersky, E., Cashmore, A. and Gruissem, W. (1987) *Mol. Gen. Genet.*, **209**, 247.

249. Pichersky, E., Bernatzky, R., Tanksley, S.D. and Cashmore, A.R. (1986) *Proc. Natl. Acad. Sci. USA*, **83**, 3880.

250. Van Haaren, M.J. and Houck, C.M. U [daily:lefspg1].

251. Lee, S.W., Robb, E.J. and Nazer, R.N. U [daily:lepal5a].

252. McGurl, B.F., Pearce, G., Orozco-Cardenas, M. and Ryan, C.A. (1992) *Science*, **255**, 1570.

253. Wingate, V.P.M. and Ryab, C.A. (1991) *J. Biol. Chem.*, **266**, 5814.

254. Tischer, E., Dassarma, S. and Goodman, H.M. (1986) *Mol. Gen. Genet.*, **203**, 221.

255. Davidowitz, E.J., Creissen, G., Lang-Unnasch, N., Vineze, E. and Kiss, G.B. U [pirembl:msleghO1].

256. Vegh, Z., Vincze, E., Kadirov, R., Toth, G., Gyorgy, G.T. and Kiss, B. U [pirembl:msnod25].

257. Gallusci, P., Dedieu, A., Journet, E.P., Huguet, T. and Barker, D.G. (1991) *Plant Mol. Biol.*, **17**, 335.

258. Cushman, J.C. and Bohnert, H.J. (1989) *Nucleic Acids Res.*, **17**, 6743.

259. Boutry, M. and Chua, N.H. (1985) *EMBO J.*, **4**, 2159.

260. Gheysen, G.D., Inze, D., Soetaert, P., Van Montagu, M. and Castresana, C. (1990) *Nucleic Acids Res.*, **18**, 6685.

261. Castresana, C., de Carvalho, F., Gheysen, G., Habets, M., Inze, D. and Van Montagu, M. (1990) *Plant Cell*, **2**, 1131.

262. Perez, C., Michelet, B., Ferrant, V., Bogaerts, P. and Boutry, M. U [pirembl:nppma1a].

263. Perez, C., Michelet, B., Ferrant, V., Bogaerts, P. and Boutry, M. U [pirembl:nppma2a].

264. Perez, C., Michelet, B., Ferrant, V., Bogaerts, P. and Boutry, M. U [pirembl:nppma3b].

265. Van Der Zaal, E.J., Droog, F.N.J., Boot, C.J.M., Hensgens, L.A.M., Hoge, J.H.C., Schilperoort, R.A. and Libbenga, K.R. (1991) *Plant Mol. Biol.*, **16**, 983.

266. Kavanagh, T.A. and Bevan, M.W. U [pirembl:ntcab36].

267. Fukuda, Y., Ohme, M. and Shinshi, H. (1991) *Plant Mol. Biol.*, **16**, 1.

268. Shinshi, H., Neuhaus, J.M., Ryals, J. and Meins Jr, F. (1990) *Plant Mol. Biol.*, **14**, 357.

269. Sperisen, C., Ryals, J. and Meins, F. (1991) *Proc. Natl. Acad. Sci. USA*, **88**, 1820.

270. Linthorst, H.J., Melchers, L.S., Mayer, A., Van Roekel, J.S., Cornelissen, B.J.C. and Bol, J.F. (1990) *Proc. Natl. Acad. Sci. USA*, **87**, 8756.

271. Van Kan, J.A.L., Cornelissen, B.J.C. and Boll, J.F. *Mol. Plant Microb. Interact.* [pirembl:ntgrpa].

272. Fleming, A., Mandel, T., Hoffmann, S., Sterk, P., De Vries, S. and Kuhlemeier, C. U [pirembl:ntltp1].

273. Chamot, D. and Kuhlemeter, C. (1992) *Nucleic Acids Res.*, **20**, 665.

274. Hua, S., Dube, S.K., Barnett, N.M. and Kung, S.D. (1991) *Plant Mol. Biol.*, **17**, 551.

275. Fromm, H., Katagiri, F. and Chua, N.H. (1991) *Mol. Gen. Genet.*, **229**, 181.

276. Rogers, H.J. and Lonsdale, D.M. U [pirembl:ntptlg].

277. Mazur, B.J. and Chui, C.F. (1985) *Nucleic Acids Res.*, **13**, 2373.

278. Seurinck, J., Truettner, J. and Goldberg, R.B. (1990) *Nucleic Acids Res.*, **18**, 3403.

279. Fukasawa-Akada, T. U [pirembl:nttpa1a].

280. Van Buuren, M., Neuhaus, J.M., Shinshi, H., Ryals, J. and Meins, F. *Mol. Gen. Genet.* [daily:ntchn14g].

281. Van Buuren, M., Neuhaus, J.M., Shinshi, H., Ryals, J. and Meins, F. *Mol. Gen. Genet.* [daily:ntchn50g].

282. Nessler, C.L., Kurz, W.G.W. and Pelcher, L.E. (1990) *Plant Mol. Biol.*, **15**, 951.

283. Landsmann, J., Dennis, E.S., Higgins, T.J.V., Appleby, C.A., Kortt, A.A., Peacock, W.J. (1986) *Nature*, **324**, 166.

284. Cass, L.G., Kirven, K.A. and Christofferen, R.E. (1990) *Mol. Gen. Genet.*, **223**, 76.

285. Van de Loecht, U., Meier, I., Hahlbrock, K. and Somssich, I.E. (1990) *EMBO J.*, **9**, 2945.

286. Tumer, N.E., Clark, W.G., Tabor, G.J., Hironaka, C.M., Frayley, R.T. and Shah, D.M. (1986) *Nucleic Acids Res.*, **14**, 3325.

287. Gregerson, R., McLean, M., Beld, M., Gerats, A. and Strommer, J. (1991) *Plant Mol. Biol.*, **17**, 37.

288. Stayton, M.M., Black, M., Bedbrook, J. and Dunsmuir, P. (1986) *Nucleic Acids Res.*, **14**, 9781.

289. Koes, R.E., Spelt, C.E., van Den Elzen, P.J.M. and Mol, J.N.M. (1989) *Gene*, **81**, 245.

290. Winter, J., Wright, R., Duck, N., Gasser, C., Fraley, R. and Shah, D. (1988) *Mol. Gen. Genet.*, 211, 315.

291. Coleman, C.E. and Kao, T.H. U [pirembl: pis1alle].

292. Coleman, C.E. and Kao, T.H. U [pirembl: pis3alle].

293. Ogushi, Y., Tetsutaro, T., Yamauchi, D. and Minamikawa, T. U [pirembl:pvendo].

294. Lee, J.S. and Verma, D.P. (1984) *EMBO J.*, **3**, 2745.

295. Slusarenko, A.J. U [pirembl:pvloxa].

296. Slightom, J.L., Sun, S.M. and Hall, T.C. (1983) *Proc. Natl. Acad. Soc. USA*, **80**, 1897.

297. Knight, M.R. U [pirembl:pvss15bco].

298. Forde, B.G., Freeman, J., Oliver, J.E. and Pineda, M. (1990) *Plant Cell*, **2**, 925.

299. Hoff, T., Stummann, B.M. and Henningsen, K.W. (1991) *Physiol. Plantarum*, **82**, 197.

300. Llewellyn, D.J., Finnegan, E.J., Ellis, J.G., Dennis, E.S. and Peacock, W.J. (1987) *J. Mol. Biol.*, **195**, 115.

301. Higgins, T.J. U [pirembl:psalbumi1].

302. Falconet, D.R., White, M.J., Fristensky, B.W., Dobres, M.S. and Thompson, W.F. (1991) *Plant Mol. Biol.*, **17**, 135.

303. Gantt, J.S., Baldauf, S.L., Calie, P.J., Weeden, N.F. and Palmer, J.D. U [pirembl: pscl22].

304. Newbigin, E.J., deLumen, B.O., Chandler, P.M., Gould, A., Blagrove, R.J., March, J.F., Kortt, A.A. and Higgins, T.J. (1980) *Planta*, **180**, 461.

305. Kolanus, W., Scharnhorst, C., Kuehne, U. and Herzfeld, F. (1987) *Mol. Gen. Genet.*, **209**, 234.

306. Lycett, G.W., Croy, R.R.D., Shirsat, A.H., Richards, D.M. and Boulter, D. (1984) *Nucleic Acids Res.*, **13**, 6733.

307. Gatehouse, J.A., Brown, D., Gilroy, J., Levasseur, M., Castleton, J. and Ellis, T.H.N. (1988) *Biochemical J.*, **250**, 15.

308. Fluhr, R., Moses, P., Morelli, G., Coruzzi, G. and Chua, N.H. (1986) *EMBO J.*, **5**, 2063.

309. Coruzzi, G., Broglie, R., Edwards, C. and Chua, N.H. (1984) *EMBO J.*, **3**, 1671.

310. Bassuener, R., Evans, I.M., Fordham-Skelton, T. and Croy, R.R.D. *Plant Mol. Biol.* [pirembl:psusp43].

311. Liaud, M.F., Brinkmann, H. and Cerff, R. (1992) *Plant Mol. Biol.*, **18**, 639.

312. Kardailsky, I., Yang, W.C., Zalensky, A., Van Kammen, A. and Bisseling, T. U [daily:psnod6g].

313. Davis, J.M., Clarke, H.R.G., Bradshaw, H.D. and Gordon, M.P. (1991) *Plant Mol. Biol.*, **17**, 631.

314. Welters, P., Metz, B., Schell, J. and de Bruijn, F.J. (1989) *Nucleic Acids Res.*, **17**, 1253.

315. Welters, P., Metz, B., Schell, J. and de Bruijn, F.J. U [daily:srglb2].

316. Van Der Leij, F.R., Visser, R.G.F., Ponstein, A.S., Jacobsen, E. and Feenstra, W.J. (1991) *Mol. Gen. Genet.*, **288**, 240.

317. Thornburg, R.W., An, G., Cleveland, T.E., Johnson, R. and Ryan, C.A. (1987) *Proc. Natl. Acad. Sci. USA*, **84**, 744.

318. Cleveland, T.E., Thornburg, R.W. and Ryan, C.A. (1987) *Plant Mol. Biol.*, **8**, 199.

319. Eckes, P., Rosahl, S., Schell, J. and Willmitzer, L. (1986) *Mol. Gen. Genet.*, **205**, 14.

320. Joos, H.J. and Hahlbrock, K. U [pirembl:stpal1].

321. Mignery, G.A., Pikaard, C.S. and Park, W.D. (1988) *Gene*, **62**, 27.

322. Rosahl, S., Schmidt, R., Schell, J. and Willmitzer, L. (1986) *Mol. Gen. Genet.*, **203**, 214.

323. Bevan, M., Barker, R., Goldsbrough, A., Jarvis, M. and Kavanagh, T. (1986) *Nucleic Acids Res.*, **14**, 4625.

324. Keil, M., Sanchez-Serrano, J., Schell, J. and Willmitzer, L. (1986) *Nucleic Acids Res.*, **14**, 5641.

325. Drouin, G. and Dover, G.A. (1990) *J. Mol. Evol.*, **31**, 132.

326. Taylor, J.L., Frizemeier, K.H., Haeuser, I., Kombrink, E., Rohwer, F., Schroeder, M., Strittmatter, G. and Hahlbrock, K. *Mol. Plant Microb. Interact.* [pirembl:stpr1a].

327. Hahlbrock, K., Schulze-Lefert, P. and Becker-Andre, M. (1991) *J. Biol. Chem.*, **266**, 8551.

328. Matton, D.P., Camirand, A. and Brisson, N. U [pirembl:ststha].

329. Stanford, A., Bevan, M. and Northcote, D. *Mol. Gen. Genet.* [pirembl:stwini2g].

330. Back, E., Dunne, W., Schneiderbauer, A., De Framond, A., Rastogi, R. and Rothstein, S.J. (1991) *Plant Mol. Biol.*, **17**, 9.

331. Mason, J.G. and Whitfield, P.R. (1990) *Plant Mol. Biol.*, **14**, 1007.

332. Bichler, J. and Herrmann, R.G. (1990) *Eur. J. Biochem.*, **190**, 415.

333. McHenry, L. and Fritz, P.J. U [pirembl:tccsv].

334. Bogusz, D., Appleby, C., Landsmann, J., Dennis, E.S., Trinick, M. and Peacock, W.J. (1988) *Nature*, **331**, 178.

335. Baeumlein, H., Wobus, U., Pustell, J. and Kafatos, F.C. (1986) *Nucleic Acids Res.*, **14**, 2707.

336. Baeumlein, H., Boerjan, W., Nagy, I., Bassuener, R., Van Montagu, M., Inze, D. and Wobus, U. (1991) *Mol. Gen. Genet.*, **225**, 459.

337. Bassuener, R., Bauemlein, H., Huth, A., Jung, R., Wobus, U., Rapoport, T.A., Saalbach, G., Muentz, K. (1988) *Plant Mol. Biol.*, **11**, 321.

338. Weschke, W., Bauemlein, H. and Wobus, U. (1987) *Nucleic Acids Res.*, **15**, 10065.

339. Akasofu, H., Yamauchi, D. and Minamikawa, T. (1990) *Nucleic Acids Res.*, **18**, 1892.

340. Botella, J.R., Schlagnhaufer, C.D., Arteca, R.N. and Phillips, A.T. U [daily:vraccsy01].

341. Botella, J.R., Schlagnhaufer, C.D., Arteca, R.N. and Phillips, A.T. U [daily:vraccsy02].

342. Botella, J.R., Schlagnhaufer, C.D., Arteca, R.N. and Phillips, A.T. U [daily:vraccsy03].

343. Baker, B., Thompson, D.V., Talbot, D.R., Swanson, J. and Bennetzen, J.L. (1984) *Nucleic Acids Res.*, **12**, 5955.

344. Bhattacharyya, M.K., Smith, A.M., Ellis, T. H.N., Hedley, C. and Martin, C. (1990) *Cell*, **60**, 115.

345. Bonas, U., Sommer, H. Harrison, B.J. and Saedler, H. (1984) *Mol. Gen. Genet.*, **194**, 138.

346. Brown, J.J., Mattes, M.G., O'Reilly, C. and Shepherd, N.S. (1989) *Mol. Gen. Genet.*, **215**, 239.

347. Fedoroff, N.V., Wessler, S. and Shure, M. (1983) *Proc. Natl Acad. Sci. USA* ., **81**, 3825.

348. Franklin, T.M. (1990) in *The Isolation of a Transposable Element Inserted in an A1 Allele of* Zea mays. PhD thesis, Sunderland Polytechnic.

349. Freeling, M. (1988) in *Plant Transposable Elements* (O. Nelson, ed.). Plenum Press, New York, p. 279.

350. Gonella, J. and Peterson, P.A. (1975) *Maize Genet. Coop. Newsl.*, **49**, 73.

351. Gonella, J. and Peterson, P.A. (1977) *Genetics*, **85**, 629.

352. Grandbastien, M.-A., Spielmann, A. and Caboche, M. (1989) *Nature*, **337**, 376.

353. Harberd, N.P., Flavell, R.B. and Thompson, R.D. (1987) *Mol. Gen. Genet.*, **209**, 326.

354. Herrmann, A., Schulz, W. and Hahlbrock, K. (1988) *Mol. Gen. Genet.*, **212**, 93.

355. Jin, Y.-K. and Bennetzen, J.L. (1989) *Proc. Natl Acad. Sci. USA* , **86**, 6235.

356. Johns, M.A., Babcock, M.S., Fuerstenberg, S.M., Freeling, M. and Simpson, R.B. (1989) *Plant Mol. Biol.*, **12**, 633.

357. Masson, P., Surosky, R., Kingsbury, J.A. and Fedoroff, N.V. (1987) *Genetics*, **177**, 117.

358. Merckelbach, A., Doring, H.P. and Starlinger, P. (1986) *Maydica*, **31**, 109.

359. Montanelli, C., Di Fonzo, N., Marotta, R., Motto, M., Soave, C. and Salamini, F. (1984) *Mol. Gen. Genet.*, **197**, 209.

360. Muller-Newmann, M., Yoder, J.I. and Starlinger, P. (1984) *Mol. Gen. Genet.*, **198**, 19.

361. Oishi, K.K. and Freeling, M. (1988) in *Plant Transposable Elements* (O. Nelson, ed.). Plenum Press, New York, p. 289.

362. Peacock, W.J., Dennis, E.S., Gerlach, W.L., Llewellyn, D., Lorz, H., Pryor, A.J., Sachs, M.M., Schwartz, D. and Sutton, W.D. (1983) in *Proceedings of the Miami Winter Symposium*, Vol. 20 (K. Downer, ed.). Academic Press, New York, p. 311.

363. Peleman, J., Cottyn, B., van Camp, W., van Montagu, M. and Inze, D. (1991) *Proc. Natl Acad. Sci. USA* , **88**, 3618.

364. Pereira, A., Schwarz-Sommer, Z., Gierl, A., Bertram, I., Peterson, P.A. and Seadler, H. (1985) *EMBO J.*, **5**, 835.

365. Pohlman, R.F., Fedoroff, N.V. and Messing, J. (1984) *Cell*, **37**, 635.

366. Pryor, A.J. (1985) in *Plant Genetics* (M. Freeling, ed.). UCLA Symposium on Molecular Cell Biology, no. 35, p. 419.

367. Rhoades, M.M. and Dempsey, E. (1982) *Maize Genet. Coop. Newsl.*, **56**, 21.

368. Rhoades, M.M. and Dempsey, E. (1982) *Maize Genet. Coop. Newsl.*, **57**, 14.

369. Rhodes, P.R. and Vodkin, L.O. (1988) *Genetics*, **120**, 597.

370. Schnable, P.S. and Peterson, P.A. (1986) *Maydica*, **31**, 59.

371. Schnable, P.S., Peterson, P.A. and Seadler, H. (1989) *Mol. Gen. Genet.*, **217**, 459.

372. Schwarz-Sommer, Z., Gierl, A., Berndtgen, R. and Seadler, H. (1985) *EMBO J.*, **4**, 2439.

373. Schwarz-Sommer, Z., Shepherd, N., Tacke, E., Gierl, A., Rohde, W., Leclercq, L., Mattes, M., Berndtgen, R., Peterson, P.A. and Saedler, H. (1987) *EMBO J.*, **6**, 287.

374. Schwarz-Sommer, Z., Leclercq, L. and Saedler, H. (1987) in *Plant Molecular Biology* (D. Wettstein and N.-H. Chua, eds). Plenum Press, New York, p. 191.

375. Shirsat, A.H. (1988) *Mol. Gen. Genet.*, **212**, 129.

376. Sommer, H., Carpenter, R., Harrison, B.J. and Saedler, H. (1985) *Mol. Gen. Genet.*, **199**, 225.

377. Upadhyaya, K.C., Sommer, H., Krebbers, E. and Saedler, H. (1985) *Mol. Gen. Genet.*, **212**, 259.

CHAPTER 7

PLANT TISSUE CULTURE AND GENETIC TRANSFORMATION

J.R. Ellis*

Plant tissue culture is used for many purposes, for example, propagating axenic tissue, reducing the structural complexity of tissue and manipulating developmental patterns. The principal application of tissue culture in molecular biology is for genetic transformation, whereby the *in vitro* environment is used to manipulate the development of transgenic cells to form callus or whole plants.

1. TISSUE CULTURE MEDIA

Innumerable permutations of ingredients exist for tissue culture media and this chapter is necessarily limited to the more common formulations. Readers seeking comprehensive data on plant tissue culture media are referred to specialist texts (1, 2).

1.1. Common basal media

Relatively few basal media (comprising macro- and microelements) form the basis of the most widely used recipes, and the ingredients of MS, B5, N6 and CM are given in *Table 1.* The popularity of media such as MS and B5 is partly attributable to their commercial availability as ready-formulated powders (available from, for example, Flow Labs, Gibco, Imperial Labs). The vitamin and organic compositions of six commonly used recipes are listed in *Table 2.* Note that formulations of inorganic salts and of vitamins can be independently combined, e.g. MS salts with B5 vitamins. Recipes created from basic ingredients are best prepared as stock solutions as indicated in *Table 3,* and stored at 4°C or −20°C in darkness.

1.2. Media supplements

A range of auxins and cytokinins is used in culture media, and common examples are listed in *Table 4,* together with their molecular weights and factors for conversion between molarity and concentration (mg l^{-1}). The molecular structures of these growth regulators are shown in *Figure 1.* Whether to use auxin or cytokinin for a specific purpose should be determined empirically, but efficacy, stability and cost are the key factors. A variety of additives optionally included in culture media is presented in *Table 5.* Semisolid media are often employed as a substratum for cultured plant tissue, and a range of gelling agents and their effective concentrations are given in *Table 6.*

*See *Preface.*

Table 1. Composition of common basal media — inorganic salts (mg l^{-1})

	Medium code			
	MS[a]	**B5**[b]	**N6**[c]	**CM**[d]
Reference	3	4	5	6
Macroelements				
KNO_3	1900	2500	2830	—
NH_4NO_3	1650	—	—	400
$CaCl_2 \cdot 2H_2O$	440	150	166	293
$MgSO_4 \cdot 7H_2O$	370	250	185	246
KH_2PO_4	170	—	400	68
$NaH_2PO_4 \cdot 2H_2O$	—	150	—	—
$(NH_4)_2SO_4$	—	134	463	—
Iron				
$FeSO_4 \cdot 7H_2O$	27.8	—	27.8	27.0
$Na_2EDTA \cdot 2H_2O$[e]	37.3	—	37.3	37.0
FeDTPA[f]	—	28.0	—	—
Microelements				
$MnSO_4 \cdot 4H_2O$	22.3	—	4.4	—
$MnSO_4 \cdot H_2O$	—	10.0	—	0.17
$ZnSO_4 \cdot 7H_2O$	8.6	2.0	1.5	0.28
H_3BO_3	6.2	3.0	1.6	6.2
KI	0.83	0.75	0.8	—
$Na_2MoO_4 \cdot 2H_2O$	0.25	0.25	—	0.024
$CuSO_4 \cdot 5H_2O$	0.025	0.025	—	0.025
$CoCl_2 \cdot 6H_2O$	0.025	0.025	—	0.024
Medium pH is usually set to 5.6–6.0				

[a] Murashige and Skoog medium.
[b] Gamborg B5 medium.
[c] Chu N6 medium.
[d] Caboche medium.
[e] EDTA, ethylenediaminetetraacetic acid.
[f] DTPA, diethylenetriaminepentaacetic acid (Sequestrene 330 Fe).

Table 2. Composition of common basal media — vitamins and organics ($mg\,l^{-1}$)

	Medium code					
	MS	**LS**[a]	**UM**[b]	**B5**	**N6**	**CM**
Reference	—	7	8	—	—	—
Myo-inositol	100	100	100	100	—	180
Casein hydrolysate	—	—	2000	—	—	—
Glycine	2	—	2	—	2.0	—
Glutamine[c]	—	—	—	—	—	146.2
Pyridoxine-HCl	0.5	—	10	1	0.5	0.5
Nicotinic acid	0.5	—	5	1	0.5	—
Thiamine-HCl[c]	0.1	0.4	10	10	1.0	—

[a] Linsmaier and Skoog medium.
[b] Uchimiya and Murashige medium.
[c] Some vitamins and amino acids may be heat-labile and filter sterilization is recommended.

Table 3. Media stock concentrations

Stock	**Concentration**	**Solvent**
Macroelements	10 ×	H_2O
Microelements	100 ×	H_2O
Iron[a]	100 ×	H_2O
Vitamins	100 ×	H_2O
Auxins	1000 ×	0.1 M NaOH, ethanol or DMSO
Cytokinins	1000 ×	0.1 M HCl, ethanol or DMSO

[a] Chelated iron is less prone to precipitation and is conveniently added as NaFeEDTA to provide the required molar quantity of iron. Alternatively, stocks are prepared by adding $FeSO_4 \cdot 7H_2O$ to a hot solution of $Na_2EDTA \cdot 2H_2O$ and stored in darkness.
DMSO, dimethylsulfoxide.

Table 4. Growth regulators: chemical names and molecular weights

Common name	Chemical name	Mol. wt	mg l^{-1} per μM
Auxins			
IAA[ab]	3-indoleacetic acid	175.2	5.71
IBA[b]	3-indolebutyric aicd	203.2	4.92
NAA	1-naphthylacetic acid	186.2	5.37
NOA	2-naphthyloxyacetic acid	202.2	4.95
CPA	4-chlorophenoxyacetic acid	186.2	5.37
2,4-D	2,4-dichlorophenoxyacetic acid	221.0	4.52
2,4,5-T	2,4,5-trichlorophenoxyacetic acid	255.5	3.91
Dicamba	3,6-dichloroanisic acid	221.0	4.52
Picloram	4-amino-3,5,6-trichloropicolinic acid	241.5	4.14
Chloramben	3-amino-2,5-dichlorobenzoic acid	206.0	4.85
Cytokinins			
Zeatin[bc]	4-hydroxy-3-methyl-trans-2-butenylaminopurine	219.3	4.56
2iP[b]	N^6-(2-isopentyl)adenine	203.3	4.92
Adenine[b]	adenine sulfate	368.3	2.72
Kinetin	6-furfurylaminopurine	215.2	4.65
BAP	6-benzylaminopurine	225.3	4.44
Gibberellin			
GA_3[b]	gibberellic acid	346.4	2.89

[a] Also used as amino-acid-conjugated IAA, e.g. 3-indoleacetyl-L-aspartic acid.
[b] Heat-labile; filter-sterilize through 0.22 μm pores.
[c] Also used as zeatin riboside.

Table 5. Other media additives

Additive	Examples
Amino acids	Arginine, asparagine, glutamine
Carbohydrates	Glucose, maltose, mannitol
Vitamins	Ca-D-pantothenate, biotin, folic acid
Anti-ethylene agents	$AgNO_3$, aminoethoxyvinylglycine
Buffers	MES[a]
Undefined additives	Coconut water, casein hydrolysate
Gelling agent	(See *Table 6*)

[a] MES, 2-morpholinoethanesulfonic acid.

Table 6. Gelling agents and concentrations

Gelling agent	Concentration ($g\,l^{-1}$)	Supplier
Phytagel	1–3	Sigma
Bacto-agar	6–8	Difco
Noble-agar	6–8	Difco
Phytagar	6–8	Gibco
Agarose	6–8	(various)
SeaPlaque agarose[a]	6–12	FMC
Alginate[b]	5–20	Sigma
Gelrite[c]	3–4	Merck & Co.

[a] Low melting point allows embedding of cells at physiological temperatures.
[b] Na-alginate solution is reversibly gelled by addition of 5–50 mM Ca^{2+}.
[c] Gelrite cannot be recycled between gel and solution.

1.3. Media and methods for the culture of tobacco

Nicotiana tabacum serves as the most popular model species for tissue culture because of its responsiveness *in vitro* and its amenability to genetic transformation. The cultivars most frequently used for tissue culture are listed in *Table 7.* A range of sterilants and fungicides are available to decontaminate organ explants of tobacco (or any other species) prior to axenic culture and these are presented in *Table 8* with recommended treatment regimes.

The culture medium adopted for tobacco depends principally on the type of explant and the course of development required. *Table 9* lists basal media and additives recommended for the culture of protoplasts, cells, callus, shoots and roots. The specific media compositions required for the sequential steps of protoplast isolation, transformation and culture are recorded in *Table 10.* The inorganic composition of CPW, KF and E media used for protoplast manipulations are detailed in *Table 11.*

Tissue Culture, Transformation

Figure 1. Structures of growth regulators. (a) Auxins, (b) cytokinins, (c) gibberellic acid.

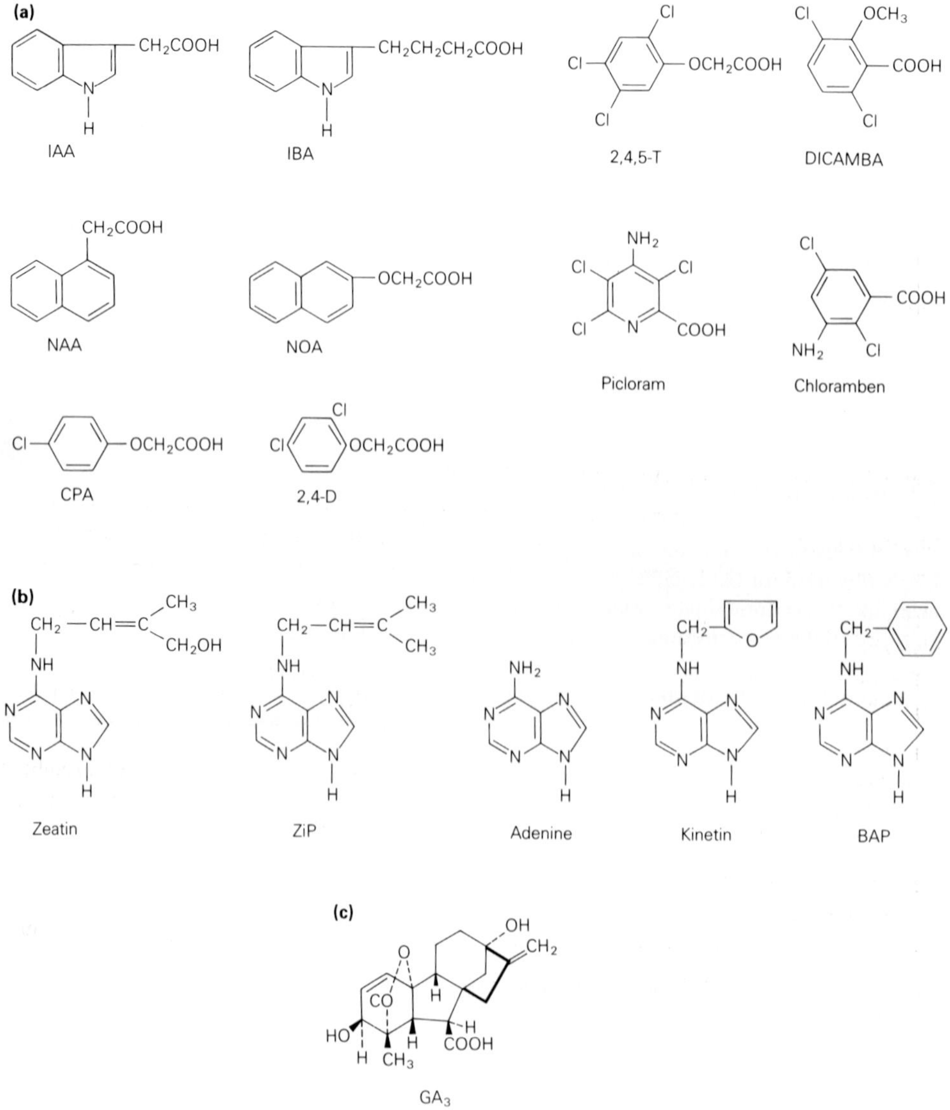

Table 7. Cultivars of tobacco commonly used for tissue culture

	Cultivar
Nicotiana tabacum cv.	Petite Havana SR1[a] Samsun White Burley[b] Wisconsin 38[c] Xanthi

[a] Streptomycin-resistant plastid marker.
[b] Popular for anther culture.
[c] Popular for *Agrobacterium*-induced tumorigenesis tests.

Table 8. Sterilant concentrations for explants

Sterilizing agent	Concentration ($g\ l^{-1}$)	Time (min)
NaOCl[a]	2–20	15
$Ca(OCl)_2$	50	15
Ethanol	70% (v/v)	1
$HgCl_2$[b]	0.1	10
Fungicide		
Benomyl	0.1–10	Overnight
Nystatin	0.04	In medium
Amphotericin B	0.005	In medium
Bacteriocidal antibiotics see *Tables 22* and *23*.		

[a] Commercial or household bleach is commonly used as 5–30%.
[b] Highly toxic.

Tissue Culture, Transformation

Table 9. Media for manipulation of tobacco in culture

Culture type	Salts + vits[a]	Sucrose ($g\ l^{-1}$)	Auxin[b] ($mg\ l^{-1}$)	Cytokinin[b] ($mg\ l^{-1}$)	Supplements ($g\ l^{-1}$)
Cell suspension	MS	30	1.0 2,4-D	0.2 KIN	—
Protoplast regeneration	MS	20	1.0 2,4-D (2.0 NAA)	0.5 BAP	80 mannitol
Protoplast (low density)	CM	20	0.1 NAA	1.0 BAP	80 mannitol
Callus	MS	30	2.0 IAA (2.0 NAA)	0.2 KIN (0.2 BAP)	—
Shoot induction	MS	20	0.2 IAA	2.0 KIN (2.0 BAP)	—
Shoot propagation	MS	20	—	—	—
Root induction[c] or seed germination	$\frac{1}{2}\times$MS	15	—	—	—

Set pH to 5.8 (before autoclaving) and solidify with 8 $mg\ l^{-1}$ agar or agarose

[a] Vitamin recipe of LS or B5 may replace that of MS.
[b] Alternative auxin or cytokinin is shown in parenthesis.
[c] A weak auxin (e.g. IBA) can be used to encourage rooting if necessary.
BAP, 6-benzylaminopurine; CM, Caboche medium; 2,4-D, 2,4-dichlorophenoxyacetic acid; IAA, 3-indoleacetic acid; KIN, 6-furfurylaminopurine; MS, Murashige and Skoog medium; NAA, 1-naphthylacetic acid.

Table 10. Media for tobacco protoplast isolation, transformation and culture

Culture type	Salts	Sugars
Plasmolysis medium	CPW	130 g l^{-1} mannitol
Flotation medium	CPW	200 g l^{-1} sucrose
Washing medium	CPW	30 g l^{-1} sucrose + 90 g l^{-1} mannitol
PEG transformation medium[a]	KF	0.9 g l^{-1} glucose
Electroporation medium	EM	73 g l^{-1} mannitol
Regeneration medium[a]	CM	20 g l^{-1} sucrose, reduce mannitol stepwise from 80 g l^{-1}

[a] See *Table 9* for growth regulators.
CPW, EM, KF, see *Table 11*.
PEG, polyethylene glycol; CM, Caboche medium.

Table 11. Salts for tobacco protoplast culture media (mg l^{-1})

Nutrients[a]	CPW	KF	EM
Reference	9	10	11
$CaCl_2 \cdot 2H_2O$	1470	18 378	—
NaCl	—	8182	—
$MgSO_4 \cdot 7H_2O$	246	—	—
$MgCl_2$	—	—	952
KNO_3	101	—	—
KCl	—	373	—
KH_2PO_4	27.2	—	—
Na_2HPO_4	—	106	—
KI	0.16	—	—
$CuSO_4 \cdot 5H_2O$	0.025	—	—
pH	5.7	7.0	5.6[b]

[a] Plus sugars indicated in *Table 10*.
[b] Include 1 g l^{-1} MES buffer.

Tissue Culture, Transformation

2. GENETIC TRANSFORMATION

The term 'genetic transformation' encompasses a variety of gene transfer events, characterized by the stability of transformation, the subcellular compartment transformed (nuclear, mitochondrial or plastid) and whether the transferred DNA is integrated into the host genome. *Table 12* documents the generally accepted definitions of these alternative transformation events.

2.1. Gene transfer methods and their evaluation

An expanding repertoire of transformation approaches is available, ranging from well-proven techniques to highly experimental (and sometimes controversial) methodologies. In *Table 13* these alternative approaches to gene delivery are listed with brief comments on their application, efficiency and limitations. Stable genetic transformation usually requires the integration of cloned DNA into the nuclear genome of the host (see *Table 12*). Strict criteria must be satisfied to authenticate the transgenic status of a tissue, based on molecular, phenotypic and inheritance evidence. *Table 14* documents the criteria necessary to validate nuclear integration of a transgene. The pattern of inheritance of a transgene is crucial for confirming its nuclear integration, and *Table 15* presents the predicted segregation ratios for a transformed phenotype consistent with Mendelian inheritance.

Table 12. Definitions of transformation

Term	Definitions
Stable transformation	Transgene and novel genetic characteristics are maintained during the life of the culture or plant. The transgene is usually, but not necessarily always, integrated into the host genome
Transient expression	Expression of the transgene is detected in the first few days after its introduction into cells. A subsequent decline in gene activity indicates that expression results largely from nonintegrated DNA
Integrative transformation	The transgene is covalently integrated into the genome of the host cell. In fertile plants the transgene is inherited by the next generation
Nuclear transformation	Gene transfer into the nuclear genome of the host cell, as confirmed by cellular fractionation, eukaryotic-type expression or Mendelian inheritance
Organellar transformation	Gene transfer into the plastid or mitochondrial genome of the host cell, as confirmed by cellular fractionation, prokaryotic-type expression or maternal inheritance
Episomal transformation	Viral genomes or 'mini-chromosomes' which replicate independently from the host genome

Table 13. Status of alternative gene delivery approaches

Gene delivery method	Characteristics
Agrobacterium	Well-established transformation vector for many dicots, and a promising vector for gymnosperms. A wide range of oncogenic and disarmed Ti- or Ri-derived plasmid vectors are available (see *Chapter 4*). Restricted use with monocots, but valuable for delivery of viral genomes to graminaceous hosts by agroinfection. See *Table 16* for references
Direct gene transfer to protoplasts	Well-established transformation technique with no host range limitation. Plasmalemma permeabilized to DNA by chemical agents or electroporation (see *Table 16* for references). Alternatively, genes can be delivered to protoplasts by fusion with DNA in encapsulated liposomes (12)
Microprojectile bombardment	A recently introduced but widely authenticated technique for accelerating DNA-coated particles into walled cells. No intrinsic host-range limitation. Gene transfer to *in situ* chloroplasts has been achieved. See *Table 16* for references
Microinjection	Effective gene delivery technique offering visual targeting to cell type and intracellular compartment. Labor-intensive to process relatively few cells, and requiring specialist skills and equipment (13, 14)
Macroinjection of inflorescence	Technically simple approach to deliver DNA to developing floral tissue by hypodermic needle. Germline transformation has not proved reproducible (15)
Impregnation by whiskers	Suspension cells mixed with DNA and micron-sized whiskers exhibit transient expression and stable transformation (16, 17)
Laser perforation	Transient expression from cells targeted with a laser microbeam in DNA solution (18)
Imbibition of tissues	Transient expression from tissues bathed in DNA solution (19)
Pollen-tube pathway	Claims of germline transformation by treating pollen or carpels with DNA remain controversial (20–23)
Ultrasonication	Stable transformation reported by ultrasonication of explants with DNA. Molecular confirmation is required (24)

Table 14. Criteria for assessing transformation

Criteria	Comment/examples
Molecular genetic evidence	
Southern hybridization analysis	$\geqslant 1$ copy of internal fragment per diploid genome and single copy of each border fragment per diploid genome
Dot blots or slot blots	Less satisfactory alternatives, and should reveal $\geqslant 1$ copy of hybridizing fragments per genome
PCR amplification	Correct size fragment provides supporting evidence for presence of the transgene
Gene expression and phenotypic evidence	
Reporter gene activity (see *Chapter 5*)	e.g. GUS, LUX, CAT or nopaline enzyme activity or immunodetection of reporter protein
Northern hybridization analysis	e.g. transcripts from transgene
Phenotype conferred by selectable marker	e.g. antibiotic resistance, herbicide resistance, hairy root morphology
Screenable phenotypes	e.g. anthocyanin pigmentation, hormone imbalance
Inheritance of transgenes or their activity	
Mendelian inheritance of transgenes	Out-crossed progeny segregate 1:1 and selfed progeny segregate 3:1 (1 dominant allele in parent); reciprocal crosses generate consistent segregation patterns

PCR, polymerase chain reaction.
CAT, chloramphenicol acetyltransferase; GUS, β-glucuronidase; LUX, luciferase (see *Chapter 5*).

Table 15. Segregation ratios for transformed phenotype resulting from transgenes at one to four loci[a]

Number of loci	Out-crossed parent	Selfed parent
1	1:1	3:1
2	3:1	15:1
3	7:1	63:1
4	15:1	255:1

[a] The progeny ratios shown are transgenic phenotype:null phenotype, assuming all loci segregate independently and the trait is dominant. These ratios may be distorted by partial linkage of two or more loci, or by transgene expression influencing gametophyte or embryo fitness.

2.2. Stably transformed plant species

Table 16 lists species which have been transformed and regenerated to plants, grouped according to their taxonomic family. This table was compiled from a comprehensive survey of the literature, but references quoted are restricted to original publications and alternative gene delivery approaches. Excluded from the species list are transformed cells, calli, crown-galls and hairy roots which were not regenerated to whole plants. Also excluded are claims of transformation which lack definitive proof or whose authenticity remains controversial.

Table 16. List of stably transformed plant species

Family and species	Common name	Method[a]	References
Actinidiaceae			
Actinidia deliciosa	Kiwi fruit	A	25
Apocynaceae			
Catharanthus roseus	Rose periwinkle	A	26
Caricaceae			
Carica papaya	Papaya	B	27
Caryophyllaceae			
Dianthus caryophyllus	Carnation	A	28
Casuarinaceae			
Allocasuarina verticillata		A	29
Chenopodiaceae			
Beta vulgaris	Sugar beet	A	30, 31
Compositae			
Bidens alba		A	32
Cichorium intybus	Chicory	A	33
Dendranthema indicum	Chrysanthemum	A	34
Helianthus annuus	Sunflower	A	35
Lactuca sativa	Lettuce	A, D	36, 37
Parthenium argentatum	Guayule	A	38
Convolvulaceae			
Convolvulus arvensis	Morning glory	A	39
Ipomoea batatas	Sweet potato	A	40
Crassulaceae			
Kalanchoe laciniata		A	41
Cruciferae			
Arabidopsis thaliana	Thale cress	A, D	42–46
Armoracia rusticana	Horseradish	A	47
Brassica carinata		A	48
B. juncea	Indian mustard	A	49
B. napus	Oilseed rape	A, C, D	50–55
B. nigra	Black mustard	A	56
B. oleracea	Cauliflower, kale, broccoli	A	57–61
Nasturtium indicum		A	62

Tissue Culture, Transformation

Table 16. Continued

Family and species	Common name	Method[a]	References
Cucurbitaceae			
Cucumis melo	Melon	A	63, 64
C. sativus	Cucumber	A	65, 66
Ericaceae			
Vaccinium macrocarpon	Cranberry	B	67
Gentianaceae			
Eustoma grandiflorum	Lisianthus	A	68
Gramineae			
Dactylis glomerata	Orchardgrass	D	69
Festuca arundinacea	Tall fescue	D	70
Hordeum vulgare	Barley	E	71
Oryza sativa	Rice	D	72–75
Secale cereale	Rye	E	76
Triticum aestivum	Wheat	B	77
Zea mays	Maize	B, D	78–80
Grossulariaceae			
Ribes nigrum	Blackcurrant	A	81
Juglandaceae			
Juglans regia	Walnut	A	82
Labiatae			
Mentha citrata	Mint	A	83
Leguminosae			
Arachis hypogaea	Peanut	B	84
Glycine argyrea	Wild soybean	A	85
G. canescens	Wild soybean	A	86
G. max	Soybean	A, B	87–89
Lotonotis bainesii		A	90
Lotus corniculatus	Birdsfoot trefoil	A	91, 92
Medicago arborea		A	93
M. sativa	Alfalfa	A	94, 95
M. truncatula		A	96
M. varia	Hybrid lucerne	A	97
Onobrychis viciifolia	Sainfoin	A	98
Pisum sativum	Pea	A	99, 100
Psophocarpus tetragonolobus	Winged bean	A	101
Stylosanthes humilis		A	102, 103
Trifolium repens	White clover	A	104
Vicia narbonensis		A	105
Vigna aconitifolia	Moth bean	D	106
V. unguiculata	Cowpea	A	107
Liliaceae			
Asparagus officinalis	Asparagus	A	108

Table 16. Continued

Family and species	Common name	Method[a]	References
Linaceae			
Linum usitatissimum	Flax	A	109, 110
Loganiaceae			
Fagraea fragrans	Tembusu tree	A	111
Magnoliaceae			
Liriodendron tulipifera	Yellow poplar	B	112
Malvaceae			
Gossypium hirsutum	Cotton	A, B	113–115
Moraceae			
Broussonetia kazinoki	Paper mulberry	A	116
Papaveraceae			
Papaver somniferum	Opium poppy	A	117
Pinaceae			
Larix decidua	European larch	A	118
Polygonaceae			
Fagopyrum esculentum	Buckwheat	A	119
Primulaceae			
Anagallis arvensis	Pimpernel	A	120
Rosaceae			
Fragaria ananassa	Strawberry	A, D	121–123
Malus domestica	Cultivated apple	A	124
M. pumila	Downy crab apple	A	125
Prunus armeniaca	Apricot	A	126
P. domestica	Plum	A	127, 128
P. persica	Peach	A	129
Rosa hybrida	Rose	A	130
Rubus spp.	Blackberry, raspberry	A	131, 132
Rutaceae			
Citrus spp.	Orange	A	133, 134
C. jambhiri		D	135
Salicaceae			
Populus spp.	Poplar, aspen	A	136–138
Scrophulariaceae			
Antirrhinum majus	Snapdragon	A	139
Solanaceae			
Atropa belladonna	Deadly nightshade	A	140
Capsicum annuum	Bell pepper	A	141
C. frutescens	Sweet pepper	A	142
Datura innoxia		A	143

Table 16. Continued

Family and species	Common name	Method[a]	References
Hyoscyamus muticus	Egyptian henbane	A	144
Lycopersicon esculentum	Tomato	A	145–147
L. esculentum × pinnellii		A	148
L. peruvianum		D	149
Nicotiana alata	Ornamental tobacco	A	150
N. benthamiana		A	151
N. bigelovii		A	152
N. clevelandii		A	153
N. debneyi		A	154
N. edwardsonii		A	155
N. glauca		A	156, 157
N. hesperis		A	158
N. plumbaginifolia		A	159–161
N. repanda		A	162
N. rustica		A	163
N. sylvestris		A	164
N. tabacum	Tobacco	A, B, D	165–168
Petunia hybrida	Petunia	A	169, 170
Solanum dulcamara	Woody nightshade	A	171, 172
S. integrifolium		A	173
S. melongena	Eggplant	A	174, 175
S. muricatum	Pepino	A	176
S. nigrum	Black nightshade	A	177, 178
S. tuberosum	Potato	A	179–182
Umbelliferae			
Apium graveolens	Celery	A	183
Daucus carota	Carrot	A	184–187
Foeniculum vulgare	Fennel	A	188
Vitaceae			
Vitis vinifera	Grapevine	A	189, 190

[a] A, *Agrobacterium*; B, bombardment; C, microinjection; D, direct gene transfer to protoplasts; E, macroinjection into tillers.

2.3. *Agrobacterium* strains, media and methods

Wild-type strains of *Agrobacterium tumefaciens* (harboring Ti plasmids) and *Agrobacterium rhizogenes* (harboring Ri plasmids) commonly used for gene transfer to plants are listed in *Table 17*. The strains are grouped by plasmid type and listed according to the opine synthetic genes present. Genetically defined C58 strains harboring either pTiC58 or transconjugant plasmids are widely employed and *Table 18* records the antibiotic-resistance markers on the chromosome and the plasmid type associated with each strain. Disarmed Ti and Ri plasmids which lack the oncogenes (causing tumorigenesis or rhizogenesis, respectively) are exploited either as co-integrate vectors or as virulence 'helper' plasmids for binary vectors (see *Chapter 4*). *Table 19* presents information on the antibiotic-resistance markers and the

Table 17. Wild-type (oncogenic) strains of *Agrobacterium*

Opine group[a]	**Strain**	**Plasmid**
Nopaline	C58	pTiC58
	T37	pTiT37
Octopine	Ach5	pTiAch5
	A6	pTiA6
	B6	pTiB6
Succinamopine	Bo542	pTiBo542
Agropine	A4	pRiA4
	1855	pRi1855
	15834	pRi15834
Cucumopine	2659	pRi2659
Mannopine	8196	pRi8196

[a] Some strains encode the synthesis of two or more opines.

Table 18. Transconjugant C58 *Agrobacterium* strains

Strain	**Resistance marker**[a]	**Plasmid**
A136 or LBA288	rif	(none)
GV3101 or LBA958	rif	pTiC58
A208	rif, nal	pTiT37
A348	rif	pTiA6
LBA1010	rif	pTiB6
GV3111	chl	pTiB6S3
A281	rif, nal	pTiBo542
R1000	rif	pRiA4b

[a] rif, rifampicin; nal, nalidixic acid; chl, chloramphenicol.

number of T-DNA borders associated with a range of disarmed plasmids. The optional methods for introducing vector plasmids into *Agrobacterium* are listed in *Table 20*, including details of the conjugation plasmids required for triparental mating. *Table 4* in *Chapter 2* lists methods used for the isolation of Ti plasmid DNA.

The media most frequently used for culture of *Agrobacterium* are detailed in *Table 21*. LB, YEB and YMB are rich media adopted for routine culture, whereas AB minimal medium is useful to favor *Agrobacterium* growth after conjugation with *E. coli*. Antibiotics recommended for controlling *Agrobacterium* overgrowth of plant cultures following co-cultivation techniques are listed in *Table 23*. For information on antibiotic stocks see *Table* 22.

Table 19. Derivation of disarmed Ti and Ri plasmids

Oncogenic parent	Disarmed plasmid	Resistance marker[a]	Number of T-borders	Ref.
pTiC58	pGV3850	amp	2	191
pTiC58	pMP90	gen	0	192
pTiC58	pTiC58-Z707	kan	0	193
pTiT37	pTiT37-SE	kan	1	194
pTiAch5	pAL4404[b]	str	0	195
pTiB6	pGV2260	amp	0	196
pTiB6S3	pTiB6S3-SE	kan	1	197
pTiBo542	pEHA101	kan	0	198
pRiA4	pRiB278b	kan	0	199

[a] amp, ampicillin; gen, gentamycin; kan, kanamycin; str, streptomycin.
[b] Strain LBA4404 has an Ach5 chromosome with a rifampicin-resistance marker.

Table 20. Introducing vector plasmids into *Agrobacterium*

Freeze/thaw: Freeze/thaw solution, 20 mM $CaCl_2$ (200)

Electroporation: Electroporation buffer, 10% (v/v) glycerol (201, 202)

Conjugation: Triparental mating between donor strain (harboring vector plasmid), mobilizing strain and recipient *Agrobacterium* strain

Vector type	Replicon	Conjugation plasmids	Mobilizing strain	Ref.
Intermediate	ColE1	pGJ28 + pR64*drd*11	*E. coli* GJ23	203
Binary	RK2	pRK2013	*E. coli* HB101	204

Physical and chemical conditions which influence the induction of virulence functions in *Agrobacterium* are presented in *Table 24*. The stringency of these requirements is highly variable and should be optimized by empirical experimentation. The molecular structure of the inducer acetosyringone is shown in *Figure 2*. *Agrobacterium* vectors have been exploited to introduce viral genomes into host plants by a process termed 'agroinfection', and has proved particularly useful for introducing recombinant DNA into cereals and grasses which are difficult to transform. *Table 25* records gemini viruses and graminaceous hosts used in agroinfection.

Table 21. Culture media for *Agrobacterium* (see also *Table 22*)

Medium ingredients	Concentration ($g\,l^{-1}$)
LB[a]	
Tryptone	10
Yeast extract	5
NaCl	5
YEB	
Beef extract	5
Peptone	5
Sucrose	5
Yeast extract	1
$MgSO_4 \cdot 7H_2O$	0.5
YMB[b]	
Mannitol	10
K_2HPO_4	0.5
Yeast extract	0.4
$MgSO_4 \cdot 7H_2O$	0.2
NaCl	0.1
AB minimal[c]	
K_2HPO_4	3
NaH_2PO_4	1
NH_4Cl	1
$MgSO_4 \cdot 7H_2O$	0.3
KCl	0.15
$CaCl_2$	0.01
$FeSO_4 \cdot 7H_2O$	0.0025
Glucose	5
Set pH of each medium to 7.0–7.2 and gel with 15 $g\,l^{-1}$ bacto-agar if required	

[a] Liquid cultures may benefit from 1 $g\,l^{-1}$ glucose.
[b] Commonly used for *A. rhizogenes*.
[c] Autoclave 20-fold stocks individually before dilution.

Tissue Culture, Transformation

Table 22. Antibiotics for plant and bacterial culture work (compiled by R.R.D. Croy)

Antibiotic	Salt	Mol. wt	Type	Working concentrations (μg ml^{-1})*				Stock concentration and solvent**	Sterilization and storage conditions***	Comments
				General range	*E. coli*	*A. tumefaciens*	Transgenic plant tissue[a]			
Ampicillin	Na	349.4	Penicillin (β-lactam)	10–100	50–100	—	—	25–50 mg ml^{-1} H_2O	Filter/−20°C	Bacterial selection (*E. coli*, *Agrobacterium*); removal of *Agrobacterium* from transformed plant tissues
Augmentin – mixture of (i) potassium clavulanate[b] and (ii) amoxycillin sodium (see also *Table 23*)	K + Na	—	Penicillin (β-lactam), Clavam	100–500	—	—	100–500	1 mg ml^{-1} H_2O fresh	Store dry powder in sealed containers at 4°C	Broad-spectrum antibiotic with particularly efficient use for the elimination of *Agrobacterium* from plant transformations. Clavulanate is a potent inhibitor of β-lactamase and greatly increases the effectiveness of the amoxycillin
Bleomycin A2/B2	SO_4	1415.6 1425.5	Glycopeptide	5–10	—	—	5–10	50 mg ml^{-1} H_2O	Filter/−20°C	Selection of transgenic plant tissues (see *Chapter 5*)
Carbenicillin	Di-Na	378.4	Penicillin (β-lactam)	100–500	100–200	100	250–1000	50–100 mg ml^{-1} H_2O	Filter/4°C or −20°C	Protoplast medium; plant growth medium; removal of *Agrobacterium* from transformed plant tissues

Cefotaxime	Na	455.5	Cephalosporin (β-lactam)	100–250	—	250	250	50–250 mg ml^{-1} H_2O	Filter/-20°C, dark	Removal of *Agrobacterium* from transformed plant tissues
Chloramphenicol	—	323.1	Chloram-phenicol	10–200	25–30 200[c]	25–170	—	10–200 mg ml^{-1} EtOH	Solvent/-20°C	Bacterial selection (*E. coli, Agrobacterium*) amplification
Cycloheximide	—	281.4	Cycloheximide	50–100	50	75–100	—	20 mg ml^{-1} H_2O	Filter/-20°C	Effective antibiotic against yeasts and molds
Erythromycin	—	733.9	Macrolide	25–50	—	50	—	25 mg ml^{-1} EtOH	Solvent/-20°C	Bacterial selection
Gentamicin A	SO_4	468.5	Amino-glycoside	10–15	5–10	2–40	—	10 mg ml^{-1} H_2O	Filter/-20°C	Protoplast medium; bacterial selection
Gentamicin G418 (Geneticin)	Di-SO_4	692.7	Amino-glycoside	10–15	—	—	10–2000	10 mg ml^{-1} H_2O	Filter/-20°C	Used as an alternative to kanamycin. Levels required for selection of different plant tissues may vary over a wide range – determine experimentally
Hygromycin B	—	527.5	Amino-glycoside	10–400	—	20–50	10–400	10 mg ml^{-1} H_2O	Filter/4°C or -20°C	Selection of transgenic plant tissues (see *Chapter 5*); bacterial selection
Kanamycin A	SO_4	484.5	Amino-glycoside	25–100	20–100	10–100	75–2000	10–100 mg ml^{-1} H_2O 50% EtOH	Filter/4°C or -20°C Solvent/-20°C	Selection of transgenic plant tissues (see *Chapter 5*)
Kasugamycin	HCl or SO_4	415.8	—	100–1000	—	—	—	10 mg ml^{-1} H_2O	Filter/-20°C	General bacteriocidal antibiotic
Methotrexate (amethopterin)	—	454.5	Folic acid antagonist	0.05–1	—	0.5–1.0	0.05–1.0	1 mg ml^{-1} H_2O (acid)	Filter/-20°C	Selection of transgenic plant tissues (see *Chapter 5*)

Table 22. Continued

Antibiotic	Salt	Mol. wt	Type	Working concentrations ($\mu g\ ml^{-1}$)*				Stock concentration and solvent**	Sterilization and storage conditions***	Comments
				General range	*E. coli*	*A. tumefaciens*	Transgenic plant tissue[a]			
Nalidixic acid	Na	254.2	Quinolone	15–20	20	15–20	—	5–20 mg ml^{-1} H_2O 10 mg ml^{-1} 0.1 M NaOH	Filter/−20°C	Selection of *Agrobacterium*
Neomycin B	SO_4	908.9	Amino-glycoside	25–250	25–50	40–50	50–250	12.5 mg ml^{-1} H_2O	Filter/−20°C	Selection of transgenic plant tissues (see *Chapter 5*)
Nystatin (mycostatin)	—	926.1	Polyene	6–40	—	6	6	10 mg ml^{-1} MeOH slightly soluble in water	Solvent, use freshly. Sterilize powder surface with DMSO then suspend in sterile water	Powerful anti-fungal agent
Rifampicin	—	823.0	Ansamycin	10–150	100	10–150	20–50	20–50 mg ml^{-1} MeOH 50 mg ml^{-1} DMSO	Use freshly. Solvent/−20°C, dark	Broad-range anti-bacterial antibiotic; *Agrobacterium* selection
Spectinomycin	Di-HCl	332.4	Amino-glycoside	50–100	50–100	100	—	10–50 mg ml^{-1} H_2O	Filter/−20°C	Bacterial selection (*E. coli*, *Agrobacterium*); selection of transgenic plant tissues
Streptomycin	SO_4	581.6	Amino-glycoside	10–300	10–100	100–300	—	10 mg ml^{-1} H_2O	Filter/−20°C	Bacterial selection (*E. coli*, *Agrobacterium*); selection of transgenic plant tissues

Sulphonamide	—	247	Sulphathiazol	100–500	250	250	100–500	125 mg ml^{-1} DMSO	Solvent/ − 20°C	Selection of transgenic plant tissues (see *Chapter 5*)
Tetracycline	HCl	444.4	Tetracycline	3–20	3–15	2.5–20	10	10 mg ml^{-1} H_2O 5 mg ml^{-1} DMSO 15 mg ml^{-1} 50% EtOH	Filter/4°C or − 20°C Solvent/ − 20°C Filter/ − 20°C, dark	Protoplast medium; bacterial selection
Trimethoprim	—	290.3	Pteridine analog	5–1000	10	1000	5–50	10 mg ml^{-1} H_2O (acidic)	Filter/ − 20°C	Selection of transgenic plant tissues (see *Chapter 5*)
Vancomycin	HCl	1485.7	Glycopeptide	50–100	—	100	—	10 mg ml^{-1} H_2O	Filter/ − 20°C	Effective against Gram + ve bacterial pathogens; removal of Agrobacteria

*Concentration ranges represent minimum and maximum levels quoted in the literature.
**Solvents: DMSO, dimethyl sulfoxide; MeOH, methanol, EtOH, ethanol.
***Sterilization: filter; filtration through 0.22 μm filter: solvent; solvent system is adequate to sterilize the solution without any further treatment. Storage: 4°C, store stock solution at 4°C for up to 1 month; − 20°C, store stock solution frozen at − 20°C, most antibiotic solutions stable for several months in this form.
[a] Antibiotics are used to remove *Agrobacterium* from insensitive transformed plant tissues or to select plant tissues transformed with genes encoding antibiotic resistance; the level of antibiotic selection in this latter case will depend on the strength of the promoter expressing the resistance gene.
[b] Clavulanate is an inhibitor of β-lactamase which destroys β-lactam antibiotics; its inclusion in this mixture extends the range and effectiveness of the penicillin antibiotic.
[c] For plasmid amplification.

Table 23. Antibiotics to control *Agrobacterium* (see also *Table 22*)

Antibiotic	Concentration (mg l^{-1})	Supplier
Penicillins		
Carbenicillin	250–500	Duchefa, Sigma
Ampicillin	500	Duchefa, Sigma
Cloxacillin[b]	500	Sigma
Cephalosporins		
Cefotaxime[b]	250–500	Duchefa, Sigma
Cefoxitin[b]	500	Sigma
Penicillin + clavulanic acid[c]		
Augmentin = Amoxycillin + clavulanic acid	200–500	Duchefa
Timentin = Ticarcillin + clavulanic acid	200–500	Duchefa
Others		
Vancomycin	100	Duchefa, Sigma

[a] The concentrations presented are those commonly used to inhibit *Agrobacterium* growth immediately after co-cultivation. At these high concentrations the antibiotics sometimes have side-effects on plant culture responses. After one or two subcultures the concentration can be reduced stepwise to eliminate any remaining bacteria.

[b] Exhibit partial resistance to β-lactamase.

[c] Clavulanic acid is an inhibitor of β-lactamase. The inclusion of clavulanic acid can shift the pH of the medium unless it is adequately buffered.

Table 24. Conditions for *Agrobacterium* virulence induction

Factor	Range	Ref.
Acetosyringone ($M_r = 196.2$)	10–200 μM	205
or α-hydroxyacetosyringone ($M_r = 212.2$)		205
pH	<5.7	206
Temperature	20–25°C	207
Glucuronic acid, xylose (plus others)	1–10 mM	208

Figure 2. Structure of acetosyringone.

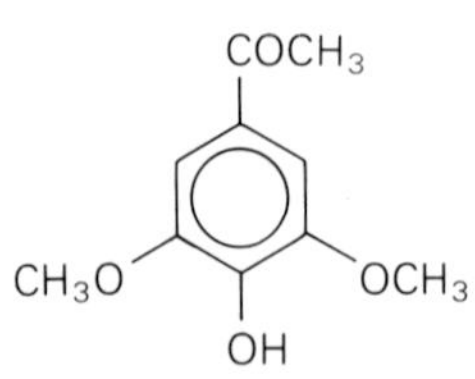

Table 25. Agroinfection of Gramineae with gemini viruses

Viral genome	Plant host	Refs
MSV	*Zea mays*	209, 210
(maize streak)	*Avena sativa*	211
	Digitaria sanguinalis	211
	Hordeum vulgare	211
	Lolium temulentum	211
	Panicum milaceum	211
WDV	*Triticum aestivum*	212–214
(wheat dwarf)	*T. durum*	215
	T. monococcum	215
	Aegilops speltoides	215
	Hordeum vulgare	216
	Zea mays	217
DSV	*Avena sativa*	218
(*Digitaria* streak)	*Digitaria sanguinalis*	218
	Zea mays	218
MiSV	*Panicum milaceum*	219
(*Miscanthus* streak)	*Zea mays*	219

2.4. Protoplast transformation

The cellulase and pectinase enzymes listed in *Table 26* are widely used to generate protoplasts from plant tissues. DNA uptake by protoplasts is generally facilitated by electroporation or by the chemicals listed in *Table 27.*

2.5. Particle bombardment of tissues

Genes can be introduced into walled cells by high-velocity bombardment with DNA-coated microprojectiles using propulsion methods presented in *Table 28.* Particle guns designed specifically for plant transformation are commercially available from Bio-Rad. A standard recipe for coating microprojectiles with DNA is given in *Table 29.*

Table 26. Enzymes for protoplast production

Enzymes[a]	Supplier
Cellulases	
Onozuka R-10	Yakult Honsha
Onozuka SS	All Japan Biochemicals
Cellulysin	Calbiochem
Cellulase CEL	Worthington Diagnostics
Driselase	Unwin
Hemicellulase	Sigma
Pectinases	
Macerozyme R-10	Yakult Honsha
Macerase	Calbiochem
Pectolyase Y-23	Unwin
Pectinase SS	Yakult Honsha
Pectinol R-10	Rohm and Haas
Meicelase P	Meiji Seika Kaisha

[a] Enzymes normally used at 1–20 mg ml^{-1} at pH 5.6. Concentration and time should be determined empirically.

Table 27. Chemicals facilitating DNA uptake by protoplasts

Chemicals	Concentration ($g\,l^{-1}$)	Ref.
Polyethyleneglycol 6000 ($+Ca^{2+}$)	130	220
Polyethyleneglycol 4000 ($+Mg^{2+}$)	200–280	221
Polyvinylalcohol ($+Ca^{2+}$)	100	222
Lipofectin	0.025	223
Polybrene	0.03	224

Table 28. Propulsion methods of particle guns

Explosive mechanism	Examples	Ref.
Gunpowder	Biolistics/DuPont	225
Electric arc H_2O vaporization	Agracetus	226
Pressurized gas (e.g. N_2)	Pneumatic	227
Pressurized gas (e.g. He)	Bio-Rad/DuPont PDS-1000/He	228

2.6. Examples of tissue-specific transient expression

Transient expression of marker genes has been demonstrated using a wide range of tissues and delivery systems, and examples of these are presented in *Table 30*.

Table 29. DNA coating of microprojectiles

Recipe	Volume (μl)
50 mg ml^{-1} tungsten particles[a] in H_2O	25
4 μg DNA in TE buffer	5
2.5 M $CaCl_2$	10
0.1 M spermidine (free base)	2.5
Ethanol[b]	30

[a] 1 μm diameter particles are recommended for most purposes; gold can substitute for tungsten.
[b] Optional wash and resuspend.

Table 30. Tissue-specific transient expression

Tissue/organ	Delivery system	Ref.
Suspension	Dgt[a] protoplasts	229
Mesophyll	Dgt protoplasts	230
Leaf base	Tissue electroporation	231
Cotyledon	Dgt protoplasts	232
Embryo	Particle bombardment	233
Aleurone	Dgt protoplasts	234
Endosperm	Dgt protoplasts	235
Epidermis	Particle bombardment	236
Floral	Dgt protoplasts	237
Pollen	Particle bombardment	238

[a] Dgt, direct gene transfer.

3. REFERENCES

1. George, E.F. and Sherrington, P.D. (1984) *Plant Propagation by Tissue Culture*. Exegetics, Basingstoke.

2. George, E.F., Puttock, D.J.M. and George, H.J. (1987) *Plant Culture Media: Vol. 1, Formulation and Uses*. Exegetics, Westbury.

3. Murashige, T. and Skoog, F. (1962) *Physiol. Plant.*, **15**, 473.

4. Gamborg, O.L., Miller, R.A. and Ojima, K. (1968) *Exp. Cell Res.*, **50**, 151.

5. Chu, C., Wang, C. and Sun, C. (1975) *Scientia Sinica*, **18**, 659.

6. Caboche, M. (1980) *Planta*, **149**, 7.

7. Linsmaier, E.M. and Skoog, F. (1965) *Physiol. Plant.*, **18**, 100.

8. Uchimiya, H. and Murashige, T. (1974) *Plant Physiol.*, **54**, 936.

9. Banks, M.S. and Evans, P.K. (1976) *Plant Science Lett.*, **7**, 409.

10. Krens, F.A., Molendijk, L., Wullems, G.J. and Schilperoort, R.A. (1982) *Nature*, **296**, 72.

11. Shillito, R.D., Saul, M.W., Paszkowski, J., Muller, M. and Potrykus, I. (1985) *Bio/technol.*, **3**, 1099.

12. Caboche, M. and Deshayes, A. (1984) *Comptes Rendus Acad. Sci.*, **299**, series 3, 663.

13. Crossway, A., Oakes, J.V., Irvine, J.M., Ward, B., Knauf, V.C. and Shewmaker, C.K. (1986) *Mol. Gen. Genet.*, **202**, 179.

14. Reich, T.J., Iyer, V.N. and Miki, B.L. (1986) *Bio/technol.*, **4**, 1001.

15. De la Pena, A., Lorz, H. and Schell, J. (1987) *Nature*, **325**, 274.

16. Kaeppler, H.F., Gu, W., Somers, D.A., Rines, H.W. and Cockburn, A.F. (1990) *Plant Cell Rep.*, **9**, 415.

17. Kaeppler, H.F., Somers, D.A., Rines, H.W. and Cockburn, A.F. (1992) *Theor. Appl. Genet.*, **84**, 560.

18. Weber, G. (1988) *Naturwissenschaften*, **75**, 35.

19. Topfer, R., Gronenborn, B., Schell, J. and Steinbiss, H.-H. (1989) *Plant Cell*, **1**, 133.

20. Zhou, G.-Y., Weng, J., Zeng, Y., Huang, J., Qian, S. and Liu, G. (1983) *Methods Enzymol.*, **101**, 433.

21. De Wet, J.M.J., Bergquist, R.R., Halan, J.R., Brink, D.E., Cohen, C.E., Newell, C.A. and De Wet, A.E. (1985) in *The Experimental Manipulation of Ovule Tissues* (G.P. Chapman, S.H. Mantell and R.W. Daniels eds). Longman, New York, p. 197.

22. Hess, D. (1987) *Intern. Rev. Cytol.*, **107**, 367.

23. Luo, Z.-X. and Wu, R. (1989) *Plant Mol. Biol. Rep.*, **7**, 69.

24. Zhang, L.-J., Cheng, L.-M., Xu, N., Zhao, N.-M., Li, C.-G., Yuan, J. and Jia, S.-R. (1991) *Bio/technol.*, **9**, 994.

25. Uematsu, C., Murase, M., Ichikawa, H. and Imamura, J. (1991) Plant Cell Rep., **10**, 286.

26. Brillanceau, M.-H., David, C. and Tempe, J. (1989) *Plant Cell Rep.*, **8**, 63.

27. Fitch, M.M.M., Manshardt, R.M., Gonsalves, D., Slightom, J.L. and Sanford, J.C. (1990) *Plant Cell Rep.*, **9**, 189.

28. Lu, C.-Y., Nugent, G., Wardley-Richardson, T., Chandler, S.F., Young, R. and Dalling, M.J. (1991) *Bio/technol.*, **9**, 864.

29. Phelep, M., Petit, A., Martin, L., Duhoux, E. and Tempe, J. (1991) *Bio/technol.*, **9**, 461.

30. Lindsey, K. and Gallois, P. (1990) *J. Exp. Bot.*, **41**, 529.

31. D'Halluin, K., Bossut, M., Bonne, E., Mazur, B., Leemans, J. and Botterman, J. (1992) *Bio/technol.*, **10**, 309.

32. Norton, R.A. and Towers, G.H.N. (1984) *Can. J. Bot.*, **62**, 408.

33. Tepfer, D. (1990) *Physiol. Plant.*, **79**, 140.

34. Ledger, S.E., Deroles, S.C. and Given, N.K. (1991) *Plant Cell Rep.*, **10**, 195.

35. Everett, N.P., Robinson, K.E.P. and Mascarenhas, D. (1987) *Bio/technol.*, **5**, 201.

36. Michelmore, R., Marsh, E., Seely, S. and Landry, B. (1987) *Plant Cell Rep.*, **6**, 439.

37. Chupeau, M.-C., Bellini, C., Guerche, P., Maisonneuve, B., Vastra, G. and Chupeau, Y. (1989) *Bio/technol.*, **7**, 503.

38. Backhaus, R.A. and Ho, J. (1989) *J. Cell. Biochem. Suppl.*, **13**, 252.

39. Tepfer, D. (1984) *Cell*, **37**, 959.

40. Prakash, C.S., Varadarajan, U. and Kumar, A.S. (1991) *Hortscience*, **26**, 492.

41. Jia, S.-R., Yang, M.-Z., Ott, R. and Chua, N.-H. (1989) *Plant Cell Rep.*, **8**, 336.

42. Feldmann, K.A. and Marks, M.D. (1987) *Mol. Gen. Genet.*, **208**, 1.

43. Sheikholeslam, S.N. and Weeks, D.P. (1987) *Plant Mol. Biol.*, **8**, 291.

44. Schmidt, R. and Willmitzer, L. (1988) *Plant Cell Rep.*, **7**, 583.

45. Valvekens, D., Van Montagu, M. and Van Lijsebettens, M. (1988) *Proc. Natl. Acad. Sci. USA*, **85**, 5536.

46. Damm, B., Schmidt, R. and Willmitzer, L. (1989) *Mol. Gen. Genet.*, **217**, 6.

47. Noda, T., Tanaka, N., Mano, Y., Nabeshima, S., Ohkawa, H. and Matsui, C. (1987) *Plant Cell Rep.*, **6**, 283.

48. Narasimhulu, S.B., Kirti, P.B., Mohaptra, T., Prakash, S. and Chopra, V.L. (1992) *Plant Cell Rep.*, **11**, 359.

49. Mathews, H., Bharathan, N., Litz, R.E., Narayanan, K.R., Rao, P.S. and Bhatia, C.R. (1990) *Plant Science*, **72**, 245.

50. Ooms, G., Bains, A., Burrell, M., Karp, A., Twell, D. and Wilcox, E. (1985) *Theor. Appl. Genet.*, **71**, 325.

51. Fry, J., Barnason, A. and Horsch, R.B. (1987) *Plant Cell Rep.*, **6**, 321.

52. Guerche, P., Charbonnier, M., Jouanin, L., Tourneur, C., Paszkowski, J. and Pelletier, G. (1987) *Plant Science*, **52**, 111.

53. Neuhaus, G., Spangenberg, G., Mittelsten Scheid, O. and Schweiger, H.-G. (1987) *Theor. Appl. Genet.*, **75**, 30.

54. Pua, E.-C., Mehra-Palta, A., Nagy, F. and Chua, N.-H. (1987) *Bio/technol.*, **5**, 815.

55. Moloney, M.M., Walker, J.M. and Sharma, K.K. (1989) *Plant Cell Rep.*, **8**, 238.

56. Sacristan, M.D., Gerdemann-Knorck, M. and Schieder, O. (1989) *Theor. Appl. Genet.*, **78**, 194.

57. David, C. and Tempe, J. (1988) *Plant Cell Rep.*, **7**, 88.

58. Srivastava, V., Reddy, A.S. and Guha-Mukherjee, S. (1988) *Plant Cell Rep.*, **7**, 504.

59. De Block, M., De Brouwer, D. and Tenning, P. (1989) *Plant Physiol.*, **91**, 694.

60. Hosoki, T., Shiraishi, K., Kigo, T. and Ando, M. (1989) *Scientia Horticult.*, **40**, 259.

61. Hosoki, T., Kigo, T. and Shiraishi, K. (1991) *J. Japan Soc. Hortic. Sci.*, **60**, 71.

62. Ghosh, S.K., Dey, N., Bhattacharya, S. and Mandal, R.K. (1992) *Current Sci.*, **63**, 44.

63. Fang, G. and Grumet, R. (1990) *Plant Cell Rep.*, **9**, 160.

64. Dong, J.-Z., Yang, M.-Z., Jia, S.-R. and Chua, N.-H. (1991) *Bio/technol.*, **9**, 858.

65. Trulson, A.J., Simpson, R.B. and Shahin, E.A. (1986) *Theor. Appl. Genet.*, **73**, 11.

66. Chee, P.P. (1990) *Plant Cell Rep.*, **9**, 245.

67. Serres, R., McCown, B., McCabe, D., Stang, E., Russell, D. and Martinell, B. (1990) *Hortscience*, **25**, 1130.

68. Ledger, S.E., Given, N.K., Butcher, S.M. and Gardner, R.C. (1990) *J. Cell. Biochem. Suppl.*, **14**, 283.

69. Horn, M.E., Shillito, R.D., Conger, B.V. and Harms, C.T. (1988) *Plant Cell Rep.*, **7**, 469.

70. Wang, Z., Takamizo, T., Iglesias, V.A., Osusky, M., Nagel, J. and Spangenberg, G. (1992) *Bio/technol.*, **10**, 691.

71. Rogers, S.W. and Rogers, J.C. (1992) *Plant Mol. Biol.*, **18**, 945.

72. Toriyama, K., Arimoto, Y., Uchimiya, H. and Hinata, K. (1988) *Bio/technol.*, **6**, 1072.

73. Zhang, H.M., Yang, H., Rech, E.L., Golds, T.J., Davis, A.S., Mulligan, B.J., Cocking, E.C. and Davey, M.R. (1988) *Plant Cell Rep.*, **7**, 379.

74. Zhang, W. and Wu, R. (1988) *Theor. Appl. Genet.*, **76**, 835.

75. Shimamoto, K., Terada, R., Izawa, T. and Fujimoto, H. (1989) *Nature* **338**, 274.

76. De la Pena, A., Lorz, H. and Schell, J. (1987) *Nature*, **325**, 274.

77. Vasil, V., Castillo, A.M., Fromm, M.E. and Vasil, I.K. (1992) *Bio/technol.*, **10**, 667.

78. Rhodes, C.A., Pierce, D.A., Mettler, I.J., Mascarenhas, D. and Detmer, J.J. (1988) *Science*, **240**, 204.

79. Fromm, M.E., Morrish, F., Armstrong, C., Williams, R., Thomas, J. and Klein, T.M. (1990) *Bio/technol.*, **8**, 833.

80. Gordon-Kamm, W.J., Spencer, T.M., Mangano, M.L., Adams, T.R., Daines, R.J., Start, W.G., O'Brien, J.V., Chambers, S.A., Adams, W.R., Willetts, N.G., Rice, T.B., Mackey, C.J., Krueger, R.W., Kausch, A.P. and Lemaux, P.G. (1990) *Plant Cell*, **2**, 603.

81. Graham, J. and McNicol, R.J. (1991) *Plant Cell Tissue Organ Culture*, **24**, 91.

82. McGranahan, G.H., Leslie, C.A., Uratsu, S.L., Martin, L.A. and Dandekar, A.M. (1988) *Bio/technol.*, **6**, 800.

83. Spencer, A., Hamill, J.D. and Rhodes, M.J.C. (1990) *Plant Cell Rep.*, **8**, 601.

84. Schnall, J.A. and Weissinger, A.K. (1992) *In Vitro*, **28**, 122A.

85. Kumar, V., Jones, B. and Davey, M.R. (1991) *Plant Cell Rep.*, **10**, 135.

86. Rech, E.L., Golds, T.J., Hammatt, N., Mulligan, B.J. and Davey, M.R. (1988) *J. Exp. Bot.*, **39**, 1275.

87. Hinchee, M.A.W., Connor-Ward, D.V., Newell, C.A., McDonnell, R.E., Sato, S.J., Gasser, C.S., Fischhoff, D.A., Re, D.B., Fraley, R.T. and Horsch, R.B. (1988) *Bio/technol.*, **6**, 915.

88. McCabe, D.E., Swain, W.F., Martinell, B.J. and Christou, P. (1988) *Bio/technol.*, **6**, 923.

89. Chee, P.P., Fober, K.A. and Slightom, J.L. (1989) *Plant Physiol.*, **91**, 1212.

90. Wier, A.T., Thro, A.M., Flores, H.E. and Janyes, J.M. (1988) *Intl. J. Exp. Bot.*, **48**, 123.

91. Jensen, J.S., Marcker, K.A., Otten, L. and Schell, J. (1986) *Nature*, **321**, 669.

92. Petit, A., Stougaard, J., Kuhle, A., Marcker, K.A. and Tempe, J. (1987) *Mol. Gen. Genet.*, **207**, 245.

93. Damiani, F. and Arcioni, S. (1991) *Plant Cell Rep.*, **10**, 300.

94. Shahin, E.A., Spielmann, A., Sukhapinda, K., Simpson, R.B. and Yashar, M. (1986) *Crop Science*, **26**, 1235.

95. Sukhapinda, K., Spivey, R. and Shahin, E.A. (1987) *Plant Mol. Biol.*, **8**, 209.

96. Thomas, M.R., Rose, R.J. and Nolan, K.E. (1992) *Plant Cell Rep.*, **11**, 113.

97. Deak, M., Kiss, G.B., Koncz, C. and Dudits, D. (1986) *Plant Cell Rep.*, **5**, 97.

98. Golds, T.J., Lee, J.Y., Husnain, T., Ghose, T.K. and Davey, M.R. (1991) *J. Exp. Bot.*, **42**, 1147.

99. De Kathen, A. and Jacobsen, H.-J. (1990) *Plant Cell Rep.*, **9**, 276.

100. Puonti-Kaerlas, J., Eriksson, T. and Engstrom, P. (1990) *Theor. Appl. Genet.*, **80**, 246.

101. Tepfer, D. (1990) *Physiol. Plant.*, **79**, 140.

102. Manners, J.M. (1988) *Plant Science*, **55**, 61.

103. Manners, J.M. and Way, H. (1989) *Plant Cell Rep.*, **8**, 341.

104. White, D.W.R. and Greenwood, D. (1987) *Plant Mol. Biol.*, **8**, 461.

105. Pickardt, T., Meixner, M., Schade, V. and Schieder, O. (1991) *Plant Cell Rep.*, **9**, 535.

106. Kohler, F., Golz, C., Eapen, S., Kohn, H. and Schieder, O. (1987) *Plant Cell Rep.*, **6**, 313.

107. Penza, R., Lurquin, P.F. and Filippone, E. (1991) *J. Plant Physiol.*, **138**, 39.

108. Bytebier, B., Deboeck, F., DeGreve, H., van Montagu, M. and Hernalsteens, J.P. (1987) *Proc. Natl. Acad. Sci. USA* , **84**, 5345.

109. Basiran, N., Armitage, P., Scott, R.J. and Draper, J. (1987) *Plant Cell Rep.*, **6**, 396.

110. Zhan, X.-C., Jones, D.A. and Kerr, A. (1988) *Plant Mol. Biol.*, **11**, 551.

111. Loh, C.S. and Rao, A.N. (1988) *Experientia*, **44**, 72.

112. Wilde, H.D., Meagher, R.B. and Merkle, S.A. (1992) *Plant Physiol.*, **98**, 114.

113. Firoozabady, E., DeBoer, D.L., Merlo, D.J., Halk, E.L., Amerson, L.N., Rashka, K.E. and Murray, E.E. (1987) *Plant Mol. Biol.*, **10**, 105.

114. Umbeck, P., Johnson, G., Barton, K. and Swain, W. (1987) *Bio/technol.*, **5**, 263.

115. Finer, J.J. and McMullen, M.D. (1990) *Plant Cell Rep.*, **8**, 586.

116. Enomoto, S., Itoh, H., Ohshima, M., Oka, S. and Ohashi, Y. (1992) *J. Seric. Sci. Japan*, **61**, 201.

117. Yoshimatsu, K. and Shimomura, K. (1992) *Plant Cell Rep.*, **11**, 132.

118. Huang, Y., Diner, A.M. and Karnosky, D.F. (1991) *In Vitro Cell. Dev. Biol.*, **27**, 201.

119. Miljus-Djukic, J., Neskovic, M., Ninkovik, S. and Crkvenjakov, R. (1992) *Plant Cell Tissue Organ Cult.*, **29**, 101.

120. Mugnier, J. (1988) *Plant Cell Rep.*, **7**, 9.

121. James, D.J., Passey, A.J. and Barbara, D.J. (1990) *Plant Science*, **69**, 79.

122. Nehra, N.S., Chibbar, R.N., Kartha, K.K., Datla, R.S.S., Crosby, W.L. and Stushnoff, C. (1990) *Plant Cell Rep.*, **9**, 10.

123. Nyman, M. and Wallin, A. (1992) *Plant Cell Rep.*, **11**, 105.

124. Maheswaran, G., Welander, M., Hutchinson, J.F., Graham, M.W. and Richards, D. (1992) *J. Plant Physiol.*, **139**, 560.

125. James, D.J., Passey, A.J., Barbara, D.J. and Bevan, M. (1989) *Plant Cell Rep.*, **7**, 658.

126. Laimer da Camara Machado, M., da Camara Machado, A., Hanzer, V., Weiss, H., Regner, F. and Steinkellner, H. (1992) *Plant Cell Rep.*, **11**, 25.

127. Scorza, R., Cordts, J.M., Mante, S., Gonsalves, D., Damsteegt, V.D. and Slightom, J.L. (1991) *Hortscience*, **26**, 786.

128. Mante, S., Morgens, P.H., Scorza, R., Cordts, J.M. and Callahan, A.M. (1991) *Bio/technol.*, **9**, 853.

129. Smigocki, A.C. and Hammerschlag, F.A. (1991) *J. Am. Soc. Hortic. Sci.*, **116**, 1092.

130. Firoozabady, E., Noriega, C., Sondahl, M.R. and Robinson, K.E.P. (1991) *In Vitro*, **27**, 154A.

131. Hassan, M.A. and Swartz, H.J. (1989) *Hortscience*, **24**, suppl., 112.

132. Graham, J., McNicol, R.J. and Kumar, A. (1990) *Plant Cell Tissue Organ Culture*, **20**, 35.

133. Hidaka, T., Omura, M., Ugaki, M., Tomiyama, M., Kato, A., Ohshima, M. and Motoyoshi, F. (1990) *Ikushugaku Zasshi*, **40**, 199.

134. Moore, G.A., Jacono, C.C., Neidigh, J.L., Lawrence, S.D. and Cline, K. (1992) *Plant Cell Rep.*, **11**, 238.

135. Vardi, A., Bleichman, S. and Aviv, D. (1990) *Plant Science*, **69**, 199.

136. Fillatti, J.J., Sellmer, J., McCown, B., Haissig, B. and Comai, L. (1987) *Mol. Gen. Genet.*, **206**, 192.

137. Pythoud, F., Sinkar, V.P., Nester, E.W. and Gordon, M.T. (1987) *Bio/technol.*, **5**, 1323.

138. De Block, M. (1990) *Plant Physiol.*, **93**, 1110.

139. Mugnier, J. (1988) *Plant Cell Rep.*, **7**, 9.

140. Mathews, H., Bharathan, N., Litz, R.E., Narayanan, K.R., Rao, P.S. and Bhatia, C.R. (1990) *J. Plant Physiol.*, **136**, 404.

141. Liu, W., Parrott, W.A., Hildebrand, D.F., Collins, G.B. and Williams, E.G. (1990) *Plant Cell Rep.*, **9**, 360.

142. Wang, Y.-W., Yang, M.-Z., Pan, N.-S. and Chen, Z.-L. (1991) *Acta Bot. Sin.*, **33**, 780.

143. Sangwan, R.S., Ducrocq, C. and Sangwan-Norreel, B.S. (1991) *Plant Cell Rep.*, **10**, 90.

144. Oksman-Caldentey, K.-M., Parkkinen, O. and Hiltunen, R. (1989) *Planta Medica*, **55**, 686.

145. Horsch, R.B., Fry, J.E., Hoffmann, N.L., Eichholtz, D., Rogers, S.G. and Fraley, R.T. (1985) *Science*, **227**, 1229.

146. McCormick, S., Niedermeyer, J., Fry, J., Barnason, A., Horsch, R. and Fraley, R. (1986) *Plant Cell Rep.*, **5**, 81.

147. Shahin, E.A., Sukhapinda, K., Simpson, R.B. and Spivey, R. (1986) *Theor. Appl. Genet.*, **72**, 770.

148. Chyi, Y.-S. and Phillips, G.C. (1987) *Plant Cell Rep.*, **6**, 105.

149. Bellini, C., Chupeau, M.-C., Guerche, P., Vastra, G. and Chupeau, Y. (1989) *Plant Science*, **65**, 63.

150. Ebert, P.R. and Clark, A.E. (1990) *Plant Mol. Biol.*, **14**, 815.

151. Elmer, J.S., Sunter, G., Gardiner, W.E., Brand, L., Browning, C.K., Bisaro, D.M. and Rogers, S.G. (1988) *Plant Mol. Biol.*, **10**, 225.

152. Schoelz, J.E., Goldberg, K.-B. and Kiernan, J.M. (1990) *Phytopathol.*, **80**, 1018.

153. Tavazza, R., Ordas, R.J., Tavazza, M., Ancora, G. and Benvenuto, E. (1988) *J. Plant Physiol.*, **133**, 640.

154. Davey, M.R., Mulligan, B.J., Gartland, K.M.A., Peel, E., Sargent, A.W. and Morgan, A.J. (1987) *J. Exp. Bot.*, **38**, 1507.

155. Kiernan, J.M., Goldberg, K.-B., Schoelz, J.E., Gowda, S. and Shepherd, R.J. (1988) *Phytopath.*, **78**, 1600.

156. Amasino, R., Taylor, B., White, F., Gordon, M. and Nester, E. (1983) *Plant Physiol.*, **72**, 147.

157. Taylor, B.H., Amasino, R.M., White, F.F., Nester, E.W. and Gordon, M.P. (1985) *Mol. Gen. Genet.*, **201**, 554.

158. Hamill, J.D. and Rhodes, M.J.C. (1988) *J. Plant Physiol.*, **133**, 506.

159. Bevan, M. (1984) *Nucleic Acids Res.*, **12**, 8711.

160. Horsch, R.B., Fraley, R.T., Rogers, S.G., Sanders, P.R., Lloyd, A. and Hoffmann, N. (1984) *Science*, **223**, 496.

161. Trinh, T.H., Mante, S., Pua, E.-C. and Chua, N.-H. (1987) *Bio/technol.*, **5**, 1081.

162. Bates, G.W. (1990) *Theor. Appl. Genet.*, **80**, 481.

163. Dijak, M., Sproule, A., Keller, W., Pandeya, R. and Gleddie, S. (1991) *Plant Cell Tissue Organ Culture*, **25**, 189.

164. Neuhaus, J.-M., Ahl-Goy, P., Hinz, U., Flores, S. and Meins, F. (1991) *Plant Mol. Biol.*, **16**, 141.

165. Zambryski, P., Joos, H., Genetello, C., Leemans, J., Van Montagu, M. and Schell, J. (1983) *EMBO J.*, **2**, 2143.

166. Paszkowski, J., Shillito, R.D., Saul, M., Mandak, V., Hohn, T., Hohn, B. and Potrykus, I. (1984) *EMBO J.*, **3**, 2717.

167. Horsch, R.B., Fry, J.E., Hoffmann, N.L., Eichholtz, D., Rogers, S.G. and Fraley, R.T. (1985) *Science*, **227**, 1229.

168. Tomes, D.T., Weissinger, A.K., Ross, M., Higgins, R., Drummond, B.J., Schaaf, S., Malone-Schoneberg, J., Staebell, M., Flynn, P., Anderson, J. and Howard, J. (1990) *Plant Mol. Biol.*, **14**, 261.

169. Horsch, R.B., Fry, J.E., Hoffmann, N.L., Eichholtz, D., Rogers, S.G. and Fraley, R.T. (1985) *Science*, **227**, 1229.

170. Ulian, E.C., Smith, R.H., Gould, J.H. and McKnight, T.D. (1988) *In Vitro Cell. Develop. Biol.*, **24**, 951.

171. Lee, J.-Y. and Davey, M.R. (1988) *Research Reports of the Rural Development Administration Suweon (South Korea)*, **30**, 21.

172. Chand, P.K., Rech, E.L., Golds, T.J., Power, J.B. and Davey, M.R. (1989) *Plant Cell Rep.*, **8**, 86.

173. Rotino, G.L., Perrone, D., Ajmone-Marsan, P. and Lupotto, E. (1992) *Plant Cell Rep.*, **11**, 11.

174. Guri, A. and Sink, K.C. (1988) *J. Plant Physiol.*, **133**, 52.

175. Rotino, G.L. and Gleddie, S. (1990) *Plant Cell Rep.*, **9**, 26.

176. Atkinson, R.G. and Gardner, R.C. (1991) *Plant Cell Rep.*, **10**, 208.

177. Davey, M.R., Mulligan, B.J., Gartland, K.M.A., Peel, E., Sargent, A.W. and Morgan, A.J. (1987) *J. Exp. Bot.*, **38**, 1507.

178. Johnson, R., Lee, J.S. and Ryan, C.A. (1990) *Plant Mol. Biol.*, **14**, 349.

179. Ooms, G., Karp, A., Burrell, M.M., Twell, D. and Roberts, J. (1985) *Theor. Appl. Genet.*, **70**, 440.

180. Shahin, E.A. and Simpson, R.B. (1986) *Hortscience*, **21**, 1199.

181. De Block, M. (1988) *Intl. Zeit. Theor. Ang. Genet.*, **76**, 767.

182. Sheerman, S. and Bevan, M.W. (1988) *Plant Cell Rep.*, **7**, 13.

183. Catlin, D., Ochoa, O., McCormick, S. and Quiros, C.F. (1988) *Plant Cell Rep.*, **7**, 100.

184. David, C., Chilton, M.-D. and Tempe, J. (1984) *Bio/technol.*, **2**, 73.

185. Tepfer, D. (1984) *Cell*, **37**, 959.

186. Scott, R.J. and Draper, J. (1987) *Plant Mol. Biol.*, **8**, 265.

187. Wurtele, E.S. and Bulka, K. (1989) *Plant Science*, **61**, 253.

188. Mugnier, J. (1988) *Plant Cell Rep.*, **7**, 9.

189. Colby, S.M., Stamp, J.A. and Meredith, C.P. (1989) *J. Cell. Biochem. Suppl.*, **13**, 255.

190. Baribault, T.J., Skene, K.G.M., Cain, P.A. and Scott, N.S. (1990) *J. Exp. Bot.*, **41**, 1045.

191. Zambryski, P., Joos, H., Genetello, C., Leemans, J., Van Montagu, M. and Schell, J. (1983) *EMBO J.*, **2**, 2143.

192. Koncz, C. and Schell, J. (1986) *Mol. Gen. Genet.*, **204**, 383.

193. Hepburn, A.G., White, J., Pearson, L., Maunders, M.J., Clarke, L.E., Prescott, A.G. and Blundy, K.S. (1985) *J. Gen. Microbiol.*, **131**, 2961.

194. Rogers, S.G., Klee, H.J., Horsch, R.B. and Fraley, R.T. (1987) *Methods Enzymol.*, **153**, 253.

195. Ooms, G., Hooykaas, P.J.J., Van Veen, R.J.M., Van Beelen, P., Regensburg-Tuink, T.J.G. and Schilperoort, R.A. (1982) *Plasmid*, **7**, 15.

196. Deblaere, R., Bytebier, B., De Greve, H., Deboeck, F., Schell, J., Van Montagu, M. and Leemans, J. (1985) *Nucleic Acids Res.*, **13**, 4777.

197. Fraley, R.T., Rogers, S.G., Horsch, R.B., Eichholtz, D.A., Flick, J.S., Fink, C.L., Hoffmann, N.L. and Sanders, P.R. (1985) *Bio/technol.*, **3**, 629.

198. Hood, E.E., Helmer, G.L., Fraley, R.T. and Chilton, M.-D. (1986) *J. Bacteriol.*, **168**, 1291.

199. Vilaine, F. and Casse-Delbart, F. (1987) *Mol. Gen. Genet.*, **206**, 17.

200. An, G, Ebert, P.R., Mitra, A. and Ha, S.M. (1988) in *Plant Molecular Biology Manual* (S.B. Gelvin and R.A. Schilperoort eds). Kluwer Academic Publishers, Dordrecht, Part A3, p. 1.

201. Shen Wen-jun and Forde, B.G. (1989) *Nucleic Acids Res.*, **17**, 8385.

202. Nagel, R., Elliot, A., Masel, A., Birch, R.G. and Manners, J.M. (1990) *FEMS Microbiol. Lett.*, **67**, 325.

203. Van Haute, E., Joos, H., Maes, M., Warren, G., Van Montagu, M. and Schell, J. (1983) *EMBO J.*, **2**, 411.

204. Ditta, G., Stanfield, S., Corbin, D. and Helinski, D.R. (1990) *Proc. Natl. Acad. Sci. USA*, **77**, 7347.

205. Stachel, S.E., Messens, E., Van Montagu, M. and Zambryski, P. (1985) *Nature*, **318**, 624.

206. Bolton, G.W., Nester, E.W. and Gordon, M.P. (1986) *Science*, **232**, 983.

207. Alt-Moerbe, J., Neddermann, P., Von Lintig, J., Weiler, E.W. and Schroder, J. (1988) *Mol. Gen. Genet.*, **213**, 1.

208. Ankenbauer, R.G. and Nester, E.W. (1990) *J. Bacteriol.*, **172**, 6442.

209. Grimsley, N., Hohn, T., Davies, J.W. and Hohn, B. (1987) *Nature*, **325**, 177.

210. Lazarovitz, S.G. (1988) *Nucleic Acids Res.*, **16**, 229.

211. Boulton, M.I., Buchholz, W.G., Marks, M.S., Markham, P.G. and Davies, J.W. (1989) *Plant Mol. Biol.*, **12**, 31.

212. Hayes, R.J., MacDonald, H., Coutts, R.H.A. and Buck, K.W. (1988) *J. Gen. Virol.*, **69**, 891.

213. Woolston, C.J., Barker, R., Gunn, H., Boulton, M.I. and Mullineaux, P.M. (1988) *Plant Mol. Biol.*, **11**, 35.

214. Marks, M.S., Kemp, J., Woolston, C. and Dale, P.J. (1989) *Plant Science*, **63**, 247.

215. Dale, P., Marks, M., Brown, M., Woolston, C., Gunn, H., Mullineaux, P., Lewis, M., Kemp, J., Chen, D., Gilmour, M. and Flavell, R. (1989) *Plant Science*, **63**, 237.

216. Creissen, G., Smith, C., Francis, R., Reynolds, H. and Mullineaux, P. (1990) *Plant Cell Rep.*, **8**, 680.

217. Ugaki, M., Ueda, T., Timmermans, M.C.P., Vieira, J., Elliston, K.O. and Messing, J. (1991) *Nucleic Acids Res.*, **19**, 371.

218. Donson, J., Gunn, H.V., Woolston, C.J., Pinner, M.S., Boulton, M.I., Mullineaux, P.M. and Davies, J.W. (1988) *Virology*, **162**, 248.

219. Chatani, M., Matsumoto, Y., Mizuta, H., Ikegami, M., Boulton, M.I. and Davies, J.W. (1991) *J. Gen. Virol.*, **72**, 2325.

220. Krens, F.A., Molendijk, L., Wullems, G.J. and Schilperoort, R.A. (1982) *Nature*, **296**, 72.

221. Negrutiu, I., Shillito, R., Potrykus, I., Biasini, G. and Sala, F. (1987) *Plant Mol. Biol.*, **8**, 363.

222. Hain, R., Stabel, P., Czernilofsky, A.P., Steinbiss, H.H., Herrera-Estrella, L. and Schell, J. (1985) *Mol. Gen. Genet.*, **199**, 161.

223. Spörlein, B. and Koop, H.-U. (1991) *Int. Z. Theor. Angew. Genet.*, **83**, 1.

224. Antonelli, N.M. and Stadler, J. (1990) *Theor. Appl. Genet.*, **80**, 395.

225. Klein, T.M., Fromm, M., Weissinger, A., Tomes, D., Schaaf, S., Sletten, M. and Sanford, J.C. (1988) *Proc. Natl. Acad. Sci. USA*, **85**, 4305.

226. Christou, P., McCabe, D.E. and Swain, W.F. (1988) *Plant Physiol.*, **87**, 671.

227. Morikawa, H., Iida, A. and Yamada, Y. (1989) *Appl. Microbiol. Biotechnol.*, **31**, 320.

228. Ye, G.-N., Daniell, H. and Sanford, J.C. (1990) *Plant Mol. Biol.*, **15**, 809.

229. Fromm, M., Taylor, L.P. and Walbot, V. (1985) *Proc. Natl. Acad. Sci. USA*, **82**, 5824.

230. Harkins, K.R., Jefferson, R.A., Kavanagh, T.A., Bevan, M.W. and Galbraith, D.W. (1990) *Proc. Natl. Acad. Sci. USA*, **87**, 816.

231. Dekeyser, R.A., Claes, B., De Rycke, R.M.U., Habets, M.E., Van Montagu, M.C. and Caplan, A.B. (1990) *Plant Cell*, **2**, 591.

232. Bustos, M.M., Battraw, M.J., Kalkan, F.A. and Hall, T.C. (1990) *J. Cell. Biochem., Suppl.*, **14**, 330.

233. Klein, T.M., Gradziel, T., Fromm, M.E. and Sanford, J.C. (1988) *Bio/technol.*, **6**, 559.

234. Huttly, A.K. and Baulcombe, D.C. (1989) *EMBO J.*, **8**, 1907.

235. Lorens, G.F., Blechl, A.E., Anderson, O.D. and Greene, F.C. (1990) *J. Cell. Biochem.*, **14**, 353.

236. Ludwig, S.R., Bowen, B., Beach, L. and Wessler, S.R. (1990) *Science*, **247**, 449.

237. Van der Meer, I.M., Spelt, C.E., Mol, J.N.M. and Stuitje, A.R. (1990) *Plant Mol. Biol.*, **15**, 95.

238. Twell, D., Klein, T.M., Fromm, M.E. and McCormick, S. (1989) *Plant Physiol.*, **91**, 1270.

CHAPTER 8

PLANT ORGANELLAR TARGETING SEQUENCES

T.P. Wallace and C.J. Howe

1. NUCLEUS

Most information comes from yeast, *Drosophila*, *Xenopus* and mammalian systems. Small proteins (less than 20–40 kDa) are thought to diffuse through nuclear pores. Larger proteins contain a nuclear localization signal, which can probably be located almost anywhere within the polypeptide and is not cleaved. The signal is thought to be a 7–13 amino acid stretch with an abundance of basic amino acids and a proline or glycine residue (1,2), although a more detailed bipartite motif has been suggested which consists of two basic residues followed by a stretch of any 10 residues and a basic cluster with three basic residues out of five (3). Relatively little information is available for plant systems, but similar features may be present. Putative signals are:

opaque-2 regulatory gene from maize (4,5):	VVGDPMEYNAILRRK	(residues 116–130)
	GFKMPTEERVRKRK	(residues 220–233)

Similar signals may function in dicots and monocots, as the maize opaque-2 protein is localized in the nucleus in transgenic tobacco plants (5). See also ref. 6 for a discussion of nuclear localization of plant viral proteins.

2. PEROXISOMES/GLYOXYSOMES (MICROBODIES)

Mammalian, yeast and plant targeting information can be represented by a C-terminal tripeptide SKL, although S may be changed to C or A, K to H or R (7,8). A more general representation may be: –uncharged (may be polar)–basic–hydrophobic–. Examples are:

catalase (8) (near C-terminus):	QKL (castor bean, maize, cotton)
	TRL (castor bean, maize)
	QKV (sweet potato)
	SRL (maize, sweet potato, cotton)
malate synthase (at C-terminus):	SRL (*Brassica napus* (9), castor bean (10))
	SKL (cucumber (11))
glycolate oxidase (at C-terminus)	AKL (spinach (12))
uricase II (at C-terminus)	SRLWSKL (soybean nodules (13))
isocitrate lyase (at C-terminus)	ARM (castor bean (14))
	SRM (*Brassica napus* (15))

Glyoxysomal malate dehydrogenase from watermelon (16) lacks the –SKL sequence (it has –IRS at the C-terminus), but possesses a cleavable 37 amino acid N-terminal transit peptide:

MQPIPDVNQRIARISAHLHPPKSQMEESSALRRANCR

Note that this contains the sequence AHL (which may be equivalent to SKL). The transit peptide shows similarities to the 26 amino acid extension of rat peroxisomal 3-ketoacyl-CoA thiolase (16):

(i) net positive charge;
(ii) glutamate and serine cluster;
(iii) lack of a long stretch of hydrophobic residues;
(iv) putative topogenic tripeptide AHL (GHL for rat protein) in similar positions.

Peroxisomal NADH-dependent hydroxypyruvate reductase from cucumber has neither a cleavable N-terminal transit sequence nor a C-terminal SKL tripeptide (17).

3. SECRETION PATHWAY

This includes rough and smooth endoplasmic reticulum (ER), Golgi, *trans*-Golgi, endosomes, secretory vesicles, vacuoles, tonoplast, plasma membrane and cell wall. For a general review, see ref. 18. Transport to extracellular and vacuolar destinations is probably always through the Golgi, although bypass may be possible in the endosperm development of some cereal seeds (18).

Entry into the system requires a signal sequence of 13–30 amino acids, with a basic N-terminus followed by an uninterrupted run of seven or eight apolar, largely hydrophobic residues and a polar region, analogous to the signal sequences of other eukaryotes (18,19). The signal sequence itself does not specify the final destination, and the nature of subsequent sorting is rather poorly understood, with specific targeting sequences generally not yet recognized. The following points may be helpful.

(i) Integral membrane proteins, such as the plasma membrane H^+-ATPase, may not have cleavable signal sequences, but may rely on internal hydrophobic domains (20).
(ii) N-linked glycosylation may take place at the sequence –N–X–S/T– (but see (v) below).
(iii) Passage to the Golgi and then the plasma membrane is the 'default' pathway. A motif I, Q/S/T, P, L, T, G, N/T/S, Q/N/S, G within the last 22 residues of the C-terminus of four extracellular peroxidases has been reported, but its significance is not clear (8).
(iv) Retention in the ER requires a C-terminal sequence KDEL (auxin-binding protein of maize (21)) or HDEL (tomato and maize BiP (18)), although other structural features of the protein may also be important. RDEL may also be suitable (22).
(v) Targeting to the vacuole does not appear to depend on correct glycosylation (23,24). For bean phytohemagglutinin and possibly other proteins, the sequence LQRD within the amino-terminal region is important (25). For some proteins, N- and C-terminal cleavable extensions are used (18,26).

4. MITOCHONDRIA

Polypeptides are generally synthesized with amino-terminal extensions termed transit peptides, which are removed during or shortly after uptake. Mitochondrial transit peptides where the cleavage site has been demonstrated from plants have been analyzed in conjunction with those from other organisms (27).

Features are:

(i) Enriched in A, L, R, S.
(ii) Depleted in D, E, I, K.
(iii) N-terminal domain with amphiphilic α-helical properties.

(iv) C-terminal domain with much lower tendency to form amphiphilic α-helices, possibly able to form an α-helix with a 'clockwise spiralling ridge' either of polar or apolar amino acids. Similar amino acid composition overall to the N-terminal domain.
(v) R residues at −2 or −10 from the cleavage site. R at −10 is often accompanied by an apolar residue (especially F, I, L) at −8.
(vi) First 10–15 residues of mature proteins are relatively rich in P and S.

For example, superoxide dismutase from maize (28):

MALRTLASKKVLSFPFGGAGRPLAAAASARG^VTTVTLPDLS

and ATP synthase β-subunit (*atp2-1*) from *Nicotiana plumbaginifolia* (29):

MASRRLLASLLRQSAQRGGGLISRSLGNSIPKSASRASSRASPKGFLLNRAVQY

ATSAAAPASQPSTPPKSGSEPSGKITDEFTGAGSIG (cleavage site not determined)

Polypeptides of the intermembrane space may be imported into the matrix, and then re-exported through the inner membrane, generally using a typical signal peptide (basic N-terminus–hydrophobic stretch–C-terminal region with cleavage site) although the cleavage site may not always obey the 'typical' A–X–A^ rule.

5. CHLOROPLASTS

As with mitochondria, polypeptides are generally synthesized with N-terminal transit sequences. Early analysis (30) of the light-harvesting chlorophyll-*a*/*b*-binding protein and the small subunit of ribulose bisphosphate carboxylase transit sequences indicated the presence of three conserved 'homology blocks' (MAXSXMXSS, PXFXGXK, GXGRV) but the analysis of von Heijne *et al.* (27) suggests that these are not general features.

Features of transit peptides identified by von Heijne *et al.* (27, 31) include:

(i) Rich in S, T.
(ii) Deficient in D, E, Y.
(iii) The N-terminal 10 residues constitute a separate domain with very few R, K, D, E, G, P residues. The second residue is usually A. There is no indication of amphiphilic structures by hydrophobic-moment analysis.
(iv) The central domain is positively charged and lacks acidic residues.
(v) The last 10 residues are richer than the first 10 in R, and more depleted in L residues. They have the potential to form an amphiphilic β-strand.
(vi) The cleavage site is loosely defined by V/I–X–A/C^A, with R frequently found at −6 to −10.

A database of chloroplast transit peptides has been established (32). At present this contains over 260 transit peptide sequences. A listing of this, together with details of how to submit entries and obtain copies of the database, is given in ref. 32.

Polypeptides of the thylakoid lumen and some polypeptides of the thylakoid membrane are also synthesized with a two-stage N-terminal extension. The first stage (missing from chloroplast-encoded proteins such as cytochrome *f* (33)) is a transit peptide, and the second is similar to prokaryotic signal peptides (basic N-terminus–central hydrophobic region–C-terminal region with cleavage site). A signal peptide similar to this second stage is also used by cyanobacteria (34). Examples where the cleavage site has been demonstrated directly are reviewed in ref. 34 and include:

photosystem II 33 kDa extrinsic polypeptide (pea):

MAASLQAAATLMQPTKLRSNTLQLKSNQSVSKAFGLEHYGAKVTCSLQSDFKEL
AHKCVEASKIAGFALATSALVVSGASA^EGAPK

photosystem I subunit IV (spinach):

MSFTIPTNLYKPLATKPKHLSSSSFAPRSKIVCQQENDQQQPKKLELAKVGANA
AAALALSSVLLSSWSVAPDAAMA^DIAGL

cytochrome c_{552} (*Chlamydomonas*):

MLQLANRSVRAKAARASQSARSVSCAAAKRGADVAPLTSALAVTASILLTTGAA
SASAA^DLALG

photosystem II 9 kDa extrinsic polypeptide (*Phormidium laminosum*, a cyanobacterium):

MKRLVGVLMILGLMLTSWGLLGSPQTAIAASLSPLSFNPSPVLA^EQQFR

The detailed nature of the cleavage site may be slightly different from the typical prokaryotic cleavage site, although it is nevertheless of the general form –AXA^E–. A weight matrix for scoring sequences is given in ref. 34. Cleavage sites for chloroplast-encoded polypeptides are less well predicted using this matrix.

The phosphate translocator of the inner envelope membrane of spinach has a transit peptide of 10–11 kDa, which is slightly enriched in S and T residues (35). Insufficient sequences are available to draw reliable conclusions.

There are several exceptions to the principles outlined above, involving C-terminal targeting information, apparent lack of targeting information, or extra 'layers' of information.

A few polypeptides may require a C-terminal sequence for insertion into the thylakoid membrane. These include:

(i) light-harvesting chlorophyll-*a*/*b*-binding protein: a putative membrane-spanning α-helix (e.g. residues 223–246 of the *Lemna gibba* LHCP) near the C-terminus (36);
(ii) the 10 kDa extrinsic polypeptide of photosystem II: a C-terminal membrane-spanning sequence of 17 amino acids (37).

The β-subunit of phycoerythrin of the cryptophyte alga *Cryptomonas* ϕ, which is a plastid-encoded soluble polypeptide of the thylakoid lumen, lacks any obvious targeting sequence (38).

A 6.7 kDa chloroplast outer-membrane protein of spinach, named E6.7, has no cleavable transit peptide and is integrated directly into the outer envelope membrane (39) independently of protease-sensitive membrane components and without ATP hydrolysis. It has a single membrane-spanning region.

Nuclear-encoded plastid polypeptides from organisms with more than two membranes round the plastid, such as *Euglena gracilis*, may have more complex transit sequences, incorporating elements resembling signal sequences (40,41). In addition, some proteins are synthesized as polyproteins and presumably cleaved into monomers after entry into the plastid (40).

6. OTHER PLASTIDS

The following observations indicate that targeting to other plastids is broadly similar to chloroplast targeting.

(i) Pea etioplasts can import and process precursors of the small subunit of ribulose bisphosphate carboxylase (42).

(ii) A 72 amino acid cleaved transit sequence directs the *waxy* gene product into maize amyloplasts (43):

MAALATSQLVATRAGLGVPDASTFRRGAAQGLRGARASAAADTLSMRTSARAAP

RHQQQARRGGRFPSLVVC^ASAGMNVVFV

A fusion protein composed of the transit sequence, the first 34 amino acids of the mature protein and *E. coli* GUS was imported into isolated chloroplasts (44).

(iii) A constitutively expressed *Silene* plastocyanin gene introduced into tomato plants resulted in the presence of mature plastocyanin in chromoplasts and leukoplasts (45).

7. REFERENCES

1. Dingwall, C. and Laskey, R.A. (1986) *Ann. Rev. Cell Biol.*, **2**, 367.

2. Silver, P.A. and Hall, M.N. (1988) in *Protein Transfer and Organelle Biogenesis* (R.C. Das and P.W. Robbins eds). Academic Press, San Diego, p. 749.

3. Dingwall, C. and Laskey, R.A. (1991) *Trends in Biochem. Sci.*, **16**, 478.

4. Schmidt, R.J., Burr, F.A., Aukerman, M.J. and Burr, B. (1990) *Proc. Natl Acad. Sci. USA*, **87**, 46.

5. Varagona, M.J., Schmidt, R.J. and Raikhel, N.V. (1991) *The Plant Cell*, **3**, 105.

6. Restrepo, M.A., Freed, D.D. and Carrington, J.C. (1990) *The Plant Cell*, **2**, 987.

7. Gould, S.J., Keller, G.-A., Hosken, N., Wilkinson, J. and Subramani, S. (1989) *J. Cell Biol.*, **108**, 1657.

8. Gonzalez, E. (1991) *Eur. J. Biochem.*, **199**, 211.

9. Comai, L.C. (1989) *J. Biol. Chem.*, **264**, 2778.

10. Rodriguez, D., Ginger, R.S., Baker, A. and Northcote, D.H. (1990) *Plant Mol. Biol.*, **15**, 501.

11. Graham, I.A., Smith, L.M., Brown, J.W.S., Leaver, C.J. and Smith, S.M. (1989) *Plant Mol. Biol.*, **13**, 673.

12. Volokita, M. and Somerville, C.R. (1987) *J. Biol. Chem.*, **262**, 15825.

13. Nguyen, T., Zelechowska, M., Foster, V., Bergmann, H. and Verma, D.P.S. (1985) *Proc. Natl Acad. Sci. USA*, **82**, 5040.

14. Beeching, J.R. and Northcote, D.H. (1987) *Plant Mol. Biol.*, **8**, 471.

15. Comai, L., Dietrich, R.A., Maslyar, D.J., Baden, C.S. and Harada, J.J. (1989) *The Plant Cell*, **1**, 293.

16. Gietl, C. (1990) *Proc. Natl Acad. Sci. USA*, **87**, 5773.

17. Greenler, J.McC., Sloan, J.S., Schwartz, B.W. and Becker, W.M. (1989) *Plant Mol. Biol.*, **13**, 139.

18. Chrispeels, M. (1991) *Ann. Rev. Plant Physiol. Plant Mol. Biol.*, **42**, 21.

19. von Heijne, G. (1986) *J. Mol. Biol.*, **189**, 239.

20. Harper, J.F., Surowy, T.K. and Sussman, M.R. (1989) *Proc. Natl. Acad. Sci. USA*, **86**, 1234.

21. Inohara, N., Shimomura, S., Fukui, T. and Futai, M. (1989) *Proc. Natl Acad. Sci. USA*, **86**, 3564.

22. Denecke, J., de Rycke, R. and Botterman, J. (1992) *EMBO J.*, **11**, 2345.

23. Voelker, T.A., Herman, E.M. and Chrispeels, M.J. (1989) *The Plant Cell*, **1**, 95.

24. Wilkins, T.A., Bednarek, S.Y. and Raikhel, N.V. (1990) *The Plant Cell*, **2**, 301.

25. Tague, B.W., Dickinson, C.D. and Chrispeels, M.J. (1990) *The Plant Cell*, **2**, 533.

26. Chrispeels, M.J. and Raikhel, N.V. (1992) *Cell*, **68**, 613.

27. von Heijne, G., Steppuhn, J. and Herrmann, R.G. (1989) *Eur. J. Biochem.*, **180**, 535.

28. White, J.A. and Scandalios, J.G. (1988) *Biochim. Biophys. Acta*, **951**, 61.

29. Boutry, M. and Chua, N.-H. (1985) *EMBO J.*, **4,** 2159.

30. Karlin-Neumann, G.A. and Tobin, E.M. (1986) *EMBO J.*, **5,** 9.

31. Gavel, Y. and von Heijne, G. (1990) *FEBS Lett.*, **261,** 455.

32. von Heijne, G., Hirai, T., Klosgen, R.-B., Steppuhn, J., Bruce, B., Keegstra, K. and Herrmann, R. (1991) *Plant Mol. Biol. Rep.*, **9,** 104.

33. Willey, D.L., Auffret, A.D. and Gray, J.C. (1984) *Cell,* **36,** 555.

34. Howe, C.J. and Wallace, T.P. (1990) *Nucleic Acids Res.*, **18,** 3417.

35. Flügge, U.I., Fischer, K., Gross A., Sebald, W., Lottspeich, F. and Eckerskorn, C. (1989) *EMBO J.*, **8,** 39.

36. Kohorn, B.D. and Tobin, E.M. (1989) *The Plant Cell,* **1,** 159.

37. Lautner, A., Klein, R., Ljungberg, U., Reilander, H., Bartling, D., Andersson, B., Reinke, H., Beyreuther, K. and Herrmann, R.G. (1988) *J. Biol. Chem.*, **263,** 10077.

38. Reith, M. and Douglas, S. (1990) *Plant Mol. Biol.*, **15,** 585.

39. Salomon, M., Fisher, K., Flügge, U.I. and Soll, J. (1990) *Proc. Natl Acad. Sci. USA,* **87,** 5778.

40. Chan, R.L., Keller, M., Canaday, T., Weil, J.-H. and Imbault, P. (1990) *EMBO J.*, **9,** 333.

41. Sharif, A.L., Smith, A.G. and Abell, C. (1989) *Eur. J. Biochem.*, **184,** 353.

42. Schindler, C. and Soll, J. (1986) *Arch. Biochem. Biophys.*, **247,** 211.

43. Klösgen, R.B., Gierl, A., Schwarz-Sommer, Z. and Saedler, H. (1986) *Mol. Gen. Genet.*, **203,** 237.

44. Klösgen, R.B., Saedler, H. and Weil, J.-H. (1989) *Mol. Gen. Genet.*, **217,** 155.

45. de Boer, D., Cremers, F., Teetstra, R., Smiths, L., Hille, J., Smeekens, S. and Weisbeek, P. (1988) *EMBO J.*, **7,** 2631.

CHAPTER 9

CHLOROPLAST MOLECULAR BIOLOGY

T. Dyer, A. Thompson and D. Herrin

1. INTRODUCTION

Many hundreds of genes are involved in the synthesis, maintenance and degradation of chloroplasts. Although quite a number of these genes are located in the chloroplast genome, by far the greater proportion is in the nucleus. The sizes of a great number of chloroplast DNAs (cpDNA) are now known from restriction enzyme mapping studies (1) and the complete sequences of three have been determined (1–5). They all appear to be circular and the great majority, at least in higher plants, contain between 135 000 and 160 000 base pairs. There is considerable uniformity, especially in higher plants, in what the chloroplast genome codes for in different plants, and the orientation of genes relative to one another is also quite similar and, in many cases, is the same as in their putative prokaryotic progenitors. In addition, coding sequences themselves are, in general, highly conserved, with most heterogeneity being in the spacer sequences between genes. An important question still being considered is whether the chloroplasts of all different photosynthetic eukaryotes may be regarded as monophyletic or whether they might have had diverse origins.

1.1. What cpDNA encodes

The main function of cpDNA is to code for all the core components of the proteinaceous thylakoid complexes involved in the conversion of light energy to chemical energy. Another important function is to code for all the RNA components of the genetic apparatus, such as the rRNAs and tRNAs and for the mRNAs for the proteins synthesized in the chloroplast. It also codes for core components of an RNA polymerase and for about a third of the chloroplast ribosomal proteins. Of the numerous metabolic proteins present in green plants, only the gene which codes for the large subunit of Rubisco (*rbcL*) and perhaps also the gene *dedB* (6) for an enzyme involved in C_1 metabolism (possibly propionyl-CoA carboxylase) have so far been identified in the cpDNA. No angiosperm chloroplast gene for any of the envelope proteins has yet been identified, but two *Marchantia* coding sequences (*mbpX* and *mbpY*) have extensive homology with proteins of a sulfate transport system in cyanobacteria (7). The sequencing of cpDNA has revealed the presence of numerous 'open reading frames' (ORFs) which may have coding functions.

1.2. Nomenclature of genes and their products

Unfortunately no standard system of nomenclature has been devised yet for all the genes for chloroplast components and for their products. Consequently, there is much confusion and ambiguity in the literature as to what they should be called. Where possible, the proposals for naming chloroplast genes, as set out by Hallick and Bottomley (8) and Hallick (9) have been adopted here. However, these proposals mainly cover thylakoid proteins involved in photosynthesis. Because many of the remaining sequences have been provisionally identified by alignment with bacterial (especially *Escherichia coli* sequences), they have in many cases been given the same identifiers as the equivalent bacterial genes.

2. GENES FOR THE POLYPEPTIDES OF MULTI-COMPONENT THYLAKOID COMPLEXES INVOLVED IN ELECTRON TRANSPORT

The lipid membranes of thylakoids contain a number of embedded proteinaceous structures which are involved in photosynthesis (*Figure 1*).

2.1. Photosystem I (PSI) plastocyanin : ferredoxin oxidoreductase

PSI contains at least 11 different core polypeptides (10, 11). Their nomenclature is extremely confused and confusing because there is considerable heterogeneity both in the size of the proteins and also because their behavior in gel electrophoretic systems is anomalous and variable. Consequently, the only certain way of identifying all subunits is by their sequences. Of the core polypeptides, five are encoded in the chloroplast genome (PSA-A, -B, -C, -I, -J) and the rest in the nuclear genome (*Table 1*). The complex also has nuclear-encoded chlorophyll *a*/*b*-binding antenna components in green plants which are distinct from similar components of PSII (12).

2.2. Photosystem II (PSII) water : plastoquinone oxidoreductase

PSII contains more than 20 different types of polypeptide (*Table 2*) as well as chlorophyll, phaeophytin, carotenoids, iron and quinone. The protein components of the complex can be subdivided into three main categories. Some are the inner-core components, which are essential for the reaction center of the complex to function. Others are outer-core components, which are essential for the stability of the core but not its basic activity. Some of these have antenna functions, as have the third category of components (12) which are chlorophyll *a*/*b*-binding proteins of the light-harvesting complex (LHCII). In addition to these PSII components there are a few extrinsic proteins which bind to the lumenal side of the complex, probably helping to create and stabilize the appropriate ionic environment necessary for water oxidation to take place. The chloroplast genome encodes all the core components of the reaction center.

Figure 1. Block diagram to show the sequential flow of electrons through the four protein complexes of the photosynthetic apparatus which are embedded in the thylakoid membrane.

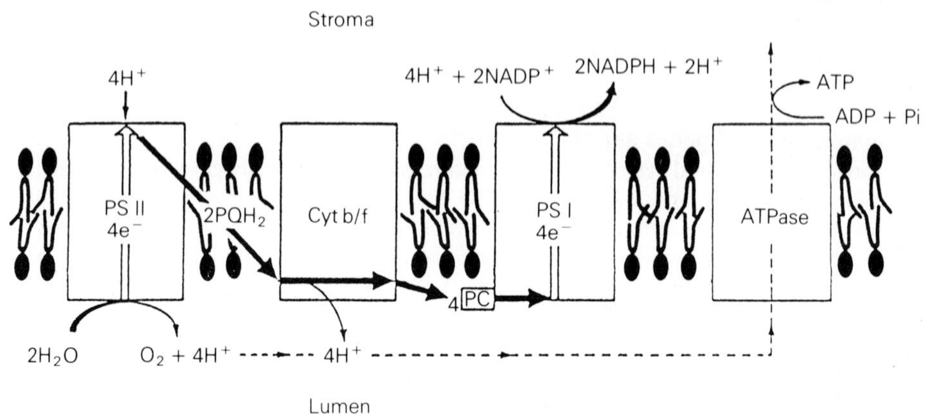

Table 1. Genes for PSI polypeptide components

Gene					Protein					References
Current name	Genome	Alternative names	Codons		Name	Alternative names	M_{calc} (kDa)	Location	Function	
			Precursor	Mature						
psaA	chl	ps1-A1	—	750	PSI-A	PSI-A1 Subunit I	83	Core; intrinsic	Reaction center	13
psaB	chl	ps1-A2	—	734	PSI-B	PSI-A2 Subunit I	82	Core; intrinsic	Reaction center	13
psaC	chl	frxA/ORF81	—	81	PSI-C	Subunit VII	9	Core; stromal face	Binds iron – sulfur centers AB	14
PsaD	nuc	—	212	162	PSI-D	Subunit II	17.9	Core; stromal face	Ferredoxin binding	15
PsaE	nuc	—	125	91	PSI-E	Subunit IV; P30	9.7	Core; stromal face	Dispensible	16
PsaF	nuc	—	231	154	PSI-F	Subunit III; P21	17	Core; lumenal face	Plastocyanin binding	17
PsaG	nuc	—	167	98	PSI-G	P35 (was SuV)	10	?	—	18
PsaH	nuc	—	144	95	PSI-H	Subunit VI; P28	10.2	Core; stromal face	—	19
psaI	chl	ORF 36	—	36	PSI-I	—	4	?	Binds early electron acceptor	20
psaJ	chl	ORF42/44	—	44	PSI-J	—	4.1	PSI/PSII interaction?	—	21
PsaK	nuc	?	113	87	PSI-K	P37	8.4	?	—	22
PsaL	nuc	—	209	109	PSI-L	Subunit V	18	Intrinsic	—	18
PsaM	nuc	—	?	?	PSI-M	—	17.5	Core	—	23

Abbreviations: chl, chloroplast; nuc, nucleus.

Table 2. Genes for PSII polypeptide components

Gene				Product					References
Name	Genome	Codons		Name	Alternative names	M_{calc} (kDa)	Location	Function	
		Precursor	Mature						
psbA	chl	360/353	344	D1 protein	Q_B polypeptide; 32 kDa protein	39.0	Inner core; intrinsic	Reaction center	24
psbB	chl	—	508	CP 47 apoprotein	47 kDa protein; P5	56.2	Outer core; intrinsic	Inner antenna	25
psbC	chl	—	473	CP 43 apoprotein	43 kDa protein; P6	51.8	Outer core; intrinsic	Inner antenna	26
psbD	chl	—	353	D2 protein		39.5	Inner core; intrinsic	Reaction center	26
psbE	chl	83	82	cyt b559 apoprotein	9 kDa subunit	9	Inner core	Inner core	27
psbF	chl	39	38		4 kDa subunit	4			
psbH	chl	73	72	10 kDa phosphoprotein		7.6	Outer core	?	28
psbI	chl	—	52	I protein	—	4.8	Inner core	Reaction center	29
psbJ	chl	—	40	J protein	—	4.2	?	Protein not yet found in plants	30
psbK	chl	61	37	K protein	—	4.3	Outer core	Oxygen evolution	31
psbL	chl	—	37	L protein	—	4.3	Outer core	Oxygen evolution	30
psbM	chl	—	34	M protein	—	3.8	Outer core	Oxygen evolution	32
psbN	chl	—	43	N protein	—	4.7	Outer core	Oxygen evolution	32
PsbO	nuc	331	248	33 kDa protein	OEE1; EP 33	26.5	Lumenal; oxygen evolving center	Mn stabilization	33
PsbP	nuc	258	185	23 kDa protein	OEE2; EP 23	20.2	Lumenal; oxygen evolving center	Ion control	34
PsbQ	nuc	232	149	16 kDa protein	OEE3; EP 16	16.5	Lumenal; oxygen evolving center	Ion control	34
PsbR	nuc	140	99	10 kDa protein	—	10.8	Lumenal; oxygen evolving center	PSII modulation?	35

Abbreviations: chl, chloroplast; nuc, nucleus.

2.3. Cytochrome *b/f* complex (plastoquinol : plastocyanin oxidoreductase)

This thylakoid complex is analogous to the cytochrome bc_1 complex of mitochondria (36). However, while the mitochondrial complex contains about 10 protein subunits, the chloroplast complex has only four major subunits, three of which are chloroplast encoded (*Table 3*). In addition, the presence of a low molecular weight component has been proposed (37).

2.4. ATP synthase (F_0F_1-ATPase, EC 3.6.1.34)

The ATPases are multimeric enzyme complexes of the cytoplasmic membrane of eubacteria, the inner membrane of mitochondria and the thylakoid membranes of cyanobacteria and chloroplasts (47). Structural analysis of the genes for their protein components shows that many features of their structure are conserved. The complex is composed of an oligomeric membrane section (F_0), which acts as a proton channel, and a peripheral catalytic center (F_1) consisting of five different types of subunit. In chloroplasts, the F_1 is located on the stromal side of the thylakoids in the regions between the grana. Most of the components of this synthase are chloroplast encoded but two regulatory components (γ and δ) are nuclear encoded (*Table 4*).

2.5. NADPH : plastoquinone oxidoreductase (NADPH dehydrogenase)

One of the most unexpected findings to emerge from the complete sequencing of the three cpDNAs is that this DNA contains a number of open reading frames which have homologies with mitochondrial genes for components of the respiratory chain NADH dehydrogenase (Complex I). Northern blot analysis has shown that all of these ORFs are actively transcribed (57). However, conclusive proof of the physical existence of this complex in chloroplasts has been difficult to obtain. The reductase is a multi-component complex (58, 59) and more of the subunits may be coded for by the chloroplast genome than by the mitochondrial genome (60–62) (*Table 5*).

3. THE CHLOROPLAST GENETIC SYSTEM

One of the most obvious similarities between the chloroplasts and eubacteria is in their genetic systems. In contrast, the cytosolic genetic system of plants is clearly similar to that of other eukaryotes. The similarity between chloroplasts and bacteria occurs at all levels – in ribosomal RNA size and sequence, in RNA polymerase composition, in properties of ribosomal proteins and in gene arrangement and transcription. However, there are substantial differences too. For example, the tRNAs are radically decreased in number (5) and their mechanisms for DNA replication are not very alike (68). There are large numbers of chloroplast DNA molecules per chloroplast and per cell (69) and their numbers vary according to the developmental state of the organelle.

3.1. Ribosomal RNA

As in bacteria, the order of genes for the ribosomal RNA of chloroplasts is *16S*-(*trnI*-*trnA*)-*23S*-*5S*. These genes are co-transcribed and subsequent processing liberates the mature molecules. In higher plants the 3′ end of the 23S rRNA is cleaved giving rise to a 4.5S rRNA which is characteristic of the ribosomes of this group of organisms (70). *Chlamydomonas* also has some unique small rRNAs but these are the result of 5′ cleavage of the precursor of the 23S rRNA (71). Some of the algae which have been studied have different arrangements of the ribosomal RNA genes (72–74) and chloroplast rRNA sequences have been widely used in phylogenetic studies (75).

Table 3. Genes for chloroplast electron transport components

Gene					Protein				References
Name	Alternative name	Genome	Codons		Name	M_{calc} (kDa)	Location	Function	
			Precursor	Mature					
petA	—	chl	320	285	Cytochrome *f* apoprotein	31.9	cyt *b*/*f* complex; lumenal face	Plastocyanin reduction	38
petB	—	chl	—	215	Cytochrome b_6 apoprotein	28.5	cyt *b*/*f* complex; intrinsic	Plastoquinone reduction	39
petC	—	nuc	247	179	Rieske apoprotein	18.8	cyt *b*/*f* complex; lumenal face	Plastoquinol oxidation	40
petD	—	chl	—	160	Subunit IV	15.3	cyt *b*/*f* complex; intrinsic	Binds plastoquinone	39
PetE	—	nuc	168	99	Plastocyanin	10.4	Lumenal	Electron transport from cyt *b*/*f* complex to PSI	41
PetF	*Fed*	nuc	149	97	Ferredoxin	9.9	Stromal	PSI electron acceptor	42
petG	*petE*	chl	—	37	Subunit V	4	cyt *b*/*f* complex; intrinsic	?	43
PetH	—	nuc	369	314	Ferredoxin:NADPH reductase	35.4	Stromal	—	44, 45
petI	—	nuc	—	173	Flavodoxin apoprotein	18.9	Lumenal	Ferredoxin equivalent	46

Abbreviations: chl, chloroplast; cyt, cytochrome; nuc, nucleus.

Table 4. Genes for chloroplast ATP synthase polypeptide components

Gene				Protein					References
Name	Genome	Codons		Name	Alternative names	M_{calc} (kDa)	Location	Function	
		Precursor	Mature						
atpA	chl	—	507	alpha	—	55.4	F_1 stromal face	Nucleotide binding	48
atpB	chl	—	498	beta	—	58.8	F_1 stromal face	Active center	49
AtpC	nuc	364	323	gamma	—	—	F_1 stromal face	Light/dark regulation	50
AtpD	nuc	257	187	delta	—	20.5	F_1 stromal face	Restricts proton flow	51
atpE	chl	—	153	epsilon	—	14.7	F_1 stromal face	CF_0 binding	49
atpF	chl	—	184	F_0 subunit I	b	21	F_0 intrinsic	CF_1 binding	52
AtpG	nuc	?	?	F_0 subunit II	b′	18	F_0 intrinsic	CF_1 binding	53
atpH	chl	—	81	F_0 subunit III	c	7.9	F_0 intrinsic	Proton translocation	54, 55
atpI	chl	247	230	F_0 subunit IV	a	25	F_0 intrinsic	Proton transport ATP synthesis	56

Table 5. Chloroplast genes for NAD(P): quinone oxidoreductase components

Gene			Protein				References
Current name	Alternative names	Codons	Name	M_{calc} (kDa)	Location	Function	
ndhA	ndh1	362/368	ND1	—	Intrinsic	Quinone binding	57
ndhB	ndh2	387/510	ND2	—	Intrinsic	Inorganic carbon transport	63
ndhC	ndh3	120	ND3	13.2	Intrinsic	—	64
ndhD	ndh4	499/509	ND4	—	Intrinsic	—	57
ndhE	ndh4L	100/101	ND4L	—	Intrinsic	—	57
ndhF	ndh5	692/734	ND5	—	Intrinsic	Iron–sulfur cluster N-2	57
ndhG	ndh6	176/191	ND6	—	Intrinsic	—	65
ndhH	ndh392 ORF 392	392/393	49 kDa protein homolog	—	Extrinsic	—	61
ndhI	frxB ORF 167/178	167/183	—	23	Extrinsic	Iron–sulfur protein; DNA binding	60, 66
ndhJ	—	158/169	—	20.8	Extrinsic	—	62
ndhK	psbG	243/284	—	27.9	—	—	67

3.2. Chloroplast ribosomal proteins

The numbers of proteins in higher plant chloroplast ribosomes has been estimated to be between 56 and 60 (76). Of these 21 are coded for in the organelle genome of higher plants. The small subunit genes are *rps2, 3, 4, 7, 8, 11, 12, 14, 15, 18* and *19* and large subunit genes are *rpl2, 14, 16, 20, 22, 23, 32, 33* and *36*. *rps16* occurs in angiosperms but not liverworts and the reverse is true for *rpl21* (5).

3.3. tRNA genes

Although chloroplast coding sequences contain all 61 possible sense codons (*Table 6*), the chloroplast genome only encodes 30–31 tRNAs (5) and there is no evidence for any tRNAs being imported. It seems that in several instances a single tRNA species can read more than one codon using a 'two out of three' recognition mechanism and a 'U:N wobble' mechanism may also operate (77). In addition to its participation in protein synthesis, the glutamyl-tRNA of chloroplasts (*trnGlu* UUC) also has a vital role in the synthesis of δ-aminolevulinc acid (ALA) which is the common precursor of hemes, chlorophylls and bilins (78).

Table 6. The genetic code showing (in bold italics) those codons for which there are tRNAs which match in all three positions. The *trnI*CAU gene codes for a tRNA which would match the Met codon, but because the C residue of the anticodon is modified, it probably recognizes the Ile-AUA codon instead (100). The Arg-CGG codon, shown in italics, is recognized by the extra tRNA found in liverwort-like plants

First position (5′ end)	**Second position**				**Third position (3′ end)**
	U	**C**	**A**	**G**	
U	Phe	Ser	Tyr	Cys	U
	Phe	***Ser***	***Tyr***	***Cys***	C
	Leu	***Ser***	Stop	Stop	A
	Leu	Ser	Stop	***Trp***	G
C	Leu	Pro	His	***Arg***	U
	Leu	Pro	***His***	Arg	C
	Leu	***Pro***	***Gln***	Arg	A
	Leu	Pro	Gln	*Arg*	G
A	Ile	Thr	Asn	Ser	U
	Ile	***Thr***	***Asn***	***Ser***	C
	Ile	***Thr***	***Lys***	***Arg***	A
	Met (start)	Thr	Lys	Arg	G
G	Val	Ala	Asp	Gly	U
	Val	Ala	***Asp***	***Gly***	C
	Val	***Ala***	***Glu***	***Gly***	A
	Val	Ala	Glu	Gly	G

3.4. RNA polymerase

RNA polymerase of prokaryotes generally consists of a catalytic core of four subunits ($\beta\beta'\alpha_2$) and a dissociable sigma subunit which confers promoter specificity. However, in addition, chloroplasts and cyanobacteria contain one more subunit (79, 80). Studies of the coding sequence of this subunit have shown that its occurrence is either due to the splitting of the gene for β' protein or that the split configuration was ancestral and that the β' subunit, as found now in organisms such as *E. coli*, is the product of gene fusion. Whatever the case, these observations provide further evidence of the common evolutionary descent of the chloroplasts and cyanobacteria. All the core components of the RNA polymerase are chloroplast encoded (81). Sigma-like factors are apparently imported into chloroplasts, giving the nucleus control of transcription (82). Chloroplasts possibly also contain a nuclear-encoded RNA polymerase, as it seems that at least two polymerases are involved in the transcription of chloroplast genes (83).

4. CHLOROPLAST GENE EXPRESSION

4.1. Transcription

Chloroplast gene expression occurs as a result of a number of (at least 33; see ref. 84) discrete mutation events. Many of the primary transcripts are apparently precursors and undergo maturation processes. The promoters at which transcription is initiated are, in general, very similar to those that occur in bacteria such as *E. coli* (85), although a nucleus-like promoter motif has also been identified (86).

4.2. Introns, splicing and processing

One of the biggest differences between the chloroplast gene and those of eubacteria is that several of the chloroplast genes contain introns. Until recently, it was thought that the eubacteria were completely devoid of introns but now it has been found that the *trnL*-UAA of cyanobacteria, like that of chloroplasts, does contain an intron of a group I type (87, 88). At least some of the introns are self-splicing (89–91) while others may require auxillary factors.

5. CHLOROPLAST GENOME STRUCTURE

The chloroplast genome has a circular conformation, as shown by electron microscopy, and the maps of the fragments produced by the digestion with restriction enzymes may be circularly permuted (*Figure 2*). Monomeric and oligomeric forms of the cpDNA have been described (92) and some linear DNA molecules have also been observed (93). Some of these may have a hairpin-like structure (94) which may be involved in replication. The multiple copies of the chloroplast genome are arranged in tightly folded clusters, called nucleoids, and each of these nucleoids may contain a large number of copies of the cpDNA (95).

5.1. Gene arrangements

A typical arrangement of genes in an angiosperm cpDNA molecule is shown in the map of tobacco (*Figure 2*). In this the inverted repeat sequences are separated by a small and a large single copy region. Although this map is representative of a great number of cpDNAs, there are many other variants of this. For example, one segment of the inverted repeat may be missing, as in some legumes and pines (96), elements of the large single copy region inverted, as in the Gramineae (4), legumes (97) and *Marchantia* (3), or the proportion of the genome which is in the inverted repeat has been increased, as in *Pelargonium* (98), or decreased in amount (*Marchantia* (3)). In general, blocks of genes which are co-transcribed are kept

Figure 2. Representative map of a chloroplast genome. Genes: *psa*, for components of photosystem I; *psb*, photosystem II; *atp*, ATP synthase; *pet*, photosynthetic electron transport; *trn*, transfer RNA; *ndh*, NADP oxidoreductase; *rps*, ribosomal proteins (small subunit); *rpl*, ribosomal proteins (large subunit); *inf*, initiation factor; *rpo*, RNA polymerase; *dedB*, C_1 metabolism enzyme; *clpP*, protease subunit; genes for ribosomal proteins are identified by the sedimentation coefficients of the proteins. URF, unidentified reading frames; ORF, open reading frames (not conserved). Coding sequences containing introns are shown with asterisks (*) and segments of the inverted repeat with thickened lines. Genes shown on the outside of the circle are transcribed in an anticlockwise direction and those on the inside in a clockwise direction.

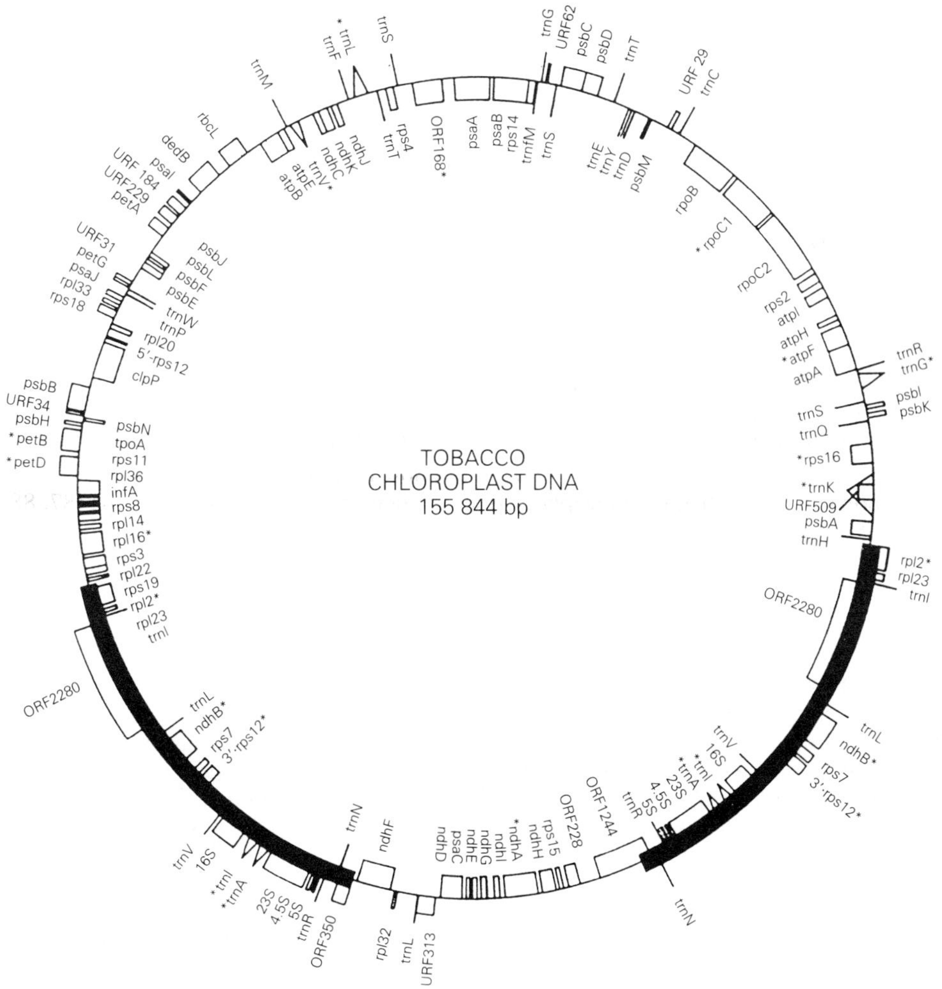

together, suggesting that their transcription is uniform, but substantial re-arrangements of cistrons have occurred in some algae (99) so this conclusion cannot be generalized to all plants.

6. CHLOROPLAST TRANSFORMATION

6.1. Transformation of plastids of higher plants

Table 7 reviews this topic. Early attempts at tobacco plastid transformation using *Agrobacterium tumefaciens* resulted in unstable chloroplast DNA integration, irreproducibility or inconclusive results due to nuclear transcription from chloroplast promoters (101, 102). A more recent preliminary report showed that *A. tumefaciens*-mediated chloroplast transformation is possible using vectors that allow homologous recombination in the rDNA region (103).

Particle bombardment techniques have also allowed the direct manipulation of the chloroplast genome by stable, homologous recombination, albeit at an extremely low transformation frequency (107, 108). Relatively efficient transient expression studies have also been achieved by particle bombardment (104–106).

Polyethylene glycol (PEG)-mediated delivery of plastid-specific reporter gene constructs into tobacco protoplasts has been reported to result in transient expression in chloroplasts (109, 110).

6.2. Transformation of the chloroplast of *Chlamydomonas reinhardtii*

This is reviewed in *Table 8*. In contrast to the low efficiency of stable plastid transformation in tobacco leaves (which contain 3000–12 000 copies of the plastid genome in up to 100 plastids per cell) (107) the single large chloroplast of *C. reinhardtii* (which occupies up to 60% of the cell volume and has a genome copy number of approximately 80) is stably transformed by particle bombardment at high frequency with vectors capable of homologous recombination. Increased transformation frequency can be achieved by growth in 0.5 mM 5-fluorodeoxyuridine (111, 112).

The *C. reinhardtii* chloroplast has also been transformed by agitation of cell-wall-deficient cells with glass beads. This method is more complex and less efficient but removes the requirement for particle bombardment apparatus (112).

6.3. Markers for selection of transformants in *C. reinhardtii*

Chlamydomonas reinhardtii can be grown heterotrophically using acetate as the carbon source, allowing isolation of photosynthetic mutants. Such mutants can then be transformed to restore photosynthetic competency which allows a strong selection by phototrophic growth on minimal medium. Alternatively, point mutations in the *rrn* or *psbA* genes confer antibiotic or herbicide resistance, respectively, and provide powerful selectable markers.

Aminoglycoside adenyl transferase, encoded by the bacterial *aadA* gene, was the first reported foreign protein stably expressed in the chloroplast. It provides the readily selectable phenotype of spectinomycin and streptomycin resistance (118). *aadA* linked to *C. reinhardtii* transcription and termination sequences has been used as a gene disruption cassette for insertional inactivation of several chloroplast genes (118, 122). In theory, any non-lethal manipulation of the chloroplast genome by homologous recombination should be possible by linking a selectable marker such as *aadA* to mutant cpDNA constructs (118).

Table 7. Transformation of higher-plant chloroplasts

Transforming DNA		Target tissue	Delivery system	Selection/expression	Ref.
Promoter	Coding				
Nopaline synthase from Ti plasmid	CAT[a]	Tobacco protoplasts	*Agrobacterium* co-cultivation	Chloramphencol resistance Unstable integration into cpDNA	101
psbA from tobacco	NPTII[a]	Tobacco protoplasts	*Agrobacterium* co-cultivation	Kanamycin resistance Inconclusive results, no evidence of cp expression	102
—	5S rDNA; $tRNA^{Arg}$ from soybean	Tobacco leaf disks regenerated into transgenic plants	*Agrobacterium* co-cultivation	Select transformants showing nuclear kanamycin resistance then screen DNA for cp recombination	103
psbA, pea, spinach	CAT[a]	Tobacco suspension culture cells	Particle gun	CAT activity assay Transient expression in cp	104
psbA, pea	GUS[a]	Wheat leaf, calli	Gunpowder gun	GUS activity assay on X-Gluc substrate Transient expression in cp	105
psbA, pea	GUS[a]	Tobacco suspension culture cells	Helium gun	As above; optimized for osmoticum and particle delivery; 300–400 transformants per shot	106

Table 7. Continued

Transforming DNA		Target tissue	Delivery system	Selection/expression	Ref.
Promoter	Coding				
—	16S rDNA; spec[r]; *rps12*; strep[r]	Tobacco leaf regenerated to plants via calli	Gunpowder gun	Spectinomycin resistance Stable gene replacement by homologous recombination 3 transformants from 50 shots (ref. 107) 2 transformants from 195 shots (ref. 108) Maternal inheritance demonstrated	107, 108
psbB and gene X from spinach	GUS	Tobacco protoplasts	PEG	Transient GUS expression on total cellular protein and cp protein	109, 110

[a]See *Chapter 5* for sequence details of the selectable and reporter genes.
Abbreviations: CAT, chloramphenicol acetyltransferase; cp, chloroplast; GUS, β-glucuronidase; PEG, polyethylene glycol; rDNA, ribosomal RNA genes; spec[r], resistance to spectinomycin; strep[r], resistance to streptomycin; X-Gluc, 5-bromo-4-chloro-3-indoyl β-D-glucuronide.

Table 8. Transformation of *Chlamydomonas* chloroplasts

Transforming DNA	Recipient strain	Delivery system	Selection/expression	Ref.
C.r. atpB gene	*atpB* deletion mutants	Gunpowder gun Glass beads Air gun	Restoration of photosynthesis by homologous recombination	111–114
Maize *rbcl* promoter + NTTII inserted 3′ to *atpB* gene	*atpB* deletion mutants	Gunpowder gun	As above, then screen for co-integration of foreign DNA	115

Table 8. Continued

Transforming DNA	**Recipient strain**	**Delivery system**	**Selection/expression**	**Ref.**
C.r. atpB, *atpA* or *rbcL* promoters + GUS inserted 3′ to *atpB* gene	*atpB* deletion mutants	Gunpowder gun	As above, but deletion mutagenesis analysis performed on promoter function; no GUS activity detected	116
C.r. atpB gene with 3′ UTR deletions	*atpB* deletion mutant	Gunpowder gun	Restoration of photosynthesis Gene replacement allowing deletion analysis of the 3′ IR of the *atpB* gene	117
C.r. tscA gene	*tscA* deletion mutant	Gunpowder gun Glass beads	Restoration of photosynthesis by homologous recombination	112, 118
C.r. psbA gene with point mutations	*C.r.* or *C.s.* wild type	Gunpowder gun	Herbicide resistance (DCMU, metribuzin); gene replacement by homologous recombination	111, 119
16S and 23S rDNA spec^r, strep^r, kan^r, eryth^r	*C.r.* or *C.s.* wild type	Gunpowder gun	Resistance to antibiotics Homologous recombination events may be followed by RFLPs between *C.r.* and *C.s.*	120
16S and 23S rDNA spec^r and eryth^r with deletion of 23S intron ORF	*C.r.* wild type	Air-gun	Resistance to antibiotics Gene replacement resulting in ORF disruption	121
rDNA, spec^r + *atpB* gene with 3′ deletion on separate plasmid	*C.r.* wild type or *atpB* deletion mutant	Gunpowder gun	Primary selection for spec^r then screen for co-transformation of *atpB* deletion by DNA analysis	112, 117
aadA gene with *C.r. psbA* or *atpB* promoter, linked to *tscA* gene	*tscA* deletion mutant	Gunpowder gun	Restoration of photosynthesis and integration of selectable *aadA* marker (spec^r/strep^r)	118
aadA gene with *C.r. atpB* promoter inserted into *psaC*, *tscA* or ORF472	*C.r.* wild type	Gunpowder gun	Selection by spec^r and/or strep^r; gene disruption by replacement with interrupted gene	118, 112

Abbreviations: *C.r.*, *Chlamydomonas reinhardtii*; *C.s.*, *Chlamydomonas smithii*; DSMU. 3-(3,4-dichlorophenyl)-1,1-dimethylurea; IR, inverted repeat; ORF, open reading frame; rDNA, ribosomal RNA genes; RFLP, restriction fragment length polymorphism; spec^r, strep^r, eryth^r, kan^r, resistance to spectinomycin, streptomycin, erythromycin and kanamycin; UTR, untranslated region.

Alternatively, co-transformation procedures in which two independent plasmids are introduced simultaneously allow primary selection for transformation (e.g. 16S spec^r), followed by screening for unselected, independent recombination events (112, 117).

6.4. Apparatus for particle bombardment

Particle guns may be obtained commercially from BioRad. The earlier model (PDS-1000) was based on a gunpowder discharge. An alternative gunpowder device has been constructed independently (123) and used for *C. reinhardtii* transformation (118, 122). The PDS-1000 has now been superseded by a helium-driven gun (PDS-1000/He) (124, 125), which has been reported to give up to 200-fold higher transformation efficiency in the transient chloroplast expression of β-glucuronidase (GUS) in tobacco suspension cells (106).

Alternatively, a simple, inexpensive device based on a ready-made air-rifle is easily constructed (126) and has been used for *C. reinhardtii* chloroplast transformation with efficiencies comparable to those using a gunpowder gun (111, 113, 121); up to 1000 transformants per shot (2.5×10^{-5} transformants per cell) were achieved when transforming the *atpB* gene into an *atpB* deletion mutant (114).

7. REFERENCES

1. Palmer, J.D. (1991) in *The Molecular Biology of Plastids* (L. Bogorad and I.K. Vasil eds). Academic Press, San Diego, Vol. 7A, p. 5.

2. Shinozaki, K., Ohme, M., Tanaka, M., Wakasugi, T., Hayashida, N., Matsubayashi, T., Zaita, N., Chungwongse, J., Obokata, J., Yamaguchi-Shinozaki, K., Ohto, C., Torazawa, K., Meng, B.Y., Sugita, M., Deno, H., Kamogashira, T., Yamada, K., Kusada, J., Takaiwa, F., Kato, A., Tohdoh, N., Shimada, H. and Sugiura, M. (1986) *EMBO Journal,* **5,** 2043.

3. Ohyama, K., Fukuzawa, H., Kohchi, T., Sano, T., Sano, S., Shirai, H., Umezomo, K., Shiki, Y., Takeuchi, M., Chang, Z., Aota, S., Inokuchi, H. and Ozeki, H. (1988) *Nature,* **322,** 572.

4. Hiratsuku, J., Shimada, H., Whittier, R., Ishibashi, T., Sakamoto, M., Mori, M., Kondo, C., Honji, Y., Sun, C.-R., Meng, B.-Y., Li, Y.-Q., Kanno, A., Nishizawa, Y., Hirai, A., Shinozaki, K. and Sugiura, M. (1989) *Mol. Gen. Genet.,* **217,** 185.

5. Shimada, H. and Sugiura, M. (1991) *Nucleic Acids Res.,* **19,** 983.

6. Smith, A.G., Wilson, R.M., Kaethner, T.M., Willey, D.L. and Gray, J.C. (1991) *Curr. Genet.,* **19,** 403.

7. Laudenbach, D.E. and Grossman, A.R. (1991) *J. Bacteriol.,* **173,** 2739.

8. Hallick, R.B. and Bottomley, W. (1983) *Plant Mol. Biol. Rep.,* **1,** 38.

9. Hallick, R.B. (1989) *Plant Mol. Biol. Reporter,* **7,** 266.

10. Scheller, H.V. and Møller, B.L. (1990) *Physiol. Plant.,* **78,** 484.

11. Ikeuchi, M. (1992) *Plant Cell Physiol.,* **33,** 669

12. Jansson, S., Pichersky, E., Bassi, R., Green, B.R., Ikeuchi, M., Melis, A., Simpson, D.J., Spangfort, M., Staelion, A. and Thornber, J.P. (1992) *Plant Mol. Biol. Rep.,* **10,** 242.

13. Kirsch, W., Seyer, P. and Herrmann, R.G. (1986) *Curr. Genet.,* **10,** 843.

14. Takahashi, Y., Goldschmidt-Clermont, M., Soen, S.-Y., Franzén, L.G. and Rochaix, J.-D. (1991) *EMBO J.,* **10,** 2033.

15. Li, N., Zhao, J., Warren, P.V., Warden, J.T., Bryant, D.A. and Goldbeck, J.H. (1991) *Biochem.,* **30,** 7863.

16. Okkels, J.S., Jepsen, L.B., Hønberg, L.S., Lehmbeck, J., Scheller, H.V., Brandt, P., Høyer-Hansen, G., Stummann, B., Henningsen, K.W., von Wettstein , D. and Møller, B.L. (1988) *FEBS Lett.,* **237,** 108.

17. Steppuhn, J., Hermans, J., Nechushtai, R., Herrmann, G.S. and Herrmann, R.G. (1989) *FEBS Lett.,* **237,** 218.

18. Okkels, J.S., Scheller, H.V., Svedsen, I. and Møller, B.L. (1991) *J. Biol. Chem.,* **266,** 6767.

19. Steppuhn, J., Hermans, J., Nechushtai, R., Herrmann, G.S. and Herrmann, R.G. (1989) *Curr. Genet.,* **16,** 99.

20. Scheller, H.V., Okkels, J.S., Høj, R.B., Svedsen, I., Roepstorff, P. and Møller, B.L. (1989) *J. Biol. Chem.,* **264,** 18402.

21. Ikeuchi, M., Nyhus, K.J., Inoue, Y. and Pakrasi, H.B. (1991) *FEBS Lett.,* **287,** 5.

22. Franzén, L.-G. Frank, G., Zuber, H. and Rochaix, J.-D. (1989) *Mol. Gen. Genet.*, **219,** 137.

23. Iwasaki, Y., Ishikawa, H., Hibino, T. and Takabe, T. (1991) *Biochim. Biophys. Acta,* **1059,** 141.

24. Zurawski, G., Bohnert, H.J., Whitfield, P.R. and Bottomley, W. (1982) *Proc. Natl. Acad. Sci. USA,* **79,** 7699.

25. Morris, J. and Herrmann, R.G. (1984) *Nucleic Acids Res.,* **12,** 2837.

26. Holschuh, K., Bottomley, W. and Whitfield, P.R. (1984) *Nucleic Acids Res.,* **12,** 8819.

27. Herrmann, R.G., Alt, J., Schiller, B., Widger, W.R. and Cramer, W.A. (1984) *FEBS Lett.,* **176,** 239.

28. Hird, S.M., Dyer, T.A. and Gray, J.C. (1986) *FEBS Lett.,* **209,** 181.

29. Ikeuchi, M. and Inoue, Y. (1988) *FEBS Lett.,* **241,** 99.

30. Webber, A.N., Hird, S.M., Packman, L.C., Dyer, T.A. and Gray, J.C. (1989) *Plant Mol. Biol.,* **12,** 141.

31. Murata, N., Miyao, M., Hayashida, N., Hidaka, T. and Sugiura, M. (1988) *FEBS Lett.,* **235,** 283.

32. Ikeuchi, M., Koike, H. and Inoue, Y. (1989) *FEBS Lett.,* **253,** 178.

33. Wales, R., Newman, B.J., Pappin, D. and Gray, J.C. (1989) *Plant Mol. Biol.,* **12,** 439.

34. Jansen, T., Rother, C., Steppuhn, J., Reinke, H., Beyreuther, K., Jansson, C., Andersson, B. and Herrmann, R.G. (1987) *FEBS Lett.,* **216,** 234.

35. Lautner, A., Klein, R., Ljungberg, U., Reiländer, H., Bartling, D., Andersson, B., Reinke, H., Beyreuther, K. and Herrmann, R.G. (1988) *J. Biol. Chem.,* **263,** 10077.

36. O'Keefe, D.P. (1988) *Photosyn. Res.,* **17,** 189.

37. Hayley, J. and Bogorad, L. (1989) *Proc. Natl. Acad. Sci. USA,* **86,** 1534.

38. Willey, D.L., Howe, C.J., Auffret, A.D., Bowman, C.M., Dyer, T.A. and Gray, J.C. (1984) *Mol. Gen. Genet.,* **194,** 416.

39. Hird, S.M., Wilson, R.J., Dyer, T.A. and Gray, J.C. (1991) *Plant Mol. Biol.,* **16,** 745.

40. Steppuhn, J., Rother, C., Hermans, J., Jansen, T., Salnikow, J., Hauska, G. and Herrmann, R.G. (1987) *Mol. Gen. Genet.,* **210,** 171.

41. Rother, C., Jansen, T., Tyagi, A., Tittgen, J. and Herrmann, R.G. (1986) *Curr. Genet.,* **11,** 171.

42. Hase, T., Kimata, Y., Yonekura, K., Matsumura, T. and Sakakibara, H. (1990) *Plant Physiol.,* **96,** 77.

43. Haley, J. and Bogorad, L. (1989) *Proc. Natl. Acad. Sci. USA,* **86,** 1534.

44. Jansen, T., Reiländer, H., Steppuhn, J. and Herrmann, R.G. (1988) *Curr. Genet.,* **13,** 517.

45. Newman, B.J. and Gray, J.C. (1988) *Plant Mol. Biol.,* **10,** 511.

46. Wakabayashi, S., Kimura, T., Fukuyama, K., Matsubara, H. and Rogers, L.J. (1989) *Biochem. J.,* **263,** 981.

47. Nelson, N. and Taiz, L. (1989) *Trends in Biochem. Sci.,* **14,** 113.

48. Deno, H., Shinozaki, K. and Sugiura, M. (1988) *Nucleic Acids Res.,* **11,** 2185.

49. Zurawski, G., Bottomley, W. and Whitfield, P.R. (1984) *Proc. Natl. Acad. Sci. USA,* **79,** 6260.

50. Werner, S., Schumann, J. and Strotmann, H. (1990) *FEBS Lett.,* **261,** 204.

51. Hermans, J., Rother, C.H., Bichler, J., Steppuhn, J. and Herrmann, R.G. (1988) *Plant Mol. Biol.,* **10,** 323.

52. Bird, C.R., Koller, B., Auffret, A.D., Huttly, A.K., Howe, C.J., Dyer, T.A. and Gray, J.C. (1985) *EMBO J.,* **4,** 1381.

53. McCarn, D.F., Whitaker, R.A., Alam, J., Vrba, J.M. and Curtis, S.E. (1988) *J. Bacteriol.,* **170,** 3448.

54. Deno, H., Shinozaki, K. and Sugiura, M. (1984) *Gene,* **32,** 195.

55. Hennig, J. and Herrmann, R.G. (1986) *Mol. Gen. Genet.,* **203,** 117.

56. Fromme, P., Gräber, P. and Salnikow, J. (1987) *FEBS Lett.,* **218,** 27.

57. Matsubayashi, T., Wakasugi, T., Shinozaki, K., Yamaguchi-Shinozaki, K., Zaita, N., Hidaka, T., Meng, B.Y., Ohto, C., Tanaka, M., Kato, A., Maruyama, T. and Sugiura, M. (1987) *Mol. Gen. Genet.,* **210,** 385.

58. Weiss, H., Friedrich, T., Hofhaus, G. and Preis, D. (1991) *Eur. J. Biochem.,* **197,** 563.

59. Arizmendi, J.M., Skehel, J.M., Runswick, M.J., Fearnley, I.M. and Walker, J.E. (1992) *FEBS Lett.,* **313,** 80.

60. Dupuis, A., Skehel, J.M. and Walker, J.E. (1991) *Biochemistry,* **30,** 3954.

61. Fearnley, I.M., Runswick, M.J. and Walker, J.E. (1989) *EMBO J.,* **8,** 665.

62. Pilkington, S.J., Skehel, J.M. and Walker, J.E. (1991) *Biochemistry,* **30,** 1901.

63. Ogawa, T. (1991) *Proc. Natl. Acad. Sci. USA,* **88,** 4275.

64. Steinmuller, K., Ley, A.C., Steinmetz, A.A., Sayre, R.T. and Bogorad, L. (1989) *Mol. Gen. Genet.,* **216,** 60.

65. Ohyama, K., Kohchi, T., Sano, T. and Yamada, Y. (1988) *Trends Biochem.*, **13**, 19.

66. Wu, M., Nie, Z.Q. and Yang, J. (1989) *The Plant Cell*, **1**, 551.

67. Nixon, P.J., Gounaris, K., Coomber, S.A., Hunter, C.N., Dyer, T.A. and Barber, J. (1989) *J. Biol. Chem.*, **264**, 14129.

68. Meeker, R. (1988) *Mol. Cell. Biol.*, **8**, 1216.

69. Miyamura, S., Kuroiwa, T. and Nagata, T. (1990) *Plant Cell Physiol.*, **31**, 597.

70. Stritmatter, G. and Kössel, H. (1984) *Nucleic Acids Res.*, **12**, 7633.

71. Lemieux, C., Boulanger, J., Otis, C. and Turmel, M. (1989) *Nucleic Acids Res.*, **17**, 7997.

72. Manhart, J.R., Hoshaw, R.W. and Palmer, J.D. (1991) *J. Phycol.*, in press.

73. Manhart, J.R., Kelly, K., Dudock, B.S. and Palmer, J.D. (1989) *Mol. Gen. Genet.*, **216**, 417.

74. Yamada, T. (1991) *Curr. Genet.*, **19**, 139.

75. Giovannoni, S.J., Turner, S., Olsen, G.J., Barns, S., Lane, D. J. and Pace, N.R. (1988) *J. Bacteriol.*, **170**, 3584.

76. Capel, M.S. and Borque, D.P. (1982) *J. Biol. Chem.*, **257**, 7746.

77. Pfitzinger, H., Weil, J.H., Pillay, D.T.N. and Guillemaut, P. (1990) *Plant Mol. Biol.*, **14**, 805.

78. Schon, A., Krupp, G., Gough, S., Berry-Lowe, S., Kannangara, C.G. and Söll, D. (1986) *Nature*, **322**, 281.

79. Bergsland, K.J. and Haselkorn, R. (1991) *J. Bacteriol.*, **173**, 3446.

80. Igloi, G.L., Meinke, A., Döry, I and Kössel, H. (1990) *Mol. Gen. Genet.*, **221**, 379.

81. Hu, J. and Bogorad, L. (1990) *Proc. Natl. Acad. Sci. USA*, **87**, 1531.

82. Bülow, S. and Link, G. (1988) *Plant Mol. Biol.*, **10**, 349.

83. Zaitlin, D., Hu, J. and Bogorad, L. (1989) *Proc. Natl. Acad. Sci. USA*, **86**, 876.

84. Woodbury, N.W., Dobres, M. and Thompson, W.F. (1989) *Curr. Genet.*, **16**, 433.

85. Gruissem, W. and Zurawski, G. (1985) *EMBO J.*, **4**, 3375.

86. Eiserman, A., Tiller, K. and Link, G. (1990) *EMBO J.*, **9**, 3981.

87. Kuhsel, M.G., Strickland, R. and Palmer, J.D. (1990) *Science*, **250**, 1570.

88. Bonnard, G., Michel, F., Weil, J.-H. and Steinmetz, A. (1984) *Mol. Gen. Genet.*, **194**, 330.

89. Herrin, D.L., Chen, Y.-F. and Schmidt, G.W. (1990) *J. Biol. Chem.*, **265**, 21134.

90. Michel, F., Umesono, K. and Ozeki, H. (1989) *Gene*, **82**, 5.

91. Shinozaki, K., Deno, H., Sugita, M., Kuramitsu, S. and Suguira, M. (1986) *Mol. Gen. Genet.*, **202**, 1.

92. Deng, X.-W., Wing, R.A. and Gruissem, W. (1989) *Proc. Natl. Acad. Sci. USA*, **86**, 4156.

93. Bendich, A.J. and Smith, S.B. (1990) *Curr. Genet.*, **17**, 421.

94. Collin, S. and Ellis, T.H.N. (1991) *Curr. Genet.*, **20**, 253.

95. Miyamura, S., Kuroiwa, T. and Nagata, T. (1990) *Plant Cell Physiol.*, **31**, 597.

96. White, E.E. (1990) *Theor. Appl. Genet.*, **79**, 119.

97. Palmer, J.D., Osorio, B. and Thompson, W.F. (1988) *Curr. Genet.*, **14**, 65.

98. Palmer, J.D., Nugent, J.M. and Herbon, L.A. (1987) *Proc. Natl. Acad. Sci. USA*, **84**, 769.

99. Woessner, J.P., Gillham, N.W. and Boynton, J.E. (1987) *Plant Mol. Biol.*, **8**, 151.

100. Kashdan, M.A. and Dudock, B.S. (1982) *J. Biol. Chem.*, **257**, 11191.

101. DeBlock, M., Schell, J. and Van Montagu, M. (1985) *EMBO J.*, **4**, 1367.

102. Cornelissen, M.J., DeBlock, M., Van Montagu, M., Leemans, J., Schreier, P.H. and Schell, J. (1987) in *Plant DNA Infectious Agents* (T.H. Hohn and J. Schell eds). Springer, Wien, p. 311.

103. Venkateswarlu, K. and Nazar, R.N. (1991) *Bio/Technol.*, **9**, 1103.

104. Daniell, H., Vivekananda, J., Nielsen, B.L., Ye, G.N., Tewari, K.K. and Sanford, J.C. (1990) *Proc. Natl. Acad. Sci. USA*, **87**, 88.

105. Daniell, H., Krishnan, M. and McFadden, B.F. (1991) *Plant Cell Rep.*, **9**, 615.

106. Ye, G.-N., Daniell, H. and Sanford, J.C. (1990) *Plant Mol. Biol.*, **15**, 809.

107. Svab, Z., Hajdukiecz, M.P. and Maliga, P. (1990) *Proc. Natl. Acad. Sci. USA*, **87**, 8526.

108. Staub, J.M. and Maliga, P. (1992) *The Plant Cell*, **4**, 39.

109. Sporlein, B., Streubel, M., Dahlfeld, G., Westhoff, P. and Koop, H.U. (1991) *Theor. Appl. Genet.*, **82**, 717.

110. Koop, H.-U., Eigel, L. and Sporlein, B. (1992) *Physiol. Plant.*, **85**, 339.

111. Boynton, J.E., Gillham, N.W., Harris, E.H., Newman, S.M., Randolph-Anderson, B.L., Johnson, A. M. and Jones, A.R. (1990) *Curr. Res. Photos.*, **III**, 509.

112. Kindle, K.L., Richards, K.L. and Stern, D.B. (1991) *Proc. Natl. Acad. Sci. USA,* **88,** 1721.

113. Boynton, J.E., Gillham, N.W., Harris, E.H., Hosler, J.P., Johnson, A.M., Jones, A.R., Randolph-Anderson, B.L., Robertson, D., Klein, T.M., Shark, K.B. and Sanford, J.C. (1988) *Science,* **240,** 1534.

114. Thompson, A.J. and Herrin, D.L. Unpublished results.

115. Blowers, A.D., Bogorad, L., Shark, K.B. and Sanford, J.C. (1989) *The Plant Cell,* **1,** 123.

116. Blowers, A.D., Ellmore, G.S., Klein, U. and Bogorad, L. (1990) *The Plant Cell,* **2,** 1059.

117. Stern, D.B., Radwanski, E.R. and Kindle, K.L. (1991) *The Plant Cell,* **3,** 285.

118. Goldschmidt-Clermont, M. (1991) *Nucleic Acids Res.,* **19,** 4083.

119. Prizibilla, E., Heiss, S., Johanningmeier, U. and Trebst, A. (1991) *The Plant Cell,* **3,** 169.

120. Newman, S.M., Boynton, J.E., Gillham, N.W., Randolph-Anderson, A.M.J. and Harris, E.H. (1990) *Genetics,* **126,** 875.

121. Thompson, A.J. and Herrin, D.L. (1991) *Nucleic Acids Res.,* **19,** 6611.

122. Takahashi, Y., Goldschmidt-Clermont, M., Soen, S.-Y., Franzen, L.G. and Rochaix, J.-D. (1991) *EMBO J.,* **8,** 2033.

123. Zumbrunn, G., Schneider, M. and Rochaix, J.-D. (1989) *Technique,* **1,** 204.

124. Johnston, S.A. (1990) *Nature,* **346,** 776.

125. Russell, J.A., Roy, M.K. and Sanford, J.C. (1992) *Plant Physiol.,* **98,** 1050.

126. Oard, J.H., Paige, D.F., Simmonds, J.A. and Gradziel, T.M. (1990) *Plant Physiol.,* **92,** 334.

CHAPTER 10
PLANT GENE EXPRESSION AND PCR TECHNIQUES

A. Fordham-Skelton, I.M. Evans and R.R.D. Croy

1. PCR STRATEGIES

This section is intended as a summary and guide to the major uses of PCR*. The reader is referred to refs 1–3 for more detailed protocols, and the chapters on PCR in refs 4 and 237. Recent reviews on PCR have been published (5, 6, 218). The journals, *PCR Methods and Applications* and *BioTechniques* are useful sources of reports on recent advances in PCR methodologies. Clontech has published two booklets: *PCR Research Tools Catalog* and *RT-PCR: Methods and Applications* which cover several aspects of PCR methodology. While the majority of PCR applications have been used for animal and prokaryotic systems, this section tries, where appropriate, to illustrate the use of PCR with plant systems. Due to the large and increasing number of reports describing PCR techniques, it is impossible to cite every paper and we apologize to those authors who have not been included. A table of some of the acronyms for PCR-based methods has been included. Since cloning of PCR products is an important, and often difficult, operation, restriction maps for some of the commercially available PCR cloning vectors have been included. Finally, all mistakes and omissions are our own.

Gene Expression/PCR

*Polymerase chain reaction. The PCR process is covered by US patent numbers 4683202, 4683195, 4800159 and 4965188 owned by Hoffman La Roche Inc. and other pending and issued patents in non-US countries owned by F. Hoffman La Roche Inc.

► p. 345

Table 1. General PCR

	Description	Comments	References
1.1 DNA polymerases Several thermostable polymerases are now available with and without 3′–5′ proofreading activity (see Section 1.9 for amplification fidelity)	*Taq* (*Thermus aquaticus*) (*rTaq*, Δ*Taq*)	Original thermostable polymerase used for PCR. No 3′–5′ proofreading activity. Has 5′–3′ exonuclease activity. Reported to have reverse transcriptase activity; may not be reproducible (discussed in ref. 12). (Available from numerous suppliers as native or recombinant form)	7–12 11, 12
	TET-z (*Thermus thermophilus*)	Has reverse transcriptase activity (Amersham)	13
	Vent_R (*Thermococcus litoralis*) and $\text{Vent}_R(\text{exo}^-)$	Available with and without 3′–5′ proofreading activity (New England Biolabs). Also supplied by Promega (*Tli* with proofreading activity)	14, 15
	Deep Vent_R (*Pyrococcus* species GB-D)	Has 3′–5′ proofreading activity (New England Biolabs)	
	Heat-TUFF	(Clontech)	
	Tth (*Thermus thermophilus* HB-8)	Has reverse transcriptase activity (United States Biochemical)	13
	Pfu (*Pyrococcus furiosus*)	Has 3′–5′ proofreading activity (Stratagene)	16
	r*Tth* (*Thermus thermophillus*)	Has reverse transcriptase activity (Perkin–Elmer)	13
	Tfl (*Thermus flavus*)	(Promega)	
	(r/R = recombinant form of polymerase)		

Table 1. Continued

	Description	Comments	References
	Proofreading activity of DNA polymerases	3′–5′ exonuclease activity will excise 3′ terminal mismatches. May be a problem if using primers which are not completely matched to target sequence (e.g. degenerate oligonucleotides).	15
		Exonucleolytic attack reduced by incorporating single phosphorothioate bonds at 3′ termini of primers, these are a less favored substrate for nuclease attack than phosphodiester bond	17
1.2 Reaction buffer additives			
Specificity enhancers	Magnesium chloride	Titration of [Mg^{2+}] may be required to optimize specific amplification of product, generally in the range 1.5–10.0 mM. Dependent on each primer pair. PCR buffers are supplied by several manufacturers without Mg^{2+} to allow for optimization	1
	TMAC (tetramethylammonium chloride)	Final concentration 1×10^{-4} M–1×10^{-5} M	18
	'Perfect match' polymerase enhancer	Specificity of primer annealing increased, supplied by Stratagene	
Amplification of G/C rich template DNAs. Formation of secondary structure may interfere with amplification efficiency	Glycerol	Addition of 20% (v/v) with 60°C annealing temperature. Cereal genomic DNA templates used to amplify α-amylase inhibitor sequences	19

Table 1. Continued

	Description	Comments	References
	7-deaza-2′-deoxyguanosine (dc^7GTP)	dc^7GTP is a structure destabilizing analog of dGTP. Optimal ratio dc^7GTP:dGTP 3:1 (150 mM:50 mM)	20
	Formamide	Addition to 1.25–5.0% (v/v)	21
	Preboiling of reaction mix	High G/C content primers with redundancy and complementary 3′ ends amplified specific product	22
Amplification of long-template DNAs	Tricine buffer	Amplification of templates up to 6.2 kb, no KCl present, low number of cycles	23
	Gene 32 protein	Single-stranded DNA binding protein from phage T_4, reported to increase yield of long amplification products	24
1.3 Procedures to increase amplification specificity			
Some of these protocols may be applicable to optimizing certain amplifications, especially if target sequences are present at a very low copy number or present in a complex template sample	Nested priming	Initial amplification followed by amplification with second set of primers internal to first set	25, 26
	Touchdown PCR	Annealing temperatures initially relatively high (65°C) then lowered by 1°C every second cycle, final 10 cycles at 55°C	27
	Pre-amplification heating/ 'Hot start'	Assemble PCR mix at 70°C before enzyme addition, this prevents primer mis-annealing. Ampliwax (Perkin–Elmer) separates reaction mix components until heated to 60–80°C. Efficient amplification of low copy number target sequences	28, 29

Table 1. Continued

	Description	Comments	References
	Size fractionation of template DNA	Enriches for target sequence/decreases template complexity (restriction fragment size must be experimentally determined beforehand)	30
	Limit number of amplification cycles	Excessive cycling reported to increase accumulation of high molecular weight products derived from PCR product	31
1.4 Primer selection and design			
Single primer species (see *Table 4* for degenerate primers)	Prediction of primer annealing temperatures	These have been determined empirically	32
		Computer programs such as OLIGO, PRIMER DETECTIVE and PROBE DESIGNER predict annealing temperatures at specified salt concentrations and inter/intra complementary. Avoid this at 3′ end of primers to minimize the posslibility of primer dimer effects	33–35
Approximately equal G/C content is desirable. Primers of 15–30 bases are mainly used	Primer specificity	Specificity of primer resides at the 3′ end (start of extension); certain base pair mismatches can be tolerated, dependent upon primer, annealing temperature and dNTP concentration	36, 37 (and references therein), 219
		Requirement for match at 3′ terminus allows for allele-specific amplification. Other sequences (e.g. restriction sites, RNA polymerase promoter sequence) can be added to 5′ end of a primer (see *Table 5*, Section 5.2)	38

Table 1. Continued

	Description	Comments	References
	Purification of primers	Polyacrylamide gel electrophoresis to size fractionate oligonucleotides Use reverse-phase column to isolate primers synthesized with trityl group on 3′ end of primer. Purifies final oligonucleotide product from prematurely terminated chains	
1.5 Minimization of contamination			
Contamination of amplification reaction mixes with foreign DNA templates (e.g. cloned or previously amplified sequences) can be a major problem and must be considered, especially when carrying out diagnostic PCR amplifications	Separate assembly of reaction mix from analysis of products		39
	Store reaction components as small aliquots		
	Use positive displacement pipet and disposable tips		
	Add template samples last to reaction mix		
	Include negative controls without template		
	UV-irradiate reaction mix prior to template addition	Irradiation with UV at 254 nm and/or 300 nm. Primers not adversely affected. Template must be added after and is a potential source of contamination. Effect of primer base composition is discussed in ref. 41	40–41

Table 1. Continued

	Description	Comments	References
Several methods described to prevent amplified DNA serving as a contaminating template	Use of exonuclease III (exoIII)	Removes nucleotides from 3′ end of duplex DNA, inactivated at 95°C. Contaminating amplified DNA preferentially degraded compared with template DNA. Alternatively, DNA can be denatured, genomic template remains mainly single stranded (not a substrate for exoIII) whereas shorter products anneal at a quicker rate and are degraded. Protocols require high molecular weight DNA and exoIII free from contaminating exonucleases. Phage T_7 exonuclease used in same manner	42 43
	Add psoralen to sample before amplification	Intercalates with double-stranded nucleic acids. UV illumination produces covalent linkages and prevents DNA acting as a template	44
	Use of isopsoralen derivatives	After amplification PCR products are photochemically reacted with isopsoralens, preventing them from serving as further templates	45, 46
	Use of uracil DNA glycosylase (UDG)	Substitute dUTP for dTTP in PCR amplifications. All PCR products then contain uracil. Further amplification reactions are incubated with UDG. UDG specifically removes uracil from DNA, i.e. any contaminating PCR products, true DNA template containing thymine not affected. On first denaturation step of thermal cycling UDG is inactivated. Kit supplied by Gibco-BRL	47

Table 1. Continued

	Description	Comments	References
1.6 Post-PCR procedures			
Removal of primers/product purification	Gel electrophoresis	e.g. low melting point agarose gel	
	Size-exclusion chromatography	e.g. Sephadex G-50 column	
	Spin column dialysis	e.g. Centricon tubes (Amicon)	
	Incubation of PCR products with exonuclease VII	*E. coli* exonuclease VII specific for single-stranded DNA. Useful if several primer sets are used sequentially with limiting amounts of template/product	48
1.7 Cloning PCR products	See *Figure 1* for examples of commercially available PCR cloning vectors		
Direct cloning into conventional plasmid vectors (e.g. pUC18, pGEM, pBluescript)	Blunt end ligation	*Pfu* and Vent DNA polymerases produce PCR products with blunt ends	49, 50
	Include restriction sites at 5′ end of each primer; allows directional cloning	Some restriction enzymes may not cut efficiently when close to the termini of the product. Concatemerization of products using T_4 DNA ligase followed by restriction reported to increase cutting efficiency (ensures primers are phosphorylated)	51

Table 1. Continued

	Description	Comments	References
	Efficiency of cloning PCR products reported to increase if treated with:		
	(i) Proteinase K or	(i) Thought to remove *Taq* polymerase from end of product	52
	(ii) T_4 DNA polymerase	(ii) Generates blunt-ended product suitable for cloning. Ensure excess dNTPs present	4
	Cloning PCR products without using ligase	Product amplified with primers containing sequences at 5′ end complementary to 3′ end of linearized vector. Several methods reported. Products annealed to plasmid vector are sufficiently stable for transformation without a ligation step	53, 54
Plasmid vectors for cloning PCR products Utilization of non-template addition of A to 3′ end of PCR products by *Taq* polymerase (see *Figure 1*)	T:A vectors	*Taq* polymerase (and *T. flavus*/*T. litoralis* DNA polymerases) commonly adds an extra residue to 3′ end of product (usually adenine). T:A vectors use this property as basis for cloning (see *Figure 1*)	55, 56
	Commercially available T:A vectors	T:A vectors supplied by Invitrogen, Stratagene, Promega and Novagen. Insertion site flanked by unique restriction enzyme sites, RNA polymerase promoters and sequencing primer sites. *lacZ* blue/white colony selection (see *Figure 1*)	57

Table 1. Continued

	Description	Comments	References
	Production of T:A vectors	Incubation of blunt end cut vector with dTTP and *Taq* polymerase or terminal transferase and ddTTP to add single T to 3′ ends of vector	58, 59
		Restrict vector containing *Xcm*I or *Hph*I sites. Restriction of these sites generates 3′ T overhangs	57, 60
1.8 Direct sequencing of PCR products			
	DNA sequence of PCR products can be determined without the requirement for cloning	Detailed protocols for the direct sequencing of PCR products are described in ref. 238	238
		Single-stranded products are generated by asymmetric PCR using either one primer in excess or strand-specific digestion using λ exonuclease	239 240
		Double-stranded PCR products can also be sequenced	
		Primers for the sequencing reactions can be the PCR primers or a third oligonucleotide internal to the flanking PCR primers	

Table 1. Continued

	Description	Comments	References
	Direct sequencing of PCR products using automated DNA sequencers	Requires specific fluorescent dye-linked primers	241, 242
		Alternative approach is to use fluorescent dye-linked chain terminators. No requirement for a labeled sequencing primer	
		See application to the sequencing of a *Brassica* self-incompatibility allele after PCR amplification	243
1.9 Potential PCR errors and artifacts			
Misincorporation of nucleotides and fidelity of DNA synthesis. References quoted describe methodologies for determining polymerase error rates	Proofreading DNA polymerases have lower error rates than *Taq* polymerase, which lacks 3′–5′ exonuclease activity, as they are able to excise mismatched 3′ nucleotides. PCR products to be used in further manipulations (e.g. where a reading frame must be maintained or a specific allele is under investigation) must be sequenced to ensure that no errors have occurred	*Taq* polymerase: Rate determined at approximately 2×10^{-4} errors per base per duplication. In a single round of DNA synthesis 1:9000 base substitutions and 1:41 000 frameshift errors	8, 61 9
		Majority of base changes A/T to G/C	61, 62
		Factors affecting the fidelity of DNA synthesis by *Taq* polymerase discussed in refs 63 and 64	63, 64
		Vent polymerase error rate reported as 2–4-fold lower than *Taq* polymerase	14, 15
		Pfu polymerase error rate reported as 10-fold lower than *Taq* polymerase	16

Table 1. Continued

	Description	Comments	References
Artifacts produced by recombination during PCR	Reported when amplifying highly similar sequences, i.e. members of multigene families. Hybrid products produced, consider possibility when amplifying alleles for diagnostic purposes	Probably produced due to pausing or termination of polymerase during extension, allowing incompletely extended products to subsequently anneal to, and amplify related sequences	65–68 69
		Incompletely extended first-strand cDNA products, produced by reverse transcriptase, reported to have acted as templates for further reverse transcription/amplification of related sequences by *Taq* polymerase. Inclusion of RNase (DNase free) after first-strand synthesis removes template	69
		Introduction of 'nicks' into template reported to increase PCR recombination	70
		Avoid old/boiled samples if possible. UV-damaged DNA, strand breaks and apurinic sites contribute to recombination	71

Table 2. Genome analysis

	Description	Comments	References
2.1 Ligation-mediated PCR (LM-PCR)			
	In vivo footprinting. Determination of protein/DNA interactions using dimethyl sulfate (DMS) methylation of guanine residues	Piperidine cleavage (Maxam and Gilbert) of genomic DNA after DMS footprinting followed by PCR with specific primer to generate blunt-end duplexes. Linker primer ligated to product, PCR with specific and generic primer. Products, extended with a 3rd, end-labeled primer, analyzed on standard sequencing gel	72
		Detection of modified guanine and adenine residues after DMS treatment	73
	DNA methylation analysis	5-methylcytosine detected	74
		Methylation analysis of *Ac* in transgenic plants	75
		Positive identification of 5-methylcytosine using permanganate cleavage or sodium bisulfite treatment	75, 220
	Improvement of LM-PCR	Biotinylated primer used to purify extension products with streptavidin-coated magnetic beads	76
		Use of Vent polymerase, blunt-ended primer extension products produced	77, 14, 15

Table 2. Continued

	Description	Comments	References
2.2 Amplification of sequences outside known regions			
Inverse PCR (IPCR)	Allows for amplification of unknown sequences flanking a characterized stretch of DNA using two specific primers. Bidirectional	Requires circularization of template DNA using ligation at a high dilution, see ref. 79, followed by amplification with primers extending into unknown region	78, 79, 221
		Modified to allow determination of linkage of exons in gene families	80
		Inverse PCR using cDNA. Double-stranded cDNA circularized and used as template for IPCR	81, 82
Chromosome walking/crawling. Anchored PCR	One specific and one generic primer used. Unidirectional	Unknown flanking sequence restricted and ligated to one end of a generic vector, i.e. M13 or double-stranded oligonucleotide cassette. Primers to known sequence and generic vector then amplify unknown sequence. Useful to have a probe for hybridization as non-specific products can be problematical. Can be used to obtain both flanking regions	82–85

Table 2. Continued

	Description	Comments	References
2.3 Amplification using limiting amounts of material			
	Single cell/DNA molecule analysis	References cited describe techniques for amplifying single or multiple alleles	86, 87, 222
		Template abundance is increased by an initial non-specific amplification using random primers	88
	Microdissected chromosomes	Microdissection followed by amplification using non-specific primers, or ligation into vector followed by amplification using universal primers. See application to construct a library from *Beta patellaris* microdissected chromosomes (92) and references therein	89–91 92
2.4 Multiplex analysis	Amplification of several target sequences in one reaction	Allows detection of different genes or allelic variants	92, 93
		Primers can be specifically dye-linked to aid product discrimination	94
2.5 Exon structure	Discrimination between genes and processed pseudogenes	PCR primers designed to span intron/exon boundaries	95–97
2.6 Genomic subtraction	Mutations caused by deletions can be cloned by subtracting wild-type genomic DNA with genomic DNA from mutant genotype	See application to cloning of the *Arabidopsis* 5.0 kbp GA1 locus (98) and references in (99)	98, 99

Table 3. PCR using cDNA after reverse transcription (RT-PCR)

	Description	Comments	References
3.1 Construction of cDNA libraries			
	Use of PCR to construct cDNA libraries from limiting amounts of tissue	cDNA population can be amplified using sequence-independent primers and subsequently cloned	100, 101
		Primers containing random hexamers used for cDNA synthesis followed by cloning into conventional vector system, e.g. phage	102
		Protocols for cDNA library production using plant tissues	103, 104
	Application to subtractive cloning of differentially expressed mRNAs	Several protocols described for producing subtractive cDNA libraries/probes	105–107 223, 224
3.2 Direct identification of differentially expressed genes			
	Determination of sequences expressed in a tissue/cell-specific manner or sequences subject to induction/repression (see also ref. 223)	Reverse transcription of poly(A)$^+$RNA using a discriminatory oligo(dT) primer selects a subset of the population which is subsequently amplified with a second random primer	108
	Products are radiolabeled and analyzed on sequencing gels	RNA populations fingerprinted using random primers and low-stringency conditions to prime cDNA synthesis	109

Table 3. Continued

	Description	Comments	References
3.3 Isolation of 5′ and 3′ ends of cDNAs			
	Protocols designed to recover the ends of cDNA clones using a gene-specific primer and a generic primer after homopolymer tailing of the cDNA	Rapid amplification of cDNA ends (RACE)	110, 111
		5′ RACE modified by producing first strand cDNA by random priming. Allows one RNA source to be used as a template for isolating several different 5′-cDNA ends	225
		Alternative to homopolymer tailing is ligation of an oligonucleotide to first strand cDNA using T_4 RNA ligase	112, 113
		Additional enrichment steps for the sequence amplified by RACE can be included but see comparison between refs 112 and 114	114
3.4 Quantification of mRNA levels			
	Alternative to Northern or dot blotting to determine transcript levels. Internal controls required to quantitate product. Either a constitutively expressed gene is used as an internal standard or a competitor sequence is used	See references in review 115 and in ref. 116	115, 116

Table 3. Continued

	Description	Comments	References
3.5 Analysis of gene expression			
	Use of PCR to distinguish expression patterns in multigene families. Transcripts can be detected without using S1 nuclease or RNase protection assays	Difference in restriction sites of gene family used as distinguishing feature, i.e. potato 4-coumarate CoA ligase and *Arachis* chitinase gene expression	117, 118
	RT-PCR used to distinguish expression of gene families using a common end-labeled primer and a gene-specific primer. Products can be sized on sequencing gels	*Phaseolus* ssRubisco and potato *USN*RNA and *U2B″* gene expression analyzed	119, 120
	Gene expression in *Volvox* embryos	Developmental control of extensin gene expression during embryo inversion determined by RT-PCR on single embryos at specific developmental stages	226
	Primers designed to detect mRNA in presence of contaminating genomic DNA	Primers can be designed to span an intron such that products of genomic and cDNA amplification can be distinguished by size difference (see *Table 2*, Section 2.5). If introns are not present then a 3′ primer containing a dT tract and the extreme 3′ terminus of the mRNA has been shown to only amplify cDNA in the presence of genomic DNA	227

Table 3. Continued

Description	Comments	References
3.6 Analysis of RNA processing		
	Mapping of polyadenylation sites	121
	Maize chloroplast tRNA processing intermediates and splice junctions mapped	122
	Analysis of wheat dwarf virus bicistronic mRNAs	123

Table 4. PCR using degenerate oligonucleotides

	Description	Comments	References
4.1 Use of degenerate primers			
Also known as mixed oligonucleotide primer amplification (MOPAC)	Primers derived from the reverse translation of amino acid sequences Primers can be completely degenerate, contain the base analog inosine (see Section 4.4) or, in some cases, codon usage data allows for the design of less redundant primers. The 3′ terminal base of each primer should not be redundant. (Methionine and tryptophan residues encoded by single codons, therefore useful for primer design)	Gene or cDNA can be cloned if some amino acid sequence is available for the gene product. Primers can also be designed from amino acid consensus sequences allowing the identification of homologs in different species or related novel genes in the same species. PCR products, even if not full-length, can be used as probes to recover corresponding genomic or cDNA clones or as a source of sequence for RACE protocols (see *Table 3*) Internal peptide sequence useful for identifying PCR products either to confirm a derived amino acid sequence of a PCR product or as an oligonucleotide hybridization probe to identify correct PCR products. Control reactions containing a single primer may be useful to distinguish non-specific products	123, 228

Table 4. Continued

	Description	Comments	References
4.2 Primers derived from reverse translation of peptide sequences			
	Partial or full amino acid sequence data can be used for primer design. Several computer programs will perform reverse translation and primer design. Selected references serve as an illustration	Porcine urate oxidase/partial N-terminal and internal amino acid (aa) sequence/cDNA	123
		Brain neurotrophic factor/partial N-terminal and internal amino acid sequence/cDNA/genomic DNA/RACE	124
		Cyanobacterial metallothionein gene *smtA*/complete aa sequence/genomic DNA/chromosome crawling with degenerate inosine primer (see *Table 2*, Section 2.2)	83
		Bean PR4 chitinase/partial aa sequence/cDNA	125
		Barley chloroplast β-ketoacyl synthase/partial aa sequence/cDNA	126
		Trichosanthin/complete aa sequence/inosine primers/genomic DNA	127
		Abrin/complete aa sequence/inosine primers/genomic DNA	128
		Saporin/N-terminal and unpositioned internal aa sequence/inosine primers/genomic DNA	129

Table 4. Continued

	Description	Comments	References
4.3 Primers derived from consensus sequences			
Use of conserved motifs in proteins, identified by multiple alignments of DNA or protein sequence, to design PCR primers (homology probing)	Examples of genes identified in animal systems using homology probing	Genes identified for pattern formation; tyrosine kinase, potassium channels. Primers designed to encode more than one residue at several positions	130
		Third member of mouse nerve growth factor/brain-derived neurotrophic factor gene family identified	131
		Identification of homologs of iron–sulfur protein of succinate dehydrogenase. Peptides conserved between *E. coli* and bovine genes	132
		Identification of new G protein-coupled receptor genes	133
		Identification of zinc finger domains	134
		Identification of putative mammalian odorant receptors. Restriction analysis of PCR products identified gene family	135

Table 4. Continued

	Description	Comments	References
	Examples of the identification of plant homologs of previously characterized genes using primers to conserved motifs	Plant homologs of p34^{cdc2} kinase in maize, alfalfa, *Arabidopsis* and pea	136–139
		Protein kinase homologs in rice and pea	140, 141, 228
		Type 1 protein phosphatases in *Arabidopsis* and *Brassica*	142, 143
		Maize receptor protein kinase	144
		Ty-1-copia retrotransposons, primers to conserved domains of reverse transcriptase	145, 146
4.4 Use of inosine as a base analog			
	The 'universal' base inosine (I) can be incorporated into ambiguous positions in a primer sequence	I will base-pair with A, T or C. Appears to act as G when present as a template for DNA synthesis, i.e. C preferentially selected on opposite strand	147–149
		Used in PCR primers to reduce complexity of primer pool	150, 127
	Other base analogs (P[a] and K[b]) function when incorporated into PCR primers		151

[a]P = 6H, 8H-3,4-dihydropyrimido[4,5-C](1,2)oxazin-7-one.
[b]K = 2-amino-6-methoxyaminopurine.

Table 5. Use of PCR to modify DNA sequences

	Description	Comments	References
5.1 Modification of sequences			
	Procedures allow insertion, deletion or replacement of template sequences. Precise regions can also be spliced together without relying on the presence of convenient restriction sites	Use relatively large amounts of template such that fewer cycles of PCR are required, less chance of errors. Verify correct changes by sequencing	232–236
5.2 Introduction of sequences for *in vitro* transcription/translation			
	5′ PCR primer incorporates an RNA polymerase promoter sequence (e.g. SP6/T7) and an in-frame initiation codon, if required. 3′ PCR primer can contain stop codon to delineate coding sequence	After amplification PCR products contain all sequences necessary for *in vitro* transcription/translation. No requirement for subcloning into expression vector	152–155

Table 6. Randomly amplified polymorphic DNA (RAPD) analysis

Description	Comments	References
Short random primers are used either singly or in combination to amplify arbitrary regions of the target genome (see *Table 3*, Section 3.2 for mRNA fingerprinting)	RAPD analysis allows the identification of DNA fragments which segregate in a Mendelian manner	156–158
	RAPDs allow the construction of linkage maps, determination of individual genotypes. Primers to tRNA genes have also been used	159 160
Amplification-fragment-length-polymorphisms (AFLPs)	Very short primers used, essentially the same as RAPD analysis. See ref. 162 for a review	161, 162
Examples of the use of RAPD analysis in plants	RAPD analysis in wheat, banana and maize/melon	163–165
	Use of RAPDs to map *Arabidopsis* genome	166
	Isolation of RAPD markers linked to *Pseudomonas* resistance gene	167
	RAPD markers used to characterize potato somatic hybrids	168
	Generation of a linkage map for white spruce using haploid DNA	169
Extraction methods for DNA template isolation for use in RAPD analysis (applicable to other PCR analyses, see *Table 7*, Section 7.2)	Quick methods for DNA isolation including leaf disks and tissue squashes	170–172
	DNA extracted from seedlings stored in microtiter plates	173
	Brassica DNA extracted from microspore-derived embryos	174
	Isolation of DNA from vine leaves, rice and maize	175, 176

Table 7. Detection and analysis of recombinant clones and transgenic plants

	Description	Comments	References
7.1 PCR analysis of recombinant clones			
	Inserts in plasmid and phage vectors can be amplified using primers flanking the cloning site		177, 178
	Screening of yeast artificial chromosome (YAC) libraries	PCR used to identify specific clones in YAC libraries by matrix pooling	179–181
		PCR used to isolate terminal sequences of YAC clones	182
	Isolation of cDNAs encoded on large cloned genomic fragments (e.g. YACs or phage P1 recombinants)	Allows mapping of regions which contain transcribed sequences	229, 230

Table 7. Continued

	Description	Comments	References
7.2 PCR analysis of transgenic plants			
	Detection of transgenic plant material using primers to part of introduced sequence	Small amounts of DNA from leaf tissue isolated using Eppendorfs/tips, suitable for PCR	183
		Numbers of integrated T-DNA copies estimated using IPCR (see *Table 2*, Section 2.2)	184–186
		Amplification of multiple target sequences in the same reaction using several primer pairs	187
	Isolation of genomic sequences flanking transposons/T-DNA insertion sites	Allows the recovery of genes inactivated by insertion events and T-DNA flanking regions using IPCR (see *Table 2*, Section 2.2)	188, 189
	Detection of DNA recombination events *in vivo* after transfection	Primers designed to only allow the amplification of specific recombination events. Studied in protoplasts. Primers amplify across recombined region	190, 191

Table 8. Evolutionary and archaeological PCR

	Description	Comments	References
8.1 PCR templates from biological/fossil remains			
	DNA has been isolated from numerous old or ancient biological samples and successfully used as a template for PCR	Dried/mummified tissue, archaeological remains	193–195
		Museum specimens	196, 197
		Fossilized animal samples (termite, saber-toothed cat)	198, 199
		Fossilized leaves from Miocene period (cypress and magnolia)	200–202
		Use of PCR to amplify ancient DNA templates discussed in several articles in ref. 203	203

Table 8. Continued

	Description	Comments	References
8.2 Use of PCR in phylogenetic/ evolutionary studies			
	Amplification of mitochondrial, chloroplastic or rRNA sequences. Selected references given to illustrate approaches taken. Amplification using consensus primers has been used with ancient templates (see Section 8.1)	Amplification of plant rRNA genes	204, 205
		Evolution and speciation investigated by amplification of mitochondrial DNAs	206–211
		Evolution of prochlorophytes investigated by amplification of 16S RNA and RNA polymerase genes	212, 213
		Design of universal primers to amplify chloroplast DNA	214
		Inosine used at 3′ terminal position of primers may be of use when amplifying DNA from several species using the same set of primers (see *Table 4*, Sections 4.3 and 4.4)	215
	RAPD analysis to investigate plant population dynamics (see *Table 6*)	RAPD analysis of *Microseries elegans* inbred populations	216
		RAPD analysis of outcrossing rates in *Datisca glomerata*	217

Table 9. Miscellaneous

	Description	Comments	References
Immuno-PCR	Exceptionally sensitive method for antigen detection	Antigens detected using an antibody–DNA conjugate. Antigen–antibody complexes are detected using a bi-specific linker (streptavidin–protein A) which binds specifically to biotinylated DNA and the Fc region of IgG. After attachment of DNA to complex, biotinylated target sequence is amplified. Sensitivity × 10^5 greater than standard alkaline phosphatase ELISA	192

Figure 1. PCR product cloning vectors. (a) pCR1000 (Invitrogen/British Biotechnology) (The pCR1000 vector has been superceded by the improved pCRII vector and is no longer available.); (b) pCRII (Invitrogen/British Biotechnology); (c) pT7Blue (T) (Novagen/AMS Biotechnology); (d) pCR-Script SK(+) (Stratagene); (e) pTOPE (T) (Novagen/AMS Biotechnology); (f) pCITE (T) (Novagen/AMS Biotechnology). *Figures 1a* and *b* reproduced with permission from Invitrogen Corporation; *Figure 1d* reproduced with permission from Stratagene; *Figure 1c, e* and *f* reproduced with permission from AMS Biotechnology (UK) Ltd.

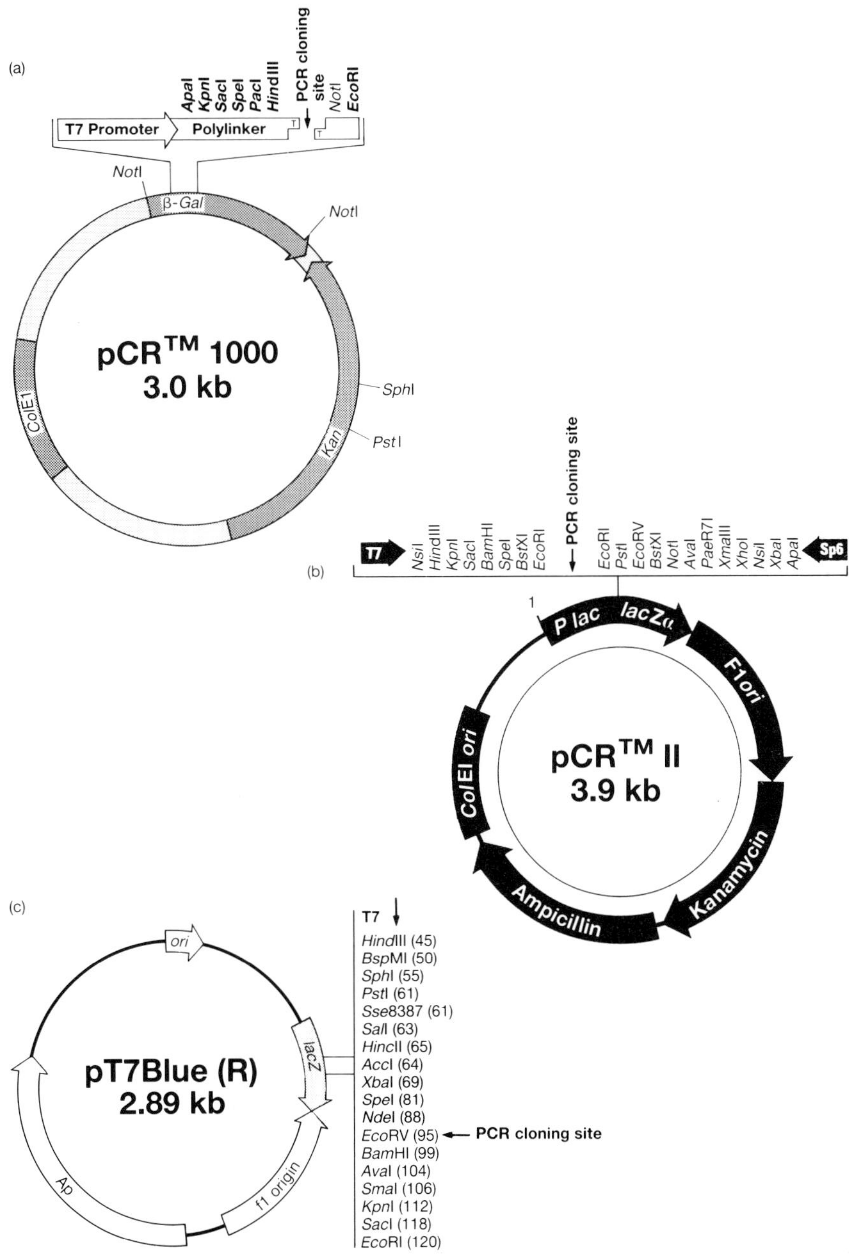

Gene Expression/PCR

Figure 1. Continued

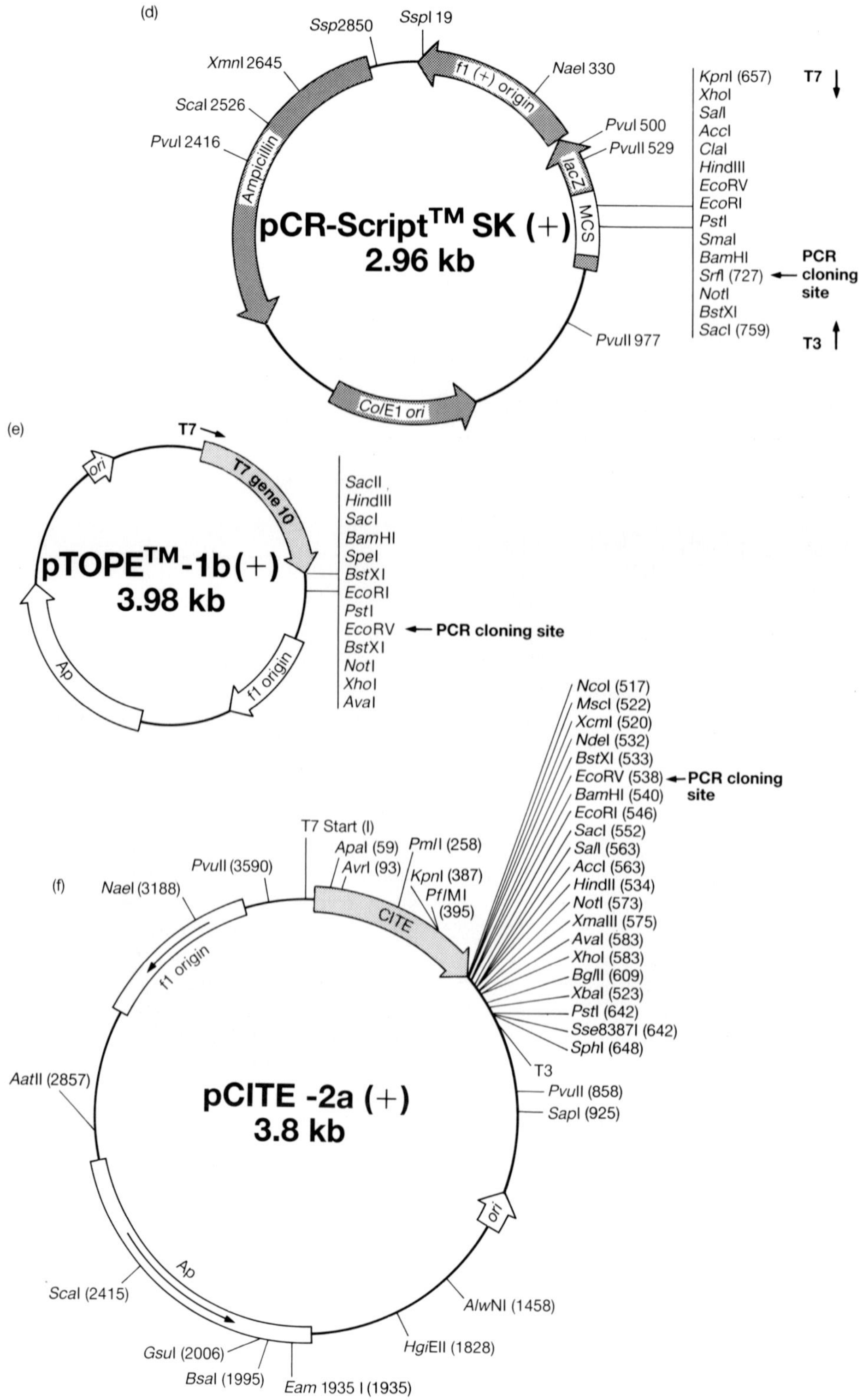

Table 10. PCR acronyms

	Description	Table/Section
AFLPs	Amplification-fragment-length-polymorphisms	6
APCR	Anchored PCR	2.2
IPCR	Inverse PCR	2.2
LM-PCR	Ligation-mediated PCR	2.1
MOPAC	Mixed oligonucleotide primer amplification	4.1
RACE	Rapid amplification of cDNA ends	3.3
RAPD	Randomly amplified polymorphic DNA	6
RT-PCR	Reverse transcription-PCR	3

2. ASSAY OF PLANT GENE EXPRESSION

2.1. Introduction

A concise and selective account of the wide variety of methods currently used in the study of gene expression in plants is given in the following tables. Several aspects of gene expression have been included: differential gene expression throughout plant development, in various organs, tissues and cells, and under the influence of some environmental signals. The study of the transcription of plant genes and processing of pre-mRNA species has long been hampered by the lack of a reproducible, homologous *in vitro* plant system, although some attempts have recently been made to develop such a system. Another serious drawback is the lack of routine plant cell cultures, although selective types can be maintained successfully. The development of *in vivo* labeling techniques in plant systems is crucial for an understanding of post-transcriptional regulation of plant gene expression. Some advantages that plants possess over other organisms include their ability to regenerate whole plants from many differentiated tissue and cell types; their photosynthetic ability and, in many cases, the existence of a diverse genetic pool that a researcher can draw from. The sizes of plant genomes vary enormously, ranging from *Arabidopsis* ($\sim 10^8$ bp), the smallest, up to *Lilium* ($\sim 10^{11}$ bp), the largest. The study of some aspects of gene expression in plants is often complicated by this feature, particularly in the important cultivated species which tend to have large genome sizes (see *Chapter 2*), and some genes are members of large multigene families. Transgenic plants have also been considered because of their increasing importance in agriculture but also for their valuable use as an experimental system for the study of genetic elements which regulate the expression of genes.

Table 11. Expression assays – RNA analyses

Techniques	Requirements	Information derived	Comments	References
11.1 Transcription start/end point mapping	Partial restriction map of gene; gene orientation; poly(A)$^+$ RNA; partial nucleotide sequence of coding region of the gene; subclones of the gene coding sequence	Organization of the transcribed region of a gene; activity of the gene; information on transcript processing	Usually used in combination with other methods in this table, such as Northern blotting (Section 11.3)	244
11.1.1 S1 nuclease assay	RNA from tissue of interest, poly(A)$^+$ RNA if possible; single-stranded DNA probe (M13 clone is useful) or synthetic oligonucleotide complementary to region of interest (see Section 11.1.3.1) – labeled at 5′ or 3′ end	Mapping of 5′ and 3′ ends of a specific transcript; quantitation of specific transcripts; direction of transcription; location and size of introns	Double-stranded RNA–DNA regions are resistant to S1 nuclease. M13 clones are useful for preparing single-stranded probes. Less sensitive than RNase protection (Section 11.1.3)	245
11.1.2 Primer extension	Oligonucleotide primer or single-stranded DNA probe (e.g. restriction fragment) complementary to the RNA region of interest – end-labeled; poly(A)$^+$ RNA; reverse transcriptase	Maps the 5′ start of a transcript; allows quantitation of specific mRNA	Primer extended sequence is analyzed on a sequencing gel; size of the sequence maps the 5′ end of the mRNA; amount of product proportional to level of mRNA in the experimental tissue. Usually used with other methods in this table to confirm transcription start points.	246

Table 11. Continued

Techniques	Requirements	Information derived	Comments	References
11.1.3 RNase protection	Fragment of gene of interest cloned in a transcription vector; ^{32}P-labeled RNA transcript from transcription vector (see Section 11.1.3.1); RNases A and T1	Sensitive, quantitative assay for presence of a specific mRNA in experimental tissue; confirms the activity of a specific gene	Double-stranded RNA–RNA regions are resistant to RNase A and T1	247
11.1.3.1 Single-stranded RNA probes	Vector containing SP6, T_7, T_3 or T_4 bacteriophage promoters upstream from cloned gene fragment of interest, e.g. Riboprobe vectors (Promega); *in vitro* transcripts produced from the DNA by RNA polymerases; RNA purified free from DNA	Production of labeled synthetic RNA for use as a gene probe in various techniques (e.g. Section 11.1.3) or unlabeled RNA for *in vitro* translation	RNA probes purified by treatment with RNase-free DNase or by electrophoresis	248
11.2 Nuclear 'run-on' transcription assays (sometimes referred to as nuclear 'run-off' assays)	Purified nuclei or isolated chromatin from tissues and stages of interest; labeled RNA transcripts; DNA probe for gene of interest	Measurement of transcription rates of specific genes active in tissues from which nuclei were isolated	Technique for assaying completion of nuclear transcripts initiated *in planta*. Provides quantitative estimates of activity of specific genes	249
11.2.1 Southern blots	DNA coding sequences for specific detection of labeled transcripts; DNA separated by gel electrophoresis and blotted onto nitrocellulose; radiolabeled transcripts	Specific nuclear transcripts. Quantitative estimation of specific nuclear transcripts	DNA applied in excess over transcript RNA for quantitative estimations	250

Table 11. Continued

Techniques	Requirements	Information derived	Comments	References
11.2.2 DNA dot/slot blots	DNA fragments applied to nitrocellulose filters as dots or slots using a template assembly; radiolabeled transcripts	Quantitative estimates of specific nuclear transcripts – activity of specific genes during development and in different tissues	Slot blot autoradiographs and images are more easily quantified by scanning densitometer	251
11.3 Northern blot analysis	Specific, defined, labeled probes for coding sequence of gene of interest; total RNA or poly(A)$^+$ RNA; RNA fractionated by electrophoresis and blotted on to nitrocellulose	Amount and size of intact mRNAs in a particular tissue. Steady-state transcription levels (abundance, number and size of transcripts); organ-specificity of gene expression	RNA blot of electrophoretic analysis gives an indication of RNA size as well as amount. Standard RNA size markers are available. Denaturing gels used, e.g. formamide, urea, formaldehyde, methyl mercury hydroxide gels. DNA probe in excess over RNA for quantitative estimates	244, 252
11.3.1 RNA dot and slot blots	RNA fragments applied to filters as dots or slots using a template assembly; labeled probe as in Section 11.3	Quantitative estimates of steady-state levels of specific mRNA. Variation in steady-state levels during development or in different tissues	Dot or slot blotting is more sensitive but specificity of hybridization must be checked with appropriate controls. Quantified by scanning densitometer	253, 254
11.4 RNase protection assay	See description above (Section 11.1.3); transcription vector clone; RNA from tissue/stage of interest	Organ-specificity and relative abundance of specific mRNA species	—	255, 256

Table 11. Continued

Techniques	Requirements	Information derived	Comments	References
11.5 RNA titration analysis	Single-stranded probes, RNA preferably (Section 11.1.3.1); RNA from tissue/stage of interest	Accurate quantitative estimate of RNA transcripts in an organ or per cell	Largely superseded by more convenient methods	257
11.6 *In situ* hybridization	Plant material fixed and embedded in an appropriate manner for this technique; thin tissue sections; labeled RNA (transcription vectors useful) or single-stranded DNA probes (M13 clones useful); labeled oligonucleotides; labels include biotin (Gibco-BRL), digoxigenin (Boehringer), fluorescent tags and radioactive isotopes (Amersham International)	Spatial and temporal patterns of specific RNA synthesis and accumulation at the tissue and cellular level. Defining the spatial control of plant gene expression	Single-stranded RNA probes are favored. Only semi-quantitative estimation; can give variable results; essential to use 'null' controls. Extremely useful in combination with immunocytochemical methods to locate sites of deposition of the corresponding protein. Can be used for normal and for transgenic plant studies	258, 259
11.7 Transgenic plants (see also *Table 12*, Section 2.5)	RNA isolates from small amounts of tissues of interest; materials as described for above methods; Northern blots or slot blots (Sections 11.3, 11.3.1)	Confirmation of activity and mode of expression of transferred genes. Studies on the stability of expression of introduced genes	Powerful technique, though labor-intensive and time-consuming. Several of the above techniques may be used to analyze the specific mRNA. Essential to use control plants in order to eliminate any interference from endogenous gene expression	260–262

Table 12. Expression assays – protein analyses

Techniques	Requirements	Information derived	Comments	References
	Appropriate extraction methods which will give quantitative yields of proteins in a stable form suitable for analysis; highly resolving reproducible analytical and detection techniques	Information on synthesis, accumulation, turnover of proteins and protein precursor processing	1D or 2D systems. Quantitative or semi-quantitative estimates of protein components by densitometric scanning of gels, autoradiographs or blots (laser densitometer), e.g. instruments by Molecular Dynamics, Bio-Rad, LKB and other companies	
12.1 Polyacrylamide gel electrophoresis (PAGE). Fluorography of radioactively labeled, native proteins	Non-denatured or denatured protein extracts from experimental tissues; labeling of proteins in tissues of interest using radioactive amino acids	Molecular mass of proteins; proteins identified and quantified by staining, radioactive label or by antibody detection methods (see Section 2.5)	As above	263–265
12.1.1 One and two-dimensional PAGE gels	Protein extracts from experimental tissues; combinations of isoelectric focusing, reducing and non-reducing, non-denaturing and denaturing electrophoretic methods; proteins identified by staining, radioactive label or labeled antibody detection	Molecular masses; isoelectric points; subunit composition disulfide-linked subunits; indications of gene complexity and/or post-translational modifications	As above. Single components can be isolated from gels and microsequenced or sequenced directly from blots (see Section 12.5.1)	266, 267

Table 12. Continued

Techniques	Requirements	Information derived	Comments	References
12.2 ***In vitro*** **translation of native or synthetic mRNA**	RNA, poly(A)$^+$ RNA or polysomes from tissues of interest; synthetic RNA transcribed from cDNA cloned in transcription vector. Purified RNAs can also be useful in these assays	Size and identification of proteins encoded by specific mRNAs; comparison of polypeptide size versus mRNA size; detection of precursor polypeptides	Synthetic mRNAs require full-length cDNAs as templates, cloned in vectors suitable for transcription. For high efficiency translation, synthetic mRNAs should be capped at their 5′ end. Detection by incorporation of radioactive amino acids and immunoprecipitation or by antibody binding on Western blots (see Section 12.5.1)	248, 268
12.2.1 Wheat germ cell-free system	Commercial wheat germ cell-free translation systems available from several sources; source of RNA (see above); radioactive amino acid(s)	As above	Homologous system for plants. Similar lysates may be derived from rice germ, pea embryonic axes, cotyledons and other tissues	248, 269
12.2.2 Reticulocyte lysate cell-free system	Commercial rabbit reticulocyte cell-free translation system available from several sources; source of RNA (see above); radio-active amino acid(s)	As above	Low molecular mass polypeptides are not resolved as well as with the wheat germ system. Endogenous mRNA may be a problem by producing large background incorporation	270

Table 12. Continued

Techniques	Requirements	Information derived	Comments	References
12.3 *Xenopus laevis* oocyte *in vivo* system	Source of oocytes; micro-injection system; cloned DNA or purified transcripts; radioactive amino acids; antibodies against encoded proteins; labeled probes for transcripts	Post-transcriptional, translational and post-translational processes including secretion of secretory proteins may be studied	Very useful experimental system, as both transcription and translation may be studied. Synthetic mRNAs may also be translated	271–273
12.4 cDNA in expression vectors – fusion proteins	Wide range of expression vectors commercially available to produce transcriptional or translational fusions with DNA of interest; inducible or constitutive expression systems	Provides cDNA encoded polypeptide for antibody production or assays. May be useful for antibody production for encoded protein and for enzyme structure/function assays	Allows synthesis of protein product encoded in any piece of DNA. Special expression systems needed if protein product is toxic to host cells	274, 275
12.4.1 Bacteriophage vectors	Bacteriophage vectors and expression in *E. coli*: e.g. λgt11 and λZAP-II	Useful for screening expression libraries with antibodies. Phagemids such as λZAP-II allow excision of insert as a plasmid suitable for transcription from viral promoters	Detection of foreign proteins with specific antibodies or DNA-binding as a ligand (both non-radioactive and radioactive methods)	276, 277

Table 12. Continued

Techniques	Requirements	Information derived	Comments	References
12.4.2 Plasmid vectors	Plasmid vectors for expression in *E. coli*; fusion protein expression vectors – e.g. pGEX-3X and pGEX-2T (Pharmacia) for translation fusions (proteins fused to C-terminus of glutathione S-transferase), pET (Novagen) transcription–translation vectors under control of T7 promoter; secretory vectors	High levels of expression of the encoded polypeptide. Translation fusions to facilitate affinity purification of product. Systems have proteolytic sites to cleave off the authentic product. Use of secretory peptide – fusion constructs promote secretion of protein into medium	Proteins can be labeled by incorporation of radioactive amino acids in the culture. Some systems are limited in ability to express genes whose products are toxic to the cell. Induceable promoters and leakproof promoters to avoid this problem	278
12.5 Transgenic plants	Experimental plants stably transformed with gene construct of interest	Wide range of experimental effects can be investigated using this technology; new phenotypic characteristics; function of gene promoters; expression and targeting of foreign proteins, etc.	Technology for plant transformation is dealt with in *Chapter 7*	See tables in *Chapter 7*
12.5.1 Western blot analysis	PAGE or SDS-PAGE (see 12.1); primary antibodies; labeled secondary antibodies, e.g. goat anti-rabbit IgG antibodies. Radioactive and non-radioactive labeling systems	Standard protein detection method using specific antibodies. Size, charge and quantity of protein synthesized in host plant	Quantitative levels of accumulated proteins may be obtained by scanning blots, autoradiographs or directly through the use of radio-imaging systems	279–281

Table 12. Continued

Techniques	Requirements	Information derived	Comments	References
12.5.1.1 Estimation of protein concentration	Protein extracts and purified antigen; slot blots frequently used (see Section 12.5.3); autoradiographic or colorimetric detection; scanning densitometer	Amount of a specific protein present in extracts from transgenic tissues	Wide variety of methods available. Direct staining of filters with procion red or amido black provides non-specific estimation of total protein	282
12.5.2 Enzyme-linked immunosorbent assay (ELISA)	Two types of detection systems are used in immunoassays: radio-iodinated reagents and enzyme-labeled reagents; biotin–streptavidin system; plate reader	As above	Highly sensitive and accurate method for detection of specific antigens	283–284
12.5.3 Dot/slot blot radio-immunoassay	Primary and labeled secondary antibodies. As for Section 12.5.1.1		Radioactive or light-emitting labels appropriate for this method	285
12.5.4 Squash-blot immuno-localization	Primary and labeled secondary antibodies (radioactive or non-radioactive labels)	Localization of sites of synthesis and deposition of a protein in various organs of transformed plants	Best results with thin organ systems or slices. Distribution of protein on blot can be checked by staining filter with amido black or procion red	285

Table 12. Continued

Techniques	Requirements	Information derived	Comments	References
12.5.5 Immunocyto-chemistry	Primary antibodies against antigen of interest and colloidal gold labeled secondary antibodies for light and electron microscopy; primary and enzyme-labeled secondary antibodies for light microscopy	Localization of specific proteins or other antigens in various organs of the transgenic plant, at the tissue, cellular and subcellular levels	Wide range of labeled secondary antibodies available commercially against primary antibodies raised in mice or rabbits	286–287
12.5.6 Reporter gene expression	Transgenic plants expressing reporter genes (see also *Chapters 4* and *5*); transient expression systems in protoplasts from suitable tissues – CAT reporter gene often used for this purpose (see *Chapter 5*)	Localization of the sites of synthesis and accumulation of reporter gene product reflecting the activity of plant promoters in different cell types. Protoplast systems may give an indication of promoter activity	Subject to the susceptibility of the expressed reporter enzyme to proteolytic degradation and to post-translational modifications, i.e. GUS inhibited by glycosylation	288
12.5.6.1 Transit peptides and targeting sequences	Transgenic plants expressing reporter gene fusions with encoded targeting or transit sequences (see also *Chapters 4* and *5*); isolated intact organelles	Function of primary sequences and protein domains in targeting proteins to or through specific subcellular compartments	Targeting of reporter or selectable gene products may be essential for correct function/activity, e.g. *epsps* gene encoding resistance to glyphosate herbicide (see *Chapter 5*). Reporter enzyme must have no endogenous targeting signals	289

Table 12. Continued

Techniques	Requirements	Information derived	Comments	References
12.5.6.2 Histochemical localization of reporter gene protein	Transgenic plants expressing reporter gene according to the promoter and any targeting sequences, e.g. glucuronidase widely used (see also *Chapters 4* and *5*). Antibodies against reporter protein can also be used (see Section 12.5.5)	Localization of sites of deposition of the reporter enzyme, thus defining the cell- and tissue-specific gene controls and also function of targeting sequences appended to the enzyme sequence	See *Chapter 4* for further details of assays. Textbook available on the use and detailed assays for the glucuronidase reporter gene	290–294

Table 13. Assays for regulation of gene expression (see ref. 295)

Techniques	Requirements	Information derived	Comments	References
13.1 Chromatin structure assays	Isolated chromatin or nuclei under conditions which maintain proteins bound to DNA; purified nuclear proteins	Organization of protein–DNA complexes in chromatin, responsible for structural organization or involved in transcriptional regulation	Structural studies including thermal denaturation; circular dichroism; X-ray diffraction; nuclease digestion patterns	296–298
13.2 *Trans*-acting protein factors	5′ gene sequence which binds the *trans*-acting protein factors; nuclear extracts containing *trans*-acting protein factors	Identification of proteins involved in the modulation of transcription	Regulation of transcriptional initiation is mediated by *trans*-acting protein factors which bind to specific DNA sequences (motifs)	299
13.3 *Cis*-acting regulatory elements	Information on the structure of gene of interest; intact 5′ gene sequence and series of 5′ deletions (substitutions and arrangements); can be fused to reporter genes for transformation or transient expression assays	Identification of 5′ sequences participating in the regulation of expression of genes of interest	Deletion studies to identify 5′ regulatory sequences are often followed by protein binding assays – DNA footprinting and gel retardation assays	300
13.3.1 DNA–protein binding assays (gel retardation assays)	Nuclear protein extracts (pure or partially purified); radioactively labeled gene target sequences	The assay is based on the differential migration of protein-bound versus free DNA in a non-denaturing gel. Protein–DNA complexes may be isolated for further characterization	For studying specific DNA-specific interactions. Success of the assay depends largely on a high specific activity to concentration ratio of the probe	301–302

Table 13. Continued

Techniques	Requirements	Information derived	Comments	References
13.3.2 DNA footprinting	Gene sequence of interest, labeled at one end only; total or partially purified nuclear proteins	The assay determines which nucleotide sequences within a DNA fragment bind specific nuclear proteins. Identification of DNA-binding proteins	Genomic DNA is protected from enzymic or chemical digestion by complexing with DNA-binding proteins. Proteins such as transcription regulation factors bind to specific DNA sequences which are revealed by this technique. Good footprinting assays require optimization of the DNA–protein binding (see Section 13.3.1)	
13.3.2.1 DNase I sensitivity	Digestion of DNA–protein complex with DNase I	DNA–protein complex protected from DNase I digestion. Indicates the region of DNA binding protein	Not particularly accurate method but quickest for identifying regions of DNA bound by specific proteins	303–304
13.3.2.2 Hydroxyl radical	Digestion of DNA–protein complex with hydroxyl radical generated from H_2O_2 and Fe(II) in FeEDTA	Regions containing DNA–protein complex protected from digestion	Very accurate for locating binding sites. Advantage of distinguishing asymmetric binding of factors to DNA	305

Table 13. Continued

Techniques	Requirements	Information derived	Comments	References
13.3.2.3 1,10 phenan-throline-Cu	Cleavage of DNA–protein complex using phenanthroline – copper ion compound	This method may be used for footprinting DNA–protein complexes directly in gel slices excised from gel retardation assays. Identifies structural changes associated with binding of low mol. wt diffusable ligands such as Mg^{2+}	Rapid and accurate. Probably the best method with respect to convenience and quality of results. Advantage of *in situ* gel digestion	306
13.4 DNA methylation assays	Genomic DNA isolated from tissues and stages of development of interest; probes for genes of interest; methylation-sensitive and -insensitive restriction enzyme isoschizomers, e.g. *Hpa*II and *Msp*I	Post-replicative methyl-modification of some cytosine residues (to 5-methylcytosine) in DNA. Indication of the methylation status of specific genes during development	Hypomethylation has been related to transcriptional activity of genes, e.g. rRNA genes	307–309
13.5 mRNA precursor splicing and small nuclear RNAs (snRNA) (see also *Chapter 6*)	Pure labeled RNA sequence containing an intron sequence – transcript from a cloned defined gene sequence (see Sections 12.4.1 and 12.4.2); purified/ partially purified cell or nuclear extracts	Purification and characterization of the factors which catalyze the splicing reactions. Sequences of intron boundaries and other important regions necessary for efficient splicing. Molecular mechanism of splicing in plants	Nuclear pre-mRNA introns constitute the largest class of introns. The abundance of snRNAs in plant nuclei is 100-fold lower than in animal nuclei. Northern blotting and *in vivo* RNA labeling can be used to detect mRNA precursor transcripts	310–311

Table 13. Continued

Techniques	Requirements	Information derived	Comments	References
13.6 mRNA half-life and cytoplasmic stability	Total or poly(A) RNA extracted from tissues of interest over a defined time-course or developmental period; *in vivo* radioactive uridine-labeled RNA and pulse-chase techniques useful in such studies	Study of post-transcriptional mechanisms which may play an important role in the modulation of cytoplasmic mRNA levels; primary transcript processing	Little is known about post-transcriptional regulatory events in plants and the stability of plant mRNAs. The poly(A) tail on some plant mRNAs has been implicated in some aspect of their stability	312–315
13.7 Developmental assays on native (non-denaturing) protein gels	Protein extracts using non-denaturing buffers, from plant materials harvested at different stages of development; PAGE or isoelectric focusing commonly used	Extensive studies have been carried out on developmental regulation of storage-protein gene expression in legume seeds and in potato tubers. Provides qualitative and quantitative data on the changes in individual protein components during development	Often used in conjunction with other electrophoretic methods in two-dimensional gels and with various detection methods, including immunochemical methods	316–318
13.8 Environmental control of gene expression	Assays for specific gene product or for its transcript (see Sections 11.3 and 12.5.1)	Effects of environmental signals including: water stress; heat shock; salt stress; heavy metal intoxication; oxygen stress; wounding (elicitors); plant growth substances	Assays for synthesis and accumulation of specific mRNAs and encoded proteins; provides basic information on the expression activity of genes during stress	319–322

Table 13. Continued

Techniques	Requirements	Information derived	Comments	References
13.9 Foreign structural genes in transgenic plants	Chimeric gene expression constructs	Phenotypic effects of the expression of specific structural genes	Widely used to alter the capability of a plant species – crop improvement or for the study of gene/protein function	323
13.9.1 Antisense RNA	Expression constructs with antisense (complement) orientation to a part of the native mRNA	Phenotypic effects of suppression or reduction of expression of specific structural genes	The technique has been used effectively to inhibit the expression of specific genes	324
13.9.2 Regulatory sequence analysis	Chimeric gene constructs with regulatory sequences linked to a reporter gene (see *Chapter 5*)	Organ- and tissue-specific promoters; enhancer(s); silencer(s); regulatory regions (deletion-mutagenesis); developmental control; environmental signals/stimuli (light, heat, stress)	In general, the data pertain to the organization and importance of the 5′ and 3′ flanking regions	325
13.9.3 Promoter trapping	Promoterless reporter or selectable gene constructs; restricted genomic DNA	Identification and isolation of organ- and tissue-specific promoters	β-Glucuronidase (GUS), luciferase (luc) and neomycin phosphotransferase (NPT) have been used (see *Chapter 5*)	326, 327

Table 13. Continued

Techniques	Requirements	Information derived	Comments	References
13.9.4 Gene targeting – homologous recombination	Gene constructs with one or two long flanking regions homologous to those of the target gene or genomic sequence; suitable system for selection of homologous recombinants	Used to promote specific gene deletion, substitution, alteration or correction	Most techniques have been developed in animal or microbiological systems; few successful systems reported for plants. Potentially eliminates position effects by targeting genes to a specific site of known activity	328
13.10 Plant transposons and gene-tagging (see also *Chapter 6*)	Plant transformation; vectors containing transposable element likely to be active or activatable in the host plant; screenable or selectable phenotype or biochemical characters for mutated genes	The method is applicable to the study of the majority of known plant characters where a mutant plant phenotype is recognizable	Transposable elements inactivate or alter gene expression by insertion in or near structural genes. Method requires combination of molecular biological and genetic techniques. Limited to active combinations of host and transposon	329, 330
13.11 Cell suspension cultures	Actively growing cell cultures from the appropriate plant tissues; synchronized culture if required	Possibility to study metabolic pathways present in actively dividing, undifferentiated plant cells; cell cycle regulated gene expression (synchronized cell culture)	Limited by the availability of a relatively homogeneous population of cells; synchronized cell cultures	331–333

Table 13. Continued

Techniques	Requirements	Information derived	Comments	References
13.12 *In vitro* transcription assays	Nuclear preparations from various eukaryotic sources, containing all the components necessary for transcription from a well-characterized gene construct	Estimates of the transcriptional efficiency of a gene promoter in the transcription assays	Cell-free transcription systems can be utilized for studying the mechanisms of transcription and for the identification of factors and nucleotide sequences which control gene expression	334
13.12.1 Isolated chromatin		Quantitative estimates of transcripts by hybridization to specific DNA probes		335
13.12.2 HeLa cell lysate system	Commercial lysate system is available; gene construct of interest	Transcriptional efficiency of a gene promoter in a heterologous system	Highly active *Drosophila* embryo extracts also available commercially for eukaryotic gene transcription	336, 337
13.12.3 Wheat germ chromatin		Transcriptional efficiency of a gene promoter in a plant system	*In vitro* system for plant genes. Needs testing for its general applicability	338, 339
13.12.4 Nuclear extract from cultured cells	Established suspension cultures derived from callus tissues, e.g. from rice (OC-2) and tobacco (NT-1)		Specialized transcription systems for plant genes. Need testing for general applicability	340

Table 13. Continued

Techniques	Requirements	Information derived	Comments	References
13.13 **Histone complement and isolation of 'active' chromatin**	Chromatin isolated from tissues of interest; specific antibodies raised against the acetylated form of histone H4; probes for genes of interest	New technique for investigating and isolating active genes directly from isolated chromatin	Chromatin containing active genes has associated histones, particularly histone H4, which is specifically acetylated. Antibodies raised specifically against the acetylated form of histone H4 can be used to select 'active' chromatin – fragments of chromosome enriched for active genes	341

ACKNOWLEDGEMENTS

Thanks are due to Drs K. Elborough, J. Gatehouse, N. Robinson, T. Fawcett, A. Shirsat, P. Urwin, V. Hilder, P. Howley and C. Brough for supplying information and references.

3. REFERENCES

1. Erlich, H.A. (ed.) (1989) *PCR Technology: Principles and Applications for DNA Amplification.* Stockton Press, New York.

2. Innis, M.A., Gelfand, D.H., Sninsky, J.J. and White, T.J. (eds) (1990) *PCR Protocols: A Guide to Methods and Applications.* Academic Press, San Diego.

3. McPherson, M.J., Quirke, P. and Taylor, G.R. (eds) (1991) *PCR: a Practical Approach.* Oxford University Press, New York.

4. Sambrook, J., Fritsch, E.F. and Maniatis, T. (eds) (1989) *Molecular Cloning: A Laboratory Manual.* Cold Spring Harbor Laboratory Press, Cold Spring Harbor, New York.

5. Bloch, W. (1991) *Biochemistry,* **30**, 2735.

6. Erlich, H.A., Gelfand, D. and Sninsky, J.J. (1991) *Science,* **252**, 1643.

7. Chien, A., Edgar, D.B. and Trela, J.M. (1976) *J. Bacteriol.,* **127**, 1550.

8. Saiki, R.K., Gelfand, D.H., Stoffel, S., Scharf, S.J., Higuchi, R., Horn, G.T., Mullis, K.B. and Erlich, H.A. (1988) *Science,* **239**, 487.

9. Tindall, K.R. and Kunkel, T.A. (1988) *Biochemistry,* **27**, 6008.

10. Longley, M.J., Bennet, S.E. and Mosbaugh, D.W. (1990) *Nucleic Acids Res.,* **18**, 7317.

11. Jones, M.D. and Foulkes, N.S. (1989) *Nucleic Acids Res.,* **17**, 8387.

12. Dilworth, D.D. and McCarrey, J.R. (1992) *PCR Methods Applications,* **1**, 279.

13. Myers, T.W. and Gelfand, D.H. (1991) *Biochemistry,* **30**, 7661.

14. Cariello, N.F., Swenberg, J.A. and Skopek, T.R. (1991) *Nucleic Acids Res.,* **19**, 4193.

15. Mattila, P., Korpela, J., Tenkanen, T. and Pitkanen, K. (1991) *Nucleic Acids Res.,* **19**, 4967.

16. Lundberg, K.S., Shoemaker, D.D., Adams, M.W.W., Short, J.M., Sorge, J.A. and Mathur, E.J. (1991) *Gene,* **108**, 1.

17. Skerra, A. (1992) *Nucleic Acids Res.,* **20**, 3551.

18. Hung, T., Mak, K. and Fong, K. (1990) *Nucleic Acids Res.,* **18**, 4953.

19. Henry, R.J. and Oono, K. (1991) *Plant Molec. Biol. Rep.,* **9**, 139.

20. McConlogue, L., Brow, M.A. and Innis, M.A. (1988) *Nucleic Acids Res.,* **16**, 9869.

21. Sarkar, G., Kapelner, S. and Sommer, S. (1990) *Nucleic Acids Res.,* **18**, 7465.

22. Kureishi, A. and Bryan, L.E. (1992) *Nucleic Acids Res.,* **20**, 1155.

23. Ponce, M.R. and Micol, J.L. (1992) *Nucleic Acids Res.,* **20**, 623.

24. Schwarz, K., Hansen-Hagge, T. and Bartram, C. (1990) *Nucleic Acids Res.,* **18**, 1079.

25. Haqqi, T.M., Sarkar, G., David, C.S. and Sommer, S.S. (1988) *Nucleic Acids Res.,* **16**, 11844.

26. Kemp, D.J., Smith, D.B., Foote, S.J., Samaras, N. and Peterson, M.G. (1989) *Proc. Natl. Acad. Sci. USA* , **86**, 2423.

27. Don, R.H., Cox, P.T., Wainwright, B.J., Barker, K. and Mattick, J.S. (1991) *Nucleic Acids Res.,* **19**, 4008.

28. D'Aquila, R.T.D., Bechtel, L.J., Videler, J.A., Eron, J.J., Gorczyca, P. and Kaplan, J.C. (1991) *Nucleic Acids Res.,* **19**, 3749.

29. Chow, Q., Russell, M., Birch, D.E., Raymond, J. and Bloch, W. (1992) *Nucleic Acids Res.,* **20**, 1717.

30. Beck, B.N. and Ho, S.N. (1988) *Nucleic Acids Res.,* **16**, 9051.

31. Bell, D.A. and DeMarini, D.M. (1991) *Nucleic Acids Res.,* **19**, 5079.

32. Rhychlick, W. and Rhoads, R.E. (1989) *Nucleic Acids Res.,* **17**, 8543.

32. Wu, D.Y., Ugozzoli, L., Pal, B.K., Qian, J. and Wallace, R.B. (1991) *DNA Cell Biol.,* **10**, 233.

33. Rychlik, W., Spencer, W.J. and Rhoads, R.E. (1990) *Nucleic Acids Res.,* **18**, 6409.

34. Lowe, T., Sharefkin, J., Yang, S.Q. and Dieffenbach, C.W. (1990) *Nucleic Acids Res.,* **18**, 1757.

36. Sommer, R. and Tautz, D. (1989) *Nucleic Acids Res.,* **17**, 6749.

37. Kwok, S., Kellog, D.E., McKinney, N., Spasic, D., Goda, L., Levenson, C. and Sninsky, J.J. (1990) *Nucleic Acids Res.,* **18**, 999.

38. Wu, D.Y., Ugozzoli, L., Pal, B.K. and Wallace, R.B. (1989) *Proc. Natl. Acad. Sci. USA*, **86**, 2757.

39. Kwok, S. and Higuchi, R. (1989) *Nature*, **339**, 237.

40. Sarkar, G. and Sommer, S.S. (1990) *Nature*, **343**, 27.

41. Cimino, G.D., Metchette, K., Isaacs, S.T. and Zhu, Y.S. (1990) *Nature*, **345**, 773.

42. Zhu, Y.S., Isaacs, S.T., Cimino, G.D. and Hearst, J.E. (1991) *Nucleic Acids Res.*, **19**, 2511.

43. Muralidhar, B. and Steinman, C.R. (1992) *Gene*, **117**, 107.

44. Jinno, Y. and Niikawa, N. (1990) *Nucleic Acids Res.*, **18**, 6739.

45. Cimino, G.D., Metchette, K.C., Tessman, J.W., Hearst, J.E. and Isaacs, S.T. (1991) *Nucleic Acids Res.*, **19**, 99.

46. Isaacs, S.T., Tessman, J.M., Metchette, K.C., Hearst, J.E. and Cimino, G.D. (1991) *Nucleic Acids Res.*, **19**, 109.

47. Longo, M.C., Berninger, M.S. and Hartley, J.L. (1990) *Gene*, **93**, 125.

48. Li, H., Cui, X. and Arnheim, N. (1991) *Nucleic Acids Res.*, **19**, 3139.

49. Lohff, C.J. and Cease, K.B. (1992) *Nucleic Acids Res.*, **20**, 144.

50. Garrity, P.A. and Wold, B.J. (1992) *Proc. Natl. Acad. Sci. USA*, **89**, 1021.

51. Jung, V., Pestka, S.B. and Pestka, S. (1990) *Nucleic Acids Res., A*, **18**, 6156.

52. Crowe, J.S., Cooper, H.J., Smith, M.A., Sims, M.J., Parker, D. and Gewert, D. (1991) *Nucleic Acids Res.*, **19**, 184.

53 Shuldiner, A.R., Scott, L.A. and Roth, J. (1990) *Nucleic Acids Res.*, **18**, 1920.

54. deJong, P.J. and Aslandis, C. (1990) *Nucleic Acids Res.*, **18**, 6069.

55. Clark, J.M. (1988) *Nucleic Acids Res.*, **16**, 9677.

56. Hemsley, A., Arnheim, M., Toney, M.D., Cortopassi, G. and Galas, D.J. (1989) *Nucleic Acids Res.*, **17**, 6545.

57. Mead, D.A., Pey, N.K., Hernstadt, C., Marcil, R.A. and Smith, L.M. (1991) *BioTechnol.*, **9**, 657.

58. Marchuk, D., Drumm, M., Sarkino, A. and Collins, F.S. (1991) *Nucleic Acids Res.*, **19**, 1154.

59. Holton, T.A. and Graham, M.W. (1991) *Nucleic Acids Res.*, **19**, 1156.

60. Kovalic, D., Kwak, J.H. and Weisblum, B. (1991) *Nucleic Acids Res.*, **19**, 4560.

61. Keohavong, P. and Thilly, W.G. (1989) *Proc. Natl. Acad. Sci. USA*, **86**, 9253.

62. Dunning, A.M., Talmud, P. and Humphries, S.E. (1988) *Nucleic Acid Res.*, **16**, 10393.

63. Eckert, K.A. and Kunkel, T.M. (1990) *Nucleic Acids Res.*, **18**, 3739.

64. Eckert, K.A. and Kunkel, T.A. (1991) in *PCR: a Practical Approach*. (M.J. McPherson, P. Quirke and G.R. Taylor, eds). Oxford University Press, New York, p. 225.

65. Meyerhans, A., Vartanian, J.P. and Wain-Hobson, S. (1990) *Nucleic Acids Res.*, **18**, 1687.

66. Jansen, R. and Ledley, F.R. (1990) *Nucleic Acids Res.*, **18**, 5153.

67. Reiss, J., Krawczak, M., Schloesser, M., Wagner, M. and Cooper, D.N. (1990) *Nucleic Acids Res.*, **18**, 973.

68. Shuldiner, A.R., Nirula, A. and Roth, J. (1989) *Nucleic Acids Res.*, **17**, 4409.

69. Brakenhoff, R.H., Schoenmakers, J.G.G. and Lubsen, N.H. (1991) *Nucleic Acids Res.*, **19**, 1949.

70. Marton, A., Delbecchi, L. and Bourgaux, P. (1991) *Nucleic Acid Res.*, **19**, 2423.

71. Paabo, S., Irwin, D.M. and Wilson, A.C. (1990) *J. Biol. Chem.*, **265**, 4718.

72. Mueller, P.R. and Wold, B. (1989) *Science*, **246**, 780.

73. Strauss, E.C., Andrews, N.C., Higgs, D.R. and Orkin, S.H. (1992) *Mol. Cell. Biol.*, **12**, 2135.

74. Pfeiffer, G.P., Steigerwald, S.D., Mueller, P.R., Wold, B. and Riggs, A.D. (1989) *Science*, **246**, 810.

75. Ott, T., Nelsen-Salz, B. and Doring, H.-P. (1992) *Plant Journal*, **2**, 705.

76. Tormanen, V.T., Swiderski, P.M., Kaplan, B.E., Pfeifer, G.P. and Riggs, A.D. (1992) *Nucleic Acids Res.*, **20**, 5487.

77. Garrity, P.A. and Wold, B.J. (1992) *Proc. Natl. Acad. Sci. USA*, **89**, 1021.

78. Triglia, T., Peterson, M.G. and Kemp, D.J. (1988) *Nucleic Acids Res.*, **16**, 8186.

79. Collins, F.S. and Weissman, S.M. (1984) *Proc. Natl. Acad. Sci. USA*, **81**, 6812.

80. Fu, P. and Evans, B. (1992) *Nucleic Acids Res.*, **20**, 2903.

81. Huang, S.H., Hu, Y., Wu, C. and Holcenberg, J. (1990) *Nucleic Acids Res.*, **18**, 1922.

82. Shyamala, V. and Ames, G.F.L. (1989) *Gene*, **84**, 1.

83. Robinson, N.J., Gupta, A., Fordham-Skelton, A.P., Croy, R.R.D., Whitton, B.A. and Huckle, J.W. (1990) *Proc. R. Soc. Lond. B.*, **242**, 241.

84. Rosenthal, A., MacKinnon, R.N. and Jones, D.S.C. (1991) *Nucleic Acids Res.*, **19**, 5395.

85. Fors, L., Saavedra, R.A. and Hood, L. (1990) *Nucleic Acids Res.*, **18**, 2793.

86. Jeffreys, A.J., Wilson, V., Neumann, R. and Keyte, J. (1988) *Nucleic Acids Res.*, **16**, 10953.

87. Li, H., Gyllensten, U., Cui, X., Saiki, R., Erlich, H. and Arnheim, N. (1988) *Nature*, **335**, 414.

88. Zhang, L., Cui, X., Schmitt, K., Hubert, R., Navidi, W. and Arnheim, N. (1992) *Proc. Natl. Acad. Sci. USA*, **89**, 5847.

89. Wesley, C.S., Ben, M., Kreitman, M., Hagag, N. and Eaves, W.F. (1990) *Nucleic Acids Res.*, **18**, 599.

90. Hadano, S., Watanabe, M., Yokoi, H., Kogi, M., Kondo, I., Tsuchiya, H., Kanawaza, I., Wakasa, K. and Ikeda, J.-E. (1991) *Genomics*, **11**, 364.

91. Han, J., Lu, C.M., Brown, G.B. and Rado, T. (1991) *Proc. Natl. Acad. Sci. USA*, **88**, 335.

92. Jung, C., Claussen, U., Horsthemke, B., Fischer, F. and Herrmann, R.G. (1992) *Plant Mol. Biol.*, **20**, 503.

93. Chowdhary, A., Oin, S., Reymez, S. and Murison, G. (1992) *Mol. Biol. Cell*, **3**, 22.

94. Chehab, F.F. and Kan, Y.W. (1989) *Proc. Natl. Acad. Sci. USA*, **86**, 9178.

95. Davies, B., Feo, S., Heard, E. and Fried, M. (1990) *Proc. Natl. Acad. Sci. USA*, **86**, 6691.

96. Feo, S., Davies, B. and Fried, M. (1990) *Nucleic Acids Res.*, **18**, 4949.

97. Nigro, J.M., Cho, K.R., Fearon, E.R., Kern, S.E., Ruppert, J.M., Oliner, J.D., Kinzler, K.W. and Vogelstein, B. (1991) *Cell*, **64**, 607.

98. Sun, T., Goodman, H.M. and Ausubel, F.M. (1992) *Plant Cell*, **4**, 119.

99. Chasan, R. (1992) *Plant Cell*, **4**, 11.

100. Belyavsky, S., Vinogradova, T. and Rajewsky, K. (1989) *Nucleic Acids Res.*, **17**, 2919.

101. Tam, A.W., Smith, M.M., Fry, K. E. and Larrick, J.W. (1989) *Nucleic Acids Res.*, **17**, 1269.

102. Froussard, P. (1992) *Nucleic Acids Res.*, **20**, 2900.

103. Jepson, I., Bray, J., Jenkins, G., Schuch, W. and Edwards, K. (1991) *Plant Molec. Biol. Rep.*, **9**, 131.

104. Gurr, S.J., McPherson, M.J., Scollan, C., Atkinson, H.J. and Bowles, D.J. (1991) *Mol. Gen. Genet.*, **226**, 361.

105. Timblin, C., Battey, J. and Kuehl, W.M. (1990) *Nucleic Acids Res.*, **18**, 1587.

106. Wieland, I., Bolger, G., Asouline, G. and Wigler, M. (1990) *Proc. Natl. Acad. Sci. USA*, **87**, 2720.

107. Hara, E., Kato, T., Nakada, S., Sekiya, S. and Odo, K. (1991) *Nucleic Acids Res.*, **19**, 7097.

108. Liang, P. and Pardee, A.B. (1992) *Science*, **257**, 967.

109. Welsh, J., Chada, K., Dalal. S.S., Cheng, R., Ralph, D. and McClelland, M. (1992) *Nucleic Acids Res.*, **20**, 4965.

110. Frohman, M.A., Dush, M.K. and Martin, G.R. (1988) *Proc. Natl. Acad. Sci. USA*, **85**, 8998.

111. Loh, E.Y., Elliot, J.F., Cwirla, S., Lanier, L.L. and Davis, M.M. (1989) *Science*, **243**, 217.

112. Dumas, J.B., Edwards, M., Delort, J. and Mallet, J. (1991) *Nucleic Acids Res.*, **19**, 5227.

113. Troutt, A.B., McHeyzer-Williams, M.G., Pulendran, B. and Nossal, G.J.V. (1992) *Proc. Natl. Acad. Sci. USA*, **89**, 9823.

114. Delort, J., Dumas, J.B., Darmon, M.C. and Mallet, J. (1989) *Nucleic Acids Res.*, **17**, 6439.

115. Siebert, P.D. and Larrick, J.W. (1992) *Nature*, **359**, 557.

116. Legoux, P., Pessgue, B., Delpech, B., Dumont, X., Piechaczyk, M., Casellas, P. and Shire, D. (1992) *J. Biol. Chem.*, **267**, 21830.

117. Becker-Andre, M., Schulze-Lefert, P. and Hahlbrock, K. (1991) *J. Biol. Chem.*, **266**, 8551.

118. Herget, T., Schell, J. and Schreier, P.H. (1990) *Mol. Gen. Genet.*, **224**, 469.

119. Simpson, C.G., Sawbridge, T.I., Jenkins, G.I. and Brown, J.W.S. (1992) *Nucleic Acids Res.*, **20**, 5861.

120. Simpson, C.G., Sinibaldi, R. and Brown, J.W.S. (1992) *Plant Journal*, **2**, 835.

121. Sadhale, P.P., Sapolsky, R., Davis, R.W., Butler, J.S. and Platt, T. (1991) *Nucleic Acids Res.*, **19**, 3683.

122. Dekker, E.L., Woolston, C.J., Cox, B. and Mullineaux, P.M. (1991) *Nucleic Acids Res.*, **19**, 4075.

123. Lee, C.C., Wu, X., Gibbs, R.A., Cook, R.G., Muzny, D.M. and Caskey, C.T. (1988) *Science*, **239**, 1288.

124. Leibrock, J., Lottspeich, F., Hohn, A., Hofer, M., Hengerer, B., Masiakowski, H., Thoenen, H. and Barde, Y.-A. (1989) *Nature*, **341**, 149.

125. Margis-Pinheiro, M., Metz-Boutigue, M.H., Awade, A., deTapia, M., leRet, M. and Burkard, G. (1991) *Plant Mol. Biol.*, **17**, 243.

126. Siggaard-Andersen, M., Kauppinen, S. and Von Wettstein-Knowles, P. (1991) *Proc. Natl. Acad. Sci. USA*, **88**, 4114.

127. Chow, T.P., Feldman, R.A., Lovett, M. and Piatak, M. (1990). *J. Biol. Chem.*, **265**, 8670.

128. Wood, K.A., Lord, J.M., Wawrzynczak, E.J. and Piatak, M. (1991) *Eur. J. Biochem.*, **198**, 723.

129. Fordham-Skelton, A.P., Yarwood, A. and Croy, R.R.D. (1990) *Mol. Gen. Genet.*, **221**, 134.

130. Kamb, A., Weir, M., Rudy, B., Varmus, H. and Kenyon, C. (1989) *Proc. Natl. Acad. Sci. USA*, **86**, 4372.

131. Hohn, H., Leibrock, J., Bailey, K. and Barde, Y.-A. (1990) *Nature*, **344**, 339.

132. Gould, S.J., Subramani, S. and Scheffler, I.E. (1989) *Proc. Natl. Acad. Sci. USA*, **86**, 1934.

133. Libert, F., Parmentier, M., Lefort, A., Dinsart, C., Van Sande, J., Maenhaut, C., Simons, M.-J., Dumont, J.E. and Vassart, G. (1989) *Science*, **244**, 569.

134. Pellegrino, G.R. and Berg, J.M. (1991) *Proc. Natl. Acad. Sci. USA*, **88**, 671.

135. Buck, L. and Axel, R. (1991) *Cell*, **65**, 175.

136. Colosanti, J., Tyers, M. and Sundresan, V. (1991) *Proc. Natl. Acad. Sci. USA*, **88**, 3377.

137. Hirt, H., Pay, A., Gyorgyey, J., Bako, L., Nemeth, K., Bogre, L., Schweyen, R.J., Heberle-Bors, E. and Dudits, D. (1991) *Proc. Natl. Acad. Sci. USA*, **88**, 1636.

138. Ferreira, P.C.G., Hemerley, A.S., Villaroel, R., Van Montagu, M. and Inze, D. (1991) *Plant Cell*, **3**, 531.

139. Feiler, H.S. and Jacobs, T.W. (1991) *Plant Mol. Biol.*, **17**, 321.

140. Feng, X.H. and Kung, S. (1991) *FEBS Lett.*, **282**, 98.

141. Anderberg, R.J. and Walker-Simmons, M.K. (1992) *Proc. Natl. Acad. Sci. USA*, **89**, 10183.

142. Nitschke, K., Fleig, U., Schell, J. and Palme, K. (1992) *EMBO J.*, **11**, 1327.

143. Rundle, S.J. and Nasrallah, J.B. (1992) *Plant Mol. Biol.*, **20**, 367.

144. Walker, J.C. and Zhang, R. (1990) *Nature*, **345**, 743.

145. Flavell, A.J., Smith, D.B. and Kumar, A. (1992) *Mol. Gen. Genet.*, **231**, 233.

146. Flavell, A.J., Dunbar, E., Anderson, R., Pearce, S.R., Hartley, R. and Kumar, A. (1992) *Nucleic Acids Res.*, **20**, 3639.

147. Ohtsuka, E., Matsuki, S., Ikehara, M., Takahashi, Y. and Matsubara, K. (1985). *J. Biol. Chem.*, **260**, 2605.

148.Takahashi, Y., Kato, K., Hayashizaki, Y., Wakabayashi, T., Ohtsuka, E., Matsuki, S., Ikehara, M. and Matsubara, K. (1985) *J. Biol. Chem.*, **82**, 1931.

149. Nordmann, P.L., Makris, J.C. and Reznikoff, W.S. (1988) *Mol. Gen. Genet.*, **214**, 62.

150. Knoth, K., Roberds, S., Poteet, C. and Tamkun, M. (1988) *Nucleic Acids Res.*, **16**, 10932.

151. Lin, P.K. T. and Brown, D.M. (1992) *Nucleic Acids Res.*, **20**, 5149.

152. Sarkar, S. and Sommer, S.S. (1989) *Science*, **244**, 331.

153. Mackow, E.R., Yamanaka, M.Y., Dang, M.N. and Greenberg, H.B. (1990) *Proc. Natl. Acad. Sci. USA*, **87**, 518.

154. Dranginis, A.M. (1990) *Nature*, **347**, 682.

155. Fordham-Skelton, A.P., Taylor, P.N., Hartley, M.R. and Croy, R.R.D. (1991) *Mol. Gen. Genet.*, **229**, 460.

156. Welsh, J. and McClelland, M. (1990) *Nucleic Acids Res.*, **18**, 7213.

157. Williams, J.G.K., Kubelik, A.R., Livak, K.J., Rafalski, J.A. and Tingey, S.V. (1990) *Nucleic Acids Res.*, **18**, 6531.

158. Welsh, J., Petersen, L. and McClelland, M. (1991) *Nucleic Acids Res.*, **19**, 303.

159. Caetano-Anolles, G., Bassam, B.J. and Gresshoff, P.M. (1991) *BioTechnol.*, **9**, 553.

160. Welsh, J. and McClelland, M. (1991) *Nucleic Acids Res.*, **19**, 861.

161. Caetano-Anolles, G., Bassam, B.J. and Gresshoff, P.M. (1992) *Mol. Gen. Genet.*, **235**, 157.

162. Caetano-Anolles, G., Bassam, B.J. and Gresshoff, P.M. (1991) *Plant Mol. Biol. Rep.*, **9**, 294.

163. D'Ovidio, R., Tanzorella, O.A. and Porceddu, E. (1990) *Plant Mol. Biol.*, **15**, 169.

164. Kaemmer, D., Afza, R., Weising, K., Kahl, G. and Novak, F.J. (1992) *BioTechnol.*, **10**, 1030.

165. Shattuckeidens, D.M., Bell, R.N., Neuhausen, S.L. and Helentjaris, T. (1990) *Genetics*, **126**, 207.

166. Reiter, R.S., Williams, J.G.K., Feldmann, K.A., Rafalski, J.A., Tingey, S.V. and Scolnik, P.A. (1992) *Proc. Natl. Acad. Sci. USA*, **89**, 1477.

167. Martin, G.B., Williams, J.G.K. and Tanksley, S.D. (1991) *Proc. Natl. Acad. Sci. USA*, **88**, 2336.

168. Baird, E., Cooper-Bland, S., Waugh, R., DeMaine, M. and Powell, W. (1992) *Mol. Gen. Genet.*, **233**, 469.

169. Tulsieram, L.K., Glaubitz, J.C., Kiss, G. and Carlson, J.E. (1992) *BioTechnol.*, **10**, 686.

170. Devagon, J.-M. and Landry, B.S. (1992) *PCR Methods Applic.*, **1**, 175.

171. Edwards, K., Johnstone, C. and Thompson, C. (1991) *Nucleic Acids Res.*, **19**, 1349.

172. Langridge, U., Schwall, M. and Langridge, P. (1991) *Nucleic Acids Res.*, **19**, 6954.

173. Brunel, D. (1992) *Nucleic Acids Res.*, **20**, 4676.

174. Horn, P. and Rafalski, A. (1992) *Plant Mol. Biol.*, **10**, 285.

175. Collins, G.G. and Symons, R.H. (1992) *Plant Mol. Biol. Rep.*, **10**, 233.

176. Oard, J.H. and Dronavalli, S. (1992) *Plant Mol. Biol. Rep.*, **10**, 236.

177. Liang, W. and Johnson, J.P. (1988) *Nucleic Acids Res.*, **16**, 3579.

178. Friedman, K.D., Rosen, N.L., Neuman, P.J. and Motgomery, R.R. (1988) *Nucleic Acids Res.*, **16**, 8718.

179. Green, E.D. and Olson, M.V. (1990) *Proc. Natl. Acad. Sci. USA* , **87**, 1213.

180. Amemiya, C.T., Alegriahartman, M.J., Aslanidis, C., Chen, C., Nikolic, J., Gingrich, J.C. and Dejong, P.J. (1992) *Nucleic Acids Res.*, **20**, 2559.

181. Kwiatkowski, T.J., Zoghbi, H.Y., Ledbetter, S.A., Ellison, K.A. and Chinault, A.C. (1990) *Nucleic Acids Res.*, **18**, 7191.

182. Riley, J., Butter, R., Ogilvie, D., Finniear, R., Jenner, D., Powell, S., Anand, R., Smith, J. C. and Markham, A.F. (1990) *Nucleic Acids Res.*, **18**, 2887.

183. Berthomieu, P. and Meyer, C. (1991) *Plant Mol. Biol.*, **17**, 555.

184. Does, M.P., Dekker, B.M.M., deGroot, M.J.A. and Offringa, R. (1991) *Plant Mol. Biol.*, **17**, 151.

185. Blake, N.K., Ditterline, R.L. and Stout, R.G. (1991) *Crop Sci.*, **31**, 1686.

186. Hammill, J.D., Rounsley, S., Spencer, A., Todd, G. and Rhodes, M.J.C. (1991) *Plant Cell Rep.*, **10**, 221.

187. Lassner, M.W., Petersen, P. and Yoder, J.I. (1989) *Plant Mol. Biol. Rep.*, **7**, 116.

188. Earp, D., Lowe, B. and Baker, B. (1990) *Nucleic Acids Res.*, **18**, 3271.

189. Deng, X.W., Matsui, M., Wei, N., Wagner, D., Chu, A.M., Feldmann, K.A. and Quail, P.H. (1992) *Cell*, **791**, 801.

190. de Groot, M.J.A., Offringa, R., Does, M.P., Hooykaas, P.J.J. and van den Elzen, P.J.M. (1992) *Nucleic Acids Res.*, **20**, 2785.

191. Puchta, H., Kocher, S. and Hohn, B. (1992) *Mol. Cell Biol.*, **12**, 3372.

192. Sano, T., Smith, C.L. and Cantor, C.R. (1992) *Science*, **258**, 120.

193. Paabo, S. (1989) *Proc. Natl. Acad. Sci. USA* , **86**, 1939.

194. Thomas, R.H., Schaffner, W., Wilson, A.C. and Paabo, S. (1989) *Nature*, **340**, 465.

195. Lawlor, D.A., Dickel, C.D., Hauswirth, W.W. and Parham, P. (1991) *Nature*, **349**, 785.

196. Diamond, J.M. (1990) *Nature*, **347**, 334.

197. Persing, D.H., Telford, S.R., Rys, P.N., Dodge, D.E., White, T.J., Malawista, S.E. and Spielman, A. (1990) *Science*, **249**, 1420.

198. DeSalle, R., Gatesy, J., Wheeler, W. and Grimaldi, D. (1992) *Science*, **257**, 1933.

199. Janczewski, D.N., Yuhki, N., Gilbert, D.A., Jefferson, G.T. and O'Brien, S.J. (1992) *Proc. Natl. Acad. Sci. USA* , **89**, 9769.

200. Golenberg, E.M., Giannasi, D.E., Clegg, M.T., Smiley, C.J., Durbin, M., Henderson, D. and Zurawski, G. (1990) *Nature*, **344**, 656.

201. Paabo, S. and Wilson, A.C. (1991) *Curr. Biol.*, **1**, 45.

202. Soltis, P.A., Soltis, D.E. and Smiley, C.J. (1992) *Proc. Natl. Acad. Sci. USA* , **89**, 449.

203. (1991) *Phil. Trans. R. Soc. Lond., Series B*, **333**, 399.

204. Bult, C., Kallersjo, M. and Suh, Y. (1992) *Plant Mol. Biol. Rep.*, **10**, 273.

205. Cox, A. V., Bennett, M.D. and Dyer, T.A. (1992) *Theor. Appl. Genet.*, **83**, 684.

206. Kocher, T.D., Thomas, W.K., Meyer, A., Edwards, S.V., Paabo, S., Villablanca, F.X. and Wilson, A.C. (1989) *Proc. Natl. Acad. Sci. USA* , **86**, 6196.

207. Meyer, A., Kocher, T.D., Basasibwaki, P. and Wilson, A.C. (1990) *Nature*, **347**, 550.

208. Hall, H.G. and Smith, D.R. (1991) *Proc. Natl. Acad. Sci. USA* , **88**, 4548.

209. Cunningham, C.W., Blackstone, C.W. and Buss, L.W. (1991) *Nature*, **355**, 539.

210. Wayne, R.K. and Jenks, S.M. (1991) *Nature*, **351**, 565.

211. Richmann, A.D. and Price, T. (1992) *Nature*, **355**, 817.

212. Palenik, B. and Haselkorn, R. (1992) *Nature*, **355**, 265.

213. Urbach, E., Robertson, D.L. and Chisholm, S.W. (1992) *Nature*, **355**, 276.

214. Taberlet, P., Gielly, L., Pautou, G. and Bouvet, J. (1991) *Plant Mol. Biol.*, **17**, 1105.

215. Batzer, M.A., Carlton, J.E. and Deininger, P.L. (1991) *Nucleic Acids Res.*, **19**, 5081.

216. Fritsch, P. and Riesberg, L.H. (1992) *Nature*, **359**, 633.

217. van Heusden, A.W. and Bachmann, K. (1992) *Plant Syst. Evol.*, **179**, 221.

218. Arnheim, N. and Erlich, H. (1992) *Ann. Rev. Biochem.*, **61**, 131.

219. Huang, M.-M., Arnheim, N. and Goodman, M.F. (1992) *Nucleic Acids Res.*, **20**, 4567.

220. Frommer, M., McDonald, L.E., Millar, D.S., Collis, C. M., Watt, F., Grigg, G.W., Molloy, P.L. and Paul, C.L. (1992) *Proc. Natl. Acad. Sci. USA*, **89**, 1827.

221. Ochman, H., Gerber, A.S. and Hartl, D.L. (1988) *Genetics*, **120**, 621.

222. Ruano, G., Kidd, K.K. and Stephens, J.C. (1990) *Proc. Natl. Acad. Sci. USA*, **87**, 6296.

223. Wang, Z. and Brown, D.D. (1991) *Proc. Natl. Acad. Sci. USA*, **88**, 11505.

224. Crompton, M.R., Bartlett, T.J. and Goodwin, G.H. (1992) *Nucleic Acids Res.*, **20**, 4107.

225. Harvey, R.J. and Darlison, M.G. (1991) *Nucleic Acids Res.*, **19**, 4002.

226. Ertl, H., Hallmann, A., Wenzl, S. and Sumper, M. (1992) *EMBO J.*, **11**, 2055.

227. Moore, R.E., Shepherd, J.W. and Hoskins, J.A. (1990) *Nucleic Acids Res.*, **18**, 1921.

228. Girgis, S.I., Alevizaki, M., Denny, P. and Ferrier, G.J.M. (1988) *Nucleic Acids Res.*, **26**, 10371.

229. Lin, X., Feng, X.-H. and Watson, J.C. (1991) *Proc. Natl. Acad. Sci. USA*, **88**, 6951.

230. Parimoo, S., Patanjali, S.R., Shukla, H., Chaplin, D.D. and Weissman, S.M. (1991) *Proc. Natl. Acad. Sci. USA*, **88**, 9623.

231. Lovett, M., Kere, J. and Hinton, L.M. (1991) *Proc. Natl. Acad. Sci. USA*, **88**, 9628.

232. Hemsley, A., Arnheim, N., Toney, M.D., Cortopassi, G. and Galas, D.J. (1989) *Nucleic Acids Res.*, **17**, 6545.

233. Clackson, T. and Winter, G. (1989) *Nucleic Acids Res.*, **17**, 10163.

234. Ho, S.N., Hunt, H.D., Horton, R.M., Pullen, J.K. and Pease, L.R. (1989) *Gene*, **77**, 51.

235. Horton, R.M., Hunt, H.D., Ho, S.N., Pullen, J.K. and Pease, L.R. (1989) *Gene*, **77**, 61.

236. Horton, R.M., Cai, Z., Ho, S.N. and Pease, L.R. (1990) *BioTechniques*, **8**, 528.

237. Ausubel, F. M., Brent, R., Kingston, R.E., Moore, D.D., Seidman, J.G., Smith, J.A. and Struhl, K. (eds) (1992) *Current Protocols in Molecular Biology.* Greene Publishing Associates and Wiley Interscience, USA.

238. Dorit, L., Ohara, O. and Hwang, C.B.-C. (1992) in *Current Protocols in Molecular Biology* (F.M. Ausubel *et al.* eds). Greene Publishing Associates and Wiley Interscience, USA, Vol. 2, p. 15.2.1.

239. Gyllenstein, U.B. and Erlich, H.A. (1988) *Proc. Natl. Acad. Sci. USA*, **85**, 7652.

240. Higuchi, R.G. and Ochman, H. (1989) *Nucleic Acids Res.*, **17**, 5865.

241. Gibbs, R.A., Nguyen, P.-N., McBride, L.J., Koepf, S.M. and Caskey, C.T. (1989) *Proc. Natl. Acad. Sci. USA*, **86**, 1919.

242. Dicker, A.P., Volkenandt, M., Adamo, A., Barreda, C. and Bertino, J.C. (1989) *BioTechniques*, **7**, 830.

243. Scutt, C.P. and Croy, R.R.D. (1992) *Mol. Gen. Genet.*, **232**, 240.

244. Williams, J.G. and Mason, P.J. (1985) in *Nucleic Acid Hybridisation: A Practical Approach* (B.D. Hames and S.J. Higgins eds). IRL Press, Oxford, p. 139.

245. Bown, D., Ellis, T.H.N. and Gatehouse, J.A. (1988) *Biochem. J.*, **251**, 717.

246. Hagen, G. *et al.* (1991) *Plant Mol. Biol.*, **17**, 567.

247. Suzuka, I. *et al.* (1991) *Eur. J. Biochem.*, **195**, 571.

248. Schreier, P.H., Reiss, B. and Kuntz, M. (1988) in *Plant Molecular Biology Manual* (S.B. Gelvin and R.A. Schilperoort eds). Kluwer Academic Publishers, Dordrecht, B5/10.

249. Gallagher, T.F. and Ellis, R.J. (1982) *EMBO J.*, **1**, 1493.

250. Evans, I.M. *et al.* (1984) *Planta*, **160**, 559.

251. Marzluff, W.F. and Huang, R.C.C. (1984) in *Transcription and Translation – A Practical Approach* (B.D. Hames and S.J. Higgins eds). IRL Pres, Oxford, p. 89.

252. Thomas, P.S. (1980) *Proc. Natl. Acad. Sci. USA*, **77**, 5201.

253. Thompson, A.J. *et al.* (1989) *Planta*, **179**, 279.

254. Nagy, F., Kay, S.A. and Chua, N.-H. (1988) in *Plant Molecular Biology Manual* (S.B. Gelvin and R.A. Schilperoort eds). Kluwer Academic Publishers, Dordrecht, B4/21.

255. Krieg, P.A. and Melton, D.A. (1987) *Methods Enzymol.*, **155**, 397.

256. Evans, I.M. *et al.* (1990) *Mol. Gen. Genet.*, **223**, 273.

257. Cox, K.H. and Goldberg, R.B. (1988) in *Plant Molecular Biology: A Practical Approach* (C.H. Shaw ed.). IRL Press, Oxford, p. 20.

258. Cox, K.H. and Goldberg, R.B. (1988) in *Plant Molecular Biology: A Practical Approach* (C.H. Shaw ed.). IRL Press, Oxford, p. 25.

259. Harris, N. *et al.* (1990) in In situ *Hybridisation; Application to Developmental Biology and Medicine*, SEB Seminar Series 40 (N. Harris and D.G. Wilkinson eds). Cambridge University Press.

260. Herrera-Estrella, L. and Simpson, J. (1988) in *Plant Molecular Biology: A Practical Approach* (C.H. Shaw ed.). IRL Press, Oxford, p. 148.

261. Ingelbrecht, I.L.W. *et al.* (1989) *The Plant Cell*, **1**, 671.

262. Dunsmuir, P. *et al.* (1988) in *Plant Molecular Biology Manual* (S.B. Gelvin and R.A. Schilperoort eds). Kluwer Academic Publishers, Dordrecht, C1/1.

263. Laemmli, U.K. (1970) *Nature*, **227**, 680.

264. Johnson, D.A. *et al.* (1984) *Gene Anal. Techn.*, **1**, 3.

265. Chamberlain, J.P. (1979) *Anal. Biochem.*, **98**, 132.

266. Hurkman, W.J. (1990) in *Environmental Injury to Plants* (F. Katterman ed.). Academic Press, Chapter 9.

267. Anderson, N.G. and Anderson, N.L. (1978) *Anal. Biochem.*, **85**, 331.

268. Evans, I.M. *et al.* (1979) *Planta*, **144**, 455.

269. Clemens, M.J. (1984) in *Transcription and Translation – A Practical Approach* (B.D. Hames and S.J. Higgins eds). IRL Press, Oxford, p. 242. 245; also Clemens, M.J. (1988) in *Plant Molecular Biology Manual* (S.B. Gelvin and R.A. Schilperoort eds). Kluwer Academic Publishers, Dordrecht, B5/13.

270. Pelham, H.R.B. and Jackson, R.J. (1976) *Eur. J.Biochem.*, **67**, 247.

271. Gurdon, J.B. *et al.* (1971) *Nature*, **233**, 177.

272. Colman, A. (1984) in *Transcription and Translation – A Practical Approach* (B.D. Hames and S.J. Higgins eds). IRL Press, Oxford, p. 271.

273. Kawata, E.E. *et al.* (1988) in *Plant Molecular Biology Manual* (S.B. Gelvin and R.A. Schilperoort eds). Kluwer Academic Publishers, Dordrecht, B7/1.

274. Johnson, K.S. *et al.* (1989) *Nature*, **338**, 585.

275. Van Etten, R.A., Jackson, P. and Baltimore, D. (1989) *Cell*, **58**, 669.

276. Vinson, C.R. *et al.* (1988) *Genes and Development*, **2**, 801.

277. Tan, T.-H. (1991) *Anal. Biochem.*, **192**, 17.

278. Rosenberg, A.H. *et al.* (1987) *Gene*, **56**, 125.

279. Burnette, W.N. (1981) *Anal. Biochem.*, **112**, 195.

280. Kyhse-Anderson, J. (1984) *J. Biochem. Biophys. Meth.*, **10**, 203.

281. Renart, J. and Sandoval, I.V. (1984) *Methods Enzymol.*, **194**, 455.

282. Bradford, M.M. (1976) *Anal. Biochem.*, **72**, 248.

283. Kang, A.S., Abbott, S.J. and Harris, N. (1988) in *Manipulating Secondary Culture* (R.J. Robbins and M.J.C. Rhodes eds). Cambridge University Press.

284. Harlow, E. and Lane, D. (1988) in *Antibodies – A Laboratory Manual.* Cold Spring Harbor Laboratory, New York.

285. Edwards, G.A. *et al.* (1991) *Plant Mol. Biol.*, **17**, 89.

286. Hawes, C. (1988) in *Plant Molecular Biology: A Practical Approach* (C.H. Shaw ed.). IRL Press, Oxford, p. 103.

287. Harris, N. and Croy, R.R.D. (1985) *Planta*, **165**, 522.

288. Herrera-Estrella, L., Teeri, T.H. and Simpson, J. (1988) in *Plant Molecular Biology Manual* (S.B. Gelvin and R.A. Schilperoort eds). Kluwer Academic Publishers, Dordrecht, B1/1.

289. Van den Broeck, G. *et al.* (1985) *Nature*, **313**, 358.

290. Pearse, A.G.E. (1972) *Histochemistry: Theory and Applied.* Churchill Livingstone Press, Edinburgh, Vol. II, p. 808.

291. Jefferson, R.A., Kavanagh, T.A. and Bevan, M.W. (1987) *EMBO J.*, **6**, 3901.

292. Jefferson, R.A. (1987) *Plant Mol. Biol. Rep.*, **5**, 387.

293. Kertbundit, S. *et al.* (1991) *Proc. Natl. Acad. Sci. USA* , **88**, 5212.

294. Gallagher, S.R. (ed.) *GUS Protocols.* Academic Press.

295. Robinson, N.J., Shirsat, A.H. and Gatehouse, J.A. (1993) in *Plant Biochemistry and Molecular Biology* (P. J. Lea and R.C. Leegood eds). John Wiley & Sons Ltd, p. 221.

296. Oudet, P., Gross-Bellard, M. and Chambon, P. (1975) *Cell*, **4**, 181.

297. Paul, A.-L. and Ferl, R.J. (1988) in *Plant Molecular Biology Manual* (S.B. Gelvin and R.A. Schilperoort eds). Kluwer Academic Publishers, Dordrecht, B2, 1.

298. Sawyer, R.M., Boulter, D. and Gatehouse, J.A. (1987) *Planta*, **171**, 254.

299. LaThangue, N.B. and Rigby, P.W.J. (1988) in *Transcription and Splicing* (B.D. Hames and D.M. Glover eds). IRL Press, Oxford, p. 1.

300. Hatzopoulos, A.K., Schlokat, U. and Gruss, P. (1988) in *Transcription and Splicing* (B.D. Hames and D.M. Glover eds). IRL Press, Oxford, p. 43.

301. Straney, D.C. and Crothers, D.M. (1985) *Cell*, **43**, 449.

302. Meakin, P.J. and Gatehouse, J.A. (1991) *Planta*, **183**, 471.

303. Treisman, R. (1986) *Cell*, **46**, 567.

304. Topol, J., Ruden, D.M. and Parker, C.S. (1985) *Cell*, **42**, 527.

305. Tullius, T.D. and Dombroski, B.A. (1986) *Proc. Natl. Acad. Sci. USA* , **83**, 5469.

306. Kuwabara, M.D. and Sigman, D.S. (1987) *Biochemistry*, **26**, 7234.

307. Razin, A. and Szyf, M. (1984) *Biochim. Biophys. Acta*, **782**, 331.

308. Razin, A. (1988) in *Plant Molecular Biology Manual* (S.B. Gelvin and R.A. Schilperoort eds). Kluwer Academic Publishers, Dordrecht, B3, 1.

309. Waterhouse, R.N., Boulter, D. and Gatehouse, J.A. (1986) *FEBS Lett.*, **209**, 223.

310. Krainer, A.R. and Maniatis, T. (1988) in *Transcription and Splicing* (B.D. Hames and D.M. Glover eds). IRL Press, Oxford, p. 131.

311. Egeland, D.B., Sturtevant, A.P. and Schuler, M.A. (1989) *The Plant Cell*, **1**, 633.

312. Shapiro, D.J., Blume, J.E. and Nielsen, D.A. (1987) *BioEssays*, **6**, 221.

313. Wreschner, D.H. and Rechavi, G. (1988) *Eur. J. Biochem.*, **172**, 333.

314. Jackson, R.J. and Standart, N. (1990) *Cell*, **62**, 15.

315. Sieliwanowicz, B. (1987) *Biochim. Biophys. Acta*, **908**, 54.

316. Goldberg, R.B., Barker, S.J. and Perez-Grau, L. (1989) *Cell*, **56**, 149.

317. Thompson, A.J. *et al.* (1989) *Planta*, **179**, 279.

318. Hendriks, T., Vreugdenhil, D. and Stiekema, W.J. (1991) *Plant Mol. Biol.*, **17**, 385.

319. Bray, E.A. (1988) *Plant Physiol.*, **88**, 1210.

320. Evans, I.M. *et al.* (1990) *FEBS Lett.*, **262**, 29.

321. Cushman, J.C., DeRocher, E.J. and Bohnert, H.J. (1990) in *Environmental Injury to Plants* (F. Katterman ed.). Academic Press, London, Chapter 8.

322. Bowles, D. (1991) *Curr. Biol.*, **1**, 165.

323. Draper, J., Scott, R., Armitage, P. and Walden, R. (eds) (1988). *Plant Genetic Transformation and Gene Expression – A Laboratory Manual.* Blackwell Scientific Publications.

324. Bird, C.R. *et al.* (1991) *Biotechnol.*, **9**, 635.

325. Shirsat, A. *et al.* (1989) *Mol. Gen. Genet.*, **215**, 326.

326. Koncz, C. *et al.* (1992) *Plant Mol. Biol.*, **20**, 963.

327. Koncz, C. *et al.* (1989) *Proc. Natl. Acad. Sci. USA* , **86**, 8467.

328. Lee, K.Y. *et al.* (1990) *Plant Cell*, **2**, 415.

329. Nevers, P., Shepherd, N.S. and Saedler, H. (1986) in *Advances in Botanical Research* (J.A. Callow, ed.). Academic Press, London, p. 103.

330. Fedoroff, N.V., Furtek, D.B. and Nelson, O.E. (1984) *Proc. Natl. Acad. Sci. USA* , **81**, 3825.

331. Zimmerman, J.L. *et al.* (1989) *The Plant Cell*, **1**, 1137.

332. Delhaize, E., Robinson, N.J. and Jackson, P.J. (1989) *Plant Mol. Biol.*, **12**, 487.

333. Gowri, G., Paiva, N.L. and Dixon, R.A. (1991) *Plant Mol. Biol.*, **17**, 415.

334. Manley, J.L. (1984) in *Transcription and Translation – A Practical Approach* (B.D. Hames and S.J. Higgins eds). IRL Press, Oxford, p. 71.

335. Gilmour, R.S. (1984) in: *Transcription and Translation – A Practical Approach* (B.D. Hames and S.J. Higgins eds). IRL Press, Oxford, p. 131.

336. Manley, J.L. *et al.* (1983) *Methods Enzymol.*, **101**, 568.

337. Evans, I.M. *et al.* (1985) *Planta*, **165**, 554.

338. Yamazaki, K.-I., Chua, N.-H. and Imaseki, H. (1990) *Plant Mol. Biol. Rep.*, **8**, 114.

339. Yamazaki, K.-I. *et al.* (1990) *Proc. Natl. Acad. Sci. USA* , **87**, 7035.

340. Roberts, M.W. and Okita, T.W. (1991) *Plant Mol. Biol.*, **16**, 771.

341. Hebbes, T.R., Thorne, A.W. and Crane-Robinson, C. (1988) *EMBO J.*, **7**, 1395.

INDEX

acetosyringone
  Agrobacterium virulence induction 276
  structure 276
acetylated histone H4 364
acridine orange 9
actin genes 116–117
agar 257
agarose 257
agglutinin genes 78–79
Agrobacterium
  antibiotic control of 276
  antibiotic selection 269, 270, 272–275
  control by antibiotics 272–275, 276
  culture media 271
  oncogenic strains 269
  plasmid DNA isolation 21, 29
  plasmids 142, 269, 270
  resistance markers 269, 270
  strains 268, 269
  transconjugant strains 269
  uptake of vector plasmids 270
  virulence induction 276
agroinfection 270
  gemini viruses 277
alginate 257
allergen genes 54
aminoglycoside adenyltransferase gene 149, 151, 304
ampicillin 272
amplification specificity, PCR 316, 317
amylase inhibitor genes 95–97
amyloplasts 291
anchored PCR (APCR) 326
aniline blue 9
anilino-naphthalene sulfonic acid (ANS) 9
ANS 9
anthocyanin regulatory genes 141
anti-ethylene agents 256
anti-viral protein genes 78–79
antibiotic resistance genes — *see also* selectable markers 150
antibiotics
  Agrobacterium selection 272–275
  for cell culture 272–275
  plant tissue culture 272–275
antigen genes 54
antisense RNA 361
assays of gene expression 346–364
ATP synthase 297, 299, 303
  genes 53–54
ATP synthesis 53–54
ATPase genes 53–54
augmentin 272
auramine 9
aurintricarboxylic acid (ATA) 34
auxins 254, 255, 256
  media supplements 253, 255
  structure 258
  -binding protein genes 104–111
  -induced genes 112–115
avenin genes 105

bacto-agar 257
bar gene 180–181
bentonite 35
bialaphos resistance 138
binary vectors 142–146
biotin-binding protein genes 104–111
bleomycin 272
bleomycin resistance gene 152
Bowman–Birk protease inhibitor genes 95–97

calcofluor 9
calculation gene copy 30–31
calmodulin genes 81
cambium 3
carbenicillin 272
caryopses 7
cat
  assay 136–137
  gene 125, 127, 132–133, 305
cauliflower mosaic virus 35S promoter 121
cefotaxime 273
cell cycle genes 57–59
cell suspension cultures 362
cell wall protein genes 116–117
cells
  cambium 3
  collenchyma 2
  culture media — *see* culture media
  embedding 8
  endodermis 1
  epidermis 1
  histology 1–3
  mesophyll 1
  pericycle 2
  phloem 3
  photosynthetic 1

Index

sclerenchyma 2
staining 8
transformation — *see* genetic transformation
wall 1
xylem 2
cellulases 277, 278
cephalosporins 276
cetyl trimethylammonium bromide (CTAB) 21
chaperonin genes 95–97
Chlamydomonas reinhardtii chloroplast transformation 306–307
chloramphenicol 273
chloramphenicol acetyltransferase gene 155–156, 305
chloroplast
codon usage 301
electron transport 294–297, 298
envelope membrane 290, 293
enzyme genes 296–300
gene expression 302
gene nomenclature 293
genes, codon usage 43–45
genetic code 301
genetic system 297–302
genetic transformation 304–308
genome 293
gene arrangements 302
size 293
structure 302
tobacco 303
introns 302
molecular biology 293–308
ribosomal proteins 301, 303
ribosomal RNA 297
RNA polymerase 302
signal sequences 289
stroma 294
targeting sequences 289–290
thylakoid 293
lumen 289, 294
membrane 290, 294
transcription 302
transformation 304–308, 305, 306
tRNA genes 301, 303
chloroplasts 293–304
chromatin
'active' 364
structural assays 357
wheatgerm 363
cis–acting factors 357
clavulanic acid 272, 276
cloning of PCR products 320, 321, 343, 344
cloxacillin 276
codon usage 37–45
chloroplast 301
cold-induced genes 112–115
collenchyma 2
common plant names 11–19
contamination in PCR 318, 319
convicilin genes 107
cpDNA 293
cruciferin genes 105
CTAB, DNA isolation 21, 22–26
culture media
Agrobacterium 271
antibiotics in 272–275
basal 253–255
gelling agents 257
protoplasts 261
sterilants for explants 259
supplements 255–256
cycloheximide 273
cysteine-rich protein genes 81
cytochrome
b 298
b/*f* complex 297
f 298
genes 53–54
cytokinins 255, 256
media supplements 253, 255
structure 258
cytoskeletal protein genes 116–117

DAPI stain 9
databases
plant gene 49, 51
sequence 125, 183
deaza-deoxyguanosine PCR 316
default pathway, secretory proteins 288
DEPC — *see* diethylpyrocarbonate 33
developmental genes 55–56
regulation 360
diamidino phenolindole (DAPI) 9
dicot
flower 6
gene introns 201–234
leaf 6
root 5
seed 8
stem 5
diethylpyrocarbonate (DEPC) 33
Digitaria streak virus (DSV) 277
dihydrofolate reductase gene 156–157
dihydropteroate synthase gene 157–158
DNA
binding protein assays 357
binding protein genes 57–59
chloroplast 297
codon usage 37–45, 301
databases 51–52
footprinting 358
gene index 49–120
isolation
Agrobacterium plasmids 29
plant 21, 22–28
methylation assays 359
polymerases thermostable 314
repetitive 60–62
satellite 60–62

sequence analysis software 51
uptake by cells
Agrobacterium 270
protoplasts 278
DNase I sensitivity 358
drought-induced genes 112–115

E. coli antibiotic selection 272–275
electron transport protein genes 53–54
embedding plant tissues 8
embryo genes for proteins 55–56
endodermis 2
endoplasmic reticulum (ER) retention sequence 288
enolpyruvylshikimate phosphate synthase gene 158–160
envelope protein genes 80
environmental gene regulation 360
enzyme
activator genes 95–97
genes 63–77
inhibitor genes 95–97
enzyme-linked immunosorbent assay (ELISA) 354
epidermis 1
erythromycin 273
EtBr, DNA isolation 21
ethidium bromide (EtBr) 21
evolutionary studies 341
expression cassettes 121
expression vectors 352
extensin genes 116–117
extracellular targeting 288

family names, plants 11–19
ferredoxin 298
genes 53–54
ferritin genes 81
flower 5
dicot histology 6
monocot histology 7
fluorescence microscopy — *see* microscopy 8, 9
fluorescent stains 9
fluorochrome stains 8, 9
F_0F_1-ATPase 297, 299
formula gene copy calculation 30
fossil remains, PCR 340
fruits 5

G418 antibiotic 273
galactosidase gene 161–164
gel retardation assays 357
gelling agents 257
gelrite 257
gene
aminoglycoside adenyltransferase 149, 151, 305
*aph*II 174–175, 305
bar 180–181
bleomycin resistance 152
chloramphenicol acetyltransferase gene 155–156
copy calculations 30–32
formula 32
nomogram 30–32
delivery systems 263
dihydrofolate reductase 156–157
dihydropteroate synthase 157–158
enolpyruvylshikimate phosphate synthase 158–160
expression assays — *see* gene expression assays
expression PCR analysis 330
galactosidase 161–164
glucuronidase (*gus*) 164–166
hygromycin phosphotransferase 166–168
index 49–120
introns 183–239
luciferase (*luc*; firefly) 168–170
A 171–172
B 172–173
nopaline synthase 175–178
*npt*II 174–175, 305
octopine synthase 178–180
pat 180–181
phosphinothricin acetyltransferase 180–181
promoter assays 361
promoters 149–181
tagging 362
targeting 362
transfer – *see* gene transfer
gene expression assays 346–364
in situ hybridization 349
antisense RNA 361
cell suspension cultures 362
chromatin
'active' 364
structure 357
wheatgerm 363
cis-acting factors 357
developmental regulation 360
DNA binding protein assays 357
DNA footprinting 358
DNA methylation assays 359
DNase I sensitivity 358
ELISA 354
environmental regulation 360
gel electrophoresis 350
gel retardation assays 357
gene promoter assays 361
gene tagging 362
gene targeting 362
in vitro translation 351
methods for RNA analysis 346–349
mRNA half-life 360
mRNA precursor splicing 359
Northern blots 348
nuclear 'run–on' assay 347
nuclear extracts 363

Index

oocyte expression system 352
primer extension 346
promoter trapping 361
protein assays 350–356
regulation 357–364
reporter gene expression 355, 356
RNA titrations 349
RNase protection assay 347, 348
S1 nuclease assay 346
Southern blots 347, 348
squash-blots 354
trans-acting factors 357
transcription assays 363
transcription mapping 346, 347
transgenic plants 349, 353, 361
Western blot analysis 353
gene transfer
Agrobacterium 263
direct gene transfer 263, 279
imbibition of tissues 263
impregnation by whiskers 263
laser perforation 263
macroinjection 263
methods 262–263
microinjection 263
microprojectile bombardment 263
particle bombardment 263
pollen tube pathway 263
ultrasonication 263
genes
actin 116–117
agglutinin 78–79
allergen 54
amylase inhibitor 95–97
anthocyanin synthesis 141
anti-viral protein 78–79
antigen 54
ATP synthases 53–54
ATP synthesis 53, 299
ATPase 53–54
auxin-binding protein 104–111
auxin-induced 112–115
avenin 105
biotin-binding protein 104–111
Bowman–Birk protease inhibitor 95–97
cell cycle 57–59
cell wall protein 116–117
chaperonin 95–97
chloroplast
ATP synthase 299
electron transport components 294–297, 298, 303
cold–induced 112–115
convicilin 107
copy number calculation 30
cruciferin 105
cysteine-rich protein 81
cytochromes 53–54, 298
cytoskeletal proteins 116–117
developmental 55–56
DNA-binding proteins 57–59
drought-induced 112–115
electron transport proteins 53–54
embryonic 55–56
envelope protein 80
enzymes 63–77
activators 95–97
inhibitors 95–97
extensin 116–117
ferredoxin 53–54
ferritin 81
gibberellin-induced 112–115
gliadin 109–110
glutelin 106–107, 110
glutenin 109
glycine-rich protein 116–117
glycinin 105–106
glycoproteins 55–56
growth regulator-induced 112–115
heat shock-induced 112–115
histones 57–59
homeogenes 57–59
hordein 106
hydroxyproline-rich protein 116–117
Kunitz trypsin inhibitor 95–97
lectins 78–79
leghemoglobin 83–84
legumin 107–108, 110
ligand-binding protein 104–111
light-influenced 85–94
luminal binding protein 104–111
membrane structural proteins 80
membrane-associated protein 80
metal-associated proteins 81
metal-binding protein 81
metalloproteins 81
metallothioneins 81
NAD(P): quinone oxidoreductase 300, 303
napin 105, 108
nitrogen-fixation protein 83–84
nitrogenase 83–84
nodulins 83–84
oleosins 80
organ-specific 55–56
patatin 108–109
pathogenesis-related 112–115
photosynthesis-associated proteins 85–94
photosynthetic 85–94, 295–296, 95–97
phytochromes 85–94
phytohemagglutinins 78–79
plastocyanins 85–94
pollen allergens 4
polymerases 57–59
prolamins 107
proline-rich proteins 116–117
protease inhibitors 95–97
proteases 95–97
protein kinase 95–97
protein phosphatase 95–97

protein modification 95–97
PSI polypeptide complexes 295
PSII polypeptide complexes 296
reporter — *see cat, gus* and *luc* 129–136, 149–181
ribonucleoproteins 98–104
ribosomal 98–104
protein 98–104
RNA 98–104
ribosome-inactivating proteins (RIPs) 78–79
rRNA 98–104
salt-induced 112–115
seed protein 105–111
selectable 149–181
small nuclear RNA 98–104
snRNP 98–104
splicing protein 98–104
storage protein 105–111
stress-induced proteins 112–115
structural protein 116–117
sulfur-rich protein 81
thioredoxin 53–54
tissue-specific 55–56
tonoplast protein 80
transcription factors 57–59
transit proteins 80
translation factor 57–59
tRNA 98–104
tuber protein 105–111
tubulin 116–117
ubiquitin 95–97
vicilin 108–111
wound-induced 112–115
zein 110–111
genetic transformation, plants 262–279
geneticin 273
genome, chloroplast 293
genome sizes 31
gentamicin 273
gibberellin 256
gibberellin-induced genes 112–115
gliadin genes 109–110
glucuronidase (*gus*) gene 164–166
glutelin genes 106–107, 110
glutenin genes 109
glycine-rich protein genes 116–117
glycinin genes 105–106
glycoprotein genes 55, 116–117
glyoxysomes targeting sequences 287–288
glyphosate resistance 158
Golgi secretion pathway 288
Gramineae gemini virus agroinfection 277
growth regulators
concentrations 256
molecular weights 256
structures 258
growth regulator induced genes 112–115
gus
enzyme assay 130
histochemical assay 131
reporter gene 121, 129, 305

heat shock-induced genes 112–115
heat shock promoter 126
HeLa cell lysate system 363
heparin 35
histochemical staining 8
histology
GUS stain 131
of cells 1
staining 8
histone genes 57–59
homeogenes 57–59
homologous recombination 362
hordein genes 106
hydroxyproline-rich protein genes 116–117
hygromycin 273
resistance 137
hygromycin phosphotransferase gene 166–168

immunoassays 354, 355
inhibitors, RNase 33–36
inorganic salts
protoplast culture media 262
tissue culture media 254
inosine PCR primers 335
insertion elements 118–120
intermembrane space, mitochondrial 289
introns 183–239
chloroplast genes 302
dicot sequences 201–234
monocot sequences 184–200
size distribution 239
splicing 235–238
inverse PCR (IPCR) 326
isolation
Agrobacterium Ti plasmids 21, 29
plant genomic DNA 21, 22–28
plant RNA 32, 36–37

kanamycin 273
resistance 136
kasugamycin 273
kinase genes 63–79, 95–97
Kunitz trypsin inhibitor genes 95–97

Latin plant names 11–19
leaf 4
dicot 6
lectin genes 78–79
leghemoglobin genes 83–84
legumin genes 107–108, 110
ligand-binding protein genes 104–111
ligation-mediated PCR 325
light-induced gene promoter 127
light-influenced genes 85–94
luc assay 135–136
luc reporter gene 132–133

luciferase (firefly) gene 168–170
α-subunit 171–172
β-subunit 172–173
A gene 171–172
B gene 172–173
luminal binding protein genes 104–111

macaloid clay 34
magnesium chloride, PCR 315
maize streak virus (MSV) 277
marker genes, *C. reinhardtii* transformants 304
matrix, mitochondrial 289
media
AB minimal 271
Agrobacterium 271
antibiotic concentrations 272–276
B5 254
CM 254
common basal 253–255
LB 271
MS 254
N6 254
plant tissue culture 253–261
stock concentrations 254
supplements 253, 256–257
tobacco protoplast
culture 262
isolation 262
transformation 262
tobacco tissue culture 257, 260
YEB 271
YMB 271
membrane
electron transport 294
electron transport, enzyme of 294–297
structural protein genes 80
membrane-associated protein genes 80
meristematic cells 2
mesophyll 1
metal-associated protein genes 81
metal-binding protein genes 81
metalloprotein genes 81
metallothionein genes 81
methotrexate 273
methotrexate resistance 141, 156
microbodies, targeting sequences 287–288
microprojectiles, DNA coating 279
microscopy
embedding 8
fluorescence 8
safety 9
Miscanthus streak virus (MSV) 277
mitochondria, targeting sequences 288–289
mitochondrial genes, codon usage 43–45
mobile genetic elements 239–242
monocot
flower 7
gene introns 184–200
seed 7
stem 4
monocotyledon — *see* monocot
mRNA half-life 360
mRNA precursor splicing 359
mycostatin 274

N-linked glycosylation 149–181, 288
NADPH dehydrogenase 297
NADPH:plastoquinone oxidoreductase 297
nalidixic acid 269, 274
napin genes 105, 108
neomycin 274
neomycin phosphotransferase gene 174–175, 305
nitrogen-fixation protein genes 83–84
nitrogenase genes 83–84
Noble-agar 257
nodulin genes 83–84
nomenclature 11
nomogram, gene copy calculation 32
nopaline synthase
gene 175–178
promoter 175, 176, 178, 305
*npt*II gene 174–175, 305
nuclear extracts 363
nuclear genes codon usage 38–42
nuclease inhibitors 33–35
nucleic acids 21–46
Agrobacterium 21, 29
plant DNA 21, 22–28
plant RNA 32, 36–37
nucleotide sequences
promoters 152, 153, 175, 178
reporter genes 155, 161, 164, 168, 171, 174, 175, 178
selectable genes 149, 152, 156, 157, 158, 166, 174, 180
nucleus, targeting sequences 287
nystatin 274

octopine synthase gene 178–180
oleosin genes 80
oligonucleotide primers, PCR 317
organ-specific genes 55–56
organelle targeting sequences 287–292

parenchyma 1
particle bombardment apparatus 308
particle guns 278, 308
propulsion systems 278
pat gene 180–181
patatin genes 105–111
pathogenesis-related genes 112–115
PCR 313
acronyms 345
amplification specificity 316, 317
buffer composition 315
cDNA libraries 328
cloning products 320, 321, 343, 344
cloning vectors 321, 343, 344
deaza-deoxyguanosine 316

degenerate primers 332–335
errors and artifacts 323, 324
evolutionary studies 341
fossil remains 340
gene expression analysis 330
ligation-mediated 324
magnesium chloride 315
minimization of contamination 318, 319
polymerase enhancer 315
primers 317
products, direct sequencing 322, 323
RACE techniques 329
RAPD analysis 337, 341
reaction buffer composition 315
recombinant clones 338
RNA processing 331
sequence modification 336
strategies 313–345
tetramethylammonium chloride 315
transgenic plants 339
pectinases 277, 278
pericycle 2
peroxidase genes 63–77
peroxisome targeting sequences 287–288
phenotypic markers, anthocyanin genes 141
phloem 3, 4
phosphatase genes 63–77, 95–97
phosphinothricin acetyltransferase gene 180–181
photoreception 85–94
photosynthesis 85–94
photosynthesis associated protein genes 85–94
photosynthetic enzymes 294–297
genes for 63–77, 294–297
photosystem I (PSI) 294, 303
photosystem II (PSII) 294, 303
phylogenetic studies, PCR 341
phytagar 257
phytochrome genes 85–94
phytohemagglutinin genes 78–79
plant
anatomy 4–8
gene expression 345–364
gene index 49–120
gene sequences 53–120
genome sizes 31, 32
histology 1–3
introns 183–239
name citation 11
names 11–19
organelles 287
secretion pathway 288
species transformed 265–268
taxonomy 10–11
tissue culture 253–279
antibiotic selection 272–275
tissue types 1–8
transformation 121–146, 253–279
transposons 362
plasma membrane 288
plasmid
pBI101 129
pBIN19 143
pC22 146
pCGN1547 144
pCGN1548 145
pCITE 343
pCR-Script 344
pCR1000 343
pCRII 343
pDH51 126
pGA482 144
pIJ4104 138
pJIT117 128
JIT118 140
pJIT134 136
pJIT161 137
pJIT26 132
pJIT27 132
pJIT52 133
pJIT53 133
pJIT6 137
pJIT60 127
pJIT72 138
pJIT74 126
pJIT75 129
pJIT82 139
pJIT92 139
pKYLX3 127
pMA406 127
pPHI443 142
pPM19 140
pPM21 141
pROK1 143
pRT103 125
pT7Blue 343
pTOPE 343
pUCD2340 145
plasmid maps 125–146
plasmid restriction maps — *see* plasmid maps
plasmid Ti DNA 21, 29
plasmids — *see also* vectors
Agrobacterium 21, 29
plastid
transformation 304
targeting sequences 291
plastocyanin 298
genes 85–94
plastocyanin:ferredoxin oxidoreductase 294
plastoquinol:plastocyanin oxidoreductase 297
ploidy level 31
pollen allergen genes 54
polymerase chain reaction (PCR) — *see also* PCR 313–345
polymerase enhancer 315
polymerase genes 57–59
precursor mRNA 183
primers
degenerate, PCR 332–335
inosine 335

Index

selection for PCR 317
prolamin genes 107
proline-rich protein genes 116–117
promoter trapping 361
promoters
cauliflower mosaic virus 35S 152–154
nopaline synthase 175–176
octopine synthase 178–180
protease genes 95–97
protease inhibitor genes 95–97
protein kinase genes 95–97
protein phosphatase genes 95–97
protein sequences
reporter gene encoded 155, 161, 164, 168, 171, 174, 175, 178
selectable gene encoded 149, 152, 156, 157, 158, 166, 174, 180
protein-modification genes 95–97
protoplast
production 278
enzymes 278
media 261
transformation 277, 278
chemical 277, 278

repetitive DNA 60–62
reporter gene assays 130, 131, 134–135
reporter gene expression
histochemical location 356
transgenic plants 355
reporter genes 129–136, 149–181
cat 132, 134, 155, 264
gus 129–131, 264
luc 132, 135, 168–173, 264
nop/nos 175
oct/ocs 178
restriction map
aad gene 151
ble gene 152
CaMV 35S promoter 152, 154
CaMV 35S terminator 154
cat gene 156
dhfr gene 157
epsps gene 160
gal gene 164
gus gene 168
hyg gene 166–168
luc gene 168–170 l
luxA gene 173
luxB gene 173
nos gene 178
npt gene 175
ocs gene 180
pat gene 181
sul gene 158
restriction maps, plasmids — *see* plasmid
Ri plasmids 268, 270
ribonucleoproteins, small nuclear genes 98–104
ribosomal protein genes 98–104
ribosomal RNA (rRNA) 98–104
genes 98–104
ribosome-inactivating protein (RIP) genes 78–79
ribulose bisphosphate carboxylase genes 85–94
rifampicin 274
RNA
isolation 32, 36–37
polymerase, chloroplast 302, 303
processing, PCR analysis 331
ribosomal (rRNA) genes 98–104, 297
small nuclear (snRNA) genes 98–104
structural 98–104
transfer (tRNA) genes 98–104
RNase block II 34
RNase inhibitors 33–36
RNasin 33
root 4
dicot 5
monocot 4
5S rRNA 98–104, 305
rRNA genes 98–104, 297
Rubisco — *see also* ribulose bisphosphate carboxylase
gene promoter 127
transit peptide 128

salt-induced genes 112–115
salts
B5 254
CM 254
CPW 262
EM 262
KF 262
MS 254
N6 254
satellite DNA 60–62
sclerenchyma 2
SDDTC, DNA isolation 21
secretion pathways 288
sectioning 8
seeds 5
dicot 8
monocot 7
protein genes 105–111
selectable genes 149–181
selectable markers 136–141
bialaphos resistance 138, 180
bleomycin resistance 152
chlorsulfuron resistance 141
glyphosate resistance 158
hygromycin resistance 137, 166
kanamycin resistance 137, 166
methotrexate resistance 141, 156
neomycin resistance 174
spectinomycin resistance 140, 149
streptomycin resistance 140, 149
sulfonamide resistance 139, 149, 157
tetracycline resistance 143

sequence databases 49, 51, 183
sequences
  cloning junctions, plasmid 123–124
  consensus splice sites 236
  plant organellar targeting 287–291
  signal peptides 287–291
sequencing, PCR products 322, 323
signal peptide sequences 287–291
small nuclear ribonucleoprotein particles (snRNPs) 183
  genes 98–104
small nuclear RNAs (snRNAs) 359
  genes 98–104
sodium diethyldithiocarbamate (SDDTC) 21
spectinomycin 274
spectinomycin resistance 141
splice site
  consensus sequences 236
  nucleotide incidence, dicots 238
  nucleotide incidence, monocots 237
spliceosome 183, 98–104
splicing
  assays 359
  chloroplast introns 302
  plant introns 183
  protein genes 98–104
staining 8
stem 4
  dicot 5
  monocot 4
sterilants tissue culture explants 259
storage protein genes 105–111
streptomycin 274
streptomycin resistance 140
stress-induced genes 112–115
structural DNA 60–62
structural protein genes 116–117
structural RNA 98–104
sulfonamide resistance 139
sulfur-rich protein genes 81
sulphonamide 275

T-DNA
  borders 270
  insertion 118
T:A vectors 321, 343, 344
targeting sequences 287–292, 355
taxonomy rules 10
terminator
  cauliflower mosaic virus 35S gene 154
  nopaline synthase gene 175–178
  octopine synthase gene 178–180
tetracycline 275
tetramethylammonium chloride, PCR 315
thioredoxin genes 53–54
thylakoids 294–297
Ti plasmids 118, 268, 269, 270
  isolation 29
timentin 276
TINS 37
tissue culture 253–279
  explant sterilants 259
  media 253–261
tissue-specific genes 55–56
tobacco
  chloroplast DNA 303
  cultivars 259
  protoplast
    culture media 262
    isolation media 262
    transformation media 262
  tissue culture cultivars 259
  tissue culture media 257, 260
tonoplast protein genes 80
trans-acting factors 357
transcription assays 363
transcription factor genes 57–59
transfer RNA (tRNA) genes 98–104
transformation 121–146
  Agrobacterium 270
  Chlamydomonas reinhardtii chloroplast 304
  chloroplasts 304–306
  criteria for assessment 264
  definitions 262
  plant phenotype segregation ratios 264
  plant species 265–268
  plant tissue culture, antibiotic selection 272–275
  plants 253–279
  plastids 304
  protoplasts 277, 278
  vectors – *see also* plasmids 121–146
transgene loci 264
transgenic plants, PCR analysis 339
transient expression, tissue-specific 279
transit peptides 288–291
transit protein genes 80
translation factor genes 57–59
translation systems
  reticulocyte lysate 351
  wheatgerm 351
transposable elements 118–120, 239–242
  homology in terminal repeats 240
  other types 242
  retrotransposon-like 241
  uncharacterized 242
transposons 118–120, 239–242
triisopropylnaphthalenedisulfonate (TINS) 37
trimethoprim 275
tRNA genes 98–104
tuber protein genes 105–111
tubulin genes 116–117

ubiquitin genes 95–97
uracil-rich small nuclear RNAs (UsnRNAs) 183

vacuolar targeting 288
vacuole 1, 2, 288
vanadyl ribonucleotide complexes (VRCs) 33
vancomycin 275, 276

vascular bundle 3
vascular tissues 2
vectors
  binary 142–146
  expression 121–146
  PCR cloning 320, 321, 343, 344
  reporter genes 131–135, 149–181
  selectable genes 136–141, 149–181
  transformation 121–146
vicilin genes 108–111
viruses
  agroinfection of *Gramineae* 277
  host plant agroinfection 277
vitamins 254, 255

water:plastoquinone oxidoreductase 294
wheat dwarf virus (WDV) 277
wound-induced genes 112–115

xylem 2

yeast genes, codon usage 40–45

zein genes 110–111